New Frontiers in Anaerobic Digestion (AD) Processes

New Frontiers in Anaerobic Digestion (AD) Processes

Editors

Sonia Heaven
Sigrid Kusch-Brandt
Charles Banks

MDPI • Basel • Beijing • Wuhan • Barcelona • Belgrade • Manchester • Tokyo • Cluj • Tianjin

Editors

Sonia Heaven
Water and Environmental
Engineering Group
University of Southampton
Southampton
United Kingdom

Sigrid Kusch-Brandt
Faculty of Mathematics,
Natural Sciences and
Management
University of Applied
Sciences Ulm
Ulm
Germany

Charles Banks
Water and Environmental
Engineering Group
University of Southampton
Southampton
United Kingdom

Editorial Office
MDPI
St. Alban-Anlage 66
4052 Basel, Switzerland

This is a reprint of articles from the Special Issue published online in the open access journal *Processes* (ISSN 2227-9717) (available at: www.mdpi.com/journal/processes/special_issues/anaerobic_digestion_frontiers).

For citation purposes, cite each article independently as indicated on the article page online and as indicated below:

LastName, A.A.; LastName, B.B.; LastName, C.C. Article Title. *Journal Name* **Year**, *Volume Number*, Page Range.

ISBN 978-3-0365-7999-3 (Hbk)
ISBN 978-3-0365-7998-6 (PDF)

Cover image courtesy of Sigrid Kusch-Brandt

Contents

About the Editors

Sonia Heaven

Prof. Dr. Sonia Heaven is an Emeritus Professor and Professorial Researcher in Environmental Engineering within Engineering and Physical Sciences at the University of Southampton, and was previously Head of the Water and Environmental Engineering Group. She is the Director of the Environmental Biotechnology Network; she worked on the EPSRC-funded SUE Waste programme, the EU FP6 CROPGEN project, and the EU FP7 All-Gas project; and was the coordinator of the EU FP7 project VALORGAS (valorisation of food waste to biogas). She is a Chartered Civil Engineer with 7 years' experience in the UK water and wastewater industry, and while employed by Southampton, she spent 7 years in central Asia working on a wide range of environmental problems. She is a Member of the Chartered Institution of Water and Environmental Management and of the Chartered Institution of Waste Management.

Sigrid Kusch-Brandt

Dr. Sigrid Kusch-Brandt is a Lecturer for the Circular Economy and Sustainable Management of Resources at the University of Applied Sciences Ulm; a Visiting Research Fellow within Engineering and Physical Sciences at the University of Southampton; and a consultant to international organisations, research institutions, public authorities, and companies. In earlier academic positions, she served as Substitute Professor for Waste Management at Technical University of Dresden and for 6 years as Visiting Professor for Sustainable and Renewable Resources at the University of Padua. With her work, she aims to contribute to the more sustainable management of environmental resources, including through a more effective integration of circular economy principles. She has a specific interest in the environmentally sound valorisation of biogenic resources.

Charles Banks

Prof. Dr. Charles Banks is an Emeritus Professor within Engineering and Physical Sciences at the University of Southampton. His research interests include innovative technology for environmental protection, including the controlled anaerobic and aerobic biodegradation of municipal and industrial solid wastes; the treatment of liquid industrial effluents using biological systems; energy production from the digestion of crops and agricultural wastes; concepts for an integrated farming system for non-competitive food and fuel production; and the provision of software tools for process energy balances, waste audit, and waste management. Research has led to the design of novel reactor systems and operating protocols to meet the challenges of new and adapted technologies for environmental protection, renewable energy production, and sustainable nutrient management systems. The development of operating protocols for anaerobic digestion, to maximise rates of substrate conversion and biogas yield, is also of special interest.

Preface to "New Frontiers in Anaerobic Digestion (AD) Processes"

Anaerobic digestion (AD) is widely used to process a variety of organic materials, but a major share of its full potential currently remains unlocked. This Special Issue of the journal *Processes* explores recent developments and advanced concepts related to the valorisation of biomass through the application of anaerobic digestion. Biological phenomena, the efficiency of the AD process, the viability of full-scale AD installations, the integration of hydrogen and biomethane concepts, and biorefineries are some of the themes covered. While AD is addressed under different perspectives, the results all strengthen AD to fulfil its important role as a sustainable and highly versatile technology. Further research and development needs are also presented. The reader is invited to study the publications in detail.

Sonia Heaven, Sigrid Kusch-Brandt, and Charles Banks
Editors

processes

Editorial

Unlocking the Full Potential: New Frontiers in Anaerobic Digestion (AD) Processes

Sigrid Kusch-Brandt [1,2,*], Sonia Heaven [1,*] and Charles J. Banks [1]

1 Water and Environmental Engineering Group, University of Southampton, Southampton SO16 7QF, UK; c.j.banks@soton.ac.uk
2 Faculty of Mathematics, Natural Sciences and Management, University of Applied Sciences Ulm, 89075 Ulm, Germany
* Correspondence: mail@sigrid-kusch.eu (S.K.-B.); s.heaven@soton.ac.uk (S.H.)

Anaerobic digestion (AD) is a bio-based solution designed to convert organic materials into renewable energy and other products, such as soil improver and organic fertiliser. AD is widely used in practice, with facilities at many thousands of sites worldwide: in Europe alone, more than 20,000 full-scale plants were in operation in 2022 [1]. The underlying biological processes are complex, and multiple options exist to steer the AD process towards optimised performance and a desired set of outputs in terms of energy and material flows. This puts AD in a prominent position with research agendas aiming for more sustainable resource management. Its ability to generate high-value products from organic wastes and residues is a key strength.

The Special Issue on "New Frontiers in Anaerobic Digestion (AD) Processes" was initiated to explore recent developments and advanced concepts related to the valorisation of biomass via the application of AD. Fourteen submissions are included in the Special Issue, and each of these publications contributes towards unlocking the full potential of AD. Five thematic clusters to advance AD can be identified based on the included publications:

- Understanding and monitoring the AD process;
- Making substrates available and increasing the efficiency of the AD process;
- Inspiring trust in non-academic stakeholders to adopt AD in practice;
- Supporting decarbonisation of the energy system through hydrogen and biomethane;
- Obtaining more value from a single unit of biomass.

All included manuscripts contribute to more than one of these five clusters (Table 1). In this paper, some selected findings reported in the publications are highlighted. These are not intended to be exhaustive, but rather to provide some first insights into the rich body of new knowledge created by the authors of the Special Issue.

For the purpose of maintaining the stability and performance of the AD process, in practice, only certain parameters can be monitored in real-time, while adequate methods are still lacking for many others. Yan et al. [2] reviewed the recent progress in applying soft sensor solutions. Some systems are available that use software-supported methods to determine the unmeasurable parameters based on measuring auxiliary variables online; but the need for more research remains high. Integration of deep learning elements into these software solutions is particularly promising.

Liu et al. [3] focused on the residual biogas potential of digestate leaving the digester and the current time-consuming standard procedures to determine this indicator through experimental laboratory testing. Residual biogas potential is a key indicator of digestate stability, which in turn is an essential requirement for spreading digestate onto agricultural land. The authors showed that kinetic modelling, in particular when supported by machine learning, could be successfully applied to reduce the testing time for residual biogas potential.

Citation: Kusch-Brandt, S.; Heaven, S.; Banks, C.J. Unlocking the Full Potential: New Frontiers in Anaerobic Digestion (AD) Processes. *Processes* **2023**, *11*, 1669. https://doi.org/10.3390/pr11061669

Received: 21 May 2023
Accepted: 29 May 2023
Published: 31 May 2023

Table 1. Publications included in this Special Issue and their relevance for the five clusters identified to unlock the full potential of anaerobic digestion.

	Cluster of AD Progress				
Publication	**Understanding and Monitoring the AD Process**	**Making Substrates Available and Increasing AD Efficiency**	**Inspiring Trust to Adopt AD in Practice**	**Supporting Energy Decarbonisation through Hydrogen and Biomethane**	**Obtaining More Value from a Single Unit of Biomass**
Review of soft sensors in anaerobic digestion process [2]	X		X		
Shortening the standard testing time for residual biogas potential (RBP) tests using biogas yield models and substrate physicochemical characteristics [3]	X		X		
Estimating the methane potential of energy crops: An overview on types of data sources and their limitations [4]		X	X		
Biogas production from residues of industrial insect protein production from black soldier fly larvae *Hermetia illucens* (L.): An evaluation of different insect frass samples [5]		X			X
Batch and semi-continuous anaerobic digestion of industrial solid citrus waste for the production of bioenergy [6]	X	X	X		
Hydrothermal pretreatment of wheat straw—Evaluating the effect of substrate disintegration on the digestibility in anaerobic digestion [7]		X			X
Anaerobic co-digestion of sewage sludge and trade wastes: Beneficial and inhibitory effects of individual constituents [8]		X	X		X
Operation of submerged anaerobic membrane bioreactors at 20 °C: Effect of solids retention time on flux, mixed liquor characteristics and performance [9]	X	X	X		
Exploring farm anaerobic digester economic viability in a time of policy change in the UK [10]			X		X
Recent advances in membrane-based biogas and biohydrogen upgrading [11]			X	X	
Potential for biomethanisation of CO2 from anaerobic digestion of organic wastes in the United Kingdom [12]				X	X
Experimental evaluation of continuous in-situ biomethanation of CO_2 in anaerobic digesters fed on sewage sludge and food waste and the influence of hydrogen gas–liquid mass transfer [13]	X			X	
Validation of two theoretically derived equations for predicting pH in CO_2 biomethanisation [14]	X		X	X	
Toward the transition of agricultural anaerobic digesters into multiproduct biorefineries [15]			X	X	X

Full-scale AD plant operators require reliable information on the methane yields of potential substrates. In agriculture, it is common to use crop-based substrates (energy crops, crop residues), either as the main AD input or to supplement biogas production from manure-fed digesters. Decisions on which crop material to use in AD can significantly impact the entirety of a farm's management. However, relying on data from the literature to estimate the methane yields of crops is not an advisable strategy. The results of Zhang et al. [4] revealed that many publications displayed deficiencies in data reporting. The transferability of the reported methane potentials was limited, because the variability in values for the same crop when tested under different experimental conditions or grown under different cultivation conditions often exceeded the variation of values between different crop species.

With a further increasing world population, food production is a major challenge to be addressed as part of the water/energy/food/climate change nexus. Valorisation of residues from the food production chain makes a valuable contribution to more sustainable food systems, and this applies both to established food production methods and to novel solutions. Among the novel solutions is the cultivation of insect biomass as an alternative feedstuff, which generates insect frass as a residue. After exploring the suitability of insect frass as an AD substrate, Wedwitschka et al. [5] reported promising results, but also highlighted some risk of instability of the process. Similarly, citrus waste is a challenging AD substrate due to the presence of toxic compounds. The work of Rosas-Mendoza et al. [6] suggests that, at an industrial scale, it might not be necessary to remove toxic D-limonene from orange peel waste when using cattle manure as inoculum; but the authors also conclude that further research is required to better understand the implications of different D-limonene concentrations under different reactor configurations. Other food production residues clearly require pre-treatment to make the material suitable for AD. One such biomass is wheat straw, a high-volume material stream. Zerback et al. [7] applied hydrothermal pre-treatment with good success, but they also observed that overly severe pre-treatment conditions had a negative impact on the degradation kinetics. For solutions to be applied at full scale, there is also a need to balance technical and economic feasibility.

Avoiding inhibition of the AD process and achieving a high gas yield are both key goals of commercial plant operators. Berzal de Frutos et al. [8] researched the co-digestion of sewage sludge and 160 different trade wastes with the aim of understanding how wastewater treatment plants can improve their AD performance by accepting trade wastes. The authors concluded that the addition of 10 percent (by volume) of trade waste can usually be recommended, but this may need to be confirmed by further experiments, for example, where inhibitory components are present or microbial acclimatisation is required.

Another approach to improve the performance of AD in the wastewater sector is the implementation of highly efficient bioreactors. Pacheco-Ruiz et al. [9] reported findings from long-term experiments conducted over 242 days with submerged anaerobic membrane bioreactors. A key result indicated that operation without chemical or external cleaning was feasible if the process conditions were adequately set by controlling solids retention and, thus, mean cell residence time. Clearly, such long-term experiments are required in order to reliably inform full-scale operators about the performance of specific operating regimes.

In practice, one of the most important factors influencing whether AD will be implemented or not is if there is sufficient confidence in its economic viability. Financial incentives directly influence the AD landscape and its development. As an example, in the United Kingdom (UK) there is currently a policy vacuum for residues-based small-scale farm AD (<150 kWe), and Bywater and Kusch-Brandt [10] showed that it is very difficult for such installations to achieve profitability, despite the currently high energy prices. An innovative policy mechanism would be to introduce financial support based on the "public goods" benefits offered by on-farm AD (e.g., greenhouse gas reduction, positive soil organic carbon impact, support of rural development).

Several of the published papers address AD as an element in decarbonising the energy system through the adoption of hydrogen and biomethane solutions. Here, AD can be applied in different ways. AD can be implemented to produce biohydrogen through dark fermentation in the digester. The concentration of hydrogen, however, is relatively low, and further processes are required to separate the hydrogen from the resulting gas mixture. As reported by Soto et al. [11], novel types of materials have become available in recent years to make gas membrane separation more effective, thus improving the competitiveness of bio-based hydrogen production. Improved membrane separation performance is of great important to other processes besides hydrogen production. A key application in the AD area is biogas upgrading to biomethane through the membrane-assisted removal of CO_2. One major advantage of biomethane is that it can directly substitute fossil natural gas in existing infrastructures.

Another approach to enhancing biomethane supply is production at biogas facilities through the biomethanation of CO_2 with hydrogen. While its commercial robustness remains to be confirmed, significant potential clearly exists. Bywater et al. [12] estimated that CO_2 biomethanation could raise the AD's contribution to bioenergy in the United Kingdom from 15 percent to 22 percent. There is, however, a relative shortage of reliable data on current UK AD feedstocks, which makes it difficult to quantify potential biomethane production from different substrates. There are also challenges related to the biomethanation process, especially when conducted in-situ, i.e., within the digester itself rather than in an external reactor. The work of Poggio et al. [13] contributed to a better understanding of the hydrogen gas–liquid mass transfer phenomena and to improving biomethanation in continuous AD. Another challenge of the process is the increase in pH because of CO_2 conversion into biomethane, potentially causing inhibition of the digestion process. Zhang et al. [14] presented a fundamentally-derived, experimentally validated approach to minimise such risks during in-situ biomethanation. These insights increase the feasibility of implementing the CO_2 biomethanation process in existing AD facilities, and, thus, of maximising the value of existing infrastructure while contributing to the decarbonisation goals [14].

A common theme across all publications of this Special Issue is making better use of infrastructure and biomass resources. This can be by increasing the efficiency of processes and performance of equipment, by reducing the risk of inhibition, by making substrates more available, or by integrating hydrogen and biomethane. As an approach with the explicit goal of making the best possible use of one unit of biomass, the concept of the biorefinery has evolved in the last decades; its main feature is to process biomass through different schemes operated widely in parallel, thus supplying a multitude of valuable outputs. Bolzonella et al. [15] present a biorefinery pilot plant based on AD that was designed to supply a set of products, namely, energy products (hydrogen, methane), chemicals (short chain volatile fatty acids, polyhydroxyalkanoates), and other materials (nutrients for agriculture, microbial proteins for food or animal feed applications). In such a biorefinery, AD becomes one integrated element of a larger system, as it is complemented by other processes (mechanical, chemical, or biological).

Clearly, applications of AD will continue to change in future, and further progress in making high-value use of this versatile bio-based technology can be expected. There are many aspects to be addressed by further research, and some current research questions are pointed out by the authors of this Special Issue. At the same time, the results of the Special Issue suggest that it is of particular relevance to inspire trust in the economic viability of AD facilities and the technical reliability of novel AD solutions.

Author Contributions: Conceptualization, S.K.-B., S.H. and C.J.B.; writing—original draft preparation, S.K.-B.; writing—review and editing, S.H. and C.J.B. All authors have read and agreed to the published version of the manuscript.

Funding: This research received no external funding.

Data Availability Statement: Not applicable.

Acknowledgments: As guest editors, we thank all the authors who have submitted their work to the Special Issue "New Frontiers in Anaerobic Digestion (AD) Processes". We are grateful to the reviewers who have invested their time to provide valuable feedback to the manuscripts, thus ensuring the high quality of the publications included. We thank the editorial team of *Processes* for the excellent support in compiling this Special Issue.

Conflicts of Interest: The authors declare no conflict of interest.

References

1. EBA. *EBA Statistical Report 2022*; European Biogas Association: Brussels, Belgium, 2022.
2. Yan, P.; Gai, M.; Wang, Y.; Gao, X. Review of soft sensors in anaerobic digestion process. *Processes* **2021**, *9*, 1434. [CrossRef]
3. Liu, Y.; Guo, W.; Longhurst, P.; Jiang, Y. Shortening the standard testing time for residual biogas potential (RBP) tests using biogas yield models and substrate physicochemical characteristics. *Processes* **2023**, *11*, 441. [CrossRef]
4. Zhang, Y.; Kusch-Brandt, S.; Salter, A.M.; Heaven, S. Estimating the methane potential of energy crops: An overview on types of data sources and their limitations. *Processes* **2021**, *9*, 1565. [CrossRef]
5. Wedwitschka, H.; Gallegos Ibanez, D.; Jáquez, D.R. Biogas production from residues of industrial insect protein production from black soldier fly larvae *Hermetia illucens* (L.): An evaluation of different insect frass samples. *Processes* **2023**, *11*, 362. [CrossRef]
6. Rosas-Mendoza, E.S.; Alvarado-Vallejo, A.; Vallejo-Cantú, N.A.; Snell-Castro, R.; Martínez-Hernández, S.; Alvarado-Lassman, A. Batch and semi-continuous anaerobic digestion of industrial solid citrus waste for the production of bioenergy. *Processes* **2021**, *9*, 648. [CrossRef]
7. Zerback, T.; Schumacher, B.; Weinrich, S.; Hülsemann, B.; Nelles, M. Hydrothermal pretreatment of wheat straw—Evaluating the effect of substrate disintegration on the digestibility in anaerobic digestion. *Processes* **2022**, *10*, 1048. [CrossRef]
8. Berzal de Frutos, O.; Götze, M.; Pidou, M.; Bajón Fernández, Y. Anaerobic Co-digestion of sewage sludge and trade wastes: Beneficial and inhibitory effects of individual constituents. *Processes* **2023**, *11*, 519. [CrossRef]
9. Pacheco-Ruiz, S.; Heaven, S.; Banks, C.J. Operation of submerged anaerobic membrane bioreactors at 20 °C: Effect of solids retention time on flux, mixed liquor characteristics and performance. *Processes* **2021**, *9*, 1525. [CrossRef]
10. Bywater, A.; Kusch-Brandt, S. Exploring farm anaerobic digester economic viability in a time of policy change in the UK. *Processes* **2022**, *10*, 212. [CrossRef]
11. Soto, C.; Palacio, L.; Muñoz, R.; Prádanos, P.; Hernandez, A. Recent advances in membrane-based biogas and biohydrogen upgrading. *Processes* **2022**, *10*, 1918. [CrossRef]
12. Bywater, A.; Heaven, S.; Zhang, Y.; Banks, C.J. Potential for biomethanisation of CO_2 from anaerobic digestion of organic wastes in the United Kingdom. *Processes* **2022**, *10*, 1202. [CrossRef]
13. Poggio, D.; Sastraatmaja, A.; Walker, M.; Michailos, S.; Nimmo, W.; Pourkashanian, M. Experimental evaluation of continuous in-situ biomethanation of CO_2 in anaerobic digesters fed on sewage sludge and food waste and the influence of hydrogen gas–liquid mass transfer. *Processes* **2023**, *11*, 604. [CrossRef]
14. Zhang, Y.; Heaven, S.; Banks, C.J. Validation of two theoretically derived equations for predicting pH in CO_2 biomethanisation. *Processes* **2023**, *11*, 113. [CrossRef]
15. Bolzonella, D.; Bertasini, D.; Lo Coco, R.; Menini, M.; Rizzioli, F.; Zuliani, A.; Battista, F.; Frison, N.; Jelic, A.; Pesante, G. Toward the transition of agricultural anaerobic digesters into multiproduct biorefineries. *Processes* **2023**, *11*, 415. [CrossRef]

processes

Review

Review of Soft Sensors in Anaerobic Digestion Process

Pengfei Yan [1], Minghui Gai [1], Yuhong Wang [1,*] and Xiaoyong Gao [2]

[1] College of Control Science and Engineering, China University of Petroleum, Qingdao 266580, China; w962995032@foxmail.com (P.Y.); g_mh0822@163.com (M.G.)
[2] Department of Automation, China University of Petroleum, Beijing 102249, China; x.gao@cup.edu.cn
* Correspondence: Y.H.Wang@upc.edu.cn

Abstract: Anaerobic digestion is associated with various crucial variables, such as biogas yield, chemical oxygen demand, and volatile fatty acid concentration. Real-time monitoring of these variables can not only reflect the process of anaerobic digestion directly but also accelerate the efficiency of resource conversion and improve the stability of the reaction process. However, the current real-time monitoring equipment on the market cannot be widely used in the industrial production process due to its defects such as expensive equipment, low accuracy, and lagging analysis. Therefore, it is essential to conduct soft sensor modeling for unmeasurable variables and use auxiliary variables to realize real-time monitoring, optimization, and control of the an-aerobic digestion process. In this paper, the basic principle and process flow of anaerobic digestion are first briefly introduced. Subsequently, the development history of the traditional soft sensor is systematically reviewed, the latest development of soft sensors was detailed, and the obstacles of the soft sensor in the industrial production process are discussed. Finally, the future development trend of deep learning in soft sensors is deeply discussed, and future research directions are provided.

Keywords: anaerobic digestion; soft sensor; deep learning

Citation: Yan, P.; Gai, M.; Wang, Y.; Gao, X. Review of Soft Sensors in Anaerobic Digestion Process. *Processes* **2021**, *9*, 1434. https://doi.org/10.3390/pr9081434

Academic Editors: Sonia Heaven, Sigrid Kusch-Brandt and Charles Banks

Received: 1 July 2021
Accepted: 16 August 2021
Published: 19 August 2021

1. Introduction

Anaerobic digestion is a highly complex biochemical reactions process, with characteristics such as multi-factor influence, dynamic change, and complex nonlinearity [1]. Anaerobic digestion can not only treat organic pollutants but also produce clean energy [2]. Therefore, anaerobic digestion technology has broad development space in the treatment of wastewater and organic solid waste [3] and is one of the practical ways to solve energy and environmental problems. However, anaerobic microorganisms of the anaerobic digestion process are intensely sensitive to changes in the digestion environment, and methanogens have extremely strict requirements on the external environment [3]. The unexpected changes in the external environment have an impact on the hydrolysis, acidification, and methanation processes of anaerobic digestion [4,5]. This will cause numerous volatile fatty acids (VFA) to accumulate in the reactor, inhibit the progress of methanation, and even result the failure of the anaerobic reactor operation [6–8]. Therefore, a more advanced online measurement system must be used to fully monitor the anaerobic digestion process in real-time to ensure that the anaerobic digestion process is stable and efficient while obtaining a higher biogas yield [9].

In terms of anaerobic digestion process variables monitoring, there is mature and reliable online monitoring equipment for temperature, pressure, flow rate, gas composition, and other variables [10,11]. However, there are still many key variables that cannot be directly measured, or the measurement equipment is expensive [12], such as biogas yield, chemical oxygen demand (COD), and VFA concentration. Online monitoring equipment for these variables cannot be widely used in industrial production due to factors such as expensive equipment, low accuracy, and lagging analysis [13–16]. Consequently, the soft sensor using online measurable auxiliary variables to estimate the unmeasurable variables in real-time has been broadly used in the anaerobic digestion process [17,18]. The soft sensor is developed based on the inference control theory proposed by *Brosilow* [19], suggesting that the mathematical relationship between auxiliary variables and target variables is established under certain optimal criteria, and the selection of auxiliary variables should be measurable and easy-to-obtain [20]. Real time monitoring of target variables is achieved through software [21]. Since the soft sensor has the advantages of fast response, low cost, easy implementation, and simple maintenance [22], it has been widely used in monitoring, optimization, and control of engineering [23]. Soft-sensor technology is broadly based on two modelling approaches: those derived mechanistically and those that are data-driven [24]. Specifically, mechanism models can be classified into common mechanism models and state estimation and system identification based on mechanism models [25]. Data-driven models can be divided into statistical machine learning models and deep learning models.

In this paper, the basic principle and process flow of anaerobic digestion are first briefly introduced. Subsequently, the soft sensors in the anaerobic digestion process are compared and analyzed, the development process of traditional soft sensors is systematically reviewed, and the defects of traditional soft sensors are presented. Next, the latest development of soft sensors is detailed, including the application of deep learning in the anaerobic digestion process. Moreover, the obstacles encountered by soft sensors in industrial production are further discussed. Finally, the future development trend of deep learning in soft sensors is deeply analyzed, and a summary and outlook are drawn.

2. Anaerobic Digestion Process

2.1. Basic Principles of Anaerobic Digestion

According to the four-stage theory of anaerobic digestion proposed by *Zeikus*, the anaerobic digestion process can be divided into four stages: hydrolysis, acidification, acetic acidification, and methanation [26]. In the hydrolysis stage, the hydrolase hydrolyzes macromolecular organics (such as protein, fat, and cellulose) into small molecular organics (such as glucose, amino acids, and long-chain fatty acids) for subsequent reactions [26]. After the initial hydrolysis, small-molecule organic substances (such as glucose and amino acids) will be further decomposed by acid-producing bacteria to produce acidified products mainly short-chain fatty acids and secondary metabolites (such as hydrogen and carbon dioxide) [27]. In the acetification stage, acetogens convert the organic acids and alcohols produced in the hydrolysis and acidification stages into acetic acid, generating carbon dioxide and hydrogen [28]. In the methanation stage, acetic acid, hydrogen, and carbon dioxide are converted into methane under the action of obligate anaerobic methanogens [29].

2.2. Process Parameters of Anaerobic Digestion

There are some essential process variables in the anaerobic digestion process, such as pH, alkalinity, temperature, VFA concentration, COD, and biogas yield. Real-time monitoring of the above variables can ensure the efficient and stable operation of the anaerobic digestion process. However, there is little widely used real-time monitoring equipment for VFA concentration, COD, and biogas yield.

1. PH: The optimal pH range of different microorganisms is different. Methanogens are extremely sensitive to pH, and the optimal pH range is 6.5–7.2 [30]. The fermenting microorganisms produce acetic acid and butyric acid when the pH is low. Acetic acid and propionic acid are formed when the pH is higher than 8.0 [31]. Therefore, reasonable monitoring of pH can ensure the maximum biological activity of microorganisms.
2. Alkalinity: Methanogens usually produce alkalinity in the form of carbon dioxide, ammonia, and bicarbonate, contributing to neutralizing VFA produced during anaerobic digestion [32]. Thus, real-time monitoring of alkalinity can improve the stability of the anaerobic digestion process when the concentration of carbon dioxide is stable.

3. Temperature: Temperature has a crucial influence on the physical and chemical properties of anaerobic digestion and fermentation substrates. It affects the growth rate and metabolism of microorganisms, which in turn influences the population dynamics of the anaerobic digestion process [33]. When the temperature changes more than 1 °C/day, the biochemical activity of methanogens will be severely affected, causing the process to fail.
4. VFA concentration: VFA concentration is an intermediate product of the anaerobic digestion process. Excessive accumulation of VFA can reduce the pH of the system and inhibit the activity of methanogens. The VFA concentration can reflect the current operating conditions of the system while being extremely sensitive to the incoming feed imbalance [34]. Hence, it is urgent to establish a soft sensor to predict the VFA concentration by monitoring the measurable and easy-to-obtain process variables in real-time.
5. COD and biogas yield: COD is an imperative indicator to measure the organic content of the effluent from the anaerobic digestion process [35]. Biogas yield is a vital indicator to measure the efficiency of anaerobic digestion [36]. Real-time monitoring of COD and biogas yield can demonstrate the operating efficiency and stability of the anaerobic digestion process and contribute to achieving the real-time calibration and optimization of production conditions and control methods.

2.3. *Anaerobic Digestion Process*

In the industrial production process, anaerobic digestion processes are usually classified according to factors such as operating temperature, feeding method, and the number of reactors [37]. It can be divided into single-phase digestion and two-phase digestion based on the number of reactors [38]. The single-phase digestion process was widely used in the immature stage of the early anaerobic digestion theory due to its low price and simple operation. Single-phase digestion suggests that the hydrolysis, acidification, acetic acidification, and methanation processes of degrading macromolecular organics are all conducted in the same digestion tank, and the inhibition of any one step will affect the overall digestion efficiency [39]. With the development of the anaerobic digestion theory, researchers and technologists have developed a two-phase digestion process to avoid acid inhibition. Two-phase anaerobic digestion suggests the hydrolysis, acidification, and acetic acid stages are conducted in the acid production tank, while the methane production stage is performed in the methane production tank [40]. This method can effectively avoid mutual inhibition between the steps, improve the efficiency of anaerobic digestion, shorten the reaction time, and increase methane production [41].

According to the biodegradability of the input materials, different two-phase anaerobic digestion devices are generally selected [42]. When industrial wastewater is treated with low solid content, the acid production tank and the methane production tank usually adopt a continuous stirred tank reactor and an up-flow anaerobic sludge blanket, respectively [43]. When organic wastewater is treated with high solid content, both the acid production tank and the methane production tank use the up-flow solid reactor [44]. When organic sludge is processed with higher solid content, both the acid production tank and the methane production tank employ the continuous stirred tank reactor [45]. The specific process flow is described as follows [28]. First, the pretreated organic materials are fed into the hydrolysis acidification tank to perform the hydrolysis reaction of macromolecular organics and the acidification reaction of small molecular organics. Then, the acidified product is input into the methane-generating tank for methane production reaction. Since the stages of acid production and methane production are performed separately, it is ensured that acid-producing bacteria and methanogens are in optimal environmental conditions and can exert maximum activity. Moreover, the acid production process improves the biochemical properties of the material, and the acidified product provides a suitable substrate for methanogens. The two-phase anaerobic digestion process is illustrated in Figure 1.

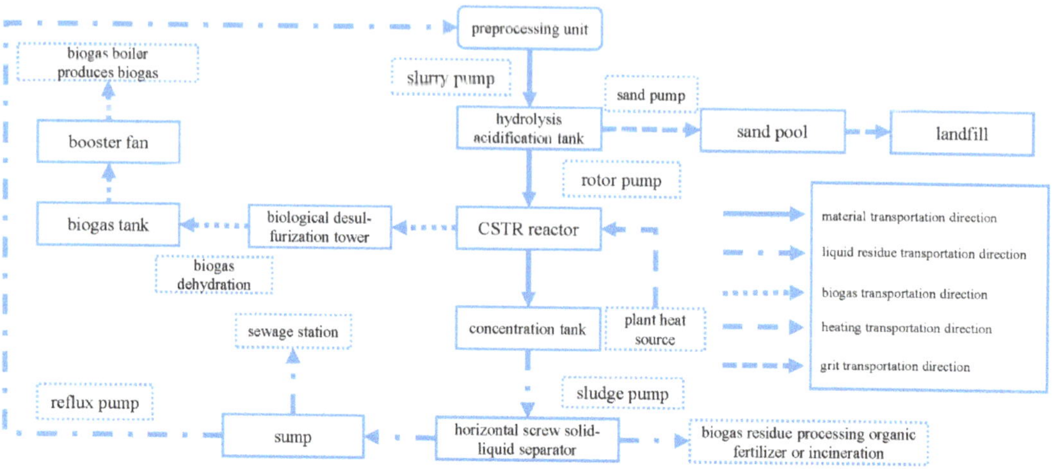

Figure 1. Two-phase anaerobic digestion process flow chart.

3. Development History of Anaerobic Digestion Soft Sensor

3.1. Soft Sensor Based on Process Mechanism

Mechanism modeling is to determine the mathematical relationship between the target variables and the auxiliary variables through the establishment of a balance equation based on a deep understanding of the process mechanism [23]. It has the advantages of high accuracy, strong interpretability, and clear industrial background. However, the biochemical reaction process of anaerobic digestion is extremely complicated, with strong nonlinearity and uncertainty, making it difficult to establish an accurate mechanism model [46,47]. Moreover, the biochemical reaction process is described by a large number of algebraic equations and differential equations. Therefore, there are defects such as large calculation amount and slow convergence, impeding it in meeting the requirements of real-time monitoring of target variables [48–50]. From another perspective, the mechanism model parameters of anaerobic digestion, such as Monod maximum specific absorption rate and the first-order decay rate in the kinetic parameters of the Anaerobic Digestion Model No.1 (ADM1), are mostly empirical values [51]. The determination of these parameters requires considerable experimental verifications, and the various indicators in the industrial production process will not be tested. Therefore, it is proposed to combine the mechanism model and the data-driven model to establish a hybrid model of anaerobic digestion [52,53]. The hybrid model fully takes advantage of the data-driven model that only pays attention to input and output and does not require a clear internal mechanism, contributing to a decrease in the difficulty of modeling the mechanism model. Moreover, the interpretability of the data-driven model is enhanced using the mechanism model. However, the prediction accuracy and generalization ability of the hybrid model need to be further improved.

3.2. Soft Sensor Based on State Estimation

In the soft sensor based on state estimation, the method of state observation and state estimation is adopted to obtain the predicted value of the state variable through auxiliary variables and then acquire the predicted value of the target variable [54,55]. With the development of anaerobic digestion soft sensors, various soft sensors based on state estimation have been proposed [56–61]. Among them, the nonlinear observer presented by *Dochain* under the improved anaerobic digestion model can estimate the VFA concentration online under different working conditions [62]. The improved anaerobic digestion model can be expressed as:

$$\begin{cases} \dot{x} = f(x,\mu) + \Delta f \\ y = Cx \end{cases} \tag{1}$$

where Δf denotes the uncertainty item related to unmodeled dynamics and load disturbance; x is the vector of dynamic states; f denotes the vector field; C = [0, 0, 1]; u and y denote the input and output of the model, respectively. For the improved anaerobic digestion model, the nonlinear observer can be expressed as:

$$\dot{\hat{x}} = f(\hat{x}, \mu) + k_l(y - \hat{y}) + k_d tanh[\gamma(y - \hat{y})] \tag{2}$$

where $\hat{x} \in R^3$ represents the state estimation vector, $\hat{y}$ indicates the predicted value of the output signal, and k_l, k_d and γ denote the observer gains. The estimated error of the model is presented in Formula (3).

$$\dot{e} = f(x, u) - f(\hat{x}, u) + \Delta f - k_l Ce - k_d tanh(\gamma Ce) \tag{3}$$

where $Ce = y - \hat{y}$. This nonlinear observer overcomes the disadvantage of the poor performance of the local observer under non-set conditions and solves the problem that the progressive observer is very sensitive to unknown load disturbances [63]. Additionally, the author has verified the convergence of the observer through Lyapunov stability. Soft sensors based on state estimation can handle situations such as dynamic characteristic differences between the variables and system lag. However, state estimation mainly applies to mature models and models that can reflect the characteristics of the measured object after approximation. Moreover, an increase in the online estimation error would be caused by simplifying the system to reduce the difficulty of modeling, and the use of this method would be restricted by the anaerobic digestion model's requirements for modeling accuracy [64–67].

3.3. Soft Sensor Based on Regression Analysis

The soft sensor of anaerobic digestion based on regression analysis majorly includes soft sensors based on multiple linear regression (MLR) and soft sensors based on partial least squares regression (PLSR).

MLR is able to establish a linear mapping between auxiliary variables and target variables through the least square method [68]. The soft sensor of anaerobic digestion based on MLR proposed by HU assumes the following linear relationship between auxiliary variables and biogas yield [69]:

$$\hat{y} = \theta_0 + \theta_1 X_1 + \theta_2 X_2 + \cdots + \theta_n X_n \tag{4}$$

where X is the auxiliary variable, θ is the parameter to be calculated, and $\hat{y}$ is the predicted value of biogas yield. The target parameter θ is solved by minimizing the error of the real biogas yield and the predicted biogas yield with the least squares method. However, the biochemical reaction process of anaerobic digestion is significantly nonlinear, and MLR cannot accurately describe the nonlinear process. Therefore, the anaerobic digestion soft sensor based on MLR has disadvantages such as low accuracy and susceptibility to external interference [70].

The anaerobic digestion soft sensor based on PLSR, which was proposed by Yang [1], can extract the principal components of auxiliary variables and target variables while maximizing the correlation between them [71]. The objective function of the soft sensor is expressed as

$$maxCov(t, y) = \sqrt{var(t)var(y)}corr(t, y) \tag{5}$$

where t represents the main component of the auxiliary variable, and y denotes the COD. The Lagrange multiplier l is introduced to solve the objective function.

$$l = p^T x^T y - \frac{\lambda}{2}\left(p^T p - 1\right) \tag{6}$$

where x and p indicate the auxiliary variable and the weight coefficient, respectively. Subsequently, the linear fitting between the principal component and the COD is realized by the MLR algorithm. This model solves the problem of the collinearity of auxiliary variables in the anaerobic digestion process. Unfortunately, the process of dimensionality reduction may eliminate the secondary principal components that are beneficial to regression and retain irrelevant noise, affecting the accuracy of the model. Meanwhile, PLSR is a linear algorithm and is only suitable for linear and weakly nonlinear models. However, there is severe nonlinearity in the anaerobic digestion process, limiting the prediction accuracy and generalization ability of the model.

3.4. Soft Sensor Based on Artificial Neural Network

Artificial neural networks can establish a non-linear mapping relationship between auxiliary variables and target variables through network learning, including back propagation (BP) neural networks and radial basis function (RBF) neural networks.

The soft sensor based on the BP neural network for the anaerobic digestion process was proposed by researchers [72–78]. In this soft sensor, the gradient descent algorithm is used to update the network weight. Therefore, the soft sensor can approximate the continuous nonlinear function with arbitrary precision and solve the highly nonlinear and uncertain problems in the anaerobic digestion process [79,80]. However, it is prone to fall into a local optimal or over-fitting state, affecting the prediction accuracy and generalization ability of the soft sensor [81].

To handle the complication that the anaerobic digestion soft sensor based on the BP neural network is prone to fall into the local minimum, Yilmaz proposed a soft sensor based on the RBF neural network to predict COD [82]. The soft sensor based on the RBF neural network has the characteristics of global best approximation and strong nonlinear mapping ability. The loss function of the soft sensor is expressed as

$$Loss = \frac{1}{2}||Y - \hat{Y}^2|| + \frac{1}{2}\lambda||D\hat{Y}||^2 \tag{7}$$

where Y and $\hat{Y}$ denote the test and predicted values of COD, respectively; λ represents the weighting factor of the regular term; D indicates the linear differential operator. With the regularization term, the curvature of the approximation function can be controlled, and the problem that the model is prone to overfitting is addressed.

The soft sensor based on the neural network can better handle the problem of non-linearity in the anaerobic digestion process. However, the performance of soft sensors is dramatically affected by the network topology and hyperparameters in practical applications. Therefore, proper hyperparameters and network topology are selected through optimization algorithms such as genetic algorithm and particle swarm optimization algorithm to improve model prediction accuracy and generalization ability [83–87].

3.5. Soft Sensor Based on Statistical Machine Learning

The soft sensor based on support-vector regression (SVR) uses the kernel function to map auxiliary variables to the high-dimensional feature space and adopts linear algorithms to analyze the nonlinear characteristics of the samples in the high-dimensional feature space. The convex quadratic programming is solved by the structural risk minimization criterion, which also addresses the high-dimensional and small-sample problems that cannot be solved by artificial neural networks [88]. Given the small sample problem caused by the difficulty of obtaining target variables in the anaerobic digestion process, Kazemi proposed the soft sensor based on SVR to predict the VFA concentration [89]. The loss function of the soft sensor is expressed as

$$Loss = \frac{1}{2}\sum_{i,j=1}^{m}(a_i - a_i^*)\left(a_j - a_j^*\right)k\left(x_i, x_j\right) + \sum_{i}^{m}a_i(\varepsilon - y_i) + a_i^*(\varepsilon + y_i) \tag{8}$$

The constraints are

$$s.t\left\{\begin{array}{c}\sum_{i=1}^{m}(a_i - a_i^*) = 0\\ a_i, a_i^* \in [0, C]\end{array}\right. \tag{9}$$

where x is the auxiliary variable; y indicates the VFA concentration; a_i and a_i^* are Lagrangian multipliers; $k(\cdot)$ represents the kernel function; ε is the insensitivity coefficient.

Given the problem of high complexity in solving SVR models, a soft sensor based on least-squares support-vector regression (LS-SVR) was proposed by Liu to monitor the VFA concentration in the anaerobic digestion process in real-time [90]. In the soft sensor based on LS-SVR, the slack variable in the optimization objective is replaced with the quadratic square term of the training error.

$$Loss = \frac{1}{2}\|w\|^2 + \frac{1}{2}\gamma\sum_{i=1}^{m}\xi_i^2 \tag{10}$$

Then, the inequality constraints are replaced with the following equality constraints.

$$y_i(wx_i + b) = 1 - \xi_i \tag{11}$$

where w and b indicates the learnable parameter of the model, γ denotes the regularization coefficient, and ξ refers to the training error. Solving the problem of convex quadratic programming is transformed into solving a set of linear equations, reducing the complexity of the model. However, the simplified soft sensor is more sensitive to abnormal values in the anaerobic digestion process, weakening the robustness of the soft sensor. Therefore, optimization algorithms are used to select the appropriate kernel function and hyperparameters to improve the prediction accuracy and generalization ability of the model [91–93].

3.6. Practical Application of Soft Sensors for Anaerobic Digestion

The soft sensor of anaerobic digestion is widely used in various industries owing to its advantages of low price, easy development, and maintenance. The soft sensor based on the process mechanism proposed by Fan [53] is employed to predict the bacterial concentration of high-temperature anaerobic digestion of cow manure. The kinetic model of anaerobic digestion of cow manure is expressed as:

$$\frac{dX}{dt} = \mu_{max}X\left(1 - \frac{X}{X_{max}}\right) \tag{12}$$

$$\frac{dP}{dt} = k_3X - k_4\frac{dX}{dt} \tag{13}$$

$$-\frac{dS}{dt} = k_1\frac{dX}{dt} + k_2\frac{dP}{dt} \tag{14}$$

where X, P, and S denote cell concentration, product concentration, and substrate concentration, respectively; μ_{max} and X_{max} indicate the maximum growth rate and concentration of the bacteria, respectively; k_1, k_2, k_3, and k_4 represent the cell growth rate, acid production rate coefficient, total enzyme activity, and cell activity coefficient, respectively, and the latter two factors can directly affect the cell growth rate and fermentation cycle. It can be observed that the cell concentration and substrate concentration are the direct factors affecting anaerobic digestion. Therefore, the cell growth rate, acid production rate, total enzyme activity, and cell activity are selected as auxiliary variables. However, the versatility of the soft sensor is poor. The prediction accuracy of the model will significantly decrease when fermentation conditions and fermentation batches change. The robust nonlinear observer proposed by *Dochain* [62] is adopted to predict the VFA concentration during the anaerobic digestion process of industrial wastewater. The mass balance equation of anaerobic digestion is expressed as:

$$\dot{S} = u\left(S_f - S\right) - k_t\mu(\cdot)X \tag{15}$$

$$\dot{X} = \mu(\cdot)X - auX \tag{16}$$

$$Q_M = k_m\mu(\cdot)X \tag{17}$$

where X, S, and Q_M indicate the methanogenic biomass, the soluble organic substrate, and the methane outflow rate, respectively; k_t and k_m represent the yield coefficient related to substrate degradation and the yield coefficient of methane production, respectively; u, a, and $\mu(\cdot)$ denote the dilution rate, the proportion of bacteria that are not attached to the support, and the growth rate of methane bacteria, respectively. Considering the limited online monitoring equipment available in the actual factory, the soft sensor only uses the methane outflow rate as an auxiliary variable to predict the VFA concentration under different working conditions and has high engineering practicability. However, the prediction accuracy of the soft sensor is generally not high when an observation model is established by simplifying the biochemical reaction and mass balance equations. *Strik* [75] employed a soft sensor based on the BP neural network to predict the content of ammonia in biogas. According to the kinetic model of anaerobic digestion, the calculation formula of related variables in biogas can be expressed as:

$$C_N = \frac{C_{TAN} \times 10^{\mathrm{pH}}}{e^{\frac{6334}{273+T}} + 10^{\mathrm{pH}}} \times \left(1 + \frac{10^{-\mathrm{pH}}}{10^{-\left(0.09+\frac{273}{T}\right)}}\right)^{-1} \tag{18}$$

$$M_t = M_0 \cdot \left(1 - e^{-Kt}\right) \tag{19}$$

where C_N, C_{TAN}, T, pH, and K denote the ammonia content, the total inorganic nitrogen concentration, the reaction temperature of anaerobic digestion, the pH value of the collected sample, and the rate constant of methane production, respectively; M_0 and M_t represent the methane production potential and the cumulative methane production at time t, respectively. As revealed from the model, pH, total inorganic nitrogen concentration, ammonium ion concentration, and temperature are the direct factors influencing ammonia content, and methane production is its indirect influence factor. Therefore, the ammonia content, ammonium ion concentration, total inorganic nitrogen concentration, nitrogen loading rate, pH, biogas production, and organic loading rate in the reactor are selected as the auxiliary variables of the model. However, the soft sensor lacks a real-time correction function. With the changes in actual working conditions and external interference factors, the prediction accuracy of the model will continue to decrease.

4. The Latest Development of Anaerobic Digestion Soft Sensor

The previous chapter introduced traditional anaerobic digestion soft sensors, reflecting the mapping relationship between auxiliary variables and target parameters to a certain extent. The characteristics of traditional soft sensors are summarized in Table 1. However, soft sensors still face many challenges in practical applications. For example:

1. The traditional soft sensor cannot extract the deep features of auxiliary variables. The performance of traditional soft sensors depends on the auxiliary variables provided, and the selection of auxiliary variables requires rich prior knowledge [94].
2. The traditional soft sensor does not consider the large number of unlabeled samples in the anaerobic digestion process. There are many unlabeled samples in the anaerobic digestion process. The semi-supervised learning mechanism, which is used to mine unlabeled sample information, can effectively improve the prediction performance of soft sensors [95].

3. The traditional soft sensor does not consider the dynamic and time lag characteristics of anaerobic digestion. The traditional soft sensor cannot adapt to changes in work and production conditions, and the prediction accuracy of the soft sensor gradually deteriorates over time [96]. Meanwhile, the slow hydrolysis process of anaerobic digestion would lead to a certain time lag between the real-time monitoring variables of the acid-producing tank and the real-time monitoring variables of the methane-producing tank.
4. The traditional soft sensor only considers the mapping relationship between auxiliary variables and target variables while ignoring the mutual influence between auxiliary variables [97]. In the actual industry, the combined auxiliary variables are generally highly correlated with the target variable while the single auxiliary variable often has a weak correlation with the target variable.

Table 1. Advantages and disadvantages of traditional soft sensors.

Soft Sensors	Advantages of Soft Sensor	Defects of Soft Sensor
Soft sensor based on process mechanism	High precision, strong interpretability, clear industrial background	It is difficult to build an accurate mechanism model
Soft sensor based on state estimation	Solve the problem of dynamic characteristic differences and system lag between variables	Simplifying the system will increase forecast errors
Soft sensor based on MLR	Only consider the mapping relationship of data; do not require a clear internal mechanism	The accuracy is not high, and it is easily affected by external interference
Soft sensor based on PLSR	Solve the problem of collinearity between auxiliary variables	Inability to handle strong nonlinear problems
Soft sensor based on BP neural network	Able to achieve an arbitrary precision approximation of nonlinear functions	Easy to fall into local optimal or over-fitting state
Soft sensor based on RBF neural network	Realize the global best approximation and solve the local optimal problem	Affected by network topology and hyperparameters
Soft sensor based on SVR	Solve the problem of high dimensions and small samples	Unable to handle large-scale data
Soft sensor based on LS-SVR	Further reduce the complexity of the model and increase the calculation speed	Very sensitive to outliers and poor robustness

In this chapter, the latest developments in anaerobic digestion soft sensors are introduced in detail. Furthermore, suitable solutions have been proposed regarding the obstacles encountered by traditional soft sensors in the industrial production process.

4.1. Soft Sensors for Extracting Deep Features

The deep belief network (DBN) achieves the approximation of complex functions through unsupervised layer-by-layer pre-training and supervised backpropagation fine-tuning [98,99]. In the process of unsupervised pre-training, the auxiliary variables are subjected to nonlinear mapping through the stacked restricted Boltzmann machine to extract the abstract features of the training samples. In the process of supervised back-propagation fine-tuning, the weights are fine-tuning through the backpropagation of the supervised signal to realize the further adjustment and optimization of the weights of the network.

To overcome the dependence of the traditional anaerobic digestion soft sensor on the features selection, Li proposed a soft sensor based on a deep belief network to predict the concentration of VFA for the anaerobic digestion process [100]. The structure diagram is illustrated in Figure 2.

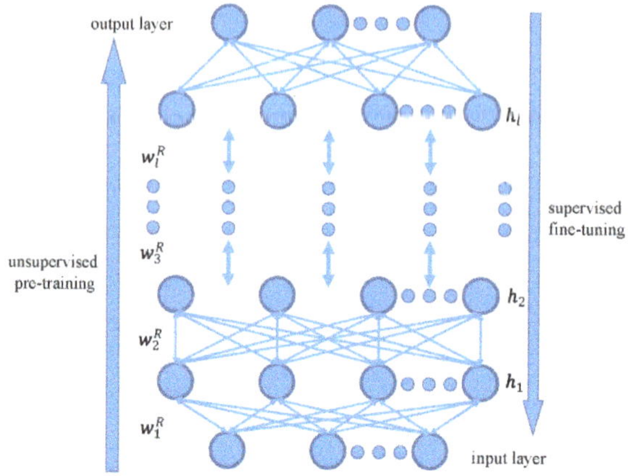

Figure 2. Deep belief network structure [100].

The gradient descent algorithm cannot effectively train the deep network. Therefore, the contrast divergence (CD) algorithm is adopted to update the weights of the restricted Boltzmann machine, layer by layer:

$$\begin{cases} w_{ij}^R = w_{ij}^R + \eta\left(v_i^{(t-1)}h_j^{(t-1)} - v_i^{(t)}h_j^{(t)}\right) \\ b_j = b_j + \eta\left(h_j^{(t-1)} - h_j^{(t)}\right) \\ a_i = a_i + \eta\left(v_i^{(t-1)} - v_i^{(t)}\right) \end{cases} \quad (20)$$

where v denotes the state vector of the visible layer, h refers to the state vector of the hidden layer, η represents the learning rate, and w and b denote the weights and biases of the network, respectively. The soft sensor, with excellent feature learning capabilities, can effectively learn the essential features from the training samples and address the defects of excessive dependence on prior knowledge in feature selection.

However, the random setting of the weights of DBN's output layer increases the randomness of the model's prediction performance. To further improve the stability of prediction performance and generalization performance, Li proposed to adopt the extreme learning machine (ELM) algorithm after the weights of the first n-1 layers were obtained using the CD algorithm to determine the weights of the output layer, and establish a soft sensor based on an improved deep belief network (IDBN) to predict the VFA concentration. IDBN structure diagram is presented in Figure 3.

$$\beta = h_{n-1}\left(w_i, \hat{b}_i\right)^+ y \quad (21)$$

where $h_{n-1}\left(w_i, \hat{b}_i\right)^+$ indicates the output of the hidden layer of the n-1 layer, β represents the weights of the output layer, and y denotes the VFA concentration. Compared with the soft sensor based on DBN, the improved soft sensor has preferable prediction accuracy and generalization performance in the experimental. However, the unsupervised layer-by-layer training process based on the CD algorithm requires a lot of iterative calculations, and the training process does not consider the mapping relationship between auxiliary variables and target variables. Therefore, Wang proposed a soft sensor based on the stacked supervised autoencoder combined with the kernel extreme learning machine (SSAE-KELM) algorithm to predict the VFA concentration [101]. The structure of SSAE-KELM is shown in Figure 4.

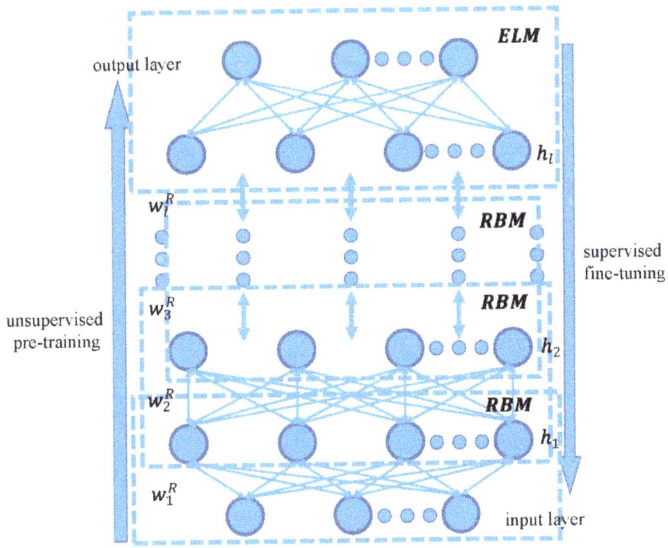

Figure 3. Improved deep belief network structure [100].

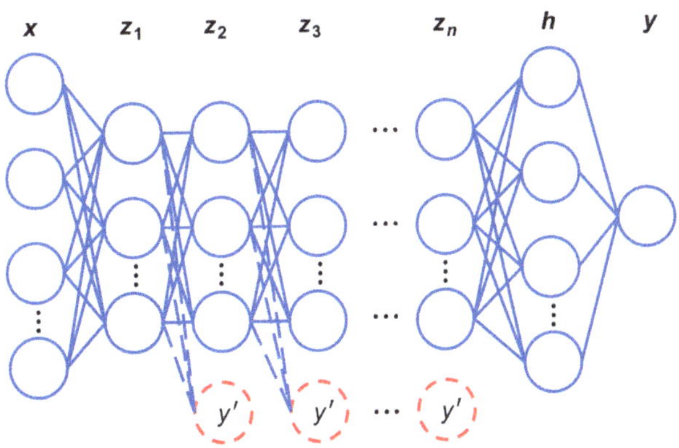

Figure 4. The stacked supervised autoencoder combined with the kernel extreme learning machine structure [101].

For the soft sensor, the ELM algorithm is employed to train supervised autoencoders (SAE), and the deep features of auxiliary variables are extracted through stacked SAE. The loss function of the training process is expressed as:

$$Loss = \frac{C_1}{2}\|X - Hr_1\|_2^2 + \frac{C_2}{2}\|Y - Hr_2\|_2^2 + \frac{1}{2}\|r\|_2^2 \tag{22}$$

By minimizing the loss function, the output weight is obtained:

$$r = [CH^T H - I_{m+1}{}^{-1} H^T Y]^T \tag{23}$$

where X refers to the auxiliary variables; Y represents the VFA concentration; H denotes the hidden layer output; r_1 and r_2 indicate the hidden layer weights and the supervised item weights, respectively; C_1 and C_2 are the weight coefficients. Finally, the kernel extreme learning machine is adopted to establish a regression model to predict the VFA concentration on the extracted deep abstract features. Compared with soft sensors based on IDBN, the soft sensor introduces supervised items by improving the loss function. As a result, the soft sensor can extract the deep features of the auxiliary variable while considering the mapping relationship between the auxiliary variable and the VFA concentration. Then, it can extract the essential features that have a greater impact on the VFA concentration. Moreover, the ELM algorithm is used to compensate for the shortcomings of the slow training speed of the traditional CD algorithm and improve the training efficiency of the model.

4.2. Soft Sensors for Extracting Information from Unlabeled Samples

In the anaerobic digestion process, the long period and high cost of target variable collection make it difficult for soft sensors to obtain sufficient labeled samples [102]. However, there are many unlabeled samples composed of process variables in the industrial process. With the semi-supervised learning mechanism, the information of unlabeled samples can be fully mined, and the prediction accuracy and generalization ability of soft sensors are improved. In recent years, semi-supervised learning mechanisms have been widely used in deep neural networks. Therefore, Yan proposed a soft sensor based on the semi-supervised hierarchical extreme learning machine to predict VFA concentration in the anaerobic digestion process [103]. The model structure of the semi-supervised hierarchical extreme learning machine is illustrated in Figure 5.

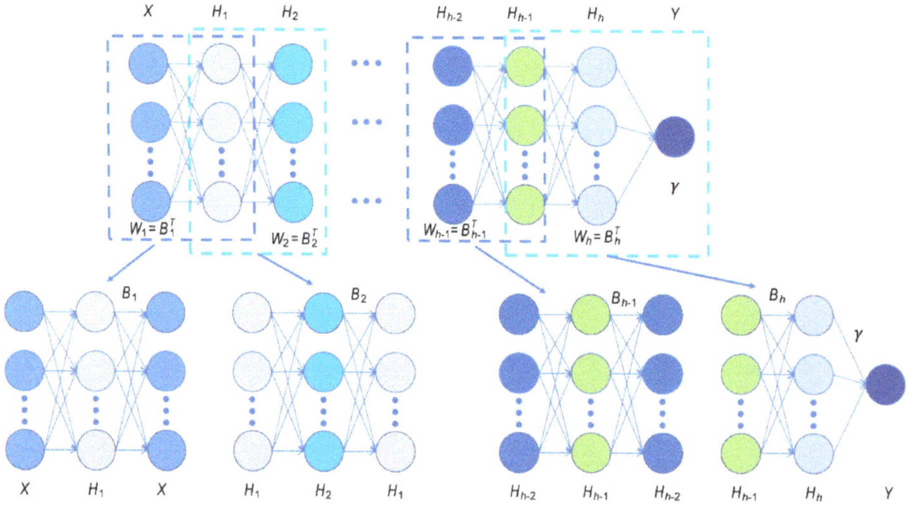

Figure 5. Semi-supervised hierarchical extreme learning machine structure [103].

Hierarchical extreme learning machine (HELM) is a multi-layer feedforward neural network composed of a multi-layer extreme learning machine-autoencoder (ELM-AE). During the training process, ELM-AE can achieve the lossless reconstruction of auxiliary variables. Therefore, the combined feature information of auxiliary variables can be extracted to a certain extent when the number of neurons in the hidden layer of ELM-AE is less than the number of neurons in the input layer [104]. The reconstruction loss function of ELM-AE is expressed as:

$$Loss = min\frac{1}{2}\|\gamma\|^2 + \frac{C}{2}\|Y - J\gamma\|^2 \tag{24}$$

The reconstruction loss function is minimized to obtain the output weight.

$$\gamma = \left(J^T J + \frac{1}{C} I_n\right)^{-1} J^T X \tag{25}$$

where γ indicates the weight of the output layer of ELM-AE; C is the weight factor; J denotes the output of the hidden layer; X and Y represent auxiliary variables and VFA concentration, respectively. Manifold regularization is used as a semi-supervised learning mechanism to learn the distribution of unlabeled samples. It can preserve the manifold domain relationship between the data vectors in the original space. The essential idea of manifold regularization is to keep the local geometric structure of the original feature space in the new projection space. The loss function of HELM that introduces the manifold regularization term is:

$$Loss = min\frac{1}{2}\|\gamma\|^2 + \frac{C}{2}\|Y - J\gamma\|^2 + \frac{\lambda}{2}Tr\left(\hat{Y}^T L \hat{Y}\right) \tag{26}$$

The loss function is minimized to acquire the output weight.

$$r = \left(I_n + CJ^T J + \lambda H^T L H\right)^{-1} CJ^T \tag{27}$$

where γ indicates the output layer weight of HELM; λ is the weight factor; $Tr(\cdot)$ represents the trace of the matrix; L refers to the graph Laplacian matrix; H and $\hat{Y}$ denote the hidden layer output and prediction output of all samples, respectively. Compared with traditional soft sensors, soft sensors based on a semi-supervised learning mechanism can learn both unlabeled sample information and label sample information. The semi-supervised learning mechanism can make full use of many unlabeled samples in the industrial process, contributing to the improvement of the prediction accuracy and generalization ability of soft sensors.

4.3. Soft Sensors for Extracting Dynamic Information

In the industrial production process of anaerobic digestion, changes in operating tasks, production materials, and production environment would cause changes in system operating conditions, making the prediction accuracy of soft sensors gradually decrease over time. Moreover, the different start-up times of the methane tank could lead to large differences in the digestion degree, substrate concentration, and biological activity, leading to inconsistent data distribution in the original data set. To handle this complication, Wang proposed to use the domain space transfer extreme learning machine (DSTELM) algorithm to adjust the data distribution [103]. The reconstruction loss function of DSTELM is:

$$loss = \frac{1}{2}\|r\|^2 + \frac{c}{2}\|X_T - H_T r\|^2 + \frac{\lambda}{2}Tr\left[r^T H^T M H r\right] \tag{28}$$

where c and λ are weighting factors; r denotes the output weight; X_T represents the auxiliary variables of the test set; $H = [H_S; H_T]$ indicates the output of the hidden layer; $Tr(\cdot)$ refers to the trace of the matrix. The M is defined as:

$$M = \begin{cases} \frac{1}{n_S^2} \; i,j \le n_S \\ \frac{1}{n_T^2} \; i,j > n_S \\ \frac{1}{n_S n_T} \; other \end{cases} \tag{29}$$

The loss function is minimized to obtain the output weight.

$$r = \left(I_L + H^T(C + \lambda M)H\right)^{-1} H^T C X_T \tag{30}$$

where $C = diag(0_{n_S \times n_S}, c, c, \ldots, c)$. The algorithm can minimize the distribution distance between the training set and the test set while retaining the essential characteristics of the test set. Moreover, it can address the problem of low model prediction accuracy caused by the inconsistent data distribution of the training set and the test set. Furthermore, a soft sensor based on the domain space migration hierarchical extreme learning machine (DSTHELM) is established by stacking DSTELM to extract the deep features of auxiliary variables. Compared with traditional soft sensors, soft sensors based on DSTHELM can better adapt to modal changes and data drift and thus present higher prediction accuracy and generalization ability.

Additionally, the hydrolysis reaction process is slow in the anaerobic digestion process, resulting in a certain time lag between the real-time monitoring variables of the acid-generating tank and the real-time monitoring variables of the methane generating tank. This suggests that the target variable is affected by the auxiliary variable in the current state, the changes in the operating conditions, and production conditions at the last moment, as well as the target variable in the current state. Therefore, *Mccormick* proposed a dynamic soft sensor based on long short-term memory (LSTM) network to predict biogas yield [105]. The LSTM structure is exhibited in Figure 6.

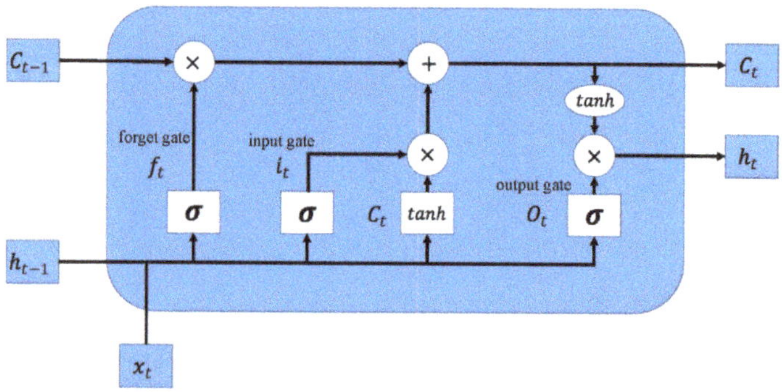

Figure 6. Long short-term memory network structure [105].

In the training process, the soft sensor realizes the retention or deletion of current information and historical information through the gate control unit. The input gate determines the extent to which the current input is retained to the current state. The forget gate determines the extent to which the state at the previous moment is retained to the current state. The output gate determines the extent to which the current state is retained to the output. The specific formulas are

$$i_t = \sigma(w_i \cdot [h_{t-1}, x_t] + b_i) \tag{31}$$

$$f_t = \sigma\left(w_f \cdot [h_{t-1}, x_t] + b_f\right) \tag{32}$$

$$h_t = \sigma(w_O \cdot [h_{t-1}, x_t] + b_O) \tag{33}$$

where i_t, f_t, h_t, and σ represent the input gate, the forget gate, the output gate, and the sigmoid activation function, respectively. The soft sensor can extract the different characteristics of the auxiliary variable at different times. Meanwhile, the soft sensor can retain historical biogas yield and its main influencing factors as auxiliary variables for current biogas yield forecasting, realizing the persistence of historical information.

The dynamic soft sensor considers the influence of historical data on the current state and overcomes the defect that the traditional soft sensor neglects the time scale information. Therefore, the dynamic soft sensor, to a certain extent, addresses the time lag caused by the slow reaction of the anaerobic digestion process. Furthermore, a dynamic soft sensor based on a combined convolutional neural network and long short-term memory network is established using the deep feature extraction ability of the convolutional neural network and the dynamic information extraction ability of LSTM to predict biogas yield. It can effectively extract the deep features of the data while using LSTM for timing error compensation. Thus, dynamic correction of the model is realized, and the prediction accuracy and generalization ability of the model are further improved.

4.4. Soft Sensors for Extracting Spatiotemporal Information

In recent years, the graph convolutional network (GCN) has been widely used, owing to its powerful feature representation ability [106]. GCN can reduce the complexity of the soft sensor through the parameter sharing of the convolution kernel in the local area. Moreover, the adjacency matrix of the GCN enables the soft sensor to quantify the mutual influence between auxiliary variables, that is, considering the degree of influence of surrounding nodes on the target node and extracting the spatial information of the sample data. In the actual industry, the combined auxiliary variables are generally highly correlated with the target variable while the single auxiliary variable often has a weak correlation with the target variable. Therefore, researchers proposed a soft sensor based on GCN to predict VFA concentration [107]. The GCN structure is exhibited in Figure 7.

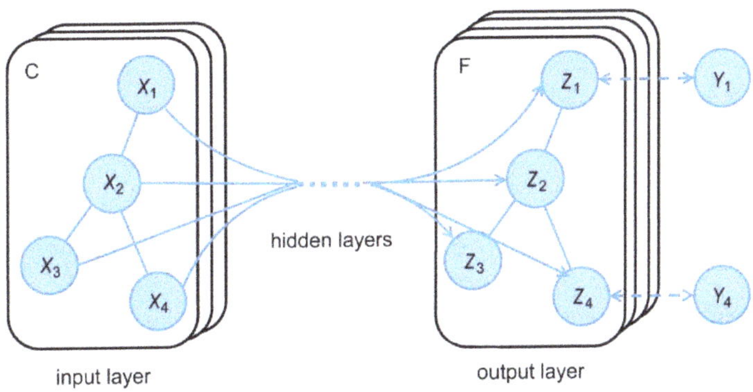

Figure 7. Graph convolutional network structure [107].

The output of the soft sensor can be expressed as:

$$Y = f(\hat{A}XW) \tag{34}$$

where X indicates the auxiliary variable; Y refers to the output of the soft sensor; f represents the nonlinear activation function; $\hat{A}$ is the normalized adjacency matrix; W denotes the learnable convolution kernel parameter. A proper adjacency matrix can be adopted to effectively extract the spatial information between auxiliary variables and improve the prediction accuracy and generalization ability of the soft sensor. Since the maximal information coefficient (MIC) can calculate the correlation between auxiliary variables, the normalized MIC is used to construct the adjacency matrix.

$$m_{ij} = MIC(x_i, x_j) \tag{35}$$

$$\alpha_{ij} = \frac{exp(m_{ij})}{\sum_{k \in N_i} exp(m_{ik})} \tag{36}$$

where m_{ij} represents the MIC between auxiliary variables i and j; α_{ij} denotes the normalized MIC between auxiliary variables i and j; $softmax$ indicates the normalization function. Compared with the traditional soft sensor, the soft sensor can learn the spatial information of the auxiliary variable by fully considering the influence of the combined feature information on the VFA concentration.

Given the dynamic characteristics and time lag characteristics of the anaerobic digestion process, a dynamic soft sensor based on the spatiotemporal graph convolutional network (STGCN) is established by introducing a gated recurrent unit (GRU). GRU can learn the dynamic changes of sample data to capture time information and consider the impact of historical sample information on current sample information. Therefore, this soft sensor can simultaneously consider the time information and spatial information of the anaerobic digestion process data. The structure of STGCN is presented in Figure 8.

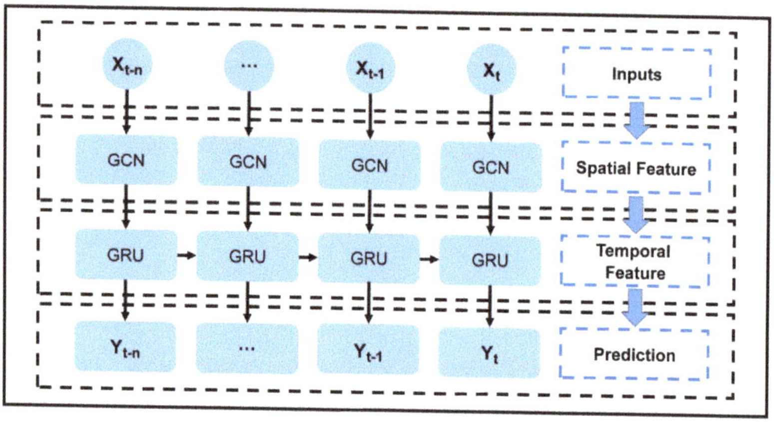

Figure 8. Spatiotemporal graph convolutional network structure [107].

During the training process, the STGCN can better handle the spatial and temporal characteristics of samples. The combined feature information of the sample is extracted using GCN to obtain its spatial dependence. Moreover, GRU is used to capture the dynamic change information of historical information and obtain temporal dependence. The specific calculation formulas are:

$$r_t = \sigma(W_r[h_{t-1}, f(X_t, A)]) \tag{37}$$

$$z_t = \sigma(W_z[h_{t-1}, f(X_t, A)]) \tag{38}$$

$$\widetilde{h}_t = tanh(W_{\widetilde{h}}[r_t \odot h_{t-1}, f(X_t, A)]) \tag{39}$$

$$h_t = (1 - z_t) \odot h_{t-1} + z_t \odot \widetilde{h}_t \tag{40}$$

where A is the adjacency matrix; $f(X_t, A)$ represents the graph convolution process; r_t denotes the reset gate; z_t represents the update gate; h refers to the state of the hidden layer; σ is the activation function; $\odot$ represents the Hadamard product. Compared with the traditional soft sensor, the dynamic soft sensor based on STGCN can effectively extract the time information and spatial information from the anaerobic digestion process data, contributing to the achievement of the accurate prediction of the current VFA concentration

5. Conclusions

The anaerobic digestion process is a time-varying, non-linear, and highly complex system with constraints. It is difficult to establish an accurate mechanism model to describe the anaerobic digestion process. The soft sensor based on regression analysis is more suitable for handling linear problems. However, there are strong nonlinear characteristics in the anaerobic digestion process. Soft sensors based on artificial neural networks are significantly affected by the network topology and the quality of training samples. They are prone to a local optimal or over-fitting state. Moreover, their generalization ability is weak. The soft sensor based on statistical learning is not suitable for processing large-scale data and is unable to monitor the anaerobic digestion process in real-time with high precision. However, soft sensors based on deep learning can learn essential features from training samples, introduce a semi-supervised learning mechanism to fully use unlabeled sample information, consider the dynamic characteristics in actual working conditions and the mutual mapping relationship between auxiliary variables, and extract the time information and space information of the sample data. Therefore, the soft sensor based on deep learning has higher prediction accuracy and generalization ability. The general idea of this paper is illustrated in Figure 9.

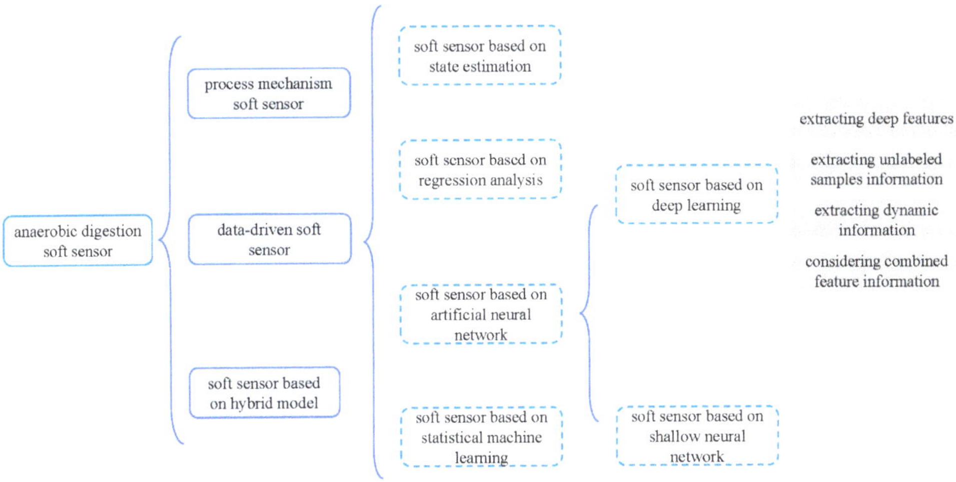

Figure 9. The general idea of this paper.

At present, a soft sensor for anaerobic digestion based on deep learning can be further developed. In the industrial production process, the mechanism model is combined with deep learning to enhance the interpretability of the soft sensor and realize the closed-loop guidance of the industrial process. Furthermore, the difficulty of sample collection during anaerobic digestion hinders researchers to obtain enough samples to train soft sensors. Therefore, constructing generated samples by the generative adversarial network is an effective solution for the shortage of soft sensor training samples.

Author Contributions: Conceptualization, P.Y. and Y.W.; methodology, P.Y. and Y.W.; software, P.Y.; validation, P.Y., Y.W., M.G. and X.G.; formal analysis, M.G.; investigation, Y.W.; resources, Y.W.; data curation, P.Y.; writing—original draft preparation, P.Y.; writing—review and editing, P.Y.; visualization, M.G. and X.G.; supervision, Y.W.; project administration, Y.W.; funding acquisition, X.G. All authors have read and agreed to the published version of the manuscript.

Funding: This research was funded by the National Natural Science Foundation of China (No. 21706282), National Key R&D Program of China (No. 2016YFC0303703), and the Research Foundation of China University of Petroleum (Beijing) (No. 2462020BJRC004, No. 2462020YXZZ023).

Institutional Review Board Statement: Not applicable.

Informed Consent Statement: Not applicable.

Conflicts of Interest: The authors declare no conflict of interest.

References

1. Yang, H.; Mo, W.-l.; Xiong, Z.-X.; Huang, M.-Z.; Liu, H.-B. Soft sensor modeling of papermaking effluent treatment processes using RPLS. *China Pulp Pap.* **2018**, *35*, 31–35. [CrossRef]
2. Yordanova, S.; Noikova, N.; Petrova, R.; Tzvetkov, P. Neuro-fuzzy modelling on experimental data in anaerobic digestion of organic waste in waters. In Proceedings of the 2005 IEEE Intelligent Data Acquisition and Advanced Computing Systems: Technology and Applications, Sofia, Bulgaria, 5–7 September 2005; pp. 84–88.

3. Liu, Z.-J.; Wan, J.-Q.; Ma, Y.-W.; Wang, Y. Online prediction of effluent COD in the anaerobic wastewater treatment system based on PCA-LSSVM algorithm. *Env. Sci. Pollut. Res. Int.* **2019**, *26*, 12828–12841. [CrossRef] [PubMed]
4. Bryant, M.P.; Wolin, E.A.; Wolin, M.J.; Wolfe, R.S. Methanobacillus omelianskii, a symbiotic association of two species of bacteria. *Arch. Mikrobiol.* **1967**, *59*, 20–31. [CrossRef] [PubMed]
5. Bryant, M.P. Microbial methane production—theoretical aspects. *J. Anim. Sci.* **1979**, *48*, 193–201. [CrossRef]
6. Franke-Whittle, I.H.; Walter, A.; Ebner, C.; Insam, H. Investigation into the effect of high concentrations of volatile fatty acids in anaerobic digestion on methanogenic communities. *Waste Manag.* **2014**, *34*, 2080–2089. [CrossRef]
7. Sbarciog, M.; Loccufier, M.; Noldus, E. Determination of appropriate operating strategies for anaerobic digestion systems. *Biochem. Eng. J.* **2010**, *51*, 180–188. [CrossRef]
8. Shen, S.; Premier, G.C.; Guwy, A.; Dinsdale, R. Bifurcation and stability analysis of an anaerobic digestion model. *Nonlinear Dyn.* **2007**, *48*, 391–408. [CrossRef]
9. Lara-Cisneros, G.; Aguilar-López, R.; Femat, R. On the dynamic optimization of methane production in anaerobic digestion via extremum-seeking control approach. *Comput. Chem. Eng.* **2015**, *75*, 49–59. [CrossRef]
10. Corona, F.; Mulas, M.; Haimi, H.; Sundell, L.; Heinonen, M.; Vahala, R. Monitoring nitrate concentrations in the denitrifying post-filtration unit of a municipal wastewater treatment plant. *J. Process Control* **2013**, *23*, 158–170. [CrossRef]
11. Jimenez, J.; Latrille, E.; Harmand, J.; Robles, A.; Ferrer, J.; Gaida, D.; Wolf, C.; Mairet, F.; Bernard, O.; Alcaraz-Gonzalez, V.; et al. Instrumentation and control of anaerobic digestion processes: A review and some research challenges. *Rev. Environ. Sci. Biotechnol.* **2015**, *14*, 615–648. [CrossRef]
12. Kawai, M.; Nagao, N.; Kawasaki, N.; Imai, A.; Toda, T. Improvement of COD removal by controlling the substrate degradability during the anaerobic digestion of recalcitrant wastewater. *J. Environ. Manag.* **2016**, *181*, 838–846. [CrossRef]
13. Gaida, D.; Wolf, C.; Meyer, C.; Stuhlsatz, A.; Lippel, J.; Bäck, T.; Bongards, M.; McLoone, S. State estimation for anaerobic digesters using the ADM1. *Water Sci. Technol.* **2012**, *66*, 1088–1095. [CrossRef] [PubMed]
14. Haimi, H.; Mulas, M.; Corona, F.; Vahala, R. Data-derived soft-sensors for biological wastewater treatment plants: An overview. *Environ. Model. Softw.* **2013**, *47*, 88–107. [CrossRef]
15. Gaida, D.; Wolf, C.; Bongards, M. Feed control of anaerobic digestion processes for renewable energy production: A review. *Renew. Sustain. Energy Rev.* **2017**, *68*, 869–875. [CrossRef]
16. Langergraber, G.; Fleischmann, N.; Hofstaedter, F.; Weingartner, A. Monitoring of a paper mill wastewater treatment plant using UV/VIS spectroscopy. *Water Sci. Technol.* **2004**, *49*, 9–14. [CrossRef] [PubMed]
17. Han, D.; Zou, Z. Soft sensor and inferential control technology. *J. Nanjing Univ. Sci. Technol.* **2005**, 212–216. [CrossRef]
18. Wade, M.J. Not just numbers: Mathematical modelling and its contribution to anaerobic digestion processes. *Processes* **2020**, *8*, 888. [CrossRef]
19. Brosilow, C.; Tong, M. Inferential control of processes: Part II. The structure and dynamics of inferential control systems. *AIChE* **1978**, *24*, 492–500. [CrossRef]
20. He, B.; Zhu, X. Soft-sensing technique based on extension method. In Proceedings of the SPIE 5253, Fifth International Symposium on Instrumentation and Control Technology, Beijing, China, 24–27 October 2003; pp. 38–42.
21. Wang, Z.-x.; Liu, Z.-w.; Xue, F.-x. Soft sensing technique for sewage treatment process. *J. Beijing Technol. Bus. Univ.* **2005**, *23*, 31–34. [CrossRef]
22. Yu, J.; Zhou, C. Soft-sensing techniques in process control. *Control Theory Appl.* **1996**, 137–144.
23. Zhu, X. Soft-sensing technique and its applications. *J. South China Univ. Technol.* **2002**, *30*, 61–67. [CrossRef]
24. James, S.C.; Legge, R.L.; Budman, H. On-line estimation in bioreactors: A review. *Rev. Chem. Eng.* **2000**, *16*, 311–340. [CrossRef]
25. Kadlec, P.; Gabrys, B.; Strandt, S. Data-driven soft sensors in the process industry. *Comput. Chem. Eng.* **2009**, *33*, 795–814. [CrossRef]
26. Zeikus, J. Microbial populations in digesters. In Proceedings of the First International Symposium on Anaerobic Digestion, London, UK, 17–21 September 1979.
27. Keymer, P.; Ruffell, I.; Pratt, S.; Lant, P. High pressure thermal hydrolysis as pre-treatment to increase the methane yield during anaerobic digestion of microalgae. *Bioresour. Technol.* **2013**, *131*, 128–133. [CrossRef]
28. Appels, L.; Baeyens, J.; Degrève, J.; Dewil, R. Principles and potential of the anaerobic digestion of waste-activated sludge. *Prog. Energy Combust. Sci.* **2008**, *34*, 755–781. [CrossRef]
29. Illi, L.; Lecker, B.; Lemmer, A.; Müller, J.; Oechsner, H. Biological methanation of injected hydrogen in a two-stage anaerobic digestion process. *Bioresour. Technol.* **2021**, *333*, 125126. [CrossRef]
30. Kazemi, P.; Bengoa, C.; Steyer, J.-P.; Giralt, J. Data-driven techniques for fault detection in anaerobic digestion process. *Process Saf. Environ. Prot.* **2021**, *146*, 905–915. [CrossRef]
31. Boe, K. *Online Monitoring and Control of the Biogas Process*; Institute of Environment & Resources, Technical University of Denmark: Copenhagen, Denmark, 2006.
32. Hwang, M.H.; Jang, N.J.; Hyun, S.H.; Kim, I.S. Anaerobic bio-hydrogen production from ethanol fermentation: The role of pH. *J. Biotechnol.* **2004**, *111*, 297–309. [CrossRef] [PubMed]
33. Stichting Toegepast Onderzoek Reiniging Afvalwater. *Optimalisatie van de Gistingsgasproduktie*; Stora: Amsterdam, The Netherlands, 1985.
34. Turovskiy, I.S.; Mathai, P. *Wastewater Sludge Processing*; John Wiley & Sons: Hoboken, NJ, USA, 2006.

35. Steyer, J.P.; Bouvier, J.C.; Conte, T.; Gras, P.; Harmand, J.; Delgenes, J.P. On-line measurements of COD, TOC, VFA, total and partial alkalinity in anaerobic digestion processes using infra-red spectrometry. *Water Sci. Technol.* **2002**, *45*, 133–138. [CrossRef] [PubMed]
36. Chae, K.J.; Jang, A.; Yim, S.K.; Kim, I.S. The effects of digestion temperature and temperature shock on the biogas yields from the mesophilic anaerobic digestion of swine manure. *Bioresour. Technol.* **2008**, *99*, 1–6. [CrossRef]
37. Massi, E. Anaerobic digestion. In *Fuel Cells in the Waste-to-Energy Chain: Distributed Generation through Non-Conventional Fuels and Fuel Cells*; McPhail, S.J., Cigolotti, V., Moreno, A., Eds.; Springer: London, UK, 2012; pp. 47–63.
38. Ren, Y.; Yu, M.; Wu, C.; Wang, Q.; Gao, M.; Huang, Q.; Liu, Y. A comprehensive review on food waste anaerobic digestion: Research updates and tendencies. *Bioresour. Technol.* **2018**, *247*, 1069–1076. [CrossRef]
39. Liu, X.; Han, Z.; Yang, J.; Ye, T.; Yang, F.; Wu, N.; Bao, Z. Review of enhanced processes for anaerobic digestion treatment of sewage sludge. *IOP Conf. Ser. Earth Environ. Sci.* **2018**, *113*, 012039. [CrossRef]
40. Ye, N.-F.; He, P.-J.; Lü, F.; Shao, L.-M. Effect of pH on microbial diversity and product distribution during anaerobic fermentation of vegetable waste. *Chin. J. Appl. Environ. Biol.* **2007**, *13*, 238–242. [CrossRef]
41. Adekunle, K.F.; Okolie, J.A. A review of biochemical process of anaerobic digestion. *Adv. Biosci. Biotechnol.* **2015**, *6*, 205–212. [CrossRef]
42. Khalid, A.; Arshad, M.; Anjum, M.; Mahmood, T.; Dawson, L. The anaerobic digestion of solid organic waste. *Waste Manag.* **2011**, *31*, 1737–1744. [CrossRef] [PubMed]
43. Lettinga, G. Anaerobic digestion and wastewater treatment systems. *Antonie Van Leeuwenhoek* **1995**, *67*, 3–28. [CrossRef] [PubMed]
44. Mumme, J.; Linke, B.; Tölle, R. Novel upflow anaerobic solid-state (UASS) reactor. *Bioresour. Technol.* **2010**, *101*, 592–599. [CrossRef]
45. Angelidaki, I.; Chen, X.; Cui, J.; Kaparaju, P.; Ellegaard, L. Thermophilic anaerobic digestion of source-sorted organic fraction of household municipal solid waste: Start-up procedure for continuously stirred tank reactor. *Water Res.* **2006**, *40*, 2621–2628. [CrossRef]
46. Tufaner, F.; Avşar, Y. Investigation of biogas production potential and adaptation to cattle manure of anaerobic flocular sludge seed. *Sigma* **2016**, *7*, 183–190.
47. Dalkılıc, K.; Ugurlu, A. Biogas production from chicken manure at different organic loading rates in a mesophilic-thermopilic two stage anaerobic system. *J. Biosci. Bioeng.* **2015**, *120*, 315–322. [CrossRef] [PubMed]
48. Moral, H.; Aksoy, A.; Gokcay, C.F. Modeling of the activated sludge process by using artificial neural networks with automated architecture screening. *Comput. Chem. Eng.* **2008**, *32*, 2471–2478. [CrossRef]
49. Güçlü, D.; Dursun, S. Amelioration of carbon removal prediction for an activated sludge process using an artificial neural network (ANN). *CLEAN–Soil Air Water* **2008**, *36*, 781–787. [CrossRef]
50. Fang, F.; Ni, B.-J.; Yu, H.-Q. Estimating the kinetic parameters of activated sludge storage using weighted non-linear least-squares and accelerating genetic algorithm. *Water Res.* **2009**, *43*, 2595–2604. [CrossRef] [PubMed]
51. Batstone, D.J.; Keller, J.; Angelidaki, I.; Kalyuzhnyi, S.V.; Pavlostathis, S.G.; Rozzi, A.; Sanders, W.T.M.; Siegrist, H.; Vavilin, V.A. The IWA anaerobic digestion model No 1 (ADM1). *Water Sci. Technol.* **2002**, *45*, 65–73. [CrossRef]
52. Bernard, O.; Hadj-Sadok, Z.; Dochain, D. Software sensors to monitor the dynamics of microbial communities: Application to anaerobic digestion. *Acta Biotheor.* **2000**, *48*, 197–205. [CrossRef]
53. Fan, Q.; Qin, G.; Zhang, L. Research and application on hybrid modeling for the monitoring of anaerobic-thermophilic fermentation of cattle manure. *Heilongjiang Sci.* **2013**, 45–47. [CrossRef]
54. Luenberger, D. An introduction to observers. *IEEE Trans. Automat. Contr.* **1971**, *16*, 596–602. [CrossRef]
55. Mohd Ali, J.; Ha Hoang, N.; Hussain, M.A.; Dochain, D. Review and classification of recent observers applied in chemical process systems. *Comput. Chem. Eng.* **2015**, *76*, 27–41. [CrossRef]
56. Bastin, G. *On-Line Estimation and Adaptive Control of Bioreactors*; Elsevier: Amsterdam, The Netherlands, 2013; Volume 1.
57. Dochain, D. State and parameter estimation in chemical and biochemical processes: A tutorial. *J. Process Control* **2003**, *13*, 801–818. [CrossRef]
58. Diop, S.; Simeonov, I. On the biomass specific growth rates estimation for anaerobic digestion using differential algebraic techniques. *Bioautomation* **2009**, *13*, 47–56.
59. Stanke, M.; Hitzmann, B. Automatic control of bioprocesses. In *Measurement, Monitoring, Modelling and Control of Bioprocesses*; Mandenius, C.-F., Titchener-Hooker, N.J., Eds.; Springer: Berlin/Heidelberg, Germany, 2013; pp. 35–63.
60. Kalchev, B.; Simeonov, I.; Christov, N. Kalman filter design for a second-order model of anaerobic digestion. *Int. J. Bioautomation* **2011**, *15*, 85–100.
61. Rodríguez, A.; Quiroz, G.; Femat, R.; Méndez-Acosta, H.O.; de León, J. An adaptive observer for operation monitoring of anaerobic digestion wastewater treatment. *Chem. Eng. J.* **2015**, *269*, 186–193. [CrossRef]
62. Lara-Cisneros, G.; Aguilar-López, R.; Dochain, D.; Femat, R. On-line estimation of VFA concentration in anaerobic digestion via methane outflow rate measurements. *Comput. Chem. Eng.* **2016**, *94*, 250–256. [CrossRef]
63. Haugen, F.; Bakke, R.; Lie, B. State estimation and model-based control of a pilot anaerobic digestion reactor. *J. Control Sci. Eng.* **2014**, *2014*, 572621. [CrossRef]
64. Benyahia, B.; Sari, T.; Cherki, B.; Harmand, J. Bifurcation and stability analysis of a two step model for monitoring anaerobic digestion processes. *J. Process Control* **2012**, *22*, 1008–1019. [CrossRef]

65. Hess, J.; Bernard, O. Design and study of a risk management criterion for an unstable anaerobic wastewater treatment process. *J. Process Control* **2008**, *18*, 71–79. [CrossRef]
66. Sbarciog, M.; Loccufier, M.; Vande Wouwer, A. On the optimization of biogas production in anaerobic digestion systems. *IFAC Proc. Vol.* **2011**, *44*, 7150–7155. [CrossRef]
67. Schaum, A.; Alvarez, J.; Garcia-Sandoval, J.P.; Gonzalez-Alvarez, V.M. On the dynamics and control of a class of continuous digesters. *J. Process Control* **2015**, *34*, 82–96. [CrossRef]
68. Eberly, L.E. Multiple linear regression. In *Topics in Biostatistics*; Ambrosius, W.T., Ed.; Humana Press: Totowa, NJ, USA, 2007; pp. 165–187.
69. Hu, K.-Q.; Li, L.-H.; Sun, Y.-M.; Kong, X.-Y.; Zhang, Y.; Yuan, Z.-H. The methane yield forecasting model of energy crops in anaerobic digestion based on feedstock components. *Adv. New Renew. Energy* **2016**, *4*, 100–104. [CrossRef]
70. Zhang, W.; Zhang, L.; Li, N.; Zhou, H. Comparing multiple regression and BP artificial nerve net model used on prediction of anaerobic co-digestion gas-producing process. *Chin. J. Environ. Eng.* **2013**, *7*, 747–752.
71. Mejdell, T.; Skogestad, S. Estimation of distillation compositions from multiple temperature measurements using partial-least-squares regression. *Ind. Eng. Chem. Res.* **1991**, *30*, 2543–2555. [CrossRef]
72. Tufaner, F.; Avşar, Y.; Gönüllü, M.T. Modeling of biogas production from cattle manure with co-digestion of different organic wastes using an artificial neural network. *Clean Techn. Environ. Policy* **2017**, *19*, 2255–2264. [CrossRef]
73. Güçlü, D.; Yılmaz, N.; Ozkan-Yucel, U.G. Application of neural network prediction model to full-scale anaerobic sludge digestion. *J. Chem. Technol. Biotechnol.* **2011**, *86*, 691–698. [CrossRef]
74. Holubar, P.; Zani, L.; Hager, M.; Fröschl, W.; Radak, Z.; Braun, R. Advanced controlling of anaerobic digestion by means of hierarchical neural networks. *Water Res.* **2002**, *36*, 2582–2588. [CrossRef]
75. Strik, D.P.B.T.B.; Domnanovich, A.M.; Zani, L.; Braun, R.; Holubar, P. Prediction of trace compounds in biogas from anaerobic digestion using the MATLAB Neural Network Toolbox. *Environ. Model. Softw.* **2005**, *20*, 803–810. [CrossRef]
76. Ozkaya, B.; Demir, A.; Bilgili, M.S. Neural network prediction model for the methane fraction in biogas from field-scale landfill bioreactors. *Environ. Model. Softw.* **2007**, *22*, 815–822. [CrossRef]
77. Sathish, S.; Vivekanandan, S. Parametric optimization for floating drum anaerobic bio-digester using response surface methodology and artificial neural network. *Alex. Eng. J.* **2016**, *55*, 3297–3307. [CrossRef]
78. Holubar, P.; Zani, L.; Hagar, M.; Fröschl, W.; Radak, Z.; Braun, R. Modelling of anaerobic digestion using self-organizing maps and artificial neural networks. *Water Sci. Technol.* **2000**, *41*, 149–156. [CrossRef]
79. Jacob, S.; Banerjee, R. Modeling and optimization of anaerobic codigestion of potato waste and aquatic weed by response surface methodology and artificial neural network coupled genetic algorithm. *Bioresour. Technol.* **2016**, *214*, 386–395. [CrossRef]
80. Abu Qdais, H.; Bani Hani, K.; Shatnawi, N. Modeling and optimization of biogas production from a waste digester using artificial neural network and genetic algorithm. *Resour. Conserv. Recycl.* **2010**, *54*, 359–363. [CrossRef]
81. Lu, J.-J.; Chen, H. Researching development on BP neural networks. *Control Eng. China* **2006**, *13*, 449–451. [CrossRef]
82. Yilmaz, T.; Seckin, G.; Yuceer, A. Modeling of effluent COD in UAF reactor treating cyanide containing wastewater using artificial neural network approaches. *Adv. Eng. Softw.* **2010**, *41*, 1005–1010. [CrossRef]
83. Han, W.; Huang, M.-z.; Ma, Y.-w.; Wan, J.-q. Multi-objective optimization in the anaerobic digestion of papermaking wastewater based on NSGA-2 and BP neural network. *Pap. Sci. Technol.* **2014**, *33*, 145–147, 165.
84. Huang, M.; Han, W.; Wan, J.; Ma, Y.; Chen, X. Multi-objective optimization for design and operation of anaerobic digestion using GA-ANN and NSGA-II. *J. Chem. Technol. Biotechnol.* **2016**, *91*, 226–233. [CrossRef]
85. Hua, Y.; Zhao, X.; Wang, X.; Teng, K. Prediction modeling for gas production of anaerobic fermentation based on improved BP neural network. *Chin. J. Environ. Eng.* **2016**, *10*, 5951–5956. [CrossRef]
86. Behera, S.K.; Meher, S.K.; Park, H.-S. Artificial neural network model for predicting methane percentage in biogas recovered from a landfill upon injection of liquid organic waste. *Clean Techn. Environ. Policy* **2015**, *17*, 443–453. [CrossRef]
87. Zhao, X.; Jiang, J. BP neural network modeling and particle swarm algorithm optimization of anaerobic fermentation process. *Appl. Energy Technol.* **2015**, 8–12. [CrossRef]
88. Liu, L.; Xie, B.; Ma, Y.; Wan, J.; Wang, Y. Hybrid model of measuring biogas yield in anaerobic digestion process based on incorporated bio-kinetic model with support vector machine model. *China Pulp Pap.* **2017**, *36*, 31–36. [CrossRef]
89. Kazemi, P.; Steyer, J.-P.; Bengoa, C.; Font, J.; Giralt, J. Robust data-driven soft sensors for online monitoring of volatile fatty acids in anaerobic digestion processes. *Processes* **2020**, *8*, 67. [CrossRef]
90. Liu, L.; Ma, Y.; Wan, J.; Wang, Y.; Xie, B.; Wu, S. An accuracy soft-sensing model for the estimation of anaerobic digestion process based on pso-SVM model. *Acta Sci. Circumstantiae* **2017**, *37*, 2122–2129. [CrossRef]
91. Cui, G.; Sun, T.; Zhang, Y. Forecast of blast furnace hot metal temperature based on least support vector machine. *Comput. Simul.* **2013**, *30*, 354–357. [CrossRef]
92. Sun, J.; Cheng, Z.; Yang, R.; Shan, S. Soft-sensor modeling for paper mill effluent COD based on PCA-PSO-LSSVM. *Comput. Appl. Chem.* **2017**, *34*, 706–710. [CrossRef]
93. Xing, Y.; Cheng, Z.; Shan, S. Dynamic soft sensing of organic pollutants in effluent from UMIC anaerobic reactor for industrial papermaking wastewater. *IOP Conf. Ser. Mater. Sci. Eng.* **2019**, *490*, 062027. [CrossRef]
94. Du, X.; Cai, Y.; Wang, S.; Zhang, L. Overview of deep learning. In Proceedings of the 31st Youth Academic Annual Conference of Chinese Association of Automation (YAC), Wuhan, China, 11–13 November 2016; pp. 159–164.

95. Yao, L.; Ge, Z. Deep learning of semisupervised process data with hierarchical extreme learning machine and soft sensor application. *ITIE* **2018**, *65*, 1490–1498. [CrossRef]
96. Yuan, X.; Li, L.; Shardt, Y.A.W.; Wang, Y.; Yang, C. Deep learning with spatiotemporal attention-based LSTM for industrial soft sensor model development. *ITIE* **2021**, *68*, 4404–4414. [CrossRef]
97. Cao, Y.; Liu, C.; Huang, Z.; Sheng, Y.; Ju, Y. Skeleton-based action recognition with temporal action graph and temporal adaptive graph convolution structure. *Multimed. Tools Appl.* **2021**. [CrossRef]
98. Bengio, Y. *Learning Deep Architectures for AI*; Now Publishers Inc.: Hanover, MA, USA, 2009.
99. Song, H.A.; Lee, S.-Y. Hierarchical representation using NMF. In Proceedings of the Neural Information Processing, Berlin, Heidelberg, 3 November 2013; pp. 466–473.
100. Wang, Y.; Li, X. Soft measurement for VFA concentration in anaerobic digestion for treating kitchen waste based on improved DBN. *IEEE Access* **2019**, *7*, 60931–60939. [CrossRef]
101. Wang, Y.; Wang, S. Soft sensor for VFA concentration in anaerobic digestion process for treating kitchen waste based on SSAE-KELM. *IEEE Access* **2021**, *9*, 36466–36474. [CrossRef]
102. Wang, X.; Liu, H. Data supplement for a soft sensor using a new generative model based on a variational autoencoder and Wasserstein GAN. *J. Process Control* **2020**, *85*, 91–99. [CrossRef]
103. Yan, P.; Shen, B.; Wang, Y. Soft sensor for VFA concentration in anaerobic digestion process for treating kitchen waste based on DSTHELM. *IEEE Access* **2020**, *8*, 223618–223625. [CrossRef]
104. Ranzato, M.; Huang, F.J.; Boureau, Y.; LeCun, Y. Unsupervised learning of invariant feature hierarchies with applications to object recognition. In Proceedings of the 2007 IEEE Conference on Computer Vision and Pattern Recognition, Minneapolis, MN, USA, 17–22 June 2007; pp. 1–8.
105. McCormick, M.; Villa, A.E.P. LSTM and 1-D convolutional neural networks for predictive monitoring of the anaerobic digestion process. In Proceedings of the Artificial Neural Networks and Machine Learning–ICANN 2019: Workshop and Special Sessions, Cham, Switzerland, 9 September 2019; pp. 725–736.
106. Zhuang, C.; Ma, Q. Dual graph convolutional networks for graph-based semi-supervised classification. In Proceedings of the Proceedings of the 2018 World Wide Web Conference, Lyon, France, 10 April 2018; pp. 499–508.
107. Wang, Y.; Yan, P.; Gai, M. Dynamic soft sensor for anaerobic digestion of kitchen waste based on SGSTGAT. *IEEE Sens. J.* **2021**, 1. [CrossRef]

processes

Article

Shortening the Standard Testing Time for Residual Biogas Potential (RBP) Tests Using Biogas Yield Models and Substrate Physicochemical Characteristics

Yanxin Liu [1], Weisi Guo [2], Philip Longhurst [1] and Ying Jiang [1,*]

1 School of Water, Energy and Environment, Cranfield University, Cranfield MK43 0AL, UK
2 School of Aerospace, Transport and Manufacturing, Cranfield University, Cranfield MK43 0AL, UK
* Correspondence: y.jiang@cranfield.ac.uk; Tel.: +44-(0)-1234-75-4492

Abstract: The residual biogas potential (RBP) test is a procedure to ensure the anaerobic digestion process performance and digestate stability. Standard protocols for RBP require a significant time for sample preparation, characterisation and testing of the rig setup followed by batch experiments of a minimum of 28 days. To reduce the experimental time to obtain the RBP result, four biogas kinetic models were evaluated for their strength of fit for biogas production data from RBP tests. It was found that the pseudo-parallel first-order model and the first-order autoregressive (AR (1)) model provide a high strength of fit and can predict the RBP result with good accuracy (absolute percentage errors < 10%) using experimental biogas production data of 15 days. Multivariate regression with decision trees (DTs) was adopted in this study to predict model parameters for the AR (1) model from substrate physicochemical parameters. The mean absolute percentage error (MAPE) of the predicted AR (1) model coefficients, the constants and the RBP test results at day 28 across DTs with 20 training set samples are 4.76%, 72.04% and 52.13%, respectively. Using five additional data points to perform the leave-one-out cross-validation method, the MAPEs decreased to 4.31%, 59.29% and 45.62%. This indicates that the prediction accuracy of DTs can be further improved with a larger training dataset. A Gaussian Process Regressor was guided by the DT-predicted AR (1) model to provide probability distribution information for the biogas yield prediction.

Keywords: RBP test; biogas yield models; decision trees; Gaussian process; regression

Citation: Liu, Y.; Guo, W.; Longhurst, P.; Jiang, Y. Shortening the Standard Testing Time for Residual Biogas Potential (RBP) Tests Using Biogas Yield Models and Substrate Physicochemical Characteristics. *Processes* **2023**, *11*, 441. https://doi.org/10.3390/pr11020441

Academic Editor: Antoni Sánchez

Received: 16 December 2022
Revised: 26 January 2023
Accepted: 27 January 2023
Published: 1 February 2023

1. Introduction

Digestate is a by-product from the anaerobic digestion (AD) process. Due to its high nutrient value, it can be used as soil improver or fertiliser if the digestate is proved to be valorised and can meet relevant quality standards [1]. The digestate stability can be evaluated with a residual biogas potential (RBP) test. The test typically is required to be carried out under mesophilic conditions for at least 28 days with an appropriate inoculum-to-substrate ratio and micro- and macronutrients supplemented to avoid the inhibition of biogas production [2]. The digestate is considered to have consistent quality if the RBP test biogas yield is below 0.25 L/g volatile solids (VS), as recommended in the Publicly Available Specification 110 (PAS110), which is a key element of the UK Government's anaerobic digestion quality protocol [3].

The 28-day continuous monitoring of biogas production in an RBP test is time-consuming and onerous for commercial AD operators. This limits the adoption of RBP tests and regulated markets for digestate. There have been many attempts to find alternative approaches to RBP that offer rapid tests result. These include assessing acid production after the inhibition of methanogenesis [4–6] and assessing the digestion of the organic fraction of the digestate after separation of the microbial cell component [7]. Nevertheless, both approaches are of great complexity and further research is needed [7].

Additionally, some researchers have attempted to relate the RBP test results to digestate physicochemical characteristics. For example, the theoretical biogas potential calculated based on the stoichiometric methane conversion from volatile fatty acid (VFA) concentrations and the soluble chemical oxygen demand (sCOD) in the digestate sample were suggested in PAS110 as preliminary pass/fail indicators for the RBP tests [8]. Although these preliminary indicators provide useful information about digestate stability, they cannot be correlated with RBP values or provide reaction kinetics that can be used to reveal further digestion performance information including inhibition. In a previous work reviewing the application of the RBP test for PAS110 [9], correlations between RBP values and various characteristics, including VFA and sCOD, were investigated. Although some low to moderate levels of correlations were found for total VFA, total solids and volatile solids, which account for 40%, 36% and 29% of variation in RBP values, respectively, none of the indicators are sufficiently reliable to predict the RBP values accurately [9].

Other researchers have evaluated using empirical biogas production models, including first-order kinetic and Gompertz models, and experimental biogas yield data from the initial stage of the RBP tests to fit specific accumulative biogas production data from RBP tests [10–12]. This has led to a promising experimental and modelling 'hybrid' approach using experimental data collected from a shorter RBP duration (3–7 days instead of 28 days) to calculate model parameters, and then predict the ultimate biogas production. However, the accuracy of prediction is not sufficient to warrant the replacement of RBP with this hybrid approach [12]; therefore, further improvement of the modelling process is required.

In this research, we evaluated the strength of fit for four biogas yield kinetic models including first-order kinetic, modified Gompertz, pseudo-parallel first-order kinetic and autoregressive (AR) time-series to describe the RBP test biogas production process. The models are then calibrated using experimental data collected from shorter RBP tests (5, 10, 15, 20 and 25 days) to calculate model parameters that are then used to predict ultimate biogas production.

In a previous work [9], although using conventional statistical methods, no significant correlations were found between key physicochemical parameters of digestate samples with RBP results. Due to the potential interplay of these parameters, which can influence the RBP results and reaction kinetics, the correlations may be deeply hidden.

Machine learning techniques including multivariate nonlinear regression analysis with decision trees (DT) were applied to predict the parameters of the biogas production model from the physicochemical characteristics of digestate samples. Compared with other multivariate nonlinear regression methods, the DT method is particularly suitable for a training dataset with limited sample size in this study [13]. The uncertainties of the predicted biogas yield were then assessed using a Gaussian process regressor (GPR).

The data processing framework described in this work can potentially have a wider application for other complex biochemical processes that are influenced by multiple physicochemical parameters of the reaction system.

2. Materials and Methods

2.1. Digestate Samples and RBP Test

The sampling point of the 25 digestate samples was the outlet of the final tank from which the biogas was collected. The AD plants involved in the study are anonymised and coded as ADP1–25. The inoculum was from the anaerobic digester at Millbrook Wastewater (WW) Treatment Plant at Southampton, UK. The RBP test followed the standard procedure described in the PAS110 [3]. Samples were tested in triplicate against two positive controls and three inoculum-only controls.

2.2. Analytical Methods

Twenty physicochemical characteristics were analysed for each digestate, including VFAs (total VFA and acetate), total ammoniacal nitrogen (TAN), total Kjeldahl nitrogen (TKN), alkalinity (total alkalinity (TA), partial alkalinity (PA) and intermediate alkalinity

(IA)), TS, VS, pH, COD (total COD and sCOD), trace element (TE) concentrations (cobalt (Co), iron (Fe), molybdenum (Mo) and nickel (Ni)), calorific value (CV) and elemental compositions (C, H, N). TAN/TKN and IA/PA were calculated as two extra metrics. TAN/TKN represents the relative contents of ammonia nitrogen and organically bonded nitrogen and thus how ready the substrate is for microorganism degradation [14]. IA/PA is an indicator of VFA accumulation in the AD process. Ripley et al. (1986) [15] suggest IA/PA < 0.3 indicates a stable state of the anaerobic process.

Determination of VFAs is based on the SCA (1979) [16] method Determination of Volatile Fatty Acids in Sewage Sludge. Supernatant layer from digestate centrifugation with 10% formic acid were quantified in a Shimadzu GC-2010 gas chromatograph with a flame ionisation detector and a capillary column type SGE BP-21. TAN and TKN were determined using a Kjeltech digestion block and steam distillation unit, according to the manufacturer's instructions (Foss Ltd., Warrington, UK). Alkalinity was measured by titration with 0.25 N H_2SO_4 to endpoints of pH 5.75 and 4.3 in order to allow calculation of TA, PA and IA [15]. TS and VS were determined with Standard Method 2540 G (APHA, 2005). Total COD and sCOD were analysed by adapting the closed reflux titrimetric method of 5220C, APHA [17]. TE concentrations were determined using ICP-MS or ICP-OES at a UKAS-accredited commercial laboratory (Severn Trent Services, Coventry, UK) after in-house hydrochloric–nitric acid digestion [18]. CV was measured with a CAL2k-ECO bomb calorimeter (CAL2k, Digital Data Systems, Gauteng, South Africa). Elemental C, H, and N analysis was performed using a Flash EA-1112 elemental analyser (Thermo Finnigan, Cheshire, UK).

2.3. Assessing the Strength of Fit for Biogas Production Kinetic Models

Four biogas-production kinetic models commonly used to estimate the kinetic constants of the AD process were compared for their abilities to fit the biogas production of RBP tests. These models include three empirical models (first-order kinetic, modified Gomperz and pseudo-parallel first-order models) and a time-series model (first-order autoregressive). The strength of a kinetic model to accurately fit the experimental biogas production data was measured using R^2 values, which indicate the percentage of the variance in the responses explained by a model.

Additionally, these models were used to predict the 28-day RBP test results by fitting the models with initial 5, 10, 15, 20 and 25 days' experimental data using the Matlab R2021b Curve Fitting Toolbox. Based on the absolute percentage error (APE) between the experimental data and model-predicted results, the accuracy of prediction and duration of experimental data required to obtain a sufficiently accurate prediction were investigated for the following four models:

(1) First-order model (FO): The FO model (Equation (1)) is derived from the assumptions that the substrate degradation is a first-order reaction with hydrolysis as the speed-limiting step and the cumulative biogas yield is proportional to the amount of substrate degraded (Equation (2)) [19,20].

$$y(t) = y_m\left(1 - e^{-kt}\right) \tag{1}$$

$$\frac{dc}{dt} = -kc\,\frac{c_0 - c}{c_0} = \frac{y}{y_m} \tag{2}$$

where $y(t)$ is the cumulative biogas yield at time t, k is the first-order rate constant, y_m is the maximum cumulative gas production, c is the concentrate of the substrate and c_0 is the initial substrate concentration.

(2) Modified Gomperz model (MG): The modified Gomperz model (Equation (3)) is derived from the Gomperz model, which is used to describe the microbial activity and has a signature sigmoid shape [21,22]. It describes the biogas production in terms of the exponential growth rates and lag phase duration of anaerobic degradation microorganisms [11].

$$y(t) = y_m \times \exp\left(-\exp\left(\frac{R}{y_m} \times e \times (\lambda - t) + 1\right)\right) \quad (3)$$

where $y(t)$ is the cumulative gas production at time t, y_m is the maximum cumulative gas production (mL CH_4/gVS), R is the maximum gas production rate (mL CH_4/gVS/d) and λ is the lag phase period or minimum time to produce biogas (days).

(3) Pseudo-parallel first-order model (PP): The pseudo-parallel first-order model (Equation (4)) is considered to be more suitable for describing the biogas yield of mixtures of substrates with different kinetic rates (rapid and slow) [23].

$$y(t) = y_m\left(1 - Pe^{-k_1 t} - (1 - P)e^{-k_2 t}\right) \quad (4)$$

where $y(t)$ is the cumulative gas production at time t, y_m is the maximum cumulative gas production (mL CH_4/g VS), P is the the proportion of the readily degradable material, k_1 is the first-order rate constant for readily degradable material, and k_2 is the first-order rate constant for less readily degradable material.

(4) First-order autoregressive model (AR (1)): AR (1) is a time-series model that predicts the present timestep based on the observations from previous timesteps. The autocorrelation function (ACF) (autocorrelation between timesteps) plots for all the RBP test biogas yield samples gradually trail off (Figure 1a, using ADP20 as an example). Therefore, the biogas production process is an AR process. Many time-series models that essentially model the randomness of the time series data need the trends in the data to be removed, in other words, to ensure the stationarity of the data. However, the application of the AR model does not intrinsically require transforming the data into stationary data. Thus, the biogas yield data were not converted to a stationary process in this study.

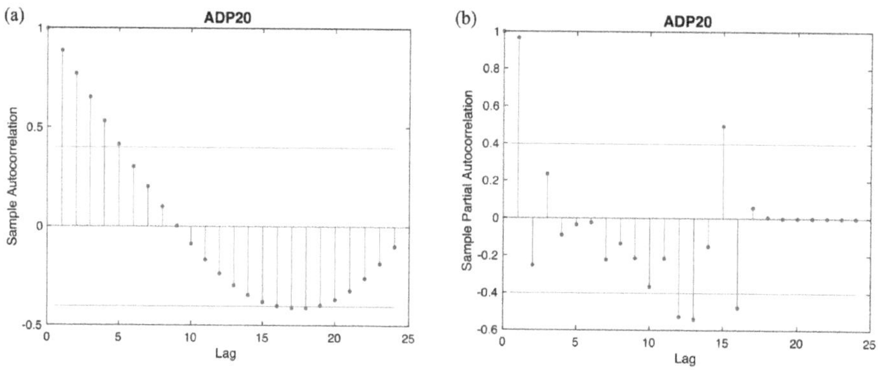

Figure 1. The ACF (**a**) and PACF (**b**) of sample ADP20 biogas production time series from day 4. The blue lines are the confidence bounds for a significant correlation. The red lines are the autocorrelation and partial autocorrelation between the current time step and the time steps of different lags.

For most of the RBP test data, the cumulative biogas production curves typically exhibit a rapid gas-production stage in the initial three days followed by a noticeable reduction in biogas production rate. This means one set of AR model parameters generally will result in a poor fitting for both the rapid-production stage and the later stage after the initial three days. In this study, to simplify the modelling process, the AR (1) model was only used to model biogas production from day 4 of the RBP tests. The partial autocorrelation function (PACF) (partial autocorrelation between timesteps) plots of data from day four for most of the biogas time series samples were cut off after one lag (Figure 1b, taking ADP20 as an example). Therefore, the biogas production from day four was an AR (1) process (Equation (5)).

The upper asymptote of an AR (1) process is determined by its unconditional mean $\mu = c/(1 - \beta)$, which corresponds to the ultimate biogas yield of a substrate.

$$X_t = \beta X_{t-1} + c + \varepsilon_t \quad (5)$$

where X_t is the response of the present timestep, the X_{t-1} is the response of the previous one timestep, β is the coefficient, c is the constant and ε_t is the white noise.

2.4. Prediction of Biogas Kinetic Model Coefficients and Constants Using Decision Tree Multivariate Regression Method

2.4.1. Decision Tree Multivariate Regression Method

The best-performing model out of the four kinetic models was further studied using decision tree (DT) multivariate regression analysis. DT is a non-parametric supervised learning technique that can be applied for both regression analysis and classification, providing piecewise constant approximations as prediction results [24].

DTs were trained using all physicochemical characteristics listed in Tables 1 and 2 as predictors to predict the coefficients and constants of the best-performing biogas production kinetic model. First, RBP results of 20 digestate samples (out of the 25-sample dataset) were used to train the DT model with the five remaining samples as the test dataset for model validation. The training process generated 53,130 groups of different combinations of 20 samples when choosing out of the 25. For practical reasons, a subset of 5000 groups were selected randomly to evaluate the absolute percentage error (APE) and mean absolute percentage error (MAPE) (Equation (6)) of the predicted biogas production model parameters and the APE of the calculated ultimate biogas yield. Then, the DT model was trained and validated using the leave-one-out (LOO) cross-validation method, which apply four additional set of data to the training data to verify if the prediction accuracy would improve.

$$APE_i = \left|\frac{y_i - \hat{y}_i}{y_i}\right| \; MAPE = \frac{1}{n}\sum_{i=1}^{n}\left|\frac{y_i - \hat{y}_i}{y_i}\right| \quad (6)$$

where y_i is the model parameter derived by fitting the model to the entire 28-day RBP test data, $\hat{y}_i$ is the model parameter inferred from the physicochemical characteristics by DT and n is the number of training set samples.

Table 1. RBP test results on day 28 (d-28) and the physicochemical characteristics of 25 samples (Part 1).

ADP No.	d-28	Acetate	Total-VFA	TAN	TKN	TAN/TKN	TA	PA	IA	IA/PA	TS	VS
	L/g VS	mg/L			g N/kg			kg/kg $CaCO_3$			g/kg	
1	0.13	178.18	193.57	1.70	3.53	0.48	12.52	9.03	3.49	0.39	49.84	37.12
2	0.18	29.08	37.35	0.62	3.28	0.19	7.97	6.42	1.55	0.24	49.11	30.39
3	0.13	846.78	1016.41	7.98	12.37	0.64	33.38	25.87	7.51	0.29	93.28	67.65
4	0.24	684.05	775.99	4.04	6.48	0.62	17.52	12.42	5.10	0.41	46.43	32.59
5	0.12	247.38	332.57	5.12	7.48	0.69	24.50	16.62	7.88	0.47	56.57	37.49
6	0.06	111.56	115.64	2.80	4.91	0.57	46.43	28.12	18.31	0.65	178.34	50.05
7	0.07	183.61	187.39	2.69	4.44	0.61	58.20	20.92	37.27	1.78	137.43	40.44
8	0.26	14.42	324.32	0.40	1.29	0.31	3.31	2.20	1.11	0.50	17.16	12.11
9	0.36	2633.79	9262.76	6.54	9.23	0.71	23.99	15.62	8.36	0.54	47.79	34.51
10	0.17	134.49	204.90	3.32	5.21	0.64	17.64	13.10	4.54	0.35	46.53	29.51
11	0.30	2662.15	3963.27	2.22	3.14	0.71	9.45	5.22	4.23	0.81	20.38	12.74
12	0.17	23.29	36.68	2.71	4.64	0.58	12.22	8.98	3.24	0.36	36.91	26.50
13	0.16	335.54	364.46	2.80	5.29	0.53	15.73	12.14	3.59	0.30	43.59	29.98
14	0.13	19.25	19.25	0.44	2.25	0.20	3.97	1.85	2.11	1.14	35.04	21.47
15	0.09	250.28	259.05	3.58	6.12	0.58	17.20	13.66	3.54	0.26	58.00	43.56
16	0.38	2706.80	3871.62	3.00	4.87	0.62	11.91	7.76	4.14	0.53	36.65	26.88
17	0.26	36.69	50.29	1.49	2.85	0.52	7.71	5.79	1.91	0.33	29.82	16.42
18	0.25	832.24	1440.64	2.42	4.27	0.57	13.70	9.98	3.72	0.37	59.04	44.97
19	0.14	39.82	53.93	4.62	6.85	0.68	21.13	16.31	4.82	0.30	52.92	38.76
20	0.19	184.69	218.06	5.76	8.55	0.67	23.56	18.34	5.23	0.29	64.64	44.93
21	0.33	89.69	260.15	3.00	5.61	0.54	24.85	14.70	10.14	0.69	211.53	106.18
22	0.22	299.39	353.29	2.25	3.11	0.72	12.78	10.07	2.71	0.27	20.86	9.85
23	0.29	270.61	411.15	4.24	6.56	0.65	17.23	13.24	3.99	0.30	52.49	32.30
24	0.28	215.32	250.62	4.47	6.98	0.64	18.19	13.87	4.33	0.31	50.25	34.87
25	0.29	241.71	272.53	4.77	7.01	0.68	19.52	15.15	4.38	0.29	46.84	31.96

Table 2. RBP test results on day 28 (d-28) and the physicochemical characteristics of 25 samples (Part 2).

ADP No.	d-28	pH	CV	C	H	N	Co	Fe	Mo	Ni	Total-COD	sCOD
	L/g VS		MJ/kg TS		%			mg/L			g O_2/L	
1	0.13	8.30	16.38	39.20	4.48	5.61	0.85	130.53	0.29	1.20	42.77	6.67
2	0.18	7.41	14.89	37.44	4.57	5.53	0.08	1144.15	0.30	0.50	40.35	2.73
3	0.13	8.35	18.61	34.80	4.52	7.34	0.24	1020.82	0.43	1.75	88.64	21.56
4	0.24	8.17	17.53	40.13	4.89	6.61	0.28	2031.83	0.19	0.60	44.04	12.14
5	0.12	8.45	16.13	38.32	4.20	6.05	0.21	90.76	0.18	0.39	40.69	15.16
6	0.06	8.14	3.29	17.08	1.71	2.20	1.08	2326.91	0.94	7.74	57.29	11.19
7	0.07	8.16	4.41	12.51	1.59	2.25	0.98	1923.48	0.69	5.06	39.32	8.45
8	0.26	7.33	20.26	46.25	5.58	6.11	0.03	87.36	0.07	0.24	28.62	3.16
9	0.36	8.35	22.26	47.75	5.32	8.51	0.09	539.07	0.14	0.50	75.39	35.94
10	0.17	8.04	14.78	36.85	4.00	5.74	0.42	174.88	0.30	0.71	36.12	7.82
11	0.30	7.62	17.02	38.87	4.43	6.30	0.04	210.67	0.08	0.19	28.69	10.00
12	0.17	8.10	18.26	42.43	4.88	6.13	0.25	234.07	0.08	0.28	31.68	10.17
13	0.16	8.37	16.21	38.47	4.47	6.65	1.04	111.12	0.30	1.02	46.88	9.53
14	0.13	7.50	15.25	32.94	4.50	5.78	0.45	1166.46	1.46	1.67	19.63	4.50
15	0.09	8.90	19.86	44.13	5.28	5.22	0.09	151.82	0.25	1.08	65.63	11.87
16	0.38	7.92	19.22	43.49	5.62	6.54	0.09	64.01	0.18	0.26	39.18	18.84
17	0.26	7.92	13.35	30.67	3.82	5.67	0.19	348.02	0.28	0.56	26.51	6.39
18	0.25	8.16	17.76	42.16	4.76	4.34	9.06	227.30	5.68	30.37	55.74	17.00
19	0.14	8.42	17.90	44.56	4.67	6.87	1.42	293.85	0.50	1.33	58.58	14.74
20	0.19	8.15	16.41	38.55	4.55	6.40	1.21	1056.09	0.45	1.55	68.52	15.57
21	0.33	8.07	12.24	33.21	2.64	2.35	2.58	4249.89	1.17	13.79	82.92	12.59
22	0.22	8.32	10.74	24.93	2.88	5.27	0.89	36.94	0.26	0.82	17.76	9.52
23	0.29	8.54	13.84	32.93	4.06	5.59	0.18	649.78	0.16	0.46	54.33	16.54
24	0.28	8.54	16.82	38.59	4.82	6.65	0.38	679.62	0.16	0.58	67.39	15.86
25	0.29	8.67	16.50	36.82	4.71	5.90	0.37	797.93	0.15	0.53	61.56	16.83

2.4.2. Assessing Prediction Uncertainty Using a Gaussian Process Regressor (GPR)

To quantify the uncertainty in the calculated biogas yield using the kinetic model predicted by DT, a Gaussian process regressor (GPR) method was applied in this study. GPR is a kernel-based Bayesian tool to perform nonlinear regression. The process is specified by a mean function m(x) and a covariance function $k(x, x' \mid \theta)$ which defines the covariances between the responses at any two input locations x and x′. θ represents the hyperparameters of the covariance function and their values are learned from the training data by maximising the log marginal likelihood. Once the hyperparameters are decided, the prediction for new input is performed by computing the marginal posterior distribution conditioning on the dimensions with known inputs [25].

The principle of GPR is to predict one timestep further each time by fitting one more datum provided by the biogas-yield kinetic model and the prediction uncertainty bands were also returned. The predictions of the GPR with zero mean and squared exponential kernel and the GPR with linear basis function and squared exponential kernel were compared.

3. Results and Discussion

3.1. Digestate Characterisation and RBP Test Results

The 28-day RBP test results for each set of RBP samples together with the physicochemical characteristics of each digestate are shown in Tables 1 and 2. Specific cumulative biogas yield data collected during the 28-day testing period were reported elsewhere [9] and were used in the model fitting and training process in this study.

3.2. Assessing Strength of Fit for Biogas Production Models

Within the 28 days of the RBP test, it was noticeable that biogas yields of some samples in this study had reached a plateau, whilst others still were showing an upward trend close to the end of the 28-day test. This is clearly due to the different concentrations of readily degradable materials in the digestate samples. To distinguish these two types of digestate samples, the 25 digestate samples were classified into two types based on the absolute average of the daily biogas production change percentages in the last four days: (1) Type I: less than 0.5%; (2) Type II: more than 0.5% (Table 3).

Table 3. The classification of two types of RBP test biogas-production time series according to the absolute average of the daily biogas production change percentages in the last four days.

	ADP	**3**	**4**	**5**	**11**	**12**	**13**	**16**	**19**	**20**	**21**					
Type I	Increase (%)	0.38	0.04	0.26	0.27	0.36	0.31	0.18	0.36	0.4	0.50					
	ADP	**1**	**2**	**6**	**7**	**8**	**9**	**10**	**14**	**15**	**17**	**18**	**22**	**23**	**24**	**25**
Type II	Increase (%)	1.07	1.13	1.23	1.56	0.70	0.58	0.58	0.94	0.95	1.13	1.00	0.92	0.77	0.86	0.80

Table 4 shows the R^2 values for the fits of the three empirical biogas production models to RBP test data. Overall, the modified Gomperz model achieved lower R^2 values across the majority of the samples, indicating poorer fitting performance. In contrast, the PP first-order model could describe the biogas production of almost all samples with an R^2 value between 97–99%. The FO model performed better at describing Type I samples, whereas the PP first-order model was more suitable for substrates mixed with materials with different reaction-rate constants, and therefore performed better with Type II samples. However, it is worth noting that there are usually multiple sets of optimal solutions of the estimates of the PP first-order model's parameters (Y_m, P, k_1 and k_2). This is because the nonlinear least-square error function of this model when fitting a particular set of biogas yield data is not always convex. Therefore, it was not chosen for the study of training DTs to predict the model parameters from the digestate physicochemical characteristics in the following section.

Table 4. R^2 values of the fits of three empirical models (FO, MP and PP) to the Type I and Type II RBP test biogas-production time series (R^2 values larger than 97% are in bold).

		Fitting R^2 Values (%)														
	ADP	**3**	**4**	**5**	**11**	**12**	**13**	**16**	**19**	**20**	**21**					
Type I	FO	**98.5**	**98.4**	**97.9**	86.8	**99.0**	**98.2**	**99.5**	**97.7**	96.4	**97.2**					
	MG	**97.6**	94.0	**96.6**	80.7	96.1	91.8	96.0	94.9	88.3	61.6					
	PP	**98.5**	**98.7**	**97.9**	94.6	**99.1**	**99.5**	**99.6**	**97.8**	**98.7**	**97.6**					
		Fitting R^2 Values (%)														
	ADP	1	2	6	7	8	9	10	14	15	17	18	22	23	24	25
Type II	FO	**97.1**	**98.6**	86.5	77.2	83.2	**97.3**	**98.4**	**97.6**	**97.5**	**97.5**	95.5	96.0	**98.8**	**99.1**	**99.5**
	MG	88.8	93.9	78.6	85.8	68.2	**99.1**	93.0	90.8	74.9	88.4	85.6	85.9	96.7	**97.3**	**99.0**
	PP	**99.9**	**99.6**	92.8	92.6	**98.0**	**97.4**	**98.9**	**99.5**	**98.9**	**97.8**	**99.1**	**98.9**	**98.8**	**99.1**	**99.5**

In addition, when fitting the PP first-order model in Matlab using the least-square algorithm, the initial value set for the parameter P should avoid 0.5 and k_1 and k_2 should not be the same. Otherwise, the partial derivatives of the error function with respect to k_1 and k_2 are the same, which means the moving direction of these two dimensions are the same and the nonlinear search for the minimum value of the error function value will settle at a local minimum point.

The fitting of the AR (1) model after the initial 3-day rapid biogas production stage was comparable to the PP first-order model, with R^2 values of 99% for the majority of samples (Table 5), regardless of Type I or Type II data. Figure 2 shows the fits of three empirical models and the AR (1) model to the cumulative biogas yield in RBP tests.

Table 5. R^2 values of the fits of the AR (1) model to the Type I and Type II RBP test biogas-production time series from day 4 (R^2 values larger than 97% are in bold).

		Fitting R^2 Values (%)														
Type I	**ADP**	**3**	**4**	**5**	**11**	**12**	**13**	**16**	**19**	**20**	**21**					
	AR (1)	**98.2**	**99.4**	**97.4**	**97.0**	**98.6**	**99.3**	**99.8**	**99.7**	**98.9**	90.1					
		Fitting R^2 Values (%)														
Type II	**ADP**	**1**	**2**	**6**	**7**	**8**	**9**	**10**	**14**	**15**	**17**	**18**	**22**	**23**	**24**	**25**
	AR (1)	**99.2**	**99.8**	85.4	80.9	95.9	**99.1**	**99.5**	**99.7**	**98.9**	**99.1**	**99.6**	**99.5**	**98.7**	**99.1**	**99.9**

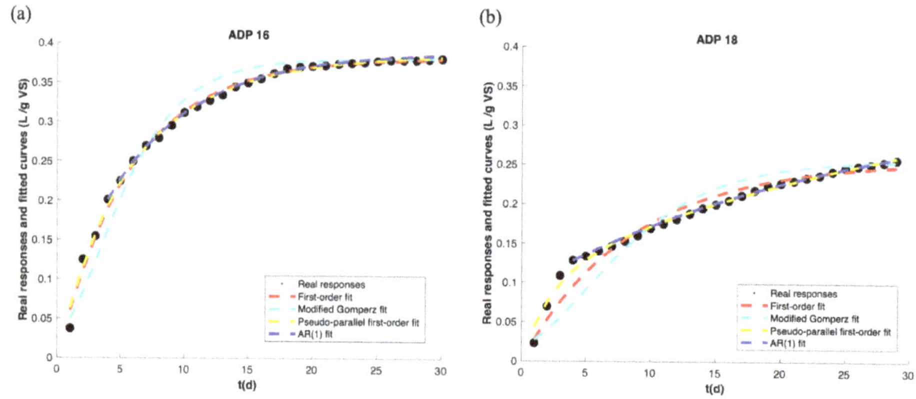

Figure 2. The fits of the three empirical models and the AR (1) model to the entire RBP test process for Type I (**a**) and Type II (**b**) data.

When using these four models to predict the final RBP results based on experimental data collected from a shorter test duration (5, 10, 15, 20 and 25 days), none of the three empirical models and the AR (1) model could achieve a sufficient level of accuracy to replace full-length RBP tests. This conclusion is in agreement with previous studies [12,26]. When using the first-order model, typically the level of prediction accuracy increases when more experimental data points are provided. FO performed moderately better than the MG model at predicting the RBP test result. This result was in agreement with the result of Nielfa et al. (2015) [26] for biomethane potential test biogas production prediction for the organic fraction of municipal solid waste and biological sludge co-digestion. However, both FO and MG models were proved to be unsuitable to predict the RBP result of Type II data.

Using 15 days of experimental data, the PP model could predict the RBP result with APE < 10% in nearly all the samples (both Type I and II). Therefore, the PP model is a preferred option for RBP test-result-prediction when the experiment is half-way through. For the AR (1) model, RBP data fitting and modelling starts from day 4, thus 10, 15, 20 and 25 days of experimental data were used for model fitting. The prediction ability of the AR (1) model was comparable to that of the PP model. Table 6 shows the predictions of RBP test results when fitting an increasing amount of experimental data to four biogas production models (due to the large sample numbers, only seven samples were randomly selected from each group of Type I and II data).

Table 6. The level of accuracy (expressed in APE%) of predicted 28-day RBP test result from fitting the mathematical models to an increasing number of experimental observations from 5 days to 25 days. APE less than 10% indicates accurate predication (in bold).

Type I	Days	FO	MG	PP	AR (1)	Type II	Days	FO	MG	PP	AR (1)
	5	77.16	24.23	35.08	/		5	46.46	51.94	28.00	/
	10	**4.98**	**9.62**	**7.26**	**1.15**		10	24.34	37.77	24.65	12.20
ADP 3	15	**1.01**	**5.14**	**2.70**	**0.47**	ADP 2	15	17.34	24.48	**6.86**	**4.52**
	20	**2.10**	**1.55**	**0.88**	**1.71**		20	10.98	14.87	**2.02**	**1.63**
	25	**2.29**	**0.07**	**3.23**	**1.81**		25	**6.40**	**6.25**	**0.43**	**0.27**
	5	57.99	26.65	52.67	/		5	**3.65**	27.78	**3.8080**	/
	10	**7.50**	15.32	**7.19**	**3.97**		10	15.24	17.56	15.13	12.57
ADP 4	15	**5.00**	**5.53**	**4.65**	**6.00**	ADP 8	15	12.03	12.02	**8.31**	**8.62**
	20	**1.84**	**1.71**	**4.26**	**2.63**		20	**9.62**	**9.62**	**5.23**	**8.20**
	25	**0.40**	**0.22**	**1.97**	**1.91**		25	**3.48**	**3.44**	**2.76**	**5.74**
	5	159.25	17.17	105.36	/		5	163.39	37.62	29.58	/
	10	**1.21**	11.80	**1.04**	**8.22**		10	**7.63**	25.74	19.18	**2.99**
ADP 5	15	**1.49**	**1.72**	**1.54**	**1.83**	ADP 10	15	**7.73**	16.41	12.75	**0.03**
	20	**0.80**	**0.86**	**1.51**	**2.33**		20	**5.40**	10.20	21.58	**0.83**
	25	**1.30**	**1.29**	**1.69**	**2.08**		25	**3.49**	**6.22**	**4.85**	**1.02**
	5	23.12	43.08	25.98	/		5	**0.16**	44.20	**5.68**	/
	10	21.74	27.40	**7.71**	**0.56**		10	27.04	32.57	27.06	44.65
ADP 13	15	11.85	12.96	22.50	**8.86**	ADP 14	15	17.04	19.33	**1.20**	**8.78**
	20	**4.95**	**4.19**	10.03	**3.70**		20	10.67	10.33	**5.62**	**3.70**
	25	**2.50**	**1.87**	**2.91**	**2.58**		25	**6.58**	**4.85**	**1.92**	**2.04**
	5	19.90	37.71	17.89	/		5	**1.6464**	45.20	11.25	/
	10	**4.66**	18.19	**4.76**	**5.69**		10	11.99	26.91	**5.02**	47.42
ADP 16	15	**3.38**	**8.20**	**3.05**	**0.57**	ADP 17	15	13.86	17.95	**0.84**	**0.25**
	20	**1.09**	**2.65**	**3.70**	**1.37**		20	10.90	11.12	**4.12**	**3.06**
	25	**0.61**	**0.95**	**1.00**	**0.89**		25	6.00	4.61	**1.53**	**1.94**
	5	183.89	24.27	146.92	/		5	**2.98**	45.76	**3.74**	/
	10	**2.33**	13.07	**2.33**	**8.62**		10	30.94	34.05	30.93	**5.6969**
ADP 19	15	**3.13**	**7.29**	**3.41**	**1.31**	ADP 18	15	22.36	22.19	**7.51**	**0.57**
	20	**2.78**	**3.23**	**2.79**	**0.36**		20	12.64	11.80	**4.16**	**1.37**
	25	**1.96**	**1.79**	**1.91**	**0.66**		25	**6.29**	**4.44**	**2.96**	**0.89**
	5	24.11	31.62	27.78			5	81.36	54.33	28.59	/
	10	11.16	21.66	**8.64**	48.84		10	36.33	23.98	**2.24**	14.84
ADP 20	15	**8.93**	13.72	**4.54**	**0.62**	ADP 25	15	**8.52**	14.08	**1.09**	**1.92**
	20	**7.15**	**9.66**	**0.06**	**1.15**		20	**3.26**	**7.75**	**7.05**	**0.48**
	25	**5.32**	**6.69**	**1.12**	**0.51**		25	**1.06**	**4.50**	**1.34**	**0.11**

3.3. Biogas Yield Prediction from Digestate Physicochemical Characteristics by DT

DTs were first trained with the physicochemical characteristics of 20 digestate samples as predictors and the fitted coefficients and constants of AR (1) models as responses. A total of 5000 groups of splits between the training set and test set were randomly chosen to evaluate the prediction accuracy. The average MAPEs of the predicted AR (1) model coefficients and constants for among 5000 groups of test sets were 4.58% and 72.04%, respectively. The MAPE of the calculated RBP test result at day 28 from the predicted AR (1) model parameters was 52.125%.

With four more samples provided for the training set, the MAPEs of the predictions for the AR (1) model coefficient and constant with the LOO cross-validation method were 4.31% and 59.29%, respectively (Table 7). The MAPE of the calculated RBP test result at day 28 was 45.620%. With four more data provided, the prediction accuracy of the RBP test result of the DTs improved 12.48%. Additionally, among 25 LOO cross-validation groups, the APEs of predicted RBP test results of five groups were smaller than 10%.

Table 7. The APEs for the AR (1) model coefficient and constant prediction and the biogas yield at 28th day prediction from the LOO cross-validation method.

Test Sample No.	Fitted AR (1) Coefficient	Predicted AR (1) Coefficient	AR (1) Coefficient APE	Fitted AR (1) Constant	Predicted AR (1) Constant	AR (1) Constant APE	Real RBP Test Result	Predicted RBP Test Result	RBP Test Result APE
ADP1	0.932	0.937	0.5%	0.010	0.010	1.1%	0.132	0.137	**4%**
ADP2	0.949	0.934	1.6%	0.011	0.019	65.6%	0.182	0.243	33.5%
ADP3	0.789	0.884	12%	0.028	0.017	40.2%	0.132	0.141	**6.9%**
ADP4	0.876	0.896	2.2%	0.030	0.035	17.5%	0.236	0.321	36%
ADP5	0.834	0.850	1.9%	0.020	0.036	84%	0.117	0.236	101.9%
ADP6	0.946	0.813	14%	0.004	0.011	174.7%	0.062	0.057	**7.9%**
ADP7	0.969	0.932	3.8%	0.003	0.011	293%	0.066	0.137	107.4%
ADP8	0.929	0.937	0.9%	0.019	0.036	84.6%	0.261	0.490	88%
ADP9	0.853	0.893	4.8%	0.055	0.032	41.2%	0.361	0.291	19.5%
ADP10	0.890	0.894	0.4%	0.020	0.023	16.1%	0.174	0.206	18.6%
ADP11	0.900	0.941	4.6%	0.030	0.035	14.6%	0.301	0.506	68.4%
ADP12	0.832	0.950	14.1%	0.028	0.035	23.5%	0.171	0.523	205.7%
ADP13	0.906	0.892	1.5%	0.016	0.009	43%	0.164	0.086	47.6%
ADP14	0.941	0.936	0.6%	0.009	0.010	14.6%	0.129	0.138	**7.1%**
ADP15	0.901	0.847	6.1%	0.009	0.020	123%	0.086	0.126	46.5%
ADP16	0.864	0.842	2.5%	0.053	0.032	38.7%	0.379	0.205	46%
ADP17	0.934	0.936	0.2%	0.019	0.009	52.8%	0.257	0.139	45.9%
ADP18	0.961	0.870	9.4%	0.013	0.041	211.6%	0.254	0.307	21%
ADP19	0.861	0.898	4.2%	0.020	0.023	16.5%	0.139	0.212	52.5%
ADP20	0.887	0.894	0.8%	0.022	0.022	1.8%	0.194	0.202	**4.3%**
ADP21	0.818	0.892	9.2%	0.006	0.010	75.2%	0.325	0.092	71.6%
ADP22	0.910	0.939	3.2%	0.021	0.022	7%	0.222	0.315	41.9%
ADP23	0.850	0.846	0.5%	0.043	0.033	21.4%	0.286	0.215	24.7%
ADP24	0.872	0.896	2.7%	0.036	0.034	5.2%	0.280	0.313	12%
ADP25	0.901	0.847	6.0%	0.030	0.035	15.3%	0.287	0.224	21.7%
MAPE			4.3%			59.3%			45.6%

3.4. AR (1) Model Prediction Guide GPR

When training the DTs with LOO cross-validation and ADP3 as the test set, the APE of the predicted biogas yield on day 28 using the inferred AR (1) model was 6.87%. This is used as an example for illustration in Figure 3. By accepting one more input each time from the inferred AR (1) model together with the experimental observations of the first three days, the predictions of every one timestep further of the GPR with zero mean and squared exponential kernel were smaller than AR (1)'s predictions for the first few timesteps and then gradually closer to the predictions of the AR (1) model. This was explained by the increased fitted GPR model length scale and vertical scale when receiving more training data. The smaller the length scale, the curvier the underlying function is, and the smaller the vertical scale, the more concentrated the underlying function is around the mean. In contrast, for the GPR with a linear basis function, the predictions surged for the first few timesteps and then approached the AR (1) model predictions. This corresponded to the decreased slope of the fitted linear mean function. In general, given that only one more timestep is predicted at a time, the selection of zero or linear mean function is not of much concern. The 95% confidence interval of the prediction of GPR narrowed when more data were provided.

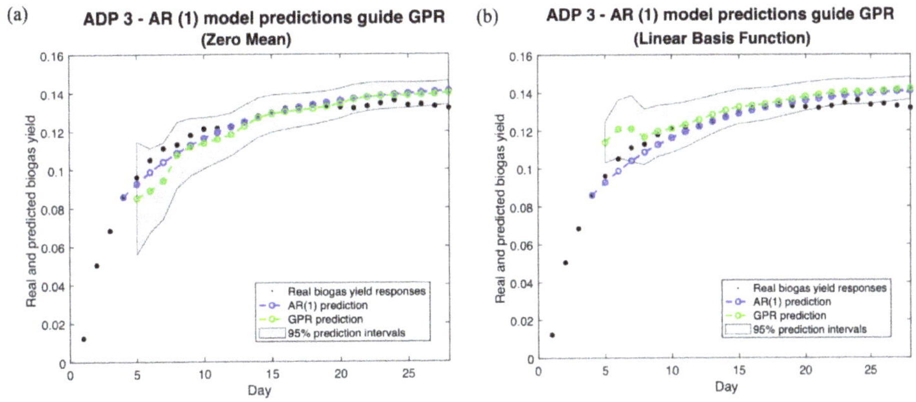

Figure 3. The prediction and its probability distribution of the GRP guided by the prediction of the inferred AR (1) model. (**a**) Zero mean and squared exponential kernel. (**b**) Linear basis function and squared exponential kernel.

4. Conclusions

This study has demonstrated that it is possible to use RBP experimental data collected in the initial stage of the test to predict the 28-day RBP result in a kinetic model fitting exercise. By fitting 15 days of experimental data from RBP tests to kinetic models, the PP first-order model and the AR (1) model achieved a promising accuracy with APE < 10%.

Further study demonstrated using the decision tree (DT) method that AR (1) model parameters can be predicted from the physicochemical characteristics of the digestate samples. This provides potential to further reduce the data requirement to four days of RBP experimental data and thereby significantly reduce the resting time of a standard 28-day RBP test to around four days. It was observed that when more training data were included in the DT machine learning model (from 20 to 24 samples), the prediction accuracy of the RBP result increased by 12.48%. This indicates that collecting more data to include in the model-training process can further improve the prediction outcome.

The framework of predicting kinetic model parameters from the physicochemical characteristics of the substrate can potentially be applied to the yield prediction of the product from other biochemical reaction processes.

Author Contributions: Conceptualization, W.G., P.L. and Y.J.; Methodology, Y.L., W.G. and Y.J.; Formal analysis, Y.L.; Writing—original draft, Y.L.; Writing—review & editing, W.G., P.L. and Y.J.; Supervision, W.G., P.L. and Y.J. All authors have read and agreed to the published version of the manuscript.

Funding: This research received no external funding.

Institutional Review Board Statement: Not applicable.

Informed Consent Statement: Not applicable.

Data Availability Statement: No new data were created or analysed in this study. Data sharing is not applicable to this article.

Conflicts of Interest: The authors declare no conflict of interest.

References

1. Lee, M.E.; Steiman, M.W.; St Angelo, S.K. Biogas digestate as a renewable fertilizer: Effects of digestate application on crop growth and nutrient composition. *Renew. Agric. Food Syst.* **2021**, *36*, 173–181. [CrossRef]
2. Walker, M.; Banks, C.; Heaven, S.; Frederickson, J. *OFW004-005. Waste and Resources Action Programme, Residual Biogas Potential Test for Digestates*; University of Southampton & Open University: Southampton, UK, 2010.

3. WRAP. PAS 110:2014. *Specification for Whole Digestate, Separated Liquor and separated Fibre Derived from the Anaerobic Digestion of Source-Segregated Biodegradable Materials*. 2014. Available online: https://wrap.org.uk/sites/default/files/2021-03/PAS110_2014.pdf (accessed on 26 June 2022).
4. Lie, E.; Welander, T. A method for determination of the readily fermentable organic fraction in municipal wastewater. *Water Res.* **1997**, *31*, 1269–1274. [CrossRef]
5. Ruel, S.M.; Comeau, Y.; Héduit, A.; Deronzier, G.; Ginestet, P.; Audic, J.M. Operating conditions for the determination of the biochemical acidogenic potential of wastewater. *Water Res.* **2002**, *36*, 2337–2341. [CrossRef] [PubMed]
6. Mohan, S.V.; Babu, V.L.; Sarma, P.N. Effect of various pretreatment methods on anaerobic mixed microflora to enhance biohydrogen production utilizing dairy wastewater as substrate. *Bioresour. Technol.* **2008**, *99*, 59–67. [CrossRef] [PubMed]
7. Banks, C.J.; Heaven, S.; Zhang, Y.; Sapp, M.; Thwaites, R. OMK002-014. Waste and Resources Action Programme. In *Review of the Application of the Residual Biogas Potential Test*; University of Southampton: Southampton, UK, 2013.
8. WRAP. OFW004-005. *Residual Biogas Potential Test for Digestates, Development and Evaluation of a Method for Testing the Residual Biogas Potential of Digestates (Section 4, Full Description of the RBP Test)*. 2010. Available online: https://www.ktbl.de/fileadmin/user_upload/Allgemeines/Download/Ringversuch-Biogas/Residual-Biogas-Potential.pdf (accessed on 20 July 2022).
9. WRAP. OMK002-014. *Review of the Application of the Residual Biogas Potential Test*. 2013. Available online: http://www.organics-recycling.org.uk/uploads/article2652/PAS110%20digestate%20stability%20review.pdf (accessed on 13 July 2022).
10. Strömberg, S.; Nistor, M.; Liu, J. Early prediction of biochemical methane potential through statistical and kinetic modelling of initial gas production. *Bioresour. Technol.* **2015**, *176*, 233–241. [CrossRef] [PubMed]
11. Pramanik, S.K.; Suja, F.B.; Porhemmat, M.; Pramanik, B.K. Performance and kinetic model of a single-stage anaerobic digestion system operated at different successive operating stages for the treatment of food waste. *Processes* **2019**, *7*, 600. [CrossRef]
12. Calabrò, P.S.; Folino, A.; Maesano, M.; Pangallo, D.; Zema, D.A. Exploring the Possibility to Shorten the Duration and Reduce the Number of Replicates in Biomethane Potential Tests (BMP). *Waste Biomass Valorization* **2022**, *2022*, 939. [CrossRef]
13. Hastie, T.; Tibshirani, R.; Friedman, J.H. Chapter 9: Additive Models, Trees, and Related Methods. In *The Elements of Statistical Learning: Data Mining, Inference, and Prediction*; Springer: New York, NY, USA, 2009; Volume 2, pp. 1–758.
14. Deng, Z.; van Linden, N.; Guillen, E.; Spanjers, H.; van Lier, J.B. Recovery and applications of ammoniacal nitrogen from nitrogen-loaded residual streams: A review. *J. Environ. Manag.* **2021**, *295*, 113096. [CrossRef] [PubMed]
15. Ripley, L.E.; Boyle, W.C.; Converse, J.C. Improved alkalimetric monitoring for anaerobic digestion of high strength wastes. *J. Water Pollut. Control Fed.* **1986**, *58*, 406–411.
16. SCA. Determination of volatile fatty acids in sewage sludge. In *Methods for the Examination of Waters and Associated Materials*; Standing Committee of Analysts; HMSO: London, UK, 1979.
17. APHA. *Standard Method for Examination of Water and Wastewater*, 21st ed.; AWWA, WPCF: Washington, DC, USA, 2005.
18. SCA. Methods for the determination of metals in soils, sediments and sewage sludge and plants by hydrochloric-nitric acid digestion. In *Methods for the Examination of Waters and Associated Materials*; Standing Committee of Analysts; HMSO: London, UK, 1986.
19. Zahan, Z.; Othman, M.Z.; Muster, T.H. Anaerobic digestion/co-digestion kinetic potentials of different agro-industrial wastes: A comparative batch study for C/N optimisation. *Waste Manag.* **2018**, *71*, 663–674. [CrossRef] [PubMed]
20. Kafle, G.K.; Chen, L. Comparison on batch anaerobic digestion of five different livestock manures and prediction of biochemical methane potential (BMP) using different statistical models. *Waste Manag.* **2016**, *48*, 492–502. [CrossRef] [PubMed]
21. Gompertz, B. XXIV. On the nature of the function expressive of the law of human mortality, and on a new mode of determining the value of life contingencies. In a letter to Francis Baily, Esq. F. R. S. &c. *Philos. Trans. R. Soc. Lond.* **1825**, *115*, 513–583.
22. Zhu, H.; Yang, J.; Cheng, X. Application of Modified Gompertz Model to Study on Biogas production from middle temperature co-digestion of pig manure and dead pigs. *E3S Web Conf.* **2019**, *118*, 1–7. Available online: https://www.e3s-conferences.org/articles/e3sconf/pdf/2019/44/e3sconf_icaeer18_03022.pdf (accessed on 10 October 2021). [CrossRef]
23. Pererva, Y.; Miller, C.D.; Sims, R.C. Existing empirical kinetic models in biochemical methane potential (BMP) testing, their selection and numerical solution. *Water* **2020**, *12*, 1831. [CrossRef]
24. Breiman, L.; Friedman, J.H.; Olshen, R.A.; Stone, C.J. Regression. In *Classification and Regression Trees*, 1st ed.; Routledge & CSC Press: Boca Raton, FL, USA, 2017; pp. 216–264.
25. Rasmussen, C.E.; Williams, C.K.I. Regression. In *Gaussian Processes for Machine Learning*; Dietterich, T., Bishop, C., Heckerman, D., Jordan, M., Kearns, M., Eds.; MIT Press: Cambridge, MA, USA, 2006; pp. 7–29.
26. Nielfa, A.; Cano, R.; Fdz-Polanco, M. Theoretical methane production generated by the co-digestion of organic fraction municipal solid waste and biological sludge. *Biotechnol. Rep.* **2015**, *5*, 14–21. [CrossRef] [PubMed]

processes
MDPI

Article

Estimating the Methane Potential of Energy Crops: An Overview on Types of Data Sources and Their Limitations

Yue Zhang, Sigrid Kusch-Brandt *, Andrew M. Salter and Sonia Heaven

Water and Environmental Engineering Group, University of Southampton, Southampton SO16 7QF, UK; Y.Zhang@soton.ac.uk (Y.Z.); amsl@soton.ac.uk (A.M.S.); S.Heaven@soton.ac.uk (S.H.)
* Correspondence: mail@sigrid-kusch.eu

Abstract: As the anaerobic digestion of energy crops and crop residues becomes more widely applied for bioenergy production, planners and operators of biogas plants, and farmers who consider growing such crops, have a need for information on potential biogas and methane yields. A rich body of literature reports methane yields for a variety of such materials. These data have been obtained with different testing methods. This work elaborates an overview on the types of data source available and the methods that are commonly applied to determine the methane yield of an agricultural biomass, with a focus on European crops. Limitations regarding the transferability and generalisation of data are explored, and crop methane values presented across the literature are compared. Large variations were found for reported values, which can only partially be explained by the methods applied. Most notably, the intra-crop variation of methane yield (reported values for a single crop type) was higher than the inter-crop variation (variation between different crops). The pronounced differences in reported methane yields indicate that relying on results from individual assays of candidate materials is a high-risk approach for planning biogas operations, and the ranges of values such as those presented here are essential to provide a robust basis for estimation.

Keywords: anaerobic digestion; biogas; methane yield; biochemical methane potential; crop material; energy crops

Citation: Zhang, Y.; Kusch-Brandt, S.; Salter, A.M.; Heaven, S. Estimating the Methane Potential of Energy Crops: An Overview on Types of Data Sources and Their Limitations. *Processes* **2021**, *9*, 1565. https://doi.org/10.3390/pr9091565

Academic Editor: Elsayed Elbeshbishy

Received: 25 July 2021
Accepted: 28 August 2021
Published: 1 September 2021

1. Introduction

Anaerobic digestion (AD) of organic materials is a proven technology to produce renewable energy in the form of biogas along with a useful soil conditioner and biofertiliser [1,2]. In agriculture, AD has traditionally been applied to treat cattle and other livestock slurries, but these are low-value substrates in terms of energy content [3,4]. To improve the energy yield, a frequently applied strategy is co-digestion, where manure is combined with energy-rich biomass such as food waste or other types of organic wastes, including crop residues or biomass grown for this purpose, i.e., energy crops [5–7]. As bioenergy production from crop-based materials and residues becomes more widespread, there is a need for information on potential biogas and methane yields of such biomass types. The information is required as a basis for the selection of crop materials to be grown and digested, in whole or part; as a reference when estimating potential energy production; and as benchmark to evaluate performance of biogas facilities [8].

Many sources of information are now available on the methane potentials of various crop materials. These include scientific journal papers, agricultural textbooks and reference works, and more recently online databases. For individual crop types, the published values have been compiled in review articles. Some address methane yields reported for lignocellulosic crops [9–12]. Others focus on a selection of the most widely grown agricultural crops across Europe [13–15], on some frequently digested energy crops [16–18], or on single biomass types such as grass silage [19]. These have established a robust knowledge base about average methane yields reported in the literature. However, there is still a lack of a comprehensive overview on the nature of works informing about methane potentials, along with an assessment of the variations in methane yields reported. Some of the quoted yields have been determined using repeatable laboratory-based tests; others come from data collected from full-scale digesters, and some are calculated values based on the elemental composition or the content of protein, fat, and carbohydrates. Thus, the outcome may be a range of values for a given crop material, due both to the nature of the substrate being tested and to differences between methods of determining the methane yield.

This paper presents the results of a review of literature values for the methane yields of various crop materials, particularly those commonly grown in European conditions. The main aim is to support practitioners and researchers in adequately placing single values reported in the literature into context and to make them aware of the risks of too strongly relying on one single value found in a publication. This work captures the diversity of information and the range of methane yields that a practitioner or researcher is likely to encounter when searching for a reference methane value for a specific crop material of interest. To explain some potential sources of variation, the results are structured into sections based on the methodology used to obtain them. The paper is not intended as a full technical review of single methods or test protocols; the focus is on clarifying the main approaches used and on highlighting potential limitations regarding the usage of literature data.

2. Materials and Methods

This work reviews the characteristics of literature data published on methane yields of crop biomass and establishes an overview of the variety of values reported. A full systematic review of all existing data for a specific type of biomass is not within the scope of this work; the methodology is explorative in so far that literature is selected which is likely to inform practitioners and researchers who are looking for reference values to estimate the potential methane yield of a crop material. In line with this goal, the focus is on data that have been effective in informing others, and thus, only results that have been quoted more than once by other authors are included. The body of literature reviewed is the outcome of applying the snowball scheme to the review articles mentioned in Section 1; i.e., literature listed in review articles was taken as a starting point. The analysis focuses on original (primary) data, i.e., by default, data are taken from the original source. Many publications, when indicating methane yields, quote earlier works of the authors or reference data from papers by others. Such secondary sources are not included here; in each case, data were tracked back to the original source. In addition, only methane potentials of commonly used crop-based biomass are considered. This includes data from biomass pre-treated using methods regularly applied at full-scale installations, such as ensiling, but not data reported after experimental pre-treatments that are not yet in widespread use. Furthermore, the focus is on crops that are commonly cultivated under European conditions. The values presented include data published in the research literature but also from the technical press, conference presentations, and web-based databases.

Literature sources are structured according to the main methodology applied to study the methane yield from a biomass. For each case, the type of method by which the results were determined is presented, and the reported methane yield is given in accompanying tables. Many studies document the results from several repetitions, and in some cases, certain results are excluded by the researchers after critical reflection. In line with the goal of this study, the value taken from each publication is that communicated by the authors as representing the methane yield determined in their work; this typically is the average of several repetitions. Each such published result is taken as one data point for the dataset of this work. Then, the range of methane yields reported for crop materials is presented and discussed.

To the extent possible, the comparability of data is facilitated by referring to a common set of standard terms. When describing both the results of analyses and the methods used, different authors use different terminology. Tests on physicochemical characteristics are usually conducted according to standard methods such as those of the International Organisation for Standardisation (ISO) or national standards. Total solids (TS) content is the material left after water has been removed, usually by drying at 103–105 °C, and it is also widely referred to as dry matter (DM) (TS is used in this work). Volatile solids (VS) may also be referred to as organic dry matter (ODM), volatile dry matter (VDM), or loss on ignition (LOI): it corresponds to the material that ignites at temperatures up to 550 °C and is the fraction from which biogas is produced. VS may be expressed as a percentage of TS or of the original fresh matter (FM), which is sometimes referred to as wet weight (WW). Biogas or methane yields may be expressed in terms of VS_{added} or $VS_{destroyed}$. VS_{added} refers to the amount of substrate VS added to the digestion, not all of which may be converted into biogas. $VS_{destroyed}$ is the amount of substrate VS that was degraded in producing the biogas, which is always less than or equal to the amount added. Some publications express biogas or methane yield in terms of TS or FM rather than VS, and sometimes, COD (chemical oxygen demand) is used (rarely employed in the case of crop biomass); where possible (i.e., where the relevant information is provided by authors), such values are converted to the VS basis in the following.

For all experimental methods employing AD, one important aspect for reporting purposes is the method used for gas collection and for the conversion of gas volumes to a standard temperature and pressure (STP). As illustrated in the next sections, the standard conditions used vary and are sometimes not stated, or no correction has been applied. This may have significant implications when comparing methane potentials, as 1 litre of methane at 1 standard atmosphere (101.325 kPa) and 0 °C equals 1.13 litres (L) at 1 atmosphere and 35 °C, which is a temperature frequently used in mesophilic AD processes. Wherever possible, in the following text, reported values have been converted to yields in terms of m^3 CH_4 kg^{-1} VS_{added} at STP of 0 °C and 101.325 kPa to facilitate direct comparison; where no information is given on the temperature and pressure conditions used, this is noted, and gas volumes are taken as reported.

3. Results

3.1. Overview on Types of Testing Methods Applied

Details of the types of methods used to determine methane yields of crops or crop residues are given in the following sections. These methods can be structured into several groups:

- Biochemical methane potential (BMP) tests and other long-retention batch assays;
- Short-retention batch tests;
- Continuous/semi-continuous tests;
- Theoretical calculations.

Most of the data reported in the literature originate from laboratory-scale experiments. Laboratory-based methods are of two main types: batch tests to determine the specific methane potential, and continuous trials, which are mostly applied to determine the specific methane production under a selected AD regime. Laboratory-based batch tests have several advantages when evaluating methane potentials. They facilitate the inclusion of control substances with a known methane yield so that the experimental setup can be validated. Multiple replicates can be used, which is seldom the case in large-scale determinations, and the tests can be operated under optimal conditions including the addition of nutrients that might otherwise be limiting. To determine methane potentials, the most widely used batch tests are BMP or other long-duration AD tests, with a retention time of 35 days or longer: this differentiation into BMP tests and other long-duration batch tests is only arbitrary in so far that it reflects that some tests are specified in the literature as being BMP assays, while others are reported more generally as being batch tests. Some authors have conducted short-retention batch tests to evaluate the methane yields of different crop materials (see Section 3.3). Others employed continuously or semi-continuously operated digesters (Section 3.4). Kinetic data from experimental studies are sometimes used in modelling to estimate potential values, as reviewed by Pererva et al. [20] and Raposo et al. [21]; these estimates may be presented with or without experimental values, but only experimental data are considered here.

In addition to experimental AD tests, some authors employed predictive methods based on biochemical composition of the biomass to determine the specific methane potential (Section 3.6). There also exists a set of literature that does not disclose which method was applied to determine the indicated methane yields (Section 3.7).

3.2. Methane Potentials Obtained in BMP and Long Retention Batch Tests

BMP tests are a form of extended batch test. The BMP of an organic material has been defined as the ultimate specific methane production under optimised digestion conditions for an indefinite degradation time [22–24], where optimised refers to the environmental conditions for microbial degradation rather than any pre-treatment of the substrate. This provides information on the energy potential of a biomass when used in AD [25,26] and thus serves as a decision basis for choosing a specific material for biogas production. Furthermore, the BMP value is often used as a benchmark to assess the efficiency of digester operation. Comprehensive reviews of factors affecting the performance of anaerobic batch tests can be found in Raposo et al. [15,21], and reviews of those affecting repeatability can be found in Mittweg et al. [27]. An inherent feature of well-conducted BMP tests is the establishment of an optimised digestion environment, in which the biochemical process will run without inhibition [28]. Different experimental protocols have been developed to determine BMP values [29–31], and some efforts at standardisation have been made (see below), but different regimes and set-ups continue to be used [30].

In practice, BMP assays are batch experiments in which a known amount of the test material is mixed with a sufficiently high quantity of inoculum that contains a mixed microbial population capable of carrying out the AD process [26]. The headspace of the test vessel is usually purged before sealing. While purging with N_2 removes oxygen and generates anaerobic conditions, an N_2/CO_2 mix containing 20–40% CO_2 is preferred in order to minimise pH changes due to loss of CO_2 from the test matrix: this is especially important if the matrix is not heavily buffered and the headspace volume is much bigger than that of the test solution [24]. Then, the vessel is maintained at a constant temperature, which may be mesophilic or thermophilic, usually depending on the source of the inoculum and/or the process under consideration. The contents of the vessel may or may not be stirred or agitated. The quantity of biogas produced is monitored, and the biogas itself is released either continuously or on an intermittent basis. In this way, a kinetic curve for gas production against time can be determined. The biogas composition is generally analysed through gas chromatography (GC) or infrared analyser (IR) to determine the percentages of methane and carbon dioxide; alternatively, the CO_2 fraction may be removed by reacting e.g., with sodium hydroxide to give methane only, in which case CO_2 content can be determined by back titration. The BMP assay usually ends when biogas production ceases, e.g., when the cumulative biogas production curve flattens, or when biogas production from the test sample is the same as from an inoculum-only control. This can be as short as 50 days for substances such as cellulose [23] or may take 100 days or more for some crop-based materials due to their lignocellulosic content [32]. In a recent attempt to standardise BMP tests, it is recommended that the test is terminated when the daily net methane production during three consecutive days is less than 1% of the accumulated methane produced from the substrate already [24,31], although the duration of the test should also be taken into consideration for unexpected degradation patterns or inhibition effects [33]. The inoculum must necessarily be digested separately (under the same conditions) to obtain a control value (amount of methane produced by the inoculum only); this control value must be deducted from the results of the test vessels to obtain the yield attributed to the actual test substrate. When using this type of data, it is important to note whether it represents methane or biogas production, as the proportion of methane contained in the biogas varies with both the material digested and the point of time within the assay.

Dividing the methane yield attributed to the test substrate by the amount of substrate added gives the specific methane yield. The specific methane yield of the test material may be expressed in units of m^3 (STP) kg^{-1} VS_{added} or m^3 (STP) kg^{-1} TS_{added}, or another unit, as appropriate.

The BMP results reported by different authors are given in Table A1 (Appendix A), and the variations are shown in Figure 1.

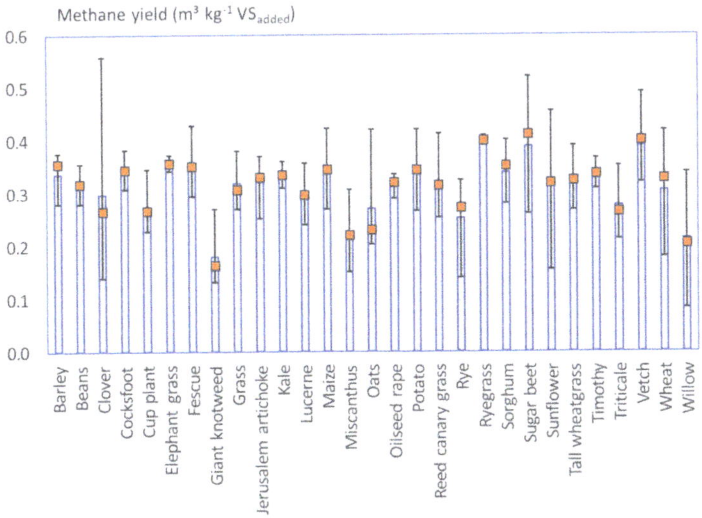

	n	Min	Max	Δ (Range)	Mean	Median
Barley	3	0.280	0.375	0.095	0.337	0.356
Beans	2	0.280	0.356	0.076	0.318	0.318
Clover	6	0.140	0.558	0.418	0.298	0.266
Cocksfoot	2	0.308	0.382	0.074	0.345	0.345
Cup plant	8	0.228	0.345	0.117	0.266	0.267
Elephant grass	2	0.342	0.372	0.030	0.357	0.357
Fescue	6	0.294	0.428	0.134	0.354	0.352
Giant knot-weed	8	0.132	0.270	0.138	0.179	0.164
Grass	13	0.270	0.380	0.110	0.319	0.307
Jerusalem artichoke	4	0.252	0.370	0.118	0.321	0.331
Kale	2	0.310	0.360	0.050	0.335	0.335
Lucerne	4	0.240	0.357	0.117	0.298	0.298
Maize	17	0.270	0.422	0.152	0.345	0.345
Miscanthus	10	0.151	0.307	0.156	0.224	0.220
Oats	4	0.203	0.420	0.217	0.271	0.230
Oilseed rape	3	0.290	0.336	0.046	0.315	0.320
Potato	2	0.267	0.420	0.153	0.344	0.344
Reed canary grass	12	0.253	0.412	0.159	0.321	0.315
Rye	4	0.140	0.324	0.184	0.252	0.273
Ryegrass	2	0.390	0.409	0.019	0.400	0.400
Sorghum	7	0.280	0.400	0.120	0.339	0.352
Sugar beet	11	0.261	0.520	0.259	0.388	0.411
Sunflower	4	0.154	0.454	0.300	0.312	0.319
Tall wheat-grass	6	0.268	0.389	0.121	0.330	0.324
Timothy	2	0.308	0.365	0.057	0.337	0.337
Triticale	3	0.212	0.351	0.139	0.276	0.265
Vetch	5	0.320	0.490	0.170	0.389	0.400
Wheat	12	0.179	0.418	0.239	0.304	0.327
Willow	8	0.082	0.339	0.257	0.213	0.204

Figure 1. BMP yields of crops (**left** side: graphical presentation, the shaded columns show the average of reported values, the orange squares indicate the median, and the lines indicate the range; **right** side: tabular presentation, n is the number of data points).

As noted, there is no single standard method for BMP determination. One of the earliest attempts to standardise the analysis was the approach put forward by Owen et al. [34]. Using 250 mL reagent bottles with a serum cap, the substrate and a broad-spectrum inoculum (e.g., from an anaerobic digester treating municipal wastewater biosolids) are added with a stock nutrient solution, and the test samples are maintained at a constant temperature; gas volumes are measured using a glass syringe in which the plunger is allowed to move horizontally until in equilibrium with atmospheric pressure. Jerger et al. [35] applied this procedure to sorghum in a 60-day incubation at 35 °C; biogas production and methane content were reported as adjusted to STP, but conditions were not stated. Gunaseelan [32] adopted the method to carry out tests on a range of crop materials in 135 mL bottles at 35 °C. Then, 500 mL test units (sealed serum bottles), maintained at 37 °C, were used by Zauner and Küntzel [36]; biogas was removed with calibrated glass syringes and volumes were converted to 0 °C and 101.29 kPa. It is not stated whether the syringes were left in place and allowed to equilibrate to ambient pressure as in Owen et al. [34], or pressure was allowed to rise in the system and then intermittently released.

In most systems, the generated biogas leaves the reactor, and its quantity is measured through gas counters, water displacement, or other methods (see below). Some systems accumulate biogas within the reactor; these employ sealed serum bottles where the biogas is occasionally released to determine the BMP of the studied substrate. The raised pressure alters the partitioning of gases between the liquid phase and the headspace and can potentially affect the degradation kinetics of digestion; careful management and reporting of both the pressure regime and the depressurisation stage is necessary to obtain reliable results. For winter rye, oilseed rape, and faba bean, Petersson et al. [37] adapted a method originally employed by Hansen et al. [38] for solid organic wastes, using sealed flasks with occasional pressure release. This method used 100 mL serum flasks that were placed in a shaking water bath at 42 °C for a 67-day period. Methane yield was calculated by taking samples of known volume with a pressure-lock syringe before and after pressure release and measuring methane concentrations by gas chromatography, thus determining the mass of methane present. Sealed 500 mL bottles with intermittent pressure release to an acidified water displacement column have been used by a number of researchers to study energy crops [39–42]. Wahid et al. [43] also used 500 mL pressurised bottles in 90-day assays on miscanthus, but it is unclear whether gas volumes were calculated from headspace pressures or gas was released to a gas bag. The method is not explicitly stated, and reference to a publication is made; however, two different methods are employed there. Jurado et al. [44] employed sealed 117 mL vials to measure methane potentials of miscanthus, wheat straw, and willow over 50 days; no details are given on how gas volumes were determined. Kakuk et al. [45] used sealed 160 mL bottles at 37 °C in a 45-day assay for willow; equipment and correction to STP are not clear.

Two attempts to standardise BMP testing were made in the German standards DIN 38414-8 [46] and VDI 4630 [47,48]; the development of these guidelines has been described elsewhere [26]. The DIN method (originally mainly applied in the wastewater sector) describes the use of eudiometer tubes for gas collection. VDI 4630 builds on DIN 38414-8 but addresses the use of small-scale digesters more generally, and it provides specific guidelines for the duration of the digestion. Batch experiments are continued until only a relatively small volume of gas (<1% of the cumulative total up to this point) is released each day. Scaled wet gas meters or precision rotor gas meters record gas production. The DIN and VDI specifications both specify the correction of biogas quantities to STP of 0 °C, 101.325 kPa, and dry gas, i.e., corrected for water vapour content. Linke et al. [49] determined gas yields according to DIN 38414-8 for various crop substrates in tests conducted at 35 °C; however, they reported only biogas yields without methane values, and thus, the findings are not included in this work. Analyses conducted according to VDI 4630 include those by Gallegos et al. [50], who used eudiometer devices at 38 °C for wheat straw, and by Amon et al. [51], who used 1-litre batch digesters operating at 38 °C to study a range of crops. Machmüller et al. [52] also employed 1-litre batch fermenters at 38 °C in a similar experimental set-up to analyse sunflower, sugar beet, maize, clover, and rye; biogas was monitored daily. Bauer et al. [53] conducted assays (maize, barley, sunflower, lucerne, sorghum, wheat) in accordance with the DIN 38414 and VDI 4630 methods but used eudiometer batch digesters of 250 mL capacity maintained at 37.5 °C, recording biogas volume on a daily basis with determination of the CH_4 content. The VDI 4630 guideline was also employed in Denmark by Heidarzadeh Vazifehkhoran et al. [54] to test sugar beet (37 °C), but digester volumes are not given. Miscanthus was tested by Schmidt et al. [55] in 2-litre batch digesters at 37 °C according to the VDI 4630 method, but although statistical analysis was conducted, no numerical BMP values are quoted, and thus, no result is included here. This overview illustrates that a high diversity of equipment and procedures continues to be applied even when accounting for existing standardisations.

A further method derived from the DIN 38414 and VDI 4630 methods is the Hohenheim Biogas Yield Test (HBT), which uses bench-scale glass syringes (100 mL) as the digester vessels combined with gas collection (in the expanding syringe volume) [56]. Methane content is periodically determined with the aid of a miniaturised infrared analyser. As the HBT uses relatively small samples, it is possible to run large numbers of tests concurrently. Examples of its application to crop materials include maize by Mittweg et al. [27], cup plant by Haag et al. [57]; cup plant, energy dock, giant knotweed, and tall wheatgrass by Mast et al. [58]; and maize and perennial energy crops by Ruf and Emmerling [59].

With the introduction of standardised procedures and guidelines prepared by the IWA task group for Anaerobic Biodegradation, Activity, and Inhibition (ABAI-Group) [24,31,60], it has become more common to operate the BMP test until gas production becomes negligible. Nevertheless, some works have continued to apply a fixed pre-defined digestion period. Chiumenti et al. [61] digested grass (38 °C) over a 40-day period; they equipped their 4-litre fermenters with bench-scale biogas meters (MilliGascounter, Ritter, Bochum, Germany) to continuously register the volume of produced biogas. Schmidt et al. [62] tested five perennial species (cup plant, giant knotweed, reed canary grass, tall wheatgrass, and virginia mallow) in 2-litre batch digesters at 37 °C with a test duration of 42 days.

Other protocols set the digestion times to reflect the point when gas production is negligible but do not record biogas generation rates during this period. A number of crop-related assays have been carried out in which biogas was collected in gas bags and then determined. Pouech et al. [63] operated 0.5-litre batch reactors at 40 °C; gas is said to have been collected in an 'inspection hole' (no further description given), stored in gas-tight bags, and then measured with a 100 mL syringe; however, it is not stated whether volumes were converted to STP. Methane potentials were measured in 0.5-litre bottles (300 mL working volume) at 37 °C in an orbital shaker water bath by Lehtomäki and Björnsson [64]; Lehtomäki et al. [65] in 2-litre glass bottles (1.5 L working volume) at 35 °C; and Seppälä et al. [66] in 1-litre glass bottles (750 mL working volume) at 35 °C over periods of 80 to 100 days, until gas production was negligible (although it is not clear how this criterion was ensured); in all cases, gas was collected in foil-lined gas bags. A similar 2-litre apparatus and protocol was used by Kaparaju et al. [67] and Lehtomäki et al. [68]. Kaparaju et al. [67] conducted experiments over 155 days on a range of crop residues. Lehtomäki et al. [68] examined crops harvested at different growth stages with the duration of the tests varying between 107 and 189 days. Parawira et al. [69] conducted assays at 37 °C in 0.5-litre flasks (working volume 300 mL) maintained in a shaking water bath. The tests ran for 50 days, being terminated when there was no significant gas production over a 2-week period; gas composition was determined by gas chromatography. Specific methane yields are expressed in terms of VS destroyed rather than added, but the percentage of degradation achieved in the test is not given. Apart from Pouech et al. [63], none of the above papers using gas bags for collection states how gas volumes were measured, but according to Lehtomäki [70], values were obtained by height difference in a water displacement column and are quoted at ambient pressure and room temperature (20–22 °C) without correction to STP or dry biogas.

Garcia et al. [71] tested a range of crop materials in 0.5 L digesters at 37 °C (45 days, checked for final daily production rate); gas volumes were determined by a water displacement method, but no information is given on any volume corrections. Measurement of biogas production in liquid displacement cylinders containing a barrier solution of 75% saturated sodium chloride at pH 2 was reported by Cornell et al. [72] for maize (37 °C, 44 days) and Rincón et al. [73,74] for wheat (35 °C, 96 days and 37 °C, 79 days). In each case, 1.5 litre stirred tank reactors maintained at constant temperature in a water bath were employed, and gas composition was measured each time the cylinders were refilled with the barrier solution, at maximum intervals of five days when gas production rates were low. Cornell et al. [72] does not indicate whether gas volumes are expressed at STP. It should be noted that the use of this type of barrier solution reduces but does not prevent CO_2 losses [75], and therefore, this method is more suitable for tests of rather short duration or where only methane yields are to be reported.

The availability of proprietary systems from various suppliers has been making BMP tests more popular in the last five years, because such solutions offer ready-to-use equipment with pre-defined specifications. Pererva et al. [20] lists several types of systems available (including YieldMaster, Nautilus BMP, Anaero Technology, AMPTS II). Most users have digested wastes, but some have studied crop-based materials. The BMP of miscanthus species harvested at different growth stages was evaluated by Peng et al. [76] using AMPTS II (Bioprocess Control, Sweden) with flow cell gas measurement in a trial ending when daily gas production was <1% of the cumulative total. Virkajärvi et al. [77] used AMPTS II for BMPs of grass and grass with clover, but the test duration is not reported. The same system was used by Nges et al. [78] and Li et al. [79] to assess the methane potential of miscanthus (50-day period), and miscanthus was further tested by Thomas et al. [80] (48 days).

A number of BMP publications are not considered in this work for methodological reasons. As an example, Pohl et al. [81] and Heeg et al. [82] carried out BMP assays on wheat straw in 2-litre gas fermenters with separate gas holders, but no detailed description of the gas measurement method is provided; and the results presented are based on modelling so are not included here. Rocha-Meneses et al. [83] used pressurised 575 mL serum bottles at 36 °C to determine the BMP of barley straw, but the pressure release regime is unclear, and as the reported values are based on a numerical model, they are not included here. Sealed 309 mL bottles at 36 °C were used by Ohlsson et al. [84] in a 94-day BMP assay for willow, with gas measurement on five occasions, but the pressure release regime is not specified, and values are available on a TS basis only, so they are omitted here. While such studies may satisfy the specific research interests of the authors, the transferability of findings is reduced by omission of the supporting data.

3.3. Methane Production Obtained from Short Retention Batch Tests

Values for methane yields have been obtained from other batch tests, which are sometimes carried out in larger-scale reactors but with gas production often measured over shorter periods. These methods may produce lower biogas or methane yields than a BMP assay, as complete digestion may not be achieved within the test period, and therefore, the full potential of the material may not be realised. On the other hand, these tests monitor the readily biodegradable components of the material, and it can be argued that the shorter duration more closely resembles the digestion time in continuous operation under standard practice. Therefore, the data may draw the attention of readers interested in methane yields obtainable in practice. The current study only includes publications where the methane yield of a specific substrate was explicitly explored and excludes, for example, research that focused on co-generation mixtures to enhance digester performance. In some cases, the equipment employed usually serves to investigate the performance of a specific reactor type (including leach-bed reactors or two-stage systems) rather than the methane potential of substrates. The review illustrates that the documentation of equipment and procedures used in many cases is even more deficient than for BMP studies.

In some cases, the methods are well-documented, but methane yields are not clear. Linke and Schelle [85] used a range of batch reactors (in accordance with the guidelines in VDI 4630) with working capacities of 1–66 kg, operated at 35 °C to digest hemp and grass. Gas production was measured with scaled wet gas meters or precision rotor gas meters, but biogas values only are reported, without methane yields, and therefore, these results are not considered here. Heiermann et al. [86,87] and Heiermann and Plöchl [88] adopted the method described by Linke and Schelle [85] for various crops (barley, rye, triticale, fodder beet, grass, hemp, ley crop, lucerne, maize), with digestion conducted at 35 °C over a period of 28 or 29 days, at a working capacity of 1.4 kg, with reactor vessels connected to scaled wet gas meters for measuring biogas production (results presented only in graphs). The methane content of biogas is reported as having been determined three and four times respectively during the digestion period. This approach may lead to errors, as gas composition may change significantly during the period. It is also difficult to read precise values from graphs. In addition, neither of the papers by Heiermann et al. [86,87] state whether the reported gas volumes were adjusted to STP. These results are not considered here.

The failure to convert gas volumes to STP, or to report whether such conversion has been carried out and what conditions were used, is a frequent problem in the analysis of literature data. This applies to studies conducted around 40 years ago, but it is a persistent issue also in more recent publications. Early studies that have been frequently cited are those by Badger et al. [89] (kale, maize, oats, sugar beet tops, and wheat straw) and Zubr [90] (cauliflower, oilseed rape, rhubarb, sugar beet, etc.). Badger et al. operated batch reactors (800 mL in 1-litre bottles) at 37 °C; the volume of biogas produced was measured using displacement of CO_2-saturated water. Methane yields were obtained after varying digestion times (17–36 days) depending on the amount of gas produced each day. Quoted gas production values are not adjusted for STP conditions. Zubr's experiments were carried out using equipment consisting of a 3-litre batch fermentation reactor, a 30-litre PVC gas collector, and a central water reservoir; fermentation was at 35 °C for durations varying from 27 to 36 days, but it is not indicated whether the quoted gas yields are adjusted to STP conditions.

Among the studies published in the last 15 years, a relatively broad range of procedures and equipment types have been used, and it is not always possible to assess the accuracy of reported values. Svensson et al. [91] examined ensiled sugar beet tops and wheat straw in batch reactors at 35 °C; the tests on sugar beet tops were conducted in a single stage batch reactor (20 days), while those on wheat straw were in a leach-bed reactor. The method of gas collection is not given, and it is also not stated whether quoted gas yields are adjusted to STP conditions. Gas collection columns were used by Nizami and Murphy [92] during investigations of the potential of ryegrass for methane production, in reactors with a working volume of 1.5 litres operated at 38 °C for 26 days; it is not stated if the measured gas volumes are corrected to STP. Raposo et al. [93] used glass vessels with a 5-litre working volume (35 °C) to digest maize over 20 days; gas volume was determined by water displacement and values corrected for STP, although the conditions are not stated. The combination of short retention time and substrate processing may account for the low values recorded. Yan et al. [94] investigated the biomethane production of various leafy vegetables over a 25-day test period in pressurised mesophilic (37 °C) 500 mL reactors; gas volumes were corrected, but the STP conditions used are not stated. Tilvikiene et al. [95] worked at a larger scale, employing 20-litre batch digesters at 38 °C, equipped with drum-type flow meters to determine gas production; the duration of the test and the conditions applied for volume correction are not stated.

In some cases, equipment and procedures are well described, but the applied digestion temperature is not clear. Kaiser et al. [96] examined a range of crop materials in 2-litre batch reactors, each with an individual small-scale gas counter (Milligascounter), in climatic test cabinets; methane yields are reported at STP, but the operating temperature of the reactors is not given. Overall, mesophilic tests are more common than thermophilic in both batch and continuous/semi-continuous experiments. Mesophilic tests with clear documentation include those by Mähnert et al. [97], who adopted the method described by Linke and Schelle [85] to digest grass (cocksfoot, among others), in 2-litre reactors at 35 °C; the volume of biogas produced was measured using calibrated wet gas meters and reported as cumulative yield after 28 days. Methane content was determined periodically. A similar method was used by Herrmann et al. [98–100] to measure methane production of various crop feedstocks in 2-litre reactors at 35 °C, reporting cumulative methane yield over 30 days corrected to STP. Kreuger et al. used 500 mL flasks incubated in a mesophilic water bath [101] to test hemp, maize, and sugar beet, and Gissén et al. [102] used them to test grass, hemp, maize, sugar beet, and triticale. Thermophilic tests were carried out by Bruni et al. [103] in batch assays with a total volume of 2140 mL at 55 °C on a range of maize varieties harvested at various times. Kreuger et al. [104] used 500 mL flasks, with an active volume of 300–350 mL, to conduct assays on hemp; flasks were incubated at 50°C in a shaker water bath and terminated after 32 or 34 days.

Since the introduction of proprietary systems on the market, such solutions have also been widely employed in the last 5 years for short-duration batch tests. Spence et al. [7] examined substrates, including grass and triticale, at 38 °C for 20 days using a proprietary system with automated normalisation to STP (Anaero Technology, Cambridge, UK). Tests on the effect of harvest date and cutting length for grass and whole-crop rye and wheat were carried out by Prade et al. [105] using the AMPTS II system, with values reported as dry gas at STP after 30 days of digestion. Allen et al. [106] also employed AMPTS II for various substrates; the test duration is not explicitly stated but appears to have been 30 days. Nges et al. [107] tested wheat straw (AMPTS II, 30-day batch test). Kolbl et al. [108] used AMPTS II modified to accommodate 5-litre vessels for a range of materials including triticale and maize over a 34-day test period. Zhao et al. [109] used the WAL-BMP-Test system 3150 (WAL, Germany) to measure the methane potential of maize stover over a 21-day test period.

The reported results from these batch tests are given in Table A2 (Appendix A) and shown in Figure 2.

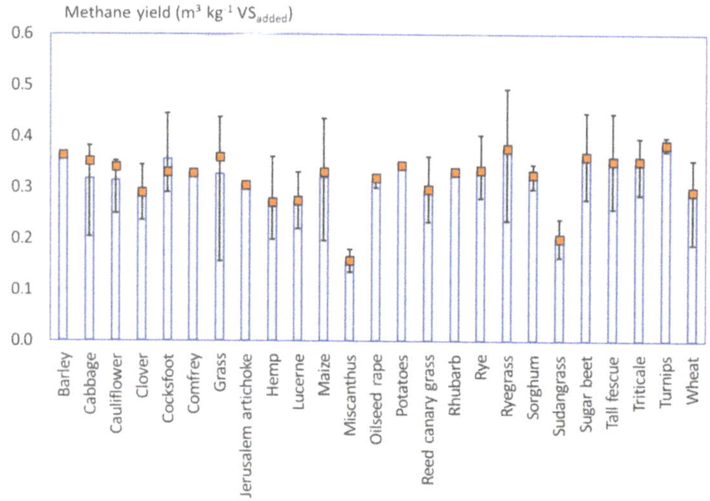

	n	Min	Max	Δ (Range)	Mean	Median
Barley	2	0.361	0.367	0.006	0.364	0.364
Cabbage	6	0.204	0.382	0.178	0.317	0.352
Cauliflower	3	0.250	0.352	0.102	0.314	0.341
Clover	2	0.236	0.344	0.108	0.290	0.290
Cocksfoot	3	0.291	0.446	0.155	0.356	0.330
Comfrey	2	0.323	0.334	0.011	0.329	0.329
Grass	9	0.156	0.438	0.282	0.326	0.359
Jerusalem artichoke	2	0.301	0.309	0.008	0.305	0.305
Hemp	8	0.199	0.360	0.161	0.276	0.271
Lucerne	4	0.220	0.330	0.110	0.275	0.275
Maize	21	0.196	0.436	0.240	0.320	0.331
Miscanthus	2	0.135	0.180	0.045	0.158	0.158
Oilseed rape	3	0.300	0.326	0.026	0.315	0.319
Potatoes	2	0.337	0.351	0.014	0.344	0.344
Reed canary grass	2	0.233	0.361	0.128	0.297	0.297
Rhubarb	2	0.330	0.334	0.004	0.332	0.332
Rye	8	0.280	0.403	0.123	0.333	0.335
Ryegrass	6	0.235	0.493	0.258	0.377	0.378
Sorghum	4	0.298	0.346	0.048	0.324	0.327
Sudangrass	2	0.164	0.239	0.075	0.202	0.202
Sugar beet	21	0.278	0.447	0.169	0.363	0.362
Tall fescue	2	0.259	0.446	0.187	0.353	0.353
Triticale	8	0.287	0.397	0.110	0.348	0.353
Turnips	2	0.372	0.399	0.027	0.386	0.386
Wheat	10	0.190	0.354	0.164	0.295	0.294

Figure 2. Methane yields from various crops obtained in short-retention batch tests (**left** side: graphical presentation, the shaded columns show the average of reported values, the orange squares indicate the median, and the lines indicate the range; **right** side: tabular presentation, n is the number of data points).

Fairly large variations can be seen for the individual crops. Some of this variation is probably because some of the tests used ground/milled feedstock. As a tendency, average methane values reported from short-duration batch tests are lower compared to BMP results, which is in line with the expectation of more complete digestion in BMP trials. However, the differences are not always high. For the relatively rapidly digesting maize, the average value in short-duration batch testing is 0.32 m^3 CH_4 kg VS_{added}, which is reasonably close to the 0.35 m^3 CH_4 kg VS_{added} obtained as an average value in the BMP assays. Similarly, ryegrass reaches on average 0.38 m^3 CH_4 kg VS_{added} in the short-duration testing compared to 0.40 m^3 CH_4 kg VS_{added} in the BMP experiments. For grass, a slightly higher average methane yield is found in the short-duration trials (0.33 CH_4 kg VS_{added}, while it was 0.32 CH_4 kg VS_{added} in the BMP testing). This contradicts the expectation that a shorter digestion produces less methane. This set of data further highlights the high variations among reported results and the limitations in the comparability and transferability of reported data.

3.4. *Methane Production from Continuous/Semi-Continuous Digestion Processes*

Experiments that more closely resemble industrial-scale processes can be conducted in continuously or semi-continuously fed laboratory-scale reactors. These are generally stirred or mixed in some way and maintained at constant temperature, and they are fed a measured amount of feedstock on a regular basis (usually daily), with digestate being removed in order to maintain the quantity of material within the digester. Biogas production may be measured by collection above a barrier solution or in gas-impermeable bags, with periodic recording of the volume collected, or by continuous monitoring via gas flow meters. The methane content of the biogas is generally determined through compositional analysis of samples taken at defined intervals.

Such experiments typically run for at least three hydraulic retention times to allow the establishment of steady-state conditions or to reveal any adverse operational symptoms. Thus, biogas production takes place in conditions similar to those in full-scale digesters. In a continuous or semi-continuous trial, the organic loading rate (OLR) in terms of kg VS_{added} m^{-3} digester day^{-1} is usually higher than the equivalent loading calculated over the duration of a BMP test; the average retention time is often less than the duration of a full BMP test; and in a completely mixed reactor, a proportion of the feedstock is removed after a very short period. For these reasons, the specific gas yield on a VS-added basis is expected to be lower than in a BMP test on the same material. Daily biogas production usually shows some fluctuation. This may be for a number of reasons, such as the heterogeneous nature of the feedstock, acclimation of the inoculum, and slight changes in operating parameters, e.g., feeding time. An average gas production rate can be determined over a period of time after the digestion trial has reached steady state.

A growing body of literature documents results from laboratory trials conducted in continuously or semi-continuously fed reactors, but the picture is less complete compared to batch testing when considering only those experiments where no process inhibition was observed and where a specific type of crop was individually tested. This criterion excludes many of the published works. As one example, in addition to carrying out BMP tests, Wahid et al. [42] attempted thermophilic (55 °C) mono-digestion of lucerne in a 15-litre working volume continuously stirred tank reactor (CSTR) over a 70-day experimental period, but they were unable to establish stable operation; therefore, results are not included here.

A relatively broad diversity exists among the scales and types of systems used. Most works employ CSTR, but other systems are also used. Similarly to the observations made above on short-retention batch tests and many BMP tests, the quality of reported values cannot always be assessed with precision because essential information is not documented in the publications.

For some published methane production results, expression on a VS basis is not possible. Stewart et al. [110] operated 20-litre CSTR digesters at 35 °C, with a retention time of 20 days, using kale, maize, oats, grass, wheat, and barley straw, among others. Biogas yields were determined daily by collection in 60-litre PVC gas collection bags from where the gas was vacuum-pumped through a gas meter; no information is given on whether the reported values are adjusted to STP. Methane yields are available on a TS basis only, and thus, the results are not considered here.

In many cases, the method for gas measurement is not clear, or information on conversion to STP is deficient. Scherer et al. [111] ran four laboratory-scale digesters at 37 °C, 45 °C, 60 °C, and 65 °C, fed with fodder beet silage; however, the method of gas collection is not given. Nges and Björnsson [112] operated two 4-litre CSTR digesters at 38 °C on sugar beet with biogas collected in gas-impermeable bags, but the method of volume measurement and any corrections to STP are not reported. Zhu et al. [113] conducted a two-stage process (1-litre CSTR followed by 5-litre CSTR) looking at H_2 and CH_4 production from homogenised potatoes; the volume of gas produced in each stage was determined using a water displacement technique, but it is not stated whether gas yields are expressed in terms of STP. Haag et al. [57] employed 20-litre horizontal digesters at 40 °C to determine the biomethane potential of cup plant; gas volumes and composition were measured automatically, but no details are reported of any corrections applied, and the duration of the experiments is not given. Lehtomäki and Björnsson [64] operated two-stage digesters (10 m^3 leach-bed hydrolytic reactor plus 2.6 m^3 leachate recirculating methanogenic reactor) under mesophilic conditions; gas volumes were measured using gas flow metres, but it is not stated whether the reported values were converted to STP. Nizami and Murphy [92] operated a two-stage CSTR system at 37 °C consisting of two digesters of 312-litre volume with 160 litres of gas headspace to examine the effects of varying loading rate on the digestion of grass; it is not reported if gas yields are corrected for STP or water vapour content. Semi-continuous digestion of maize stover at 37 °C was tested in wet

(5-litre CSTR) and dry (packed-bed) fermentation conditions by Kakuk et al. [114], with gas volumes measured by mass flow controllers, but details of STP corrections are not reported.

Veluchamy et al. [115] used a mesophilic (35 °C) plug flow digester with a 50-litre working volume working in fill-and-draw mode for the semi-continuous digestion of maize silage at a range of OLR. Hydraulic retention times ranged from 13 to 25 days, but the digester only operated at each OLR for 30 days. Gas volumes were measured using a multi-chamber rotor meter; whether values were corrected in any way is not stated.

Rincón et al. [116] ran eight CSTR reactors with working volumes of 4 litres at 35 °C, using winter wheat collected at medium milk harvest stage; digesters were fed at loading rates of 2, 3, 4, and 5 g VS L^{-1} day^{-1}. Gas production was measured using tipping-bucket gas counters with continuous data logging. Calibration was checked weekly by collecting the gas in an impermeable bag; gas volumes were corrected to dry biogas at STP as described in Walker et al. [75]. Wheat straw digestion was tested by Pohl et al. [81,117] and Heeg et al. [82] in two-stage systems consisting of an upflow anaerobic solid-state (UASS) reactor with a working volume of 39 litres and a 30-litre anaerobic filter under mesophilic (37 °C), thermophilic (55 °C), and hyperthermophilic (60 °C) conditions. Biogas volumes were measured by flow meter (Ritter, Germany) and normalised to STP and 0% humidity. Nges et al. [107] carried out semi-continuous digestion of wheat straw under various regimes of nutrient addition and digestate recycling in 15-litre CSTRs at 37 °C, with OLRs of 2–4 g VS L^{-1} day^{-1} and a fixed SRT of 30 days. Gas volumes were measured using a real-time monitoring system with built-in correction to STP (Bio-process Control AB, Lund, Sweden).

Values reported for various crops are documented in Table A3 (Appendix A). Relatively high variations are found across literature sources. For many crops shown in Table A3, including grass, ryegrass, wheat, and sugar beet, the methane yields are reasonably close to those found in BMP testing (Section 3.2), which illustrates the effectiveness of such processing, if operated in an optimised way and without inhibition occurring.

3.5. Variations of the Reported Methane Potential for Various Crops

Experimental results reported for the specific methane yields of different crops are summarised in Figures 3 and 4. Literature data from the experimental tests described above (batch and semi-continuous/continuous assays) are considered in both figures. Figure 3 contains all values for each crop, including all growth stages and crop parts. Figure 4 shows the values for various crops where these have been divided into crop parts or other specifications, e.g., straw, whole crop, or ensiled.

As can be seen in Figure 3, the range of values is particularly large for clover, grass, maize, oats, rye, ryegrass, sugar beet, sunflower, wheat, and willow, but high variations are also found for other crops. The average specific methane values for the individual crops range between 0.18 (giant knotweed) and 0.39 (ryegrass, vetch) CH_4 kg^{-1} VS_{added}. However, for many of the single crops, the range of reported values is much wider than this inter-crop variation. Therefore, the difference between reported values for a single crop (intra-crop variation) is frequently greater than the difference between crops (inter-crop variation). Thus, when relying on single publications, the interpretation of the methane yield risks being misleading, because the identified value might be particularly high or low.

Disaggregating the data according to plant components (Figure 4) removes a part of the intra-crop variation, in particular for maize, oats, sugar beet, and wheat, but not all of it. Evidently, it can be advantageous to know as many details as possible about the biomass treated, but relying on literature data still bears the risk of being confronted with a value that is particularly high or low. Figure 4 shows a selection of crops only, as disaggregated data are not available for all the crops included in Figure 3.

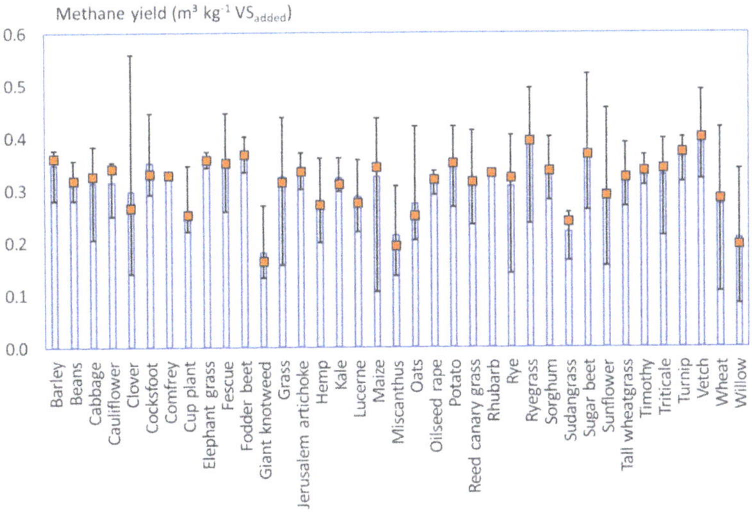

	n	Min	Max	Δ (Range)	Mean	Median
Barley	5	0.280	0.375	0.095	0.348	0.361
Beans	2	0.280	0.356	0.076	0.318	0.318
Cabbage	8	0.204	0.382	0.178	0.313	0.326
Cauliflower	3	0.250	0.352	0.102	0.314	0.341
Clover	8	0.140	0.558	0.418	0.296	0.266
Cocksfoot	5	0.291	0.446	0.155	0.351	0.330
Comfrey	2	0.323	0.334	0.011	0.329	0.329
Cup plant	10	0.220	0.345	0.125	0.260	0.253
Elephant grass	2	0.342	0.372	0.030	0.357	0.357
Fescue	8	0.259	0.446	0.187	0.354	0.352
Fodder beet	2	0.333	0.401	0.068	0.367	0.367
Giant knot-weed	8	0.132	0.270	0.138	0.179	0.164
Grass	23	0.156	0.438	0.282	0.325	0.315
Jerusalem artichoke	4	0.301	0.370	0.069	0.335	0.335
Hemp	8	0.199	0.360	0.161	0.276	0.271
Kale	3	0.296	0.360	0.064	0.322	0.310
Lucerne	8	0.220	0.357	0.137	0.287	0.275
Maize	42	0.105	0.436	0.331	0.325	0.343
Miscanthus	12	0.135	0.307	0.172	0.213	0.192
Oats	5	0.203	0.420	0.217	0.273	0.250
Oilseed rape	6	0.290	0.336	0.046	0.315	0.320
Potato	5	0.267	0.420	0.153	0.352	0.351
Reed canary grass	14	0.233	0.412	0.179	0.318	0.315
Rhubarb	2	0.330	0.334	0.004	0.332	0.332
Rye	12	0.140	0.403	0.263	0.306	0.323
Ryegrass	10	0.235	0.493	0.258	0.387	0.393
Sorghum	11	0.280	0.400	0.120	0.334	0.336
Sudangrass	3	0.164	0.256	0.092	0.220	0.239
Sugar beet	35	0.261	0.520	0.259	0.371	0.367
Sunflower	6	0.154	0.454	0.300	0.290	0.290
Tall wheat-grass	6	0.268	0.389	0.121	0.330	0.324
Timothy	2	0.308	0.365	0.057	0.337	0.337
Triticale	11	0.212	0.397	0.185	0.328	0.340
Turnip	3	0.314	0.399	0.085	0.362	0.372
Vetch	5	0.320	0.490	0.170	0.389	0.400
Wheat	31	0.105	0.418	0.313	0.271	0.283
Willow	9	0.082	0.339	0.257	0.207	0.195

Figure 3. Methane yields obtained in experimental (batch and semi-continuous/continuous) tests (**left** side: graphical presentation, the shaded columns show the average of reported values, the orange squares indicate the median, and the lines indicate the range; **right** side: tabular presentation, n is the number of data points).

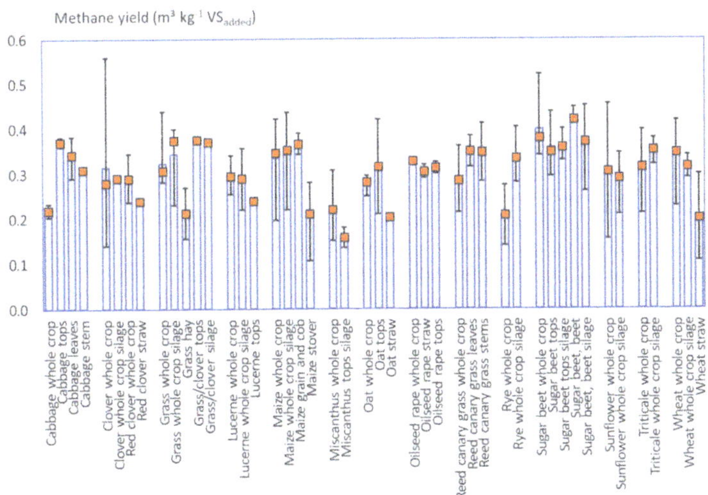

	n	Min	Max	Δ (Range)	Mean	Median
Cabbage whole crop	2	0.204	0.233	0.029	0.219	0.219
Cabbage tops	2	0.360	0.381	0.021	0.371	0.371
Cabbage leaves	3	0.291	0.382	0.091	0.339	0.343
Cabbage stem	1	0.309	0.309	0.000	0.309	0.309
Clover whole crop	4	0.140	0.558	0.418	0.315	0.280
Clover whole crop silage	1	0.291	0.291	0.000	0.291	0.291
Red clover whole crop	2	0.236	0.344	0.108	0.290	0.290
Red clover straw	1	0.240	0.240	0.000	0.240	0.240
Grass whole crop	13	0.282	0.438	0.156	0.323	0.307
Grass whole crop silage	5	0.230	0.399	0.169	0.345	0.374
Grass hay	2	0.156	0.270	0.114	0.213	0.213
Grass/clover tops	2	0.370	0.380	0.010	0.375	0.375
Grass/clover silage	1	0.370	0.370	0.000	0.370	0.370
Lucerne whole crop	4	0.255	0.340	0.085	0.296	0.295
Lucerne whole crop silage	3	0.220	0.357	0.137	0.289	0.290
Lucerne tops	1	0.240	0.240	0.000	0.240	0.240
Maize whole crop	21	0.196	0.422	0.226	0.335	0.346
Maize whole crop silage	14	0.219	0.436	0.217	0.346	0.353
Maize grain and cob	2	0.343	0.389	0.046	0.366	0.366
Maize stover	5	0.105	0.281	0.176	0.205	0.210
Miscanthus whole crop	10	0.151	0.307	0.156	0.224	0.220
Miscanthus tops silage	2	0.135	0.180	0.045	0.158	0.158
Oat whole crop	3	0.250	0.295	0.045	0.275	0.281
Oat tops	2	0.210	0.420	0.210	0.315	0.315
Oat straw	1	0.203	0.203	0.000	0.203	0.203
Oilseed rape whole crop	2	0.319	0.336	0.017	0.328	0.328
Oilseed rape straw	2	0.290	0.320	0.030	0.305	0.305
Oilseed rape tops	2	0.300	0.326	0.026	0.313	0.313
Reed canary grass whole c.	10	0.213	0.361	0.148	0.291	0.285
Reed canary grass leaves	2	0.315	0.384	0.069	0.350	0.350
Reed canary grass stems	2	0.283	0.412	0.129	0.348	0.348
Rye whole crop	2	0.140	0.275	0.135	0.208	0.208
Rye whole crop silage	7	0.280	0.403	0.123	0.333	0.334
Sugar beet whole crop	8	0.340	0.520	0.180	0.398	0.379
Sugar beet tops	12	0.294	0.437	0.143	0.349	0.349
Sugar beet tops silage	4	0.330	0.399	0.069	0.362	0.359
Sugar beet, beet	3	0.416	0.447	0.031	0.428	0.420
Sugar beet, beet silage	8	0.261	0.451	0.190	0.361	0.371
Sunflower whole crop	2	0.154	0.454	0.300	0.304	0.304
Sunflower whole c. silage	4	0.210	0.345	0.135	0.284	0.290
Triticale whole crop	7	0.212	0.397	0.185	0.315	0.314
Triticale whole crop silage	4	0.320	0.378	0.058	0.351	0.353
Wheat whole crop	11	0.228	0.418	0.190	0.339	0.346
Wheat whole crop silage	2	0.290	0.340	0.050	0.315	0.315
Wheat straw	15	0.105	0.299	0.194	0.208	0.200

Figure 4. Methane yields separated into various crop parts or other specifications (left side: graphical presentation, the shaded columns show the average of reported values, the orange squares indicate the median, and the lines indicate the range; right side: tabular presentation, n is the number of data points).

3.6. Methane Production Based on Elemental/Biochemical Composition

Another method by which methane potential can be determined is through knowledge of the elemental composition of the feedstock material, which is followed by application of an equation such as that given by Symons and Buswell [118] and Buswell and Hatfield [119].

$$C_cH_hO_oN_nS_s + 1/4(4c - h - 2o + 3n + 2s)H_2O \rightarrow 1/8(4c - h + 2o + 3n + 2s)CO_2 + 1/8(4c + h - 2o - 3n - 2s)CH_4 + nNH_3 + sH_2S \quad (1)$$

The Buswell equation has been used to calculate the maximal potential methane production of defined organic chemicals, such as different sugars and alcohols. However, this equation is not readily adapted for complex substances containing refractory components, as, for example, lignocellulosic materials. The methane yield calculated from the Buswell equation is a theoretical maximum and thus should be always higher than values measured in a biochemical assay. When applied to defined organic chemicals, such as sugars and alcohols, reasonably good agreement between experimental and theoretical values can be obtained. However, for complex substances containing refractory components (e.g., lignocellulosic materials), the results of the Buswell equation need to be adjusted; this can be done based on fibre content (estimated or measured) or ruminal digestibility as demonstrated by Czepuck et al. [120].

Alternatively, the potential biogas and methane yield can be calculated based on the biochemical composition of the material in terms of its carbohydrate, protein, and fat content, and an assumed or calculated yield for each of these constituents. This has now become a frequently-used method to estimate the methane potential of biomass. For accurate results, the carbohydrate component should refer to readily available storage materials such as starch, rather than to cellulosic material embedded in lignin, which may be less degradable. An example of this approach is found in Linke et al. [121], which gives results for a range of cereal whole crop substrates based on calculated values from protein (taken as 0.7 L biogas g TS^{-1}), fat (1.2 L biogas g TS^{-1}), and carbohydrate (0.8 L biogas g TS^{-1}) with predicted percentage of CH_4 in the biogas. Then, the values for biogas and methane yield are predicted from the measured composition of the substrates including total solids, ash content, fibre, fat, and protein as given in DLG (German Agricultural Society) [122].

A similar method is adopted as the basis for the database maintained by LfL (Bavarian State Institute for Agriculture) [123], where calculated methane potential values for a range of substrates are available, based on the content of protein (taken as 700 L biogas kg TS^{-1}), fat (1250 L biogas kg TS^{-1}), and carbohydrate (790 L biogas kg TS^{-1}) and on the substrate digestibility taken from animal fodder value tables. Values are given for biogas yield and methane composition, and these have been used to calculate the methane values given in Table A4 (Appendix B).

Hundreds of papers are available where this approach has been used. Methods and their applications have been reviewed [124]. Combining experimental testing and calculation of methane potential from proximate analysis is also common [106]. It is not within the scope of this work to review the results calculated with this approach across the literature, as the focus here is on experimental determination of methane yields. Table A4 lists some results to provide examples.

3.7. Results Published without Methods

A number of values for the methane yield of various crops have appeared without details of the methods used to determine them. Examples can be found in different media, for example in conference presentations [125], in handbooks for practitioners [126], in books [127], on web sites [128], and also in journal publications [129]. The existence of methane values provided without indicating the method by which these were determined might reflect one of the key challenges elaborated in this work—namely, the high variability of methane yields reported in the literature along with strongly differing methods applied, which make the data difficult to interpret. A detailed review of the publications that do not report the methods used is not within the scope of this work. However, some values are listed as examples in Table A5 (Appendix C). Those data tend to be close to average values found in experimental works as reported in earlier sections of this publication. This might indicate that they have been compiled by considering the data published across various works. However, a lack of transparency must be critically mentioned.

4. Discussion

4.1. Issues Potentially Affecting the Precision of Methane Potential Reported in Literature

The BMP test provides the most reproducible approach for the determination of ultimate methane potential. Its main advantage is the possibility to process the biomass under environmentally optimised conditions; for this to be fulfilled, careful attention is required to ensure adequate procedures and conditions, including a sufficiently high inoculum-to-substrate ratio [Koch et al. 2020]. The BMP test can only give a valid result for the ultimate biochemical methane potential of a substrate when no inhibition and no loss of biogas occurs.

Several specific points can be highlighted from the observations made. For some experimental set-ups, long processing times applied to determine the BMP of slowly degrading biomass can be assumed to have influenced precision of results, e.g., the methods where a barrier solution was involved [72,73] (see Section 3.2). This approach can lead to a loss of both CO_2 and CH_4, but more CO_2 because of its greater solubility [75]. In short-duration tests where substrate is rapidly degraded, this may not strongly affect the BMP result, but at longer digestion times, the impact could be significant; and the method is unsuitable where values and production kinetics for biogas rather than methane are required. However, one advantage is that the level of the barrier solution can be logged automatically at short intervals, giving a more detailed picture of gas production kinetics.

Some methods involve an extra step to determine gas volumes after storing the biogas in gas bags. However, the equipment and procedures used are not always clearly stated, hindering any attempt to evaluate the results. Gas collection bags are not completely impermeable or leak-proof. Different qualities exist (foil-lined gas collection bags are now standard), and the rate of diffusion from a bag will also vary for different gases (unpublished experimental data, University of Southampton). Other systems involve pressurised conditions as biogas accumulates in the test system, but the exact pressure regime and its management is often unclear. Depending on the method used to release pressure and in particular the time allowed for equilibration of headspace and liquid phase, pressurisation can affect the quantity of dissolved gases [24], altering the relative proportion of CH_4 and CO_2 due to the higher solubility of the latter. This may affect the pH and carbonate equilibrium of the digestion process and thus influence its outcome and kinetics. Manometric measurement can also be a source of error due to leakage as well as volume and pressure errors [130]. Especially when working at relatively small scale, gas storage over differing time periods and the use of apparatus to extract and quantify volumes can all introduce potential risks and variations due to technical or managerial challenges. These will affect both the BMP value and the gas production kinetics.

As noted above, many publications do not specify whether gas volumes are corrected to STP or, if corrected, do not give the actual temperatures and pressures used. In particular, reference temperatures can vary significantly between standards, and higher values will give larger volumes. It is also not always clear whether the measured biogas yields were corrected for water vapour content, i.e., expressed as dry gas. In some cases, the selected duration of the test is not concisely explained; sometimes, no information is provided, and sometimes, a weak criterion such as "negligible gas production" is given with insufficient detail for replication.

An increasing number of studies are now using proprietary systems, i.e., test arrangements that are available as complete systems on the market. This specifies the equipment used and the general procedures but does not necessarily improve the accuracy of testing and precision of the reported results. Proprietary systems usually automatically correct gas volumes to dry gas and STP, but this automatised feature increases the likelihood that the relevant information is not included in publications. One particularly challenging point regarding proprietary systems is that they generally use relatively small sample volumes, and thus, a high level of pre-processing is needed to ensure homogeneity. Regardless of the system used, the knowledge and skills of the users are critical to improve the quality of reported results [60].

The extent to which the variations among methane potentials reported in the literature can be attributed to the differences in methods applied cannot conclusively be quantified based on reviewing the published data. A wide range of other potential errors can affect the accuracy of the BMP assay and make comparison of results from different tests problematic. These include inadequately homogenised samples (substrate), poor choice of inoculum-to-substrate ratio, lack of buffering or nutrient deficiency in the substrate–inoculum mix, and incorrect or inadequately defined methods for monitoring and calculating biogas/methane production [23,75,131]. Furthermore, the number of repetitions conducted is likely to impact the precision of findings. The composition of crop biomass also varies significantly depending on the stage of growth in which the crop was harvested [132,133], the conditions under which it was grown (soil type, climatic conditions, year-on-year variations in weather) [134,135], and on post-harvest storage conditions [136]; in turn, these factors affect the energy potential. To produce reliable and meaningful data, it is essential to understand the limitations of BMP tests and the complex requirements for their adequate application.

4.2. Limitations Regarding Literature Data from Experiments Other Than BMP Tests

Short-duration batch tests will primarily show the gas yield obtained from readily degradable components, while other components may require longer digestion times. Therefore, with a view to estimating the ultimate methane potential of a specific crop, this type of data is less useful compared to BMP test results. Especially for slowly degradable biomass, data from this type of tests must be interpreted with care, and any pre-treatment such as grinding may also have a significant impact on the result.

As with BMP tests, some of the variation in reported methane yields is likely to originate from the differences in the methods applied to determine these, but quantification of this impact and assessment of the main factors driving it was not feasible based on the published data. However, one recurring issue across the literature reviewed is that not all values are expressed as dry gas at STP, while in some cases, it is unclear whether or not any correction has taken place.

Compared to batch tests, trials that apply continuous/semi-continuous digestion processes are closer to conditions in full-scale digesters, but they have their own limitations. A number of process parameters can influence methane production in continuous or semi-continuous digestion. For example, specific biogas production tends to be lower when intermediate products (e.g., VFA) accumulate in the digester, or when the digesters are run at very high OLR or very short retention times [65,137,138]. A simple kinetic model for indicating the effect of OLR on biogas yield for crops was derived by Mähnert and Linke [139] and applied to maize, rye, and beet silage with success, although results were provided only graphically.

It should be noted that this work only included data where no inhibition was observed during testing. Many experiments with continuous processes are designed to study potential inhibitions. As such, it is relatively challenging to identify works that inform about the methane potential of a biomass rather than its kinetic performance. Ensuring optimum environmental conditions in continuously/semi-continuously operated digesters is more demanding than running BMP tests, and minor process inhibition might not necessarily be noticed. Nevertheless, data from continuously operated digesters are also useful because they demonstrate the actual methane production under current practice in digestion technology and operating protocols. Extended studies under steady-state conditions, normally defined as operation for a minimum of three hydraulic retention times (HRT), can benefit from biomass acclimatisation to the feedstock, and well-designed studies with replicate digesters running at different OLR, HRT, and/or nutrient supplementation strategies are labour-intensive but can provide detailed insights into optimal conditions and performance.

4.3. Transparency of Published Data

This review has identified frequent deficiencies regarding the full documentation of experimental equipment and procedures applied. Such deficiencies reduce the usefulness and comparability of data. Results published without methods are particularly difficult to evaluate. On the other hand, there is the need of practitioners for reference values. As such, data published without methods might be attempts to provide practitioners with the best possible overall estimates as distilled by experts in the field to help them navigate through the jungle of existing data. Listing methane values without indicating the method used was more common some two decades ago, but this practice seems to be reducing now, which is a positive change regarding the transparency of data reported.

4.4. Alternatives to Relying on Literature Data

Well-conducted experiments can clarify the properties that a specific substrate displays during AD, but they are time-consuming and require a skilled work force and appropriate equipment. A viable alternative to relying on published methane yields or to conducting experimental AD testing is the estimation of the methane potential based on the chemical composition of the biomass (see Section 3.6). This requires a detailed knowledge of the biomass composition and thus involves some laboratory analyses. This method is also applied to complement experimental AD testing and allow evaluation of the efficacity of the AD process [140].

Experimental testing and theoretical calculations based on the chemical composition of the material both require the practical availability of the specific biomass of interest, and the results will apply to that specific substrate. However, in many cases, practitioners and researchers interested in a specific crop material will not have a sample available; for example, at the planning stage, a crop sample grown under relevant conditions might not be within reach, or the interest in the methane yield has arisen in a situation after a particular substrate has been consumed. The difficulty of obtaining representative homogeneous samples of crop biomass at the scale required for analysis can also be relevant. Thus, published literature remains an important source, and awareness about the limitations of reported data is an essential element to support sound decision making.

4.5. Which Crop Material Has the Highest Methane Potential?

A question often asked is, which crop material has the highest methane potential in anaerobic digestion? It can be seen from the above results that published data from methane potential assays do not necessarily provide a simple answer. The differences in reported values result both from the test protocols used and from the nature of the material tested, which also depends on the growth stage at which it was harvested and the method of storage, as discussed elsewhere [73,87].

It is important to know what the methane value is to be used for and to take the method of deriving the value into consideration. For example, BMP test results for a specific biomass may give the maximum methane potential, which is a value unlikely to be achieved in continuous laboratory trials or full-scale operation.

In addition to the specific energetic content of the biomass harvested, the biomass yield per hectare must be considered [62,74,141]. Thus, the actual methane yield in m^3 CH_4 ha^{-1} will be affected by climate, soil type, crop rotation regime [142,143], and many other agronomic conditions, which also lead to changes in the biomass yield in tonnes VS ha^{-1} [51,96,127]. Therefore, the 'best' crop to grow or crop residue to utilise for AD, in terms of maximising methane production per hectare per year, is likely to be one with the highest biomass yield under the particular geographic conditions rather than necessarily having a high specific methane yield during digestion.

5. Conclusions

Predicting specific methane yields for crop-based biomass is a difficult process. Selecting from published values can lead to considerable differences in the predicted outcome. Therefore, great caution is due when relying on literature data for estimates of the potential methane yield of a specific crop type in order to decide on its economic viability as an AD feedstock. Results from the literature indicate that variation between different crops is less than that within reported values for a single crop.

A wide range of techniques have been used to derive the reported values, and the results may be highly dependent both on the nature of the test and on the quality of its execution. Many publications display deficiencies in adequately documenting the equipment and procedures used, and too often, it is not clear whether reported gas volumes are expressed at STP. The casual reader looking for a methane value for a crop type might not analyse the documentation of the method applied in detail, and thus, deficiencies or low transparency might not trigger critical reflections on the usefulness of the published value. The test methods have partially been standardised, and the German VDI and DIN systems, along with the guidelines elaborated by the IWA ABAI Group, now offer a strong basis for comparability of results. Nevertheless, this does not resolve the challenge of a high variation among the published results, and therefore, this paper seeks to draw attention to the importance of assessing these aspects in the peer review process.

Where the biomass is available for analyses, theoretical calculations based on biochemical composition can provide quite accurate predictions of methane yields, and values obtained in this way appear to be compatible with those derived from experimental anaerobic digestion assays. The biochemical approach requires less practical testing compared to AD experiments; however, it still involves a substantial laboratory analytical procedure and does not avoid the issue of spatial and temporal variations in biomass properties.

Where the AD of crop-based materials is being considered as part of an integrated farming system, it may be better to consider the average biomass yield per hectare rather than the methane potential per unit of biomass, especially as relatively robust data on the former are often locally available. Whatever value is adopted for the potential methane yield and whatever the method used to predict or determine it, it must also be remembered that these are indicative values only and the methane yield actually generated in any full-scale continuous or semi-continuous process will be subject to many other factors including retention times, operating conditions, and co-digestion materials.

Author Contributions: Conceptualisation, Y.Z., S.K.-B., A.M.S. and S.H.; methodology, Y.Z., S.K.-B., A.M.S. and S.H.; formal analysis, S.K.-B. and A.M.S.; investigation, Y.Z., S.K.-B., A.M.S. and S.H.; data curation, Y.Z., S.K.-B. and A.M.S.; writing—original draft preparation, Y.Z., S.K.-B., A.M.S. and S.H.; writing—review and editing, Y.Z., S.K.-B., A.M.S. and S.H.; visualisation, S.K.-B. and A.M.S.; supervision, S.H.; project administration, A.M.S. All authors have read and agreed to the published version of the manuscript.

Funding: This work was partly carried out under the ESRC funded project RELU RES-229-25-002, and also builds on the EU project CROPGEN—Renewable Energy from Crops and Agrowastes, FP6 Grant no. 502824.

Institutional Review Board Statement: Not applicable.

Informed Consent Statement: Not applicable.

Data Availability Statement: No new data were created in this study, this is a review of published literature data, and the data used are documented in this article. Data sharing is not applicable to this article.

Acknowledgments: The authors would like to thank Charles Banks for his insightful comments during conceptualizing and conducting this study.

Conflicts of Interest: The authors declare no conflict of interest. The funders had no role in the design of the study; in the collection, analyses, or interpretation of data; in the writing of the manuscript, or in the decision to publish the results.

Appendix A. Literature Data from Different Types of Assays

Table A1 lists the data discussed in Section 3.2 (BMP tests) and Table A2 lists those examined in Section 3.3 (short-duration batch test). The data from semi-continuous and continuous tests are reported in Table A3 (discussed in Section 3.4). In Table A3, note that the data for Stewart et al. [110] are given in terms of TS_{added}, as there was not enough information given to convert them to VS_{added}.

Table A1. Methane values derived from BMP and long retention batch tests.

Reference	Crops	Crop Parts	Methane Yield (m^3 kg^{-1} VS_{added})	Comments
Amon et al. [51]	Maize	Whole crop	0.359–0.422	Range of varieties
	Maize	Whole crop	0.343–0.407	Range of harvest times
	Wheat	Whole crop	0.228–0.343	Range of harvest times
	Rye	Whole crop	0.140–0.275	Range of harvest times
	Sunflowers	Whole crop	0.154–0.454	Range of harvest times
	Triticale	Whole crop	0.212–0.265	Range of harvest times
Bauer et al. [53]	Barley	Whole crop	0.375	Silage
	Lucerne	Whole crop	0.357	Silage
	Maize	Whole crop	0.345	Silage
	Sorghum	Whole crop	0.362	Silage
	Sunflower	Whole crop	0.345	Silage
	Wheat	Straw	0.276	
Chiumenti et al. [61]	Grass	Whole crop	0.308–0.340	
Cornell et al. [72]	Maize	Whole crop	0.33	Ensiled
Feng et al. [41]	Fescue	Whole crop	0.294–0.310	Ensiled
Gallegos et al. [50]	Wheat	Straw	0.179–0.244	Ensiled and chopped
Garcia et al. [71]	Barley	Whole crop	0.280	
	Maize	Whole crop	0.289	
	Millet	Whole crop	0.253	
	Sorghum	Whole crop	0.313	
	Triticale	Whole crop	0.351	
Gunaseelan [32]	Cabbage	Stems	0.309	
		Leaves	0.291	
	Carrot	Leaves	0.241	
		Petiole	0.309	
	Elephant grass	Lamina	0.372	
		Sheath	0.342	
	Garden beet	Leaves	0.231	
	Pea	Pods (seeds removed)	0.390	
	Potato	Peels	0.267	
	Sudangrass	Whole crop	0.256	
	Turnip	Leaves	0.314	
Haag et al. [57]	Cup plant	Whole crop	0.228–0.261	Several varieties, dried and ground
Heidarzadeh Vazifehkhoran et al. [54]	Sugar beet	Beet	0.337–0.420	Silage in open silos
		Beet	0.411–0.451	Silage in closed silos
Jerger et al. [35]	Sorghum	Tops	0.28–0.40	Range of cultivars
Jurado et al. [44]	Miscanthus	Whole crop	0.249	Milled
	Wheat	Straw	0.200	Milled
	Willow	Woody component	0.082	Milled

Table A1. *Cont.*

Reference	Crops	Crop Parts	Methane Yield ($m^3\ kg^{-1}\ VS_{added}$)	Comments
Kakuk et al. [45]	Willow	Leaves	0.107–0.309	Green (< 1 year), various harvest dates and species
		Stems	0.149–0.252	
		Stems	0.195–0.212	Woody (> year)
Kandel et al. [40]	Fescue	Whole crop	0.401–0.428	Various harvest dates
Kandel et al. [39]	Reed canary grass	Leaf	0.315–0.384	Various harvest patterns
	Reed canary grass	Stem	0.283–0.412	Various harvest patterns
Kaparaju et al. [67]	Clover	Whole crop	0.14–0.21	
	Grass hay	Whole crop	0.27	
	Oats	Whole crop	0.25	
Lehtomäki and Björnsson [64]	Grass/clover	Whole crop	0.37	Silage
	Sugar beet	Leaves and beets	0.45	
	Willow	Whole crop	0.29	
Lehtomäki et al. [65]	Giant knotweed	Tops	0.17–0.27	Range of harvest dates
	Grass/clover mix	Tops	0.37–0.38	
	Jerusalem artichoke	Tops	0.36–0.37	
	Lupine	Whole crop	0.3	
	Marrow kale	Tops	0.31–0.36	
	Nettle	Tops	0.31–0.32	
	Oat	Tops	0.21–0.42	
	Oilseed rape	Straw	0.32	
	Red clover	Straw	0.24	
	Reed canary grass	Whole crop	0.28–0.34	
	Sugar beet		0.34–0.43	
	Vetch–oat mixture	Tops	0.32–0.49	
		Whole crop	0.4–0.41	
Lehtomäki et al. [68]	Grass	Whole crop	0.306	
	Oat	Straw	0.203	
	Sugar beet	Tops	0.353	
Li et al. [79]	Miscanthus	Whole crop	0.182	Particle sizes 20 and 30 mm
Machmüller et al. [52]	Clover	Whole crop	0.291	Silage
	Maize	Grain and cob	0.343	
		Whole crop	0.338	Silage
	Rye	Whole crop	0.324	Silage
	Sugar beet	Beet	0.261	Silage
	Sunflower	Whole crop	0.293	Silage
Mast et al. [58]	Cup plant	Whole crop	0.232–0.275	Various dates, dried and milled
	Giant knotweed	Whole crop	0.146–0.158	Various dates, dried and milled
	Energy dock	Whole crop	0.187–0.297	Various dates, dried and milled
	Tall wheatgrass	Whole crop	0.311–0.376	Various dates, dried and milled
Mittweg et al. [27]	Maize	Whole crop	0.346–0.362	
		Cobs	0.389	
Nges et al. [78]	Miscanthus	Whole crop	0.151–0.238	Particle sizes 0.5–20 mm
Parawira et al. [69]	Potato	Tuber waste	(0.42)	Value for VS degraded
	Sugar beet	Leaves	(0.52)	Value for VS degraded
Peng et al. [76]	Miscanthus	Whole crop	0.172–0.267	Range of types and growth stages
Petersson et al. [37]	Faba bean	Straw	(0.28) calculated	Reported: 18.9 g (100g DM)$^{-1}$
	Oilseed rape	Straw	(0.29)	18.8 g (100g DM)$^{-1}$
	Winter rye	Straw	(0.27)	18.2 g (100g DM)$^{-1}$

Table A1. *Cont.*

Reference	Crops	Crop Parts	Methane Yield (m^3 kg^{-1} VS_{added})	Comments
Pouech et al. [63]	Barley	Whole crop	0.356	
	Clover	Whole crop	0.350–0.558	Range of harvest times
	Forage sorghum	Whole crop	0.295	
	Grain sorghum	Whole crop	0.372	
	Lucerne	Whole crop	0.340	
	Maize	Whole crop	0.397	
	Oilseed rape	Whole crop	0.336	
	Ryegrass	Whole crop	0.390–0.409	Fresh and ensiled samples
	Sweet sorghum	Whole crop	0.352	
	Wheat	Whole crop	0.384–0.418	Range of harvest times
Rincón et al. [73]	Wheat	Whole crop	0.311–0.360	Various harvest dates
Rincón et al. [74]	Wheat	Whole crop	0.346–0.361	Spring and winter planting
Ruf and Emmerling [59]	Cup plant	Whole crop	0.236–0.282	2 years, poorly drained, dried
	Giant knotweed	Whole crop	0.189–0.222	2 years, poorly drained, dried
	Jerusalem artichoke	Whole crop	0.252–0.301	2 years, poorly drained, dried
	Maize	Whole crop	0.282–0.347	2 years, poorly drained, dried
	Reed canary grass	Whole crop	0.277–0.290	2 years, poorly drained, dried
	Tall wheatgrass	Whole crop	0.268–0.302	2 years, poorly drained, dried
Schmidt et al. [62]	Cup plant	Whole crop	0.272–0.345	Three sites, range of harvest dates
	Giant knotweed	Whole crop	0.132–0.147	Three sites, range of harvest dates
	Reed canary grass	Whole crop	0.315–0.355	Two sites, range of harvest dates
	Tall wheatgrass	Whole crop	0.336–0.389	Two sites, range of harvest dates
	Virginia mallow	Whole crop	0.213–0.315	Three sites, range of harvest dates
Seppälä et al. [66]	Cocksfoot	Whole crop	0.308–0.382	Range of sites and harvest dates
	Fescue	Whole crop	0.296–0.394	Range of sites and harvest dates
	Timothy	Whole crop	0.308–0.365	Range of sites and harvest dates
	Reed Canary Grass	Whole crop	0.253–0.351	Range of sites and harvest dates
Thomas et al. [80]	Miscanthus	Whole crop	0.166–0.202	Various species
Virkajärvi et al. [77]	Grass	Whole crop	0.302–0.307	Various fertiliser strategies
	Grass/clover mix	Whole crop	0.285–0.292	Various fertiliser strategies
Wahid et al. [42]	Lucerne	Whole crop	0.255	
Wahid et al. [43]	Miscanthus	Whole crop	0.302–0.307	Various harvest times
Zauner and Küntzel [36]	Horse bean	Tops	0.356	
	Lucerne	Tops	0.240	
	Maize	Whole crop	0.270–0.298	Various growth stages
	Mixed grass	Whole crop	0.298–0.315	Various fermentation periods
	Sugar beet	Tops	0.294	
	Vetch	Tops	0.323	

Table A2. Methane values derived from short duration batch tests.

Reference	Crops	Crop Parts	Methane Yield ($m^3\ kg^{-1}\ VS_{added}$)	Comments
Allen et al. [106]	Energy beet	Whole crop	0.375	
	Fodder beet	Whole crop	0.333	
	Grass	Whole crop	0.368	
		Whole crop	0.374–0.399	Silage, various harvests
		Whole crop	0.156	Hay
	Maize	Whole crop	0.354	Fresh
			0.394	Ensiled
	Oilseed rape	Whole crop	0.319	
	Potatoes	Whole crop	0.337–0.351	Various species
	Spring barley	Whole crop	0.361	
	Spring wheat	Whole crop	0.340	
	Sugar beet	Whole crop	0.344	
	Triticale	Whole crop	0.314	
	Turnips	Whole crop	0.399	
	Winter barley	Whole crop	0.367	
	Winter oats	Whole crop	0.281	
	Winter wheat	Whole crop	0.354	
Badger et al. [89]	Kale	Tops	0.296	(Methane values calculated from biogas and indicated percentage of methane)
	Maize	Tops	0.342	
	Oats	Whole crop	0.295	
	Sugar beet	Tops	0.297	
	Wheat	Straw	0.255	
Bruni et al. [103]	Maize	Whole crop	0.313–0.401	Range of varieties and harvest dates
Gissen et al. [102]	Hemp	Whole crop	0.260–0.292	Various fertiliser strategies
	Maize	Whole crop	0.327–0.382	Various fertiliser strategies
	Sugar beet	Root	0.416–0.420	Various fertiliser strategies
		Tops	0.362–0.367	Various fertiliser strategies
	Triticale	Whole crop	0.397	
Herrmann et al. [98]	Maize	Whole crop	0.331–0.378	All crops as silage; range of silage periods
	Sorghum	Whole crop	0.317–0.346	
	Forage rye	Whole crop	0.293–0.346	
	Triticale	Whole crop	0.340–0.365	
Herrmann et al. [99]	Maize	Whole crop	0.323–0.362	All crops as silage; range of chopping length and silage periods
	Sorghum	Whole crop	0.298–0.336	
	Forage rye	Whole crop	0.334–0.403	
	Triticale	Whole crop	0.320–0.378	
	Winter rye	Whole crop	0.321–0.336	
Herrmann et al. [100]	Sugar beet	Whole crop	0.350–0.399	Silage
	Sunflowers	Whole crop	0.210–0.286	Silage
	Winter wheat	Whole crop	0.269–0.328	Silage
Kaiser et al. [96]	Grass	Whole crop	0.282–0.438	Fresh
	Hemp	Tops	0.250–0.360	Ensiled, various cuts
	Lucerne	Whole crop	0.260–0.330	
	Maize	Whole crop	0.219–0.436	Silage
	Miscanthus	Tops	0.135–0.180	Ensiled
	Red clover	Whole crop	0.236–0.344	Fresh and ensiled
	Ryegrass	Whole crop	0.220–0.290	
			0.235–0.395	Fresh and ensiled
	Sudangrass	Tops	0.164–0.239	Fresh and ensiled
	Sugar beet	Beet	0.278–0.328	Silage
		Tops	0.335–0.395	
Kakuk et al. [114]	Maize	Stover	0.210–0.281	Various particles sizes and loadings

Table A2. *Cont.*

Reference	Crops	Crop Parts	Methane Yield (m^3 kg^{-1} VS_{added})	Comments
Kolbl et al. [108]	Forage turnip	Whole crop	0.372	Milled
	Maize	Whole crop	0.236	Ensiled
	Triticale	Whole crop	0.287	Chopped
Kreuger et al. [104]	Hemp	Whole crop	0.199–0.270	Range of harvest dates
Kreuger et al. [101]	Hemp	Whole crop	0.301	Fresh frozen
		Whole crop	0.272	Ensiled
	Maize	Whole crop	0.363	Fresh frozen
		Whole crop	0.367	Ensiled
	Sugar beet	Beet	0.447	Low-sugar variety, fresh frozen
		Beet	0.405	Low-sugar variety, ensiled
		Tops	0.437	Low-sugar variety, fresh frozen
		Tops	0.367	Low-sugar variety, ensiled
Mähnert et al. [97]	Cocksfoot	Whole crop	0.33	Fresh and ensiled
	Meadow foxtail	Whole crop	0.31	Fresh
	Ryegrass	Whole crop	0.36	Fresh and ensiled
Nges et al. [107]	Wheat	Straw	0.297	Dried and ground
Nizami and Murphy [92]	Ryegrass	Whole crop	0.483–0.493	Fresh
Prade et al. [105]	Grass	Whole crop	0.230–0.330	Ensiled, range of harvest dates
	Rye	Whole crop	0.280–0.350	Ensiled, range of harvest dates
	Wheat	Whole crop	0.290–0.340	Ensiled, range of harvest dates
Raposo et al. [93]	Maize	Whole crop	0.196–0.233	Range of inoculum–substrate ratios
Spence et al. [7]	Grass	Whole crop	0.359	
	Ryegrass	Whole crop	0.294	
	Triticale	Whole crop	0.380	
	Wheat	Whole crop	0.283	
Svensson et al. [91]	Sugar beet	Tops	0.33	Ensiled
	Wheat	Straw	0.19	
Tilvikiene et al. [95]	Cocksfoot	Whole crop	0.291–0.446	Various fertiliser and harvest times
	Reed canary grass	Whole crop	0.233–0.361	Various fertiliser and harvest times
	Tall fescue	Whole crop	0.259–0.446	Various fertiliser and harvest times
Yan et al. [94]	Broccoli	Whole crop		
	Cabbage	Whole crop	0.204	Ground
	Cauliflower	Whole crop	0.250	Ground
	Leek	Whole crop	0.183	Ground
	Purple cabbage	Whole crop	0.233	Ground
Zhao et al. [109]	Maize	Stover	0.250	Dried and ground
Zubr [90]	Cauliflower	Leaves	0.341–0.352	Fresh and ensiled
	Comfrey	Tops	0.323–0.334	
	Jerusalem artichoke	Tops	0.301–0.309	
	Oilseed rape	Tops	0.300–0.326	
	Rhubarb	Tops	0.330–0.334	
	Sugar beet	Tops	0.316–0.345	
	White cabbage	Tops	0.360–0.381	
		Leaves	0.343–0.382	

Table A3. Methane values derived from continuous/semi-continuous experiments.

Reference	Crops	Crop Parts	Methane Yield (m^3 kg^{-1} VS_{added})	Comments
Haag et al. [57]	Cup plant	Whole crop	0.220–0.244	Various varieties, ensiled
Heeg et al. [82]	Wheat	Straw	0.105–0.173	Two-phase, meso- and thermophilic
Kakuk et al. [114]	Maize	Stover	0.105–0.177	Wet and dry digestion
Lehtomäki and Björnsson [64]	Grass Sugar beet Willow	Whole crop Beets and tops Shoots	0.39 0.38 0.16	Silage Chopped
Nges and Björnsson [112]	Sugar beet	Roots and tops	0.343–0.383	Various loading rates
Nges et al. [107]	Wheat	Straw	0.250-0.299	Dried and ground, different loading
Nizami and Murphy [92]	Ryegrass	Whole crop	0.363–0.451	Range of loading rates
Pohl et al. [81]	Wheat	Straw	0.127–0.180	Two-phase, meso- and thermophilic
Pohl et al. [117]	Wheat	Straw	0.144–0.207	Two-phase, meso- and thermophilic
Rincón et al. [116]	Winter wheat	Whole crop	0.334	Ensiled
Scherer et al. [111]	Fodder beet	Beet	0.401	
Veluchamy et al. [115]	Maize	Whole crop	0.360–0.410	Ensiled
Zhu et al. [113]	Potato	Tubers	0.387	
The following values are not included further because they are not available (cannot be derived) on a VS basis.				
			Methane yield (m^3 kg^{-1} TS_{added})	
Stewart et al. [110]	Grass Grass/clover mix Kale Lucerne Maize Oats Wheat Barley Potato	Whole crop Whole crop Tops Whole crop Tops Whole crop Straw Straw Tuber waste	0.217–0.292 0.278 0.179–0.304 0.248–0.390 0.231 0.227–0.257 0.245 0.128–0.162 0.350–0.410	Fresh and ensiled Chopped and ground

Appendix B. Calculated Methane Yields

Table A4 shows the data discussed in Section 3.6 (calculated methane yields based on the composition of the biomass).

Table A4. Examples of calculated methane values.

Reference	Crops	Crop Parts	Methane Yield (m^3 kg^{-1} VS_{added})	Comments
Linke et al. [121]	Barley	Whole crop	0.420	
	Fodder beet	Leaves	0.430	
		Beet	0.411	
	Forage rye	Whole crop	0.431	
	Grass	Whole crop	0.433	
	Hemp	Whole crop	0.409	
	Lucerne	Whole crop	0.432	
	Maize	Whole crop	0.422	
	Rye	Grain	0.419	
		Straw	0.409	
LfL [123]	Barley	Straw	0.196	
	Beans	Whole crop	0.277	Green
		Whole crop	0.286	Silage
	Clover grass	Whole crop	0.284–0.325	Various harvest dates
	Grass	Whole crop	0.296–0.333	Various harvest dates; Wet silage
		Whole crop	0.296–0.340	Various harvest dates; Wilted silage
	Grass hay	Whole crop	0.250–0.307	Various harvest dates
	Lucerne	Whole crop	0.265–0.308	Various harvest dates
	Maize	Whole crop	0.287–0.328	Various harvest dates
	Meadow grass	Whole crop	0.290–0.322	Various harvest dates
	Mustard	Whole crop	0.358–0.368	Various harvest dates
	Oat	Straw	0.197	
		Grain	0.322	
	Oilseed rape	Whole crop	0.341–0.356	Various harvest dates; Green
		Whole crop	0.361–0.376	Various harvest dates; Silage
	Oilseed rape	Straw	0.129	
		Seed	0.504	
	Pasture grass	Whole crop	0.307–0.322	Various harvest dates
	Red clover	Whole crop	0.278–0.316	Various harvest dates
	Rye	Straw	0.179	
		Grain	0.365	
		Whole crop	0.261	Silage
	Ryegrass	Whole crop	0.287–0.320	Various harvest dates
	Sainfoin	Whole crop	0.267–0.292	Various harvest dates
	Sugar beet	Tops	0.321–0.331	Various harvest dates
	Wheat	Straw	0.187	
		Grain	0.370	
		Whole crop	0.262	Silage

Appendix C. Methane Values Reported without Methods

Table A5 lists examples of values published without providing information about the method how these were derived. Note that the methane yields for Koettner [125] were given per fresh material in the publication; using the indicated VS contents of the substrates, the methane yields per VS were calculated for inclusion in Table A5.

Table A5. Examples of methane values reported without methods.

Reference	Crops	Crop Parts	Methane Yield (m^3 kg^{-1} VS_{added})	Comments
Koettner [125]	Milled grain		0.37	
	Grass		0.32	Silage
	Maize		0.32	Silage
	Meadow grass		0.32	
	Rape seed cake		0.46	
	Vegetable residues		0.38	
KTBL [126,128]	Fodder beet	Beet	0.364–0.496	
		Tops	0.312	Silage
	Hemp	Whole crop	0.143	
	Maize	Whole crop	0.304	
	Rape seed cake		0.439	
	Rye	Whole crop	0.313–0.319	Fresh and ensiled
	Sugar beet	Beet	0.357	
		Tops	0.312	Silage
	Wheat	Grain	0.33	
		Straw	0.138	
Weiland [129]	Barley	Whole crop	0.36	Silage
	Clover	Whole crop	0.35	Silage
	Fodder beet	Beet and leaf	0.456	Silage
	Kale	Whole crop	0.255	Silage
	Lucerne	Whole crop	0.41	Silage
	Maize	Whole crop	0.41	Silage
	Oilseed rape	Whole crop	0.34	Silage
	Potato	Tubers	0.276	Silage
	Ryegrass	Whole crop	0.41	Silage
	Wheat	Whole crop	0.39	Silage

References

1. Anukam, A.; Mohammadi, A.; Naqvi, M.; Granström, K. A review of the chemistry of anaerobic digestion: Methods of accelerating and optimizing process efficiency. *Processes* **2019**, *7*, 504. [CrossRef]
2. Achinas, S.; Euverink, G.J.W. Rambling facets of manure-based biogas production in Europe: A briefing. *Renew. Sustain. Energy Rev.* **2020**, *119*, 109566. [CrossRef]
3. Ma, G.; Ndegwa, P.; Harrison, J.H.; Chen, Y. Methane yields during anaerobic co-digestion of animal manure with other feedstocks: A meta-analysis. *Sci. Total Environ.* **2020**, *728*, 138224. [CrossRef]
4. Jeung, J.H.; Chung, W.J.; Chang, S.W. Evaluation of anaerobic co-digestion to enhance the efficiency of livestock manure anaerobic digestion. *Sustainability* **2019**, *11*, 7170. [CrossRef]
5. Karki, R.; Chuenchart, W.; Surendra, K.C.; Shrestha, S.; Raskin, L.; Sung, S.; Hashimoto, A.; Kumar Khanal, S. Anaerobic co-digestion: Current status and perspectives. *Bioresour. Technol.* **2021**, *330*, 125001. [CrossRef] [PubMed]
6. Iglesias, R.; Muñoz, R.; Polanco, M.; Díaz, I.; Susmozas, A.; Moreno, A.D.; Guirado, M.; Carreras, N.; Ballesteros, M. Biogas from anaerobic digestion as an energy vector: Current upgrading development. *Energies* **2021**, *14*, 2742. [CrossRef]
7. Spence, A.; Blanco Madrigal, E.; Patil, R.; Bajón Fernández, Y. Evaluation of anaerobic digestibility of energy crops and agricultural by-products. *Bioresour. Technol. Rep.* **2019**, *5*, 243–250. [CrossRef]
8. Nwokolo, N.; Mukumba, P.; Obileke, K.; Enebe, M. Waste to energy: A focus on the impact of substrate type in biogas production. *Processes* **2020**, *8*, 1224. [CrossRef]
9. Frigon, J.C.; Guiot, S.R. Biomethane production from starch and lignocellulosic crops: A comparative review. *Biofuels Bioprod. Biorefin.* **2010**, *4*, 447–458. [CrossRef]
10. Gunaseelan, V.N. Anaerobic digestion of biomass for methane production: A review. *Biomass Bioenergy* **1997**, *13*, 83–114. [CrossRef]
11. Martínez Gutiérrez, E. Biogas production from different lignocellulosic biomass sources: Advances and perspectives. *3 Biotech* **2018**, *8*, 233. [CrossRef]
12. Dar, R.A.; Parmar, M.; Dar, E.A.; Sani, R.K.; Phutela, U.G. Biomethanation of agricultural residues: Potential, limitations and possible solutions. *Renew. Sustain. Energy Rev.* **2021**, *135*, 110217. [CrossRef]
13. Lesteur, M.; Bellon-Maurel, B.; Gonzalez, C.; Latrille, E.; Roger, J.M.; Junqua, G.; Steyera, J.P. Alternative methods for determining anaerobic biodegradability: A review. *Process Biochem.* **2010**, *45*, 431–440. [CrossRef]
14. Murphy, J.; Braun, R.; Weiland, P.; Wellinger, A. Biogas from Crop Digestion. IEA Bioenergy: Cork, Ireland, 2011.

15. Raposo, F.; De la Rubia, M.A.; Fernández-Cegrí, V.; Borja, R. Anaerobic digestion of solid organic substrates in batch mode: An overview relating to methane yields and experimental procedures. *Renew. Sustain. Energy Rev.* **2012**, *16*, 861–877. [CrossRef]
16. Kulichkova, G.I.; Ivanova, T.S.; Koettner, M.; Volodko, O.I.; Spivak, S.I.; Tsygankov, S.P.; Blume, Y.B. Plant feedstocks and their biogas production potentials. *Open Agric. J.* **2020**, *14*, 219–234. [CrossRef]
17. Weiland, P. Biogas production: Current state and perspectives. *Appl. Microbiol. Biotechnol.* **2010**, *85*, 849–860. [CrossRef] [PubMed]
18. Ward, A.J.; Hobbs, P.J.; Holliman, P.J.; Jones, D.L. Optimisation of the anaerobic digestion of agricultural resources. *Bioresour. Technol.* **2008**, *99*, 7928–7940. [CrossRef]
19. Nizami, A.-S.; Murphy, J.D. What type of digester configurations should be employed to produce biomethane from grass silage? *Renew. Sustain. Energy Rev.* **2010**, *14*, 1558–1568. [CrossRef]
20. Pererva, Y.; Miller, C.D.; Sims, R.C. Existing empirical kinetic models in biochemical methane potential (BMP) testing, their selection and numerical solution. *Water* **2020**, *12*, 1831. [CrossRef]
21. Raposo, F.; Borja, R.; Ibelli-Bianco, C. Predictive regression models for biochemical methane potential tests of biomass samples: Pitfalls and challenges of laboratory measurements. *Renew. Sustain. Energy Rev.* **2020**, *127*, 109890. [CrossRef]
22. Angelidaki, I.; Sanders, W. Assessment of the anaerobic biodegradability of macropollutants. *Rev. Environ. Sci. Biotechnol.* **2004**, *3*, 117–129. [CrossRef]
23. Angelidaki, I.; Alves, M.; Bolzonella, D.; Borzacconi, L.; Campos, J.L.; Guwy, A.J.; Kalyuzhnyi, S.; Jenicek, P.; van Lier, J.B. Defining the biomethane potential (BMP) of solid organic wastes and energy crops: A proposed protocol for batch assays. *Water Sci. Technol.* **2009**, *59*, 927–934. [CrossRef]
24. Holliger, C.; Alves, M.; Andrade, D.; Angelidaki, I.; Astals, S.; Baier, U.; Bougrier, C.; Buffière, P.; Carballa, M.; De Wilde, V.; et al. Towards a standardization of biomethane potential tests. *Water Sci. Technol.* **2016**, *74*, 2515–2522. [CrossRef] [PubMed]
25. Esposito, G.; Frunzo, L.; Liotta, F.; Panico, A.; Pirozzi, F. Bio-methane potential tests to measure the biogas production from the digestion and co-digestion of complex organic substrates. *Open Environ. Eng. J.* **2012**, *5*, 1–8. [CrossRef]
26. Weinrich, S.; Schäfer, F.; Bochmann, G.; Liebetrau, J. Value of Batch Tests for Biogas Potential Analysis. In *Method Comparison and Challenges of Substrate and Efficiency Evaluation of Biogas Plants*; IEA Bioenergy Task 37; University College Cork: Cork, Ireland, 2018; ISBN 978-1-910154-49-6.
27. Mittweg, G.; Oechsner, H.; Hahn, V.; Lemmer, A.; Reinhardt-Hanisch, A. Repeatability of a laboratory batch method to determine the specific biogas and methane yields. *Eng. Life Sci.* **2012**, *12*, 270–278. [CrossRef]
28. Koch, K.; Hafner, S.D.; Weinrich, S.; Astals, S.; Holliger, C. Power and limitations of biochemical methane potential (BMP) tests. *Front. Energy Res.* **2020**, *8*, 63. [CrossRef]
29. Amodeo, C.; Hafner, S.D.; Teixeira Franco, R.; Benbelkacem, H.; Moretti, P.; Bayard, R.; Buffière, P. How different are manometric, gravimetric, and automated volumetric BMP results? *Water* **2020**, *12*, 1839. [CrossRef]
30. Filer, J.; Ding, H.H.; Chang, S. Biochemical methane potential (BMP) assay method for anaerobic digestion research. *Water* **2019**, *11*, 921. [CrossRef]
31. Holliger, C.; Astals, S.; Fruteau de Laclos, H.; Hafner, S.D.; Koch, K.; Weinrich, S. Towards a standardization of biomethane potential tests: A commentary. *Water Sci. Technol.* **2021**, *83*, 247–250. [CrossRef]
32. Gunaseelan, V.N. Biochemical methane potential of fruits and vegetable solid waste feedstocks. *Biomass Bioenergy* **2004**, *26*, 389–399. [CrossRef]
33. Zhang, Y.; Kusch-Brandt, S.; Heaven, S.; Banks, C.J. Effect of pasteurisation on methane yield from food waste and other substrates in anaerobic digestion. *Processes* **2020**, *8*, 1351. [CrossRef]
34. Owen, W.F.; Stuckey, D.C.; Healy, J.B.; Young, L.Y.; Mc Carty, P.L. Bioassay for monitoring biochemical methane potential and anaerobic toxicity. *Water Res.* **1979**, *13*, 485–492. [CrossRef]
35. Jerger, D.E.; Chynoweth, D.P.; Isaacson, H.R. Anaerobic digestion of sorghum biomass. *Biomass* **1987**, *14*, 99–113. [CrossRef]
36. Zauner, E.; Küntzel, U. Methane production from ensiled plant material. *Biomass* **1986**, *10*, 207–223. [CrossRef]
37. Petersson, A.; Thomsen, M.H.; Hauggaard-Nielsen, H.; Thomsen, A.-B. Potential bioethanol and biogas production using lignocellulosic biomass from winter rye, oilseed rape and faba bean. *Biomass Bioenergy* **2007**, *31*, 812–819. [CrossRef]
38. Hansen, T.L.; Schmidt, J.E.; Angelidaki, I.; Marca, E.; la Cour Jansen, J.; Mosbæk, H.; Christensen, T.H. Method for determination of methane potentials of solid organic waste. *Waste Manag.* **2004**, *24*, 393–400. [CrossRef]
39. Kandel, T.P.; Sutaryo, S.; Møller, H.B.; Jørgensen, U.; Lærke, P.E. Chemical composition and methane yield of reed canary grass as influenced by harvesting time and harvest frequency. *Bioresour. Technol.* **2013**, *130*, 659–666. [CrossRef]
40. Kandel, T.P.; Ward, A.J.; Elsgaard, L.; Møller, H.B.; Lærke, P.E. Methane yield from anaerobic digestion of festulolium and tall fescue cultivated on a fen peatland under different harvest managements. *Acta Agric. Scand. Sect. B—Soil Plant Sci.* **2017**, *67*, 670–677. [CrossRef]
41. Feng, L.; Kristensen, E.F.; Moset, V.; Ward, A.J.; Møller, H.B. Ensiling of tall fescue for biogas production: Effect of storage time, additives and mechanical pretreatment. *Energy Sustain. Dev.* **2018**, *47*, 143–148. [CrossRef]
42. Wahid, R.; Feng, L.; Cong, W.-F.; Ward, A.J.; Møller, H.B.; Eriksen, J. Anaerobic mono-digestion of lucerne, grass and forbs—Influence of species and cutting frequency. *Biomass Bioenergy* **2018**, *109*, 199–208. [CrossRef]
43. Wahid, R.; Nielsen, S.F.; Hernandez, V.M.; Ward, A.J.; Gislum, R.; Jørgensen, U.; Møller, H.B. Methane production potential from Miscanthus sp.: Effect of harvesting time, genotypes and plant fractions. *Biosyst. Eng.* **2015**, *133*, 71–80. [CrossRef]

44. Jurado, E.; Gavala, H.N.; Skiadas, I.V. Enhancement of methane yield from wheat straw, miscanthus and willow using aqueous ammonia soaking. *Environ. Technol.* **2013**, *34*, 2069–2075. [CrossRef]
45. Kakuk, B.; Bagi, Z.; Rákhely, G.; Maróti, G.; Dudits, D.; Kovács, K.L. Methane production from green and woody biomass using short rotation willow genotypes for bioenergy generation. *Bioresour. Technol.* **2021**, *333*, 125223. [CrossRef]
46. *German Standard Methods for the Examination of Water, Waste Water and Sludge; Sludge and Sediments (Group S)*; DIN 38414-8; Determination of the Amenability to Anaerobic Digestion (S 8); Deutsches Institut Fuer Normung e.V.: Berlin, Germany, 1985.
47. *Fermentation of Organic Materials. Characterisation of the Substrates, Sampling, Collection of Material Data, Fermentation Tests*; VDI 4630; VDI-Handbuch Energietechnik, Issue 2006; VDI: Duesseldorf, Germany, 2006.
48. *Fermentation of Organic Materials. Characterisation of the Substrate, Sampling, Collection of Material Data, Fermentation Tests*; VDI 4630 (updated 2016); VDI-Handbuch Energietechnik, Issue 2016; VDI: Duesseldorf, Germany, 2016.
49. Linke, B.; Heiermann, M.; Grundmann, P.; Hertwig, P. Grundlagen, Verfahren und Potenzial der Biogasgewinnung im Land Brandenburg. In *Biogas in der Landwirtschaft—Leitfaden für Landwirte und Investoren im Land Brandenburg*, 2nd ed.; Heiermann, M., Plöchl, M., Eds.; Ministerium für Landwirtschaft Umweltschutz und Raumordnung des Landes Brandenburg: Potsdam, Germany, 2003; pp. 10–23.
50. Gallegos, D.; Wedwitschka, H.; Moeller, L.; Zehnsdorf, A.; Stinner, W. Effect of particle size reduction and ensiling fermentation on biogas formation and silage quality of wheat straw. *Bioresour. Technol.* **2017**, *245*, 216–224. [CrossRef] [PubMed]
51. Amon, T.; Amon, B.; Kryvoruchko, V.; Machmüller, A.; Hopfner-Sixt, K.; Bodiroza, V.; Hrbek, R.; Friedel, J.; Pötsch, E.; Wagentristl, H.; et al. Methane production through anaerobic digestion of various energy crops grown in sustainable crop rotations. *Bioresour. Technol.* **2007**, *98*, 3204–3212. [CrossRef]
52. Machmüller, A.; Hrbek, R.; Kryvoruchko, V.; Amon, T. Biogas Production from Energy Crops produced on Sustainable Crop Rotations. In Proceedings of the 15th European Biomass Conference, Berlin, Germany, 7–11 May 2007.
53. Bauer, A.; Leonhartsberger, C.; Bösch, P.; Amon, B.; Friedl, A.; Amon, T. Analysis of methane yields from energy crops and agricultural by-products and estimation of energy potential from sustainable crop rotation systems in EU-27. *Clean Technol. Environ. Policy* **2010**, *12*, 153–161. [CrossRef]
54. Heidarzadeh Vazifehkhoran, A.; Triolo, J.M.; Larsen, S.U.; Stefanek, K.; Sommer, S.G. Assessment of the variability of biogas production from sugar beet silage as affected by movement and loss of the produced alcohols and organic acids. *Energies* **2016**, *9*, 368. [CrossRef]
55. Schmidt, A.; Lemaigre, S.; Ruf, T.; Delfosse, P.; Emmerling, C. Miscanthus as biogas feedstock: Influence of harvest time and stand age on the biochemical methane potential (BMP) of two different growing seasons. *Biomass Convers. Biorefinery* **2018**, *8*, 245–254. [CrossRef]
56. Helffrich, D.; Oechsner, H. Hohenheim biogas yield test—Comparing various laboratory methods on biomass fermentation. *Landtechnik* **2003**, *58*, 148–149.
57. Haag, N.L.; Nägele, H.J.; Reiss, K.; Biertümpfel, A.; Oechsner, H. Methane formation potential of cup plant (*Silphium perfoliatum*). *Biomass Bioenergy* **2015**, *75*, 126–133. [CrossRef]
58. Mast, B.; Lemmer, A.; Oechsner, H.; Reinhardt-Hanisch, A.; Claupein, W.; Graeff-Hönninger, S. Methane yield potential of novel perennial biogas crops influenced by harvest date. *Ind. Crops Prod.* **2014**, *58*, 194–203. [CrossRef]
59. Ruf, T.; Emmerling, C. Site-adapted production of bioenergy feedstocks on poorly drained cropland through the cultivation of perennial crops. A feasibility study on biomass yield and biochemical methane potential. *Biomass Bioenergy* **2018**, *119*, 429–435. [CrossRef]
60. Hafner, S.D.; de Laclos, H.F.; Koch, K.; Holliger, C. Improving inter-laboratory reproducibility in measurement of biochemical methane potential (BMP). *Water* **2020**, *12*, 1752. [CrossRef]
61. Chiumenti, A.; Boscaro, D.; Da Borso, F.; Sartori, L.; Pezzuolo, A. Biogas from fresh spring and summer grass: Effect of the harvesting period. *Energies* **2018**, *11*, 1466. [CrossRef]
62. Schmidt, A.; Lemaigre, S.; Delfosse, P.; von Francken-Welz, H.; Emmerling, C. Biochemical methane potential (BMP) of six perennial energy crops cultivated at three different locations in W-Germany. *Biomass Convers. Biorefin.* **2018**, *8*, 873–888. [CrossRef]
63. Pouech, P.; Fruteau, H.; Bewa, H. Agricultural crops for biogas production on anaerobic digestion plants. In *Biomass for Energy and Industry, Proceeding of the International Conference, Würzburg, Germany, 8–11 June 1998*; C.A.R.M.E.N.: Straubing, Germany, 1998; pp. 163–165.
64. Lehtomäki, A.; Björnsson, L. Two-stage anaerobic digestion of energy crops: Methane production, nitrogen mineralisation and heavy metal mobilisation. *Environ. Technol.* **2006**, *27*, 209–218. [CrossRef]
65. Lehtomäki, A.; Huttunena, S.; Rintala, J. Laboratory investigations on co-digestion of energy crops and crop residues with cow manure for methane production: Effect of crop to manure ratio. *Resour. Conserv. Recycl.* **2007**, *51*, 591–609. [CrossRef]
66. Seppala, M.; Paavola, T.; Lehtomaki, A.; Rintala, J. Biogas production from boreal herbaceous grasses—Specific methane yield and methane yield per hectare. *Bioresour. Technol.* **2009**, *100*, 2952–2958. [CrossRef]
67. Kaparaju, P.; Luostarinen, S.; Kalmari, E.; Kalmari, J.; Rintala, J. Co-digestion of energy crops and industrial confectionery by-products with cow manure: Batch-scale and farm-scale evaluation. *Water Sci. Technol.* **2002**, *45*, 275–280. [CrossRef]
68. Lehtomäki, A.; Viinikainen, T.; Rintala, J. Screening boreal energy crops and crop residues for methane biofuel production. *Biomass Bioenergy* **2008**, *32*, 541–550. [CrossRef]

69. Parawira, W.; Murto, M.; Zvauya, R.; Mattiasson, B. Anaerobic batch digestion of solid potato waste alone and in combination with sugar beet leaves. *Renew. Energy* **2004**, *29*, 1811–1823. [CrossRef]
70. Lehtomäki, A.; Department of Biological and Environmental Science, University of Jyväskylä, Jyväskylä, Finland. Personal Communication, November 2011.
71. Garcia, N.H.; Mattioli, A.; Gil, A.; Frison, N.; Battista, F.; Bolzonella, D. Evaluation of the methane potential of different agricultural and food processing substrates for improved biogas production in rural areas. *Renew. Sustain. Energy Rev.* **2019**, *112*, 1–10. [CrossRef]
72. Cornell, M.; Heaven, S.; Banks, C. Impact of the addition of maize on the anaerobic digestion of cattle slurry. In Proceedings of the 5th IWA International Symposium on Anaerobic Digestion of Solid Wastes and Energy Crops, Hammamet, Tunisia, 25–28 May 2008.
73. Rincón, B.; Banks, C.J.; Heaven, S. Biochemical methane potential of winter wheat (*Triticum aestivum* L.): Influence of growth stage and storage practice. *Bioresour. Technol.* **2010**, *101*, 8179–8184. [CrossRef]
74. Rincón, B.; Heaven, S.; Salter, A.M.; Banks, C.J. Anaerobic digestion of spring and winter wheat: Comparison of net energy yields. *J. Environ. Sci. Health Part A* **2016**, *51*, 1084–1089. [CrossRef]
75. Walker, M.; Zhang, Y.; Heaven, S.; Banks, C. Potential errors in the quantitative evaluation of biogas production in anaerobic digestion processes. *Bioresour. Technol.* **2009**, *100*, 6339–6346. [CrossRef]
76. Peng, X.; Li, C.; Liu, J.; Yi, Z.; Han, Y. Changes in composition, cellulose degradability and biochemical methane potential of Miscanthus species during the growing season. *Bioresour. Technol.* **2017**, *235*, 389–395. [CrossRef]
77. Virkajärvi, P.; Hyrkäs, M.; Kykkänen, S.; Pyykkönen, V.; Laakso, J.; Kekkonen, P.; Ervasti, S.; Luostarinen, S. Effect of digestate application, N fertilisation and forage species on biomass and biogas production. Sustainable meat and milk production from grasslands 2018. In Proceedings of the 27th General Meeting of the European Grassland Federation, Cork, Ireland, 17–21 June 2018; pp. 688–690.
78. Nges, I.A.; Li, C.; Wang, B.; Xiao, L.; Yi, Z.; Liu, J. Physio-chemical pretreatments for improved methane potential of Miscanthus lutarioriparius. *Fuel* **2016**, *166*, 29–35. [CrossRef]
79. Li, C.; Liu, G.; Nges, I.A.; Liu, J. Enhanced biomethane production from miscanthus lutarioriparius using steam explosion pretreatment. *Fuel* **2016**, *179*, 267–273. [CrossRef]
80. Thomas, H.L.; Arnoult, S.; Brancourt-Hulmel, M.; Carrère, H. Methane production variability according to miscanthus genotype and alkaline pretreatments at high solid content. *BioEnergy Res.* **2019**, *12*, 325–337. [CrossRef]
81. Pohl, M.; Mumme, J.; Heeg, K.; Nettmann, E. Thermo-and mesophilic anaerobic digestion of wheat straw by the upflow anaerobic solid-state (UASS) process. *Bioresour. Technol.* **2012**, *124*, 321–327. [CrossRef] [PubMed]
82. Heeg, K.; Pohl, M.; Sontag, M.; Mumme, J.; Klocke, M.; Nettmann, E. Microbial communities involved in biogas production from wheat straw as the sole substrate within a two-phase solid-state anaerobic digestion. *Syst. Appl. Microbiol.* **2014**, *37*, 590–600. [CrossRef]
83. Rocha-Meneses, L.; Raud, M.; Orupõld, K.; Kikas, T. Potential of bioethanol production waste for methane recovery. *Energy* **2019**, *173*, 133–139. [CrossRef]
84. Ohlsson, J.A.; Harman-Ware, A.E.; Sandgren, M.; Schnürer, A. Biomass recalcitrance in willow under two biological conversion paradigms: Enzymatic hydrolysis and anaerobic digestion. *BioEnergy Res.* **2020**, *13*, 260–270. [CrossRef]
85. Linke, B.; Schelle, H. Solid state anaerobic digestion of organic wastes. In Proceedings of the Conference Agricultural Engineering into the Third Millenium, AgEng2000, Coventry, UK, 2–7 July 2000.
86. Heiermann, M.; Plöchl, M.; Linke, B.; Schelle, H. Biogas production in agriculture by co-fermentation of energy crops. *Asp. Appl. Biol.* **2001**, *65*, 65–70.
87. Heiermann, M.; Plöchl, M.; Linke, B.; Schelle, H. Preliminary evaluation of some cereals as energy crops for biogas production. In Proceedings of the World Renewable Energy Congress VII (WREC), Cologne, Germany, 29 June–5 July 2002.
88. Heiermann, M.; Plöchl, M. Crops—A big potential for biogas production. In Proceedings of the World Renewable Energy Congress VIII, Denver, CO, USA, 28 August–3 September 2004.
89. Badger, D.M.; Bogue, M.J.; Stewart, D.J. Biogas production from crops and organic wastes 1: Results of batch digestions. *N. Z. J. Sci.* **1979**, *22*, 11–20.
90. Zubr, J. Methanogenic fermentation of fresh and ensiled plant materials. *Biomass* **1986**, *11*, 159–171. [CrossRef]
91. Svensson, L.M.; Christensson, K.; Björnsson, L. Biogas production from crop residues on a farm-scale level: Is it economically feasible under conditions in Sweden. *Bioprocess Biosyst. Eng.* **2005**, *28*, 139–148. [CrossRef]
92. Nizami, A.-S.; Murphy, J.D. Optimizing the operation of a two-phase anaerobic digestion system digesting grass silage. *Environ. Sci. Technol.* **2011**, *45*, 7561–7569. [CrossRef] [PubMed]
93. Raposo, F.; Banks, C.J.; Siegert, I.; Heaven, S.; Borja, R. Influence of inoculum to substrate ratio on the biochemical methane potential of maize in batch tests. *Process Biochem.* **2006**, *41*, 1444–1450. [CrossRef]
94. Yan, H.; Zhao, C.; Zhang, J.; Zhang, R.; Xue, C.; Liu, G.; Chen, C. Study on biomethane production and biodegradability of different leafy vegetables in anaerobic digestion. *AMB Express* **2017**, *7*, 27. [CrossRef] [PubMed]
95. Tilvikiene, V.; Kadziuliene, Z.; Dabkevicius, Z.; Venslauskas, K.; Navickas, K. Feasibility of tall fescue, cocksfoot and reed canary grass for anaerobic digestion: Analysis of productivity and energy potential. *Ind. Crops Prod.* **2016**, *84*, 87–96. [CrossRef]

96. Kaiser, F.; Diepolder, M.; Eder, J.; Hartmann, S.; Prestele, H.; Gerlach, R.; Ziehfreund, G.; Gronauer, A. Ertragspotenziale verschiedener nachwachsender Rohstoffe in landwirtschaftlichen Biogasanlagen. In *Biogas in Bayern*; Bayerische Landesanstalt für Landwirtschaft: Freising, Germany, 2004, pp. 43–56.
97. Mähnert, P.; Heiermann, M.; Linke, B. Batch- and semi-continuous biogas production from different grass species. *Agric. Eng. Int CIGR Ej* **2005**, *7*, 1–11.
98. Herrmann, C.; Heiermann, M.; Idler, C. Effects of ensiling, silage additives and storage period on methane formation of biogas crops. *Bioresour. Technol.* **2011**, *102*, 5153–5161. [CrossRef]
99. Herrmann, C.; Heiermann, M.; Idler, C.; Prochnow, A. Particle size reduction during harvesting of crop feedstock for biogas production I: Effects on ensiling process and methane yields. *Bioenergy Res.* **2012**, *5*, 926–936. [CrossRef]
100. Herrmann, C.; Idle, C.; Heiermann, M. Biogas crops grown in energy crop rotations: Linking chemical composition and methane production characteristics. *Bioresour. Technol.* **2016**, *206*, 23–35. [CrossRef]
101. Kreuger, E.; Nges, I.A.; Björnsson, L. Ensiling of crops for biogas production: Effects on methane yield and total solids determination. *Biotechnol. Biofuels* **2011**, *4*, 44. [CrossRef]
102. Gissén, C.; Prade, T.; Kreuger, E.; Nges, I.A.; Rosenqvist, H.; Svensson, S.E.; Lantz, M.; Mattsson, J.E.; Börjesson, P.; Björnsson, L. Comparing energy crops for biogas production—Yields, energy input and costs in cultivation using digestate and mineral fertilisation. *Biomass Bioenergy* **2014**, *64*, 199–210. [CrossRef]
103. Bruni, E.; Jensen, A.P.; Pedersen, E.S.; Angelidaki, I. Anaerobic digestion of maize focusing on variety, harvest time and pre-treatment. *Appl. Energy* **2010**, *87*, 2212–2217. [CrossRef]
104. Kreuger, E.; Prade, T.; Escobar, F.; Svensson, S.-E.; Englund, J.-E.; Björnsson, L. Anaerobic digestion of industrial hemp—Effect of harvest time on methane energy yield per hectare. *Biomass Bioenergy* **2011**, *35*, 893–900. [CrossRef]
105. Prade, T.; Svensson, S.E.; Hörndahl, T.; Kreuger, E. Impact of harvest date and cutting length of grass ley and whole-crop cereals on methane yield and economic viability as feedstock for biogas vehicle fuel production. *Bioenergy Res.* **2019**, *12*, 137–149. [CrossRef]
106. Allen, E.; Wall, D.M.; Herrmann, C.; Murphy, J.D. A detailed assessment of resource of biomethane from first, second and third generation substrates. *Renew. Energy* **2016**, *87*, 656–665. [CrossRef]
107. Nges, I.A.; Wang, B.; Cui, Z.; Liu, J. Digestate liquor recycle in minimal nutrients-supplemented anaerobic digestion of wheat straw. *Biochem. Eng. J.* **2015**, *94*, 106–114. [CrossRef]
108. Kolbl, S.; Paloczi, A.; Panjan, J.; Stres, B. Addressing case specific biogas plant tasks: Industry oriented methane yields derived from 5 L Automatic Methane Potential Test Systems in batch or semi-continuous tests using realistic inocula, substrate particle sizes and organic loading. *Bioresour. Technol.* **2014**, *153*, 180–188. [CrossRef]
109. Zhao, X.; Luo, K.; Zhang, Y.; Zheng, Z.; Cai, Y.; Wen, B.; Cui, Z.; Wang, X. Improving the methane yield of maize straw: Focus on the effects of pretreatment with fungi and their secreted enzymes combined with sodium hydroxide. *Bioresour. Technol.* **2018**, *250*, 204–213. [CrossRef]
110. Stewart, D.J.; Bogue, M.J.; Badger, D.M. Biogas production from crops and organic wastes 2: Results from continuous digestion tests. *N. Z. J. Sci.* **1984**, *27*, 285–294.
111. Scherer, P.A.; Dobler, S.; Rohardt, S.; Loock, R.; Büttner, B.; Nöldeke, P.; Brettschuh, A. Continuous biogas production from fodder beet silage as sole substrate. *Water Sci. Technol.* **2003**, *48*, 229–233. [CrossRef]
112. Nges, I.A.; Björnsson, L. High methane yields and stable operation during anaerobic digestion of nutrient-supplemented energy crop mixtures. *Biomass Bioenergy* **2012**, *47*, 62–70. [CrossRef]
113. Zhu, H.; Stadnyk, A.; Béland, M.; Seto, P. Co-production of hydrogen and methane from potato waste using a two-stage anaerobic digestion process. *Bioresour. Technol.* **2008**, *99*, 5078–5084. [CrossRef] [PubMed]
114. Kakuk, B.; Kovács, K.L.; Szuhaj, M.; Rákhely, G.; Bagi, Z. Adaptation of continuous biogas reactors operating under wet fermentation conditions to dry conditions with corn stover as substrate. *Anaerobe* **2017**, *46*, 78–85. [CrossRef] [PubMed]
115. Veluchamy, C.; Gilroyed, B.H.; Kalamdhad, A.S. Process performance and biogas production optimizing of mesophilic plug flow anaerobic digestion of corn silage. *Fuel* **2019**, *253*, 1097–1103. [CrossRef]
116. Rincón, B.; Heaven, S.; Banks, C.J.; Zhang, Y. Anaerobic digestion of whole-crop winter wheat silage for renewable energy production. *Energy Fuels* **2012**, *26*, 2357–2364. [CrossRef]
117. Pohl, M.; Heeg, K.; Mumme, J. Anaerobic digestion of wheat straw—Performance of continuous solid-state digestion. *Bioresour. Technol.* **2013**, *146*, 408–415. [CrossRef] [PubMed]
118. Symons, G.E.; Buswell, A.M. The methane fermentation of carbohydrates. *J. Am. Chem. Soc.* **1933**, *55*, 2028–2036. [CrossRef]
119. Buswell, A.M.; Hatfield, W.D. *Anaerobic Fermentation—Bulletin 32*; State of Illinois Department of Registration and Education: Urbana, IL, USA, 1936.
120. Czepuck, K.; Oechsner, H.; Schumacher, B.; Lemmer, A. Hohenheim biogas yield test. *Landtechnik* **2006**, *61*, 82–83.
121. Linke, B.; Baganz, K.; Schlauderer, R. Nutzung von Feldfrüchten zur Biogasgewinnung. *Agaratechnische Forsch.* **1999**, *5*, 81–90.
122. DLG (Ed.) *Futterwerttabellen—Wiederkäuer*; DLG-Verlag: Frankfurt, Germany, 1997.
123. LfL (Bayerische Landesanstalt für Landwirtschaft). Biogasausbeuten Verschiedener Substrate (Database). Available online: https://www.lfl.bayern.de/iba/energie/049711/ (accessed on 6 June 2021).

124. Rodrigues, R.P.; Rodrigues, D.P.; Klepacz-Smolka, A.; Martins, R.C.; Quina, M.J. Comparative analysis of methods and models for predicting biochemical methane potential of various organic substrates. *Sci. Total Environ.* **2019**, *649*, 1599–1608. [CrossRef] [PubMed]
125. Koettner, M. The role of biogas in German farm business development. In Proceedings of the Bioexell European Biogas Conference—Biogas in Society, Enniskillen, UK, 21–23 October 2004.
126. KTBL (Ed.) *Betriebsplanung Landwirtschaft 2008/09*; KTBL: Darmstadt, Germany, 2009.
127. Karpenstein-Machan, M. *Energiepflanzenbau für Biogasanlagenbetreiber*; DLG Verlag: Frankfurt, Germany, 2005.
128. KTBL (Ed.) Wirtschaftlichkeitsrechner Biogas. Available online: https://daten.ktbl.de/biogas/ (accessed on 6 June 2021).
129. Weiland, P. Production and energetic use of biogas from energy crops and wastes in Germany. *Appl. Biochem. Biotechnol.* **2003**, *109*, 263–274. [CrossRef]
130. Hafner, S.D.; Astals, S. Systematic error in manometric measurement of biochemical methane potential: Sources and solutions. *Waste Manag.* **2019**, *91*, 147–155. [CrossRef]
131. Müller, W.-R.; Frommert, I.; Jörg, R. Standardized methods for anaerobic biodegradability testing. *Rev. Environ. Sci. Biotechnol.* **2004**, *3*, 141–158. [CrossRef]
132. Stolarski, M.J.; Snieg, M.; Krzyżaniak, M.; Tworkowski, J.; Szczukowski, S.; Graban, L.; Lajszner, W. Short rotation coppices, grasses and other herbaceous crops: Biomass properties versus 26 genotypes and harvest time. *Ind. Crops Prod.* **2018**, *119*, 22–32. [CrossRef]
133. White, P.M.; Viator, R.P.; Webber, C.L. Temporal and varietal variation in sugarcane post-harvest residue biomass yields and chemical composition. *Ind. Crops Prod.* **2020**, *154*, 112616. [CrossRef]
134. Butkute, B.; Lemežienė, N.; Kanapeckas, J.; Navickas, K.; Dabkevičius, Z.; Venslauskas, K. Cocksfoot, tall fescue and reed canary grass: Dry matter yield, chemical composition and biomass convertibility to methane. *Biomass Bioenergy* **2014**, *66*, 1–11. [CrossRef]
135. Ierna, A.; Sortino, O.; Mauromicale, G. Biomass, seed and energy yield of *Cynara cardunculus* L. as affected by environment and season. *Agronomy* **2020**, *10*, 1548. [CrossRef]
136. Bochmann, G.; Montgomery, L.F.R. Storage and pre-treatment of substrates for biogas production. In *The Biogas Handbook*; Woodhead Publishing Series in Energy; Wellinger, A., Murphy, J., Baxter, D., Eds.; Woodhead Publishing: Oxford, UK, 2013; pp. 85–103. [CrossRef]
137. Alvarez, R.; Lidén, G. Semi-continuous co-digestion of solid slaughterhouse waste, manure, and fruit and vegetable waste. *Renew. Energy* **2008**, *33*, 726–734. [CrossRef]
138. Demirel, B.; Scherer, P. Bio-methanization of energy crops through mono-digestion for continuous production of renewable biogas. *Renew. Energy* **2009**, *34*, 2940–2945. [CrossRef]
139. Mähnert, P.; Linke, B. Kinetic study of biogas production from energy crops and animal waste slurry: Effect of organic loading rate and reactor size. *Environ. Technol.* **2009**, *30*, 93–99. [CrossRef]
140. Pizarro-Loaiza, C.A.; Torres-Lozada, P.; Illa, J.; Palatsi, J.; Bonmatí, A. Effect of harvesting age and size reduction in the performance of anaerobic digestion of Pennisetum grass. *Processes* **2020**, *8*, 1414. [CrossRef]
141. Brauer-Siebrecht, W.; Jacobs, A.; Christen, O.; Götze, P.; Koch, H.-J.; Rücknagel, J.; Märländer, B. Silage maize and sugar beet for biogas production in rotations and continuous cultivation: Dry matter and estimated methane yield. *Agronomy* **2016**, *6*, 2. [CrossRef]
142. Wannasek, L.; Ortner, M.; Kaul, H.-P.; Amon, B.; Amon, T. Double-cropping systems based on rye, maize and sorghum: Impact of variety and harvesting time on biomass and biogas yield. *Eur. J. Agron.* **2019**, *110*, 125934. [CrossRef]
143. Graß, R.; Malec, S.; Wachendorf, M. Biomass performance and competition effects in an established temperate agroforestry system of willow and grassland—Results of the 2nd rotation. *Agronomy* **2020**, *10*, 1819. [CrossRef]

processes

Article

Biogas Production from Residues of Industrial Insect Protein Production from Black Soldier Fly Larvae *Hermetia illucens* (L.): An Evaluation of Different Insect Frass Samples

Harald Wedwitschka [1,*], Daniela Gallegos Ibanez [1] and Damián Reyes Jáquez [2]

1 Department Biochemical Conversion, DBFZ Deutsches Biomasseforschungszentrum Gemeinnützige GmbH, Torgauer Straße 116, D-04347 Leipzig, Germany
2 Unidad de Posgrado, Investigación y Desarrollo Tecnológico, Tecnológico Nacional de México, Instituto Tecnológico de Durango, Victoria de Durango 34080, Mexico
* Correspondence: harald.wedwitschka@dbfz.de; Tel.: +49-341-2434-562

Abstract: Insect biomass shows promise as an alternative animal feedstuff with a low climate effect. Industrial insect rearing generates residual materials, such as feed remains and insect excrements, so-called insect frass, which exhibits a high organic content. Commonly, these residues are utilized as soil amendment. Information on the suitability of these residues for biogas production is rather scarce. The energetic utilization of insect frass as feedstock for anaerobic digestion (AD) would allow for the simultaneous residue material reduction and bioenergy production. Additionally, synergies in heat management could arise using the exhaust heat of the biogas plant in the insect farming process. In laboratory-scale anaerobic digestion trials, the specific methane yield (SMY) of six different insect frass samples from black soldier fly (Hermetia) rearing were tested in batch biochemical methane potential (BMP) tests. Further, semi continuous anaerobic digestion trials on a lab scale using continuously stirred tank reactors (CSTRs) were carried out with Hermetia insect frass from a pilot plant operation in order to determine the digestibility and process stability of the AD process. The BMP results showed SMY values of the different insect frass samples ranging from 201 ± 9 to 287 ± 37 mL/gVS that are similar to those of other animal excrements, such as cow or pig manure already been used as feedstock in agricultural biogas plants. Results of the semi-continuous digestion of insect frass from the pilot plant operation showed a SMY value of 167 ± 15 mL/gVS, suggesting no process-inhibiting effect caused by the feed material. Although, the high nitrogen content must be taken into account for stable AD performance.

Keywords: insect frass; black soldier fly; *Hermetia illucens*; anaerobic digestion; BMP tests; CSTR digestion

Citation: Wedwitschka, H.; Gallegos Ibanez, D.; Jáquez, D.R. Biogas Production from Residues of Industrial Insect Protein Production from Black Soldier Fly Larvae *Hermetia illucens* (L.): An Evaluation of Different Insect Frass Samples. *Processes* **2023**, *11*, 362. https://doi.org/10.3390/pr11020362

Academic Editors: Sonia Heaven, Sigrid Kusch-Brandt and Charles Banks

Received: 12 December 2022
Revised: 17 January 2023
Accepted: 18 January 2023
Published: 23 January 2023

1. Introduction

The continued growth in global meat production is leading to an increasing demand for high-quality protein feed. Due to the limited availability of natural resources, increasing climate change, and land-use competition between food-feed-fuel production, the importance of cost-effective and sustainably produced protein sources is growing [1].

The utilization of insects as feed animals, as food sources for human nutrition and for the production of technical products, such as silk, shellac, or bee wax, has a long tradition. Insects are the most diverse class of animals, with approximately one million described species. Due to their evolutionary development history, they are optimally adapted to a variety of habitats, environmental conditions, and feed materials. They are able to convert a wide spectrum of organic substrates and residues into high-quality raw materials.

Industrially produced insect meal represents an alternative feed protein source [2] and has been successfully tested as livestock, pet, and aquaculture feed [3–8]. Insect biomass shows a high protein content and a high-quality amino acid spectrum.

Nutritionally, insect protein is also suitable for human consumption and the climate impact of insect protein production turns out to be better compared to conventional animal protein production. Water consumption, land requirements, and required feed quantities and slaughter losses are generally lower in insect farming than in cattle and pig fattening and fish production. In Europe, however, the consumption of insects is hardly culturally anchored and consumer acceptance of insects as food is comparatively low. In contrast, there are hardly any reservations about the use of insects as animal feed [9]. Forecasts see the largest market potential for insects as feed in the aquaculture and pet food sectors, followed by livestock feed for poultry and swine [10].

In addition, insect products can be used as a bio-based alternative to conventional fossil raw materials in the production of a wide range of technical products, such as cosmetics, pharmaceuticals, surfactants, surface coatings, lubricants, and fuels. Insect farming represents a promising building block of a future bioeconomy, because against the background of limited resources, the multiple use and recycling of biomass in utilization cascades is increasingly gaining in importance. The carbon footprint of insect products is particularly advantageous when residual materials are used as insect feed and the process heat requirement is covered by exhaust heat or with the aid of renewable energies. Another advantage of insect protein production is that the water and land requirements and the amount of feed used for insect farming are relatively low, and residual materials can be returned to the nutrient cycle, for example, as biogas substrate or as agricultural fertilizer [11,12].

Insect framing of black soldier fly, respectively Hermetia in industrial scale comprises of the process steps fly rearing, larva fattening, and product processing [13]. The insect *Hermatia illucens* goes through the following development stages: egg, larvae, pupae, and fly, whereby insect protein is usually derived from the adult larvae. Fly breeding requires light and warmth and is necessary in order to provide sufficient young larvae for the fattening process. The fattening of Hermetia larvae is carried out without light, commonly in boxes and tubs of different sizes in a climate-controlled environment. In this production step, young larvae are added to a feed medium. Hermetia larvae consume organic matter, such as carbohydrates, proteins, and lipids, and increase in weight. Before entering the pupae state, larvae are separated mechanically from feed residues and insect excrement. The majority of the larvae are further processed into animal feed, while a smaller proportion is used for further fly rearing. Normally, all production steps take place in centralized farm concepts which ideally contribute from favorable heat energy and feed material supply. Decentralized concepts with outsourced larval fattening and centralized larva processing and product recovery represent an alternative approach.

The feed remains and insect excrements are residue materials of the Hermetia farming process and are usually utilized as organic fertilizers and soil additives [14]. According to European law, insect frass may only be used as agricultural fertilizer after sufficient sanitization. This requires heating to at least 70 °C for at least one hour. Possible hygienization measures that can meet the temperature requirements would be, for example, the heating, pelleting, or extrusion of the insect frass. All of these processes require a relatively high energy input but ensure that neither pathogens nor live larvae are released into the environment. Another technology that is used for waste biomass treatment and has a proven sanitization effect is the biogas process [15–18].

With regards to insect frass, anaerobic digestion is an interesting waste treatment option providing bio-methane as biofuel or an energy source for the production electricity and heat which could be reused in the insect rearing process and product processing. Additionally, residues of the digestion process still hold plant nutrients contained in the substrate material and can be utilized as organic fertilizer and soil amendment. The process combination of the biogas plant and insect farm enables various synergies. By integrating the insect farm into existing biogas plants, the exhaust heat utilization of the biogas plant could be optimized and digested residues from the biogas plant could also serve as an insects feed source [19].

Insect farming as the first stage of corresponding value chains and the utilization of residues from insect production in the biogas process could increase the efficiency of resource utilization. The large-scale production of Hermetia is a new technology. There are so far only a few companies worldwide with an insect production capacity on an industrial scale. Data on the methane potential of the residue materials are scarce and there are no data available on the long-term digestibility when this manuscript was written. One research aim of the study was the assessment of the biomethane potential of insect frass from Hermetia rearing on different feed sources. Insect frass samples were subjected to biochemical methane potential (BMP) tests in triplicate in laboratory batch scale in order to determine the specific methane yields and methane production kinetics of the sample materials. A further aim of the study was to evaluate the feedstock suitability of insect frass for AD processes in long term semi-continuous anaerobic digestion trials. Therefore, digestion experiments on a lab scale were carried out with actual insect frass from a large-scale pilot production of Hermetia in order to determine the feedstock digestibility and AD process stability. The AD characteristics presented in this study extend the data basis required for the suitability assessment of Hermetia insect frass as a raw material for biogas production. The BMP results determined can be used for an economic feasibility evaluation of the energetic utilization of insect frass in the AD process.

2. Materials and Methods

2.1. Acquisition and Characterization of the Insect Frass Samples

Five insect frass samples were obtained from previous Hermetia rearing trials in laboratory scale run between 2020 and 2021, where Hermetia larva were fed on five different substrates: corn silage (CS), brewers spend grain (BS), thin stillage from bioethanol production (ST), aquatic plants from Elodea genus (EL), and bran (BR). Larvae feeding trials were carried out in triplicate batch attempts in 550 mL plastic containers (CLIP & CLOSE Food storage container, EMSA, Germany) with a size of 16.3 × 11.3 × 5.8 cm and a working volume of approximately 250 mL. The container caps were perforated to allow gas exchange. The substrate feed load was 240 mg VS/Larvae. Containers were stored at 30.0 ± 0.25 °C in temperature chambers (New Brunswick Innova 44). After 12 days, the feed remains and excrements (insect frass) were separated from the larvae and used in the present study without prior drying. In addition, another sample of insect frass was obtained from Hermetia rearing in pilot plant operation (IF_PP) (Hermetia Baruth GmbH, Baruth Mark, Germany). In 2020 and 2021, the annual production capacity of the pilot plant was approximately 300 t Hermetia larvae which were fed on a feed mixture mainly composed of cereal grain.

Insect frass samples were tested for their material properties regarding total solids (TS) and volatile solids (VS), nitrogen, protein, fat, and fibre composition and subjected to biochemical methane potential (BMP) tests. Additionally, long term semi-continuous digestion experiments were performed with the insect frass sample from a pilot plant operation (IF_PP). Wet samples were stored at 5 °C after sampling. Dry samples were stored in air tight plastic barrels at room temperature. Sample characteristics are depicted in Table 1.

Table 1. Insect frass sample material characteristics.

Insect Frass	TS *	VS *	Ash *	Crude Protein	Raw Fat	Crude Fibre	Other Carbohydrates
(Feedstock)	[%FM]	[%TS]	[g/kgTS]	[g/kgTS]	[g/kgTS]	[g/kgTS]	[g/kgTS]
Stillage (ST)	9.6	94.1	58.5	240.5 ± 1.17	63.1 ± 3.0	315.2 ± 2.7	322.8 ± 3.4
Brewers spent grain (BS)	2.6	51.2	487.7	215.2 ± 4.27	37.9 ± 5.0	47.1 ± 2.3	307.7 ± 2.3
Corn silage (CS)	7.3	81.1	189.0	230.0 ± 7.09	30.5 ± 4.5	101.7 ± 1.0	347.9 ± 1.4
Elodea nutallii (EL)	12.9	94.5	54.8	46.4 ± 3.2	21.2 ± 6.7	533.0 ± 1.8	344.7 ± 2.5
Bran (BR)	12.4	85.7	143.1	288.8 ± 2.4	23.3 ± 3.9	338.6 ± 7.8	206.2 ± 4.3
Insect frass pilot plant (IF_PP)	84.2	91.0	89.7	228.8 ± 5.1	33.9 ± 5.2	226.5 ± 0.8	421.1 ± 5.2

TS total solids; vs. volatile solids; FM fresh matter; * single sample.

2.2. Analytical Methods

Total solids (TS) and volatile solids (VS) were measured in accordance with DIN EN 12,880 (2001) [20] and DIN EN 12,879 [21]. The pH-value of digestate samples was measured with a pH device 3310 (WTW Wissenschaftlich-Technische Werkstätten GmbH, Weilheim, Germany). The Weender feed analysis of insect frass and ammonia nitrogen (NH4-N) and the total ammonia nitrogen (TAN) of the digestate were determined, as described in [22]. Once a week, fresh digestate samples were taken from CSTR digestion and centrifuged with 10,000 rpm for 10 min at 10 °C. Filtered samples (10 mL) of the supernatant liquid were used for the quantification of all volatile organic acids (VOA) and the ratio of VOA to total inorganic carbonate to calcium carbonate (VOA/TIC, gVOA/gCACO3) measurement in a Titration Excellence T90 titrator (Mettler-Toledo GmbH, Zurich, Switzerland).

2.3. Anaerobic Digestion Trials

2.3.1. Biochemical Methane Potential (BMP) Test

The six different insect frass samples were analyzed for specific methane yields (SMY) at lab-scale using the AMPTS2 BMP test system (Bioprocesscontrol, Lund, Sweden). BMP tests were carried out in accordance with the VDI guideline 4630 (2016) [23] under mesophilic conditions (39 ± 1 °C). The inoculum to substrate (ISR) ratio was approximately 3:1 (based on mass VS). Before the batch experiments, the AD reactors' headspace was flushed with nitrogen gas for about 2 min to assure anaerobic conditions. Each reactor contained approximately 2.5 gVS of insect frass and 400 g inoculum and was analyzed in triplicate. The SMY was standardized according to DIN 1343 [24] (dry gas, 273.15 K, 101.325 kPa). The BMP test ended after 38 days; the daily methane production had reduced to just 0.5% of the total biogas production for a minimum of 5 days. The pure inoculum was measured as a blank sample to determine the specific methane yield and to subtract this from the other samples. To monitor the inoculum performance, microcrystalline cellulose (MCC) was used as a reference substrate and the reference BMP confirmed a sufficient inoculum quality with 351 ± 11 mL/gVS. As inoculum served digestate (pH 7.8, VOA/TIC, Ammonia NH4-N 1.49 g/L) which was adapted to a wide range of substrate components, such as protein, fat, fiber, and carbohydrates, at a low OLR of 0.5 gVS/(L*d) over the duration of one year.

2.3.2. Semi-Continuous Anaerobic Digestion Tests

Two continuously stirred tank reactors (CSTRs; R1 and R2) in duplicate, each with a net volume of 15 L (10 L working volume) (Bräutigam Kunststofftechnik GmbH, Mohlsdorf-Teichwolframsdorf, Germany), were used for semi-continuous AD of insect frass derived from pilot plant operation (IF_PP). The main objective was assessing the AD process performance, stability, and methane production from IF_PP. The temperature was set at 39 °C using a thermostat (JULABO GmbH, Seelbach, Germany) and kept under mesophilic conditions (38 ± 1 °C) by recirculating hot water through the double-walled reactors. The reactors were continuously stirred (100 rpm) using a Stirrer 'RZR 2102 control' (Heidolph Instruments GmbH & Co.KG, Schwabach, Germany) located in the upper part of the reactors. The biogas volume was measured with a drum-type gas meter TG05/5 (Dr.-Ing. RITTER Apparatebau GmbH & Co. KG, Bochum, Germany), and the biogas quality was determined using a AwiFLEX (Awite Bioenergie GmbH, Langenbach, Germany). CSTR tests were conducted in accordance with the VDI guideline 4630 (2016) [23] as well. Methane and biogas yields were standardized, respectively (dry gas, 273.15 K, 1013.25 kPa). The fermentation experiments were accompanied by numerous analyses that were used for process characterization and monitoring, such as the dry matter organic dry matter analysis of the substrate and digestate samples, pH, ammonium, and volatile fatty acid concentration.

The general procedure for reactor operation at the DBFZ and detailed information on the accompanying analytics can be found in the literature reference [22]. For the CSTR experiment, the same inoculum was used as for the BMP tests (see Section 3.2). The experiment was carried-out over 314 consecutive days with the same feeding frequency (once per day). After 5 days without feeding, reactors R1 and R2 were fed with DDGS (distillers' dried grains with solubles) pellets due to a delay in the supply with insect frass. On day 25, feeding of both reactors with IF_PP started. During start-up (Phase 1) the organic loading rate (OLR) was set to 1.0 g VS/L·d. Between days 52–140 (Phase 2), the OLR was increased to approximately 1.5 g VS/L·d and between day 141–173 (Phase 3) reduced to 0.7 g VS/L·d for 32 days due to process instability. Thereafter, the OLR was gradually increased from 0.7 to 1.5 and finally to 2.2 g VS/L·d until the end of the experiment between day 174–314 (Phase 4). When the final OLR of 2.2 g VS/L·d was reached, 30 g FM insect frass and 150 mL tap water were added daily to each digester. The hydraulic retention time (HRT) of ~80 days was kept constant over the first half of the experiment until day 130. Thereafter, the HRT was reduced to about 60 days until the end of the experiment. No additives, such as trace elements, were used. Detailed information about different feeding rates, OLR, and HRT are listed in Table 2.

Table 2. Overview of the reactor's setup during the AD experiment with IF_PP.

Phase	Period (Day)	HRT (Days)	OLR (g VS/L·d)
Phase I	0–52	80	1.0
Phase II	53–140	80	1.5
Phase III	141–173	60	0.7
Phase IV	174–314	60	1.5–2.2

Insect frass from pilot plant operation (IF_PP), hydraulic retention time (HRT), and organic loading rate (OLR).

2.4. Kinetic Evaluation

Two kinetic models were used to fit the experimental data of the BMP of the six different insect frass (i.e., EL, CS, BS, BR, ST, and IF_PP). These models were the first-order models and the modified Gompertz model, as given in Equations (1) and (2),

$$\beta(t) = \beta_0 \cdot \left[1 - e^{-kt}\right] \tag{1}$$

$$\beta(t) = \beta_0 \cdot e^{\left[-e^{\left(\frac{\beta_m \cdot e}{\beta_0} \cdot (\lambda - t) + 1\right)}\right]} \tag{2}$$

where $\beta(t)$ is the cumulative methane yield at time t (mLCH4/gVS), β_0 is the maximum cumulative methane production predicted at a theoretically infinite digestion time (mLCH4/gVS), k is the first order hydrolysis constant (1/days), t is the time (days), β_m is the maximum methane production rate (mLCH4/gVS·d), and λ is the lag phase (days). In addition, model parameters and their uncertainties were estimated using the negative logarithm of the likelihood LL (Equation (3)) as the objective function with constant error variance [25],

$$LL = \frac{n ln(2\pi\sigma^2)}{2} + \frac{\sum_{i=1}^{n}\left(\beta_i^{obs} - \beta_i^{est}\right)}{2\sigma^2} \tag{3}$$

where n is the total number of experimental data, i is an index, β_i^{obs} represents the observed cumulative specific methane yield at time t, β_i^{est} represents the estimated cumulative methane yield calculated with Equations (1) and (2), and σ is the standard error. The model selection for the best fit to observed data was conducted using the Akaike information criterion (AIC) (Equation (4)).

$$AIC = -2ln(L_{max}) + 2P \tag{4}$$

where L_{max} is the maximum likelihood and P is the number of parameters included in the model. All parameters and their uncertainties were estimated using the subroutine *"optim"* from the statistical package R [26] using the L-BFGS-B algorithm.

2.5. Statistical Analysis

The data recorded after 32 days of AD obtained from the BMP test was analyzed using one-way analysis of variance (ANOVA) for comparing the specific methane yield (SMY) means among the six different insect frass samples (i.e., EL, CS, BS, BR, ST, and IF_PP). After one-way ANOVA, a post-hoc analysis with a Sidak post test for multiple comparison was performed. In addition, we used the Mann–Whitney rank sum test (Normality test, $p = 0.000$) for the data obtained from the semi-continuous experiment to compare the SMY values between reactors (R1 and R2). Descriptive statistics (mean and standard deviation) were computed for all the insect frass samples. All analysis was performed with Minitab V16.0 (Minitab Inc., State College, PA, USA). The statistical significance was set at $p \leq 0.05$.

3. Results and Discussion

3.1. Insect Frass Characteristics

As shown in Table 1, the total solid (TS) content of insect frass samples from lab scale rearing experiments was in a range between 2.6 and 12.9% FM while the TS of residue material obtained from the pilot plant operation was comparably dry with a TS of 84.2% FM. During lab scale rearing experiments, the humidity of the feed medium was controlled while in the pilot plant scale a certain drying of the feed medium is wanted. The lab scale separation of larvae and insect frass by wet sieving is easily possible while large-scale dry sieving is favored, as less handling effort is required, and a more transport-worthy insect frass is produced.

3.2. Effect of the Six Different Insect Frass Samples on Specific Methane Yield (SMY) from the BMP Test

The specific methane yield (SMY) of the tested insect frass samples ranged from 201 to 287 mL/g VS (Figure 1A) and is comparable to other residues from livestock farming, such as cattle manure (110–275 mL/g VS), pig manure (180–360 mL/g VS), and chicken manure (200–360 mL/g VS) [27]. A slightly lower SMY of approximately 177 mL/g VS was reported by Bulak et al. [28], measured in BMP tests with insect frass from Hermetia reared on residues from the fruit and vegetable industry in the form of carrot-beetroot marc.

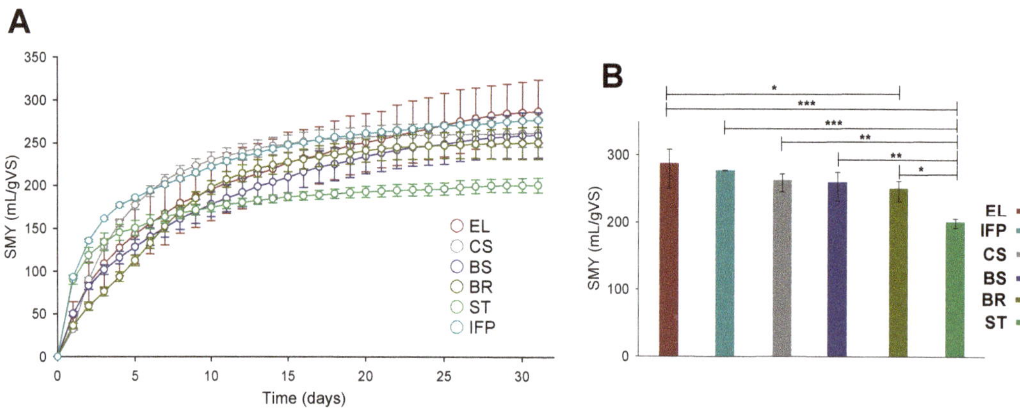

Figure 1. Specific methane yield (SMY) of the different insect frass samples. (**A**): *Elodea nutallii* (EL), corn silage (CS), brewers spent grain (BS), bran (BR), stillage (ST), and insect frass (IF_PP). Effect of different substrates on SMY (**B**). Data are mean ± standard deviation ($n = 3$) for each insect frass sample. * $p < 0.05$, ** $p < 0.01$. *** $p < 0.001$ (one-way ANOVA with Sidak post-test).

Our test results show the highest SMY for EL, followed by IF_PP, CS, BS, and BR, while the lowest SMY was shown by the insect frass sample ST with a value of 201 ± 8.6 mL/g VS. Results of the one-way ANOVA analysis indicated that at least one insect frass sample was significantly different in the SMY mean values among the six different frass samples ($F = 5.833$, $p = 0.006$; Figure 1B). A post-hoc analysis for multiple comparison (Holm-Sidak method) showed that samples EL, CS, BS, BR, and IF_PP had significantly higher SMY versus the ST sample ($p < 0.05$; Figure 1B), with increases of up to 30.07, 23.41, 22.53, 19.76, and 27.55%, respectively. These increases were probably due to differences in the characteristics of the insect frass samples used in the present study. For example, the IF_PP sample showed a lower fiber content than that of the ST (Table 1). Indeed, a negative effect on the SMY values would be expected from the increase in fiber fractions, in particular the ADF and ADL-like fraction [29].

The data also showed that only the EL sample had a significantly higher SMY compared to the BR sample, increasing SMY by about 12.85% ($p = 0.059$, considered significant due to borderline significance). We also found that statistically similar SMY was observed among the EL, CS, BS, BR, and IF_PP insect frass samples (287 ± 36.8, 262 ± 16.9, 259 ± 26.9, 250 ± 19.1 and 277 ± 0.8, respectively; $p > 0.05$) and were not significantly different from each other, except for the EL and BR samples (Figure 1B). This suggests that any change in SMY values may be attributed to differences in chemical composition among the six insect frass samples, likely due to the composition of the substrates used to previously feed the larvae. Overall, the SMY obtained from the BMP trials indicate, in the first instance, a good degradability of all insect frass samples used in the present study. After only about 30 days of the BMP test, gas formation was largely completed.

SMY from batch tests were similar for all insect frass samples and showed the potential suitability as AD feedstock. However, an economic evaluation would be necessary to assess the economic feasibility. There is, however, still very limited knowledge about methane production (i.e., SMY) from insect frass available that can support the findings of this study. Further research is required in order to validate the presented results and to extend the database of different insect frass materials.

3.3. *Effects of the Different Insect Frass Samples on Estimated Model Parameters*

Parameter estimates of the applied model structures (Equations (1) and (2)) are presented in Table 3. All models fit the observed data well with correlation coefficients (R^2) varying from 0.998–0.999; however, the first order model had the overall lowest AIC values in all the insect frass samples. Furthermore, parameter estimates of the lag phase duration (λ) in the modified Gompertz model were most often negative. Since the modified Gompertz model is only defined for positive parameter values ($\lambda \geq 0$), negative parameter estimates indicate that the model is not suitable for the description of measured methane production. Therefore, we selected the first order model as the best fit for the observed methane production of the insect frass samples: EL, CS, BS, ST, and IF_PP. The hydrolysis constant (k) obtained from the first-order kinetic model is mainly used to evaluate the substrate suitability and estimate the process rate-limiting stage. In this way, k describes the velocities of degradation and methane production; therefore, a high k represents high rates of degradation and methane production. In this study, our results provided evidence that high substrate biodegradation improved k, and thus improved the methane production rate and methane yield. The overall highest k corresponded to the ST, IF_PP, and CS samples, while the lowest k values were obtained for EL, BS, and BR. One possible explanation for the highest k values of the samples is that there is a greater proportion of more easily degradable substances in these insect frass samples.

Table 3. Estimated parameters for the insect frass samples: *Elodea nutallii* (EL), corn silage (CS), brewers spent grain (BS), bran (BR), stillage (ST), and insect frass (IF_PP) for the first-order model and modified Gompertz model.

Parameters	Insect Frass Samples					
	EL	CS	BS	BR	ST	IF_PP
Observed SMY	287 ± 36.8	262 ± 16.9	259 ± 26.9	250 ± 19.1	201 ± 8.6	277 ± 0.8
First order model						
β_0 (mL/gVS)	280 ± 4.5	261 ± 1.0	257 ± 3.8	259 ± 2.1	190 ± 2.2	262 ± 3.7
k (1/day)	0.13 ± 0.01	0.22 ± 0.00	0.13 ± 0.01	0.13 ± 0.00	0.38 ± 0.03	0.25 ± 0.02
Correlation coefficient (R^2)	0.9915	0.9985	0.9932	0.9983	0.9725	0.9761
Akaike information criterion (AIC)	247.10	183.49	236.95	191.03	243.25	266.63
Modified Gompertz model						
β_0 (mL/gVS)	269 ± 3.7	256 ± 1.2	248 ± 3.7	248 ± 1.3	190 ± 2.4	259 ± 2.7
λ (days)	−1.00 ± 0.39	−0.33 ± 0.21	−1.00 ± 0.45	−0.29 ± 0.14	−1.00 ± 0.37	−1.00 ± 0.30
β_m (mL/gVS·day)	20.35 ± 1.25	34.13 ± 1.93	18.64 ± 1.26	21.65 ± 0.52	32.75 ± 4.51	33.25 ± 2.73
Correlation coefficient (R^2)	0.9787	0.9929	0.9825	0.9972	0.9563	0.9583
Akaike information criterion (AIC)	287.40	232.97	265.93	205.35	258.46	309.52

Specific methane yield (SMY), maximum cumulative methane production predicted (β_0), first order hydrolysis constant (k), maximum methane production rate (β_m), and lag phase (λ).

3.4. Results of the Semi-Continuous Anaerobic Digestion Experiment

The purpose of the fermentation test was to determine the maximum biogas potential and the process stability when using insect frass from a pilot plant operation (IF_PP) as a sole substrate. Figure 2 illustrates the AD performance of the two reactors R1 and R2, where the whole digestion period was divided into four phases: phase I (start-up, 0–52 day), II (53–140 day), III (141–173 day), and IV (174–314 day). It can be observed that R1 and R2 produced statistically similar SMY during phase I, II, and IV. Only during phase III did the digester show statistically significant differences according to the Mann–Whitney rank sum test (p = 0.000). During the course of the experiment, it was observed that the ammonium nitrogen (NH_4-N) concentration in both reactors increased continuously, which was likely due to the degradation of the protein-containing components of the insect frass (Figure 1A). According to Yenigün et al. (2013) [30], at a NH_4-N concentration of ~3 g/L, depending on the test temperature and pH value, an inhibition of the biogas process can already occur, which leads to an increase in fermentation acids, a decrease in pH, and thus a reduction in methane formation in the long-term test. This effect is very likely to be seen in Figure 1B at the end of phase II and during phase III. Only after a short-term reduction of the daily feeding amount (day 145 to 173) and an increase in the amount of water added to the substrate material (day 140), could the process be stabilized again.

In the second half of the experiment (phase IV), the organic loading rate (OLR) was increased in two steps from 0.7 to 1.4 and thereafter to 2.2 g VS/L·d, which led again to an increase in NH4-N concentration. However, the biogas process seemed to be already adapted to the substrate and higher NH4-N concentrations, as no further process instability was observed until the end of the experiment (phase IV).

According to Lalander et al. (2019) [31] and Abduh et al. (2022) [32], insect feeds with high protein content result in a higher insect biomass and protein yield. Therefore, common insect feed materials can contain comparatively high protein concentrations and, as a result, higher nitrogen levels may also be found in insect frass. Ammonium nitrogen (NH4-N) is a degradation product of organic nitrogen, such as proteins or urea during the AD process. In the digester liquid, NH_4-N is present as ammonium ions (NH_4+) and as free ammonia (NH_3). Increasing the pH or temperature results in a higher percentage of NH_4-N present as NH_3. It has been demonstrated that NH_3 is more toxic, as it can pass through the cell membrane, causing a proton imbalance and potassium deficiency [30]. According to Jiang et al. [33], several research groups reported that the inhibitory NH_4-N concentration may range between 1.5 to 5.0 g/L. However, concentrations of up to 14 g/L are possible by the microbial adaptation of the digestion process. In the case of our study, an inhibition of the process biology can be expected during the end of phase II and beginning of phase III, which negatively affected the production of methane. Thus, the mixing of ammonia rich insect frass with other biogas substrates with a lower nitrogen concentration, such as whole plant silage could be an approach in order to stabilize the AD process in a practical application.

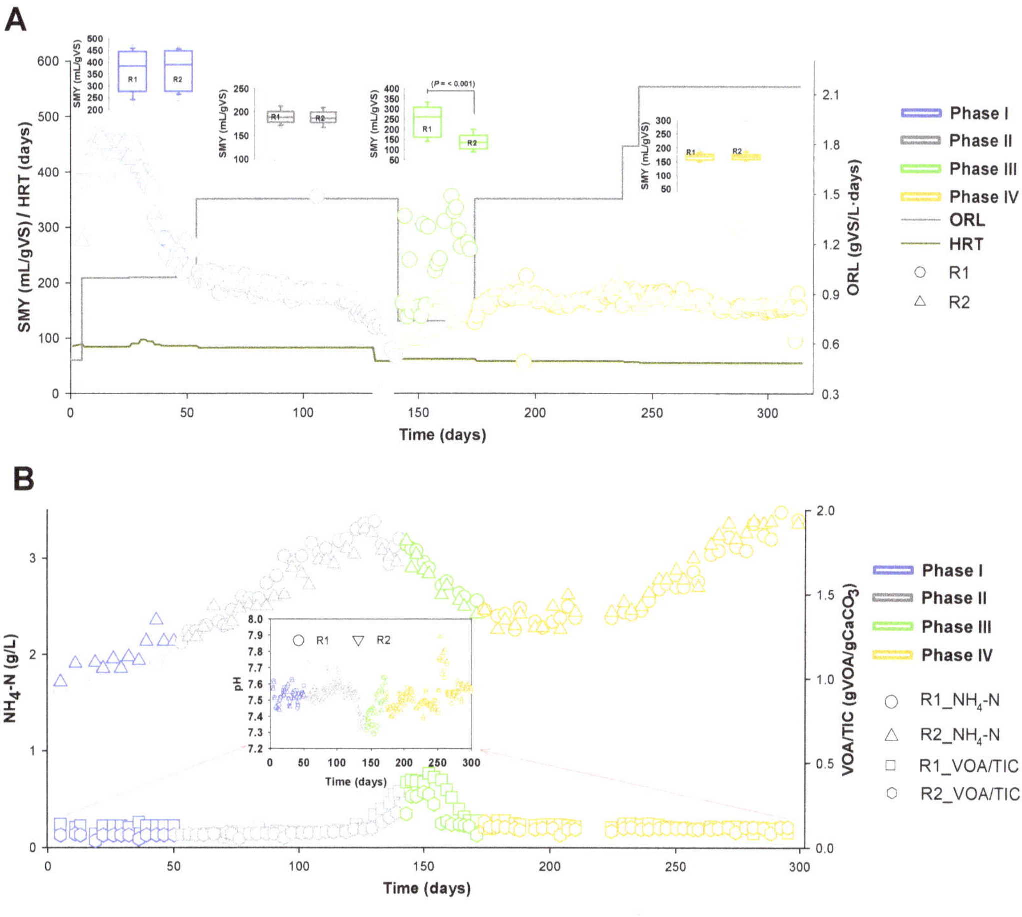

Figure 2. Anaerobic digestion (AD) process performance of the insect frass pilot plant (IF_PP) sample. Specific methane yield (SMY), organic loading rate (ORL), and hydraulic retention time (HRT) (**A**). Ammonium nitrogen (NH_4-N) concentration, ratio of volatile organic acids to total inorganic carbon (VOA/TIC) and pH (**B**).

The mean SMY measured in the semi continuous tests was in the range of 167 ± 15 mL/g VS and the mean methane content of the biogas was approximately 54%. Thus, the anaerobic digestion of insect frass in the CSTR trials also resulted in specific methane yields comparable to other agricultural residues from livestock farming (see Section 3.1). However, in comparison to the BMP tests the long-term digestion of insect frass from pilot plant operation (IF_PP) resulted in an approximately 38% lower SMY.

10. de Jong, B.; Nikolik, G. No Longer Crawling: Insect Protein to Come of Age in the 2020s. Rabo Bank Research. Available online: https://insectfeed.nl/wp-content/uploads/2021/03/Rabobank_No-Longer-Crawling-Insect-Protein-to-Come-of-Age-in-the-2020s_Feb2021-1.pdf (accessed on 11 December 2022).
11. Almeida, C.; Murta, D.; Nunes, R.; Baby, A.R.; Fernandes, Â.; Barros, L.; Rijo, P.; Rosado, C. Characterization of lipid extracts from the Hermetia illucens larvae and their bioactivities for potential use as pharmaceutical and cosmetic ingredients. *Heliyon* **2022**, *8*, e09455. [CrossRef]
12. Elsayed, M.; Li, W.; Abdalla, N.S.; Ai, P.; Zhang, Y.; Abomohra, A.E.-F. Innovative approach for rapeseed straw recycling using black solider fly larvae: Towards enhanced energy recovery. *Renew. Energy* **2022**, *188*, 211–222. [CrossRef]
13. Ites, S.; Smetana, S.; Toepfl, S.; Heinz, V. Modularity of insect production and processing as a path to efficient and sustainable food waste treatment. *J. Clean. Prod.* **2020**, *248*, 119248. [CrossRef]
14. Lopes, I.G.; Yong, J.W.; Lalander, C. Frass derived from black soldier fly larvae treatment of biodegradable wastes. A critical review and future perspectives. *Waste Manag.* **2022**, *142*, 65–76. [CrossRef]
15. Seruga, P.; Krzywonos, M.; Paluszak, Z.; Urbanowska, A.; Pawlak-Kruczek, H.; Niedźwiecki, Ł.; Pińkowska, H. Pathogen Reduction Potential in Anaerobic Digestion of Organic Fraction of Municipal Solid Waste and Food Waste. *Molecules* **2020**, *25*, 275. [CrossRef]
16. Hupfauf, S.; Winkler, A.; Wagner, A.O.; Podmirseg, S.M.; Insam, H. Biomethanation at 45 °C offers high process efficiency and supports hygienisation. *Bioresour. Technol.* **2020**, *300*, 122671. [CrossRef]
17. Wöss, D.; Ortner, M.; Mensik, J.; Kirchmayr, R.; Schumergruber, A.; Pröll, T. Implementation and long term experiences with a continuous hygienisation process in food industry—A case study. *Chem. Eng. Process. Process Intensif.* **2019**, *137*, 100–107. [CrossRef]
18. Levantesi, C.; Beimfohr, C.; Blanch, A.R.; Carducci, A.; Gianico, A.; Lucena, F.; Tomei, M.C.; Mininni, G. Hygienization performances of innovative sludge treatment solutions to assure safe land spreading. *Environ. Sci. Pollut. Res. Int.* **2015**, *22*, 7237–7247. [CrossRef]
19. Pas, C.; Brodeur, D.; Deschamps, M.-H.; Lebeuf, Y.; Adjalle, K.; Barnabé, S.; Eeckhout, M.; Vandenberg, G.; Vaneeckhaute, C. Valorization of pretreated biogas digestate with black soldier fly (Hermetia illucens, L.; Diptera: Stratiomyidae) larvae. *J. Environ. Manag.* **2022**, *319*, 115529. [CrossRef]
20. *DIN 12880:2000*; Characterization of Sludges—Determination of Dry Residue and Water Content. Deutsches Institut für Normung e. V.: Berlin, Germany, 2001.
21. *DIN Standard EN 12879*; Characterization of Sludges—Determination of the Loss on Ignition of Dry Mass. Deutsches Institut für Normung e. V.: Berlin, Germany, 2001.
22. Liebetrau, J.; Pfeiffer, D.; Thrän, D. (Eds.) *Energetische Biomassenutzung: Messmethodensammlung Biogas: Methoden zur Bestimmung von Analytischen und Prozessbeschreibenden Parametern im Biogasbereich*; Fischer Druck: Leipzig, Germany, 2015.
23. *VDI Guideline 4630*; Fermentation of Organic Materials—Characterization of the Substrate, Sampling, Collection of Material Data, Fermentation Tests. The Association of German Engineers: Berlin, Germany, 2016.
24. *DIN 1343:1990-01*; Reference Conditions, Normal Conditions, Normal Volume; Concepts and Values. Deutsches Institut für Normung e. V.: Berlin, Germany, 1990.
25. Kimura, D. Likelihood methods for the von Bertalanffy growth curve. *Fish. Bull.* **1980**, *77*, 765–776.
26. R Core Team. *R: A Language and Environment for Statistical Computing*; R Foundation for Statistical Computing: Vienna, Austria, 2019. Available online: https://www.R-project.org/ (accessed on 5 January 2022).
27. FNR. *Leitfaden Biogas: Von der Gewinnung zur Nutzung*, 7th ed.; Druckerei Weidner: Rostock, Germany, 2016; ISBN 3-00-014333-5.
28. Bulak, P.; Proc, K.; Pawłowska, M.; Kasprzycka, A.; Berus, W.; Bieganowski, A. Biogas generation from insects breeding post production wastes. *J. Clean. Prod.* **2020**, *244*, 118777. [CrossRef]
29. McEniry, J.; O'Kiely, P. Anaerobic methane production from five common grassland species at sequential stages of maturity. *Bioresour. Technol.* **2013**, *127*, 143–150. [CrossRef]
30. Yenigün, O.; Demirel, B. Ammonia inhibition in anaerobic digestion: A review. *Process Biochem.* **2013**, *48*, 901–911. [CrossRef]
31. Lalander, C.; Diener, S.; Zurbrügg, C.; Vinnerås, B. Effects of feedstock on larval development and process efficiency in waste treatment with black soldier fly (Hermetia illucens). *J. Clean. Prod.* **2019**, *208*, 211–219. [CrossRef]
32. Abduh, M.Y.; Perdana, M.P.; Bara, M.A.; Anggraeni, L.W.; Putra, R.E. Effects of aeration rate and feed on growth, productivity and nutrient composition of black soldier fly (Hermetia illucens L.) larvae. *J. Asia-Pac. Entomol.* **2022**, *25*, 101902. [CrossRef]
33. Jiang, Y.; McAdam, E.; Zhang, Y.; Heaven, S.; Banks, C.; Longhurst, P. Ammonia inhibition and toxicity in anaerobic digestion: A critical review. *J. Water Process Eng.* **2019**, *32*, 100899. [CrossRef]
34. Weinrich, S.; Murphy, J.D. *Value of Batch Tests for Biogas Potential Analysis: Method Comparison and Challenges of Substrate and Efficiency Evaluation of Biogas Plants*; IEA Bioenergy: Dublin, Ireland, 2018; ISBN 978-1-910154-49-6.
35. Ruffino, B.; Fiore, S.; Roati, C.; Campo, G.; Novarino, D.; Zanetti, M. Scale effect of anaerobic digestion tests in fed-batch and semi-continuous mode for the technical and economic feasibility of a full scale digester. *Bioresour. Technol.* **2015**, *182*, 302–313. [CrossRef] [PubMed]
36. Zhang, C.; Su, H.; Tan, T. Batch and semi-continuous anaerobic digestion of food waste in a dual solid-liquid system. *Bioresour. Technol.* **2013**, *145*, 10–16. [CrossRef]

37. Holliger, C.; Fruteau de Laclos, H.; Hack, G. Methane Production of Full-Scale Anaerobic Digestion Plants Calculated from Substrate's Biomethane Potentials Compares Well with the One Measured On-Site. *Front. Energy Res.* **2017**, *5*, 927. [CrossRef]
38. Stiborova, H.; Wolfram, J.; Demnerova, K.; Macek, T.; Uhlik, O. Bacterial community structure in treated sewage sludge with mesophilic and thermophilic anaerobic digestion. *Folia Microbiol.* **2015**, *60*, 531–539. [CrossRef]

processes

MDPI

Article

Batch and Semi-Continuous Anaerobic Digestion of Industrial Solid Citrus Waste for the Production of Bioenergy

Erik Samuel Rosas-Mendoza [1], Andrea Alvarado-Vallejo [2], Norma Alejandra Vallejo-Cantú [2], Raúl Snell-Castro [3], Sergio Martínez-Hernández [4] and Alejandro Alvarado-Lassman [2,*]

1 CONACYT-Tecnológico Nacional de México, Instituto Tecnológico de Orizaba, Av. Oriente 9, 852. Col., Emiliano Zapata, Orizaba C.P. 94320, Mexico; eriksamuelrm@hotmail.com
2 División de Estudios de Posgrado e Investigación, Tecnológico Nacional de México, Instituto Tecnológico de Orizaba, Av. Oriente 9, 852. Col., Emiliano Zapata, Orizaba C.P. 94320, Mexico; alvaradov.andrea@gmail.com (A.A.-V.); nvallejoc@prodigy.net.mx (N.A.V.-C.)
3 Departamento de Ingeniería Química, CUCEI-Universidad de Guadalajara, Blvd. Marcelino García Barragán 1421, Guadalajara C.P. 44430, Mexico; raul.snell@cucei.udg.mx
4 Instituto de Biotecnología y Ecología Aplicada, Universidad Veracruzana, Avenida de las Culturas Veracruzanas 101, Col., Emiliano Zapata, Xalapa C.P. 91090, Mexico; sermartinez@uv.mx
* Correspondence: lassman@prodigy.net.mx

Citation: Rosas-Mendoza, E.S.; Alvarado-Vallejo, A.; Vallejo-Cantú, N.A.; Snell-Castro, R.; Martínez-Hernández, S.; Alvarado-Lassman, A. Batch and Semi-Continuous Anaerobic Digestion of Industrial Solid Citrus Waste for the Production of Bioenergy. *Processes* **2021**, *9*, 648. https://doi.org/10.3390/pr9040648

Academic Editors: Sigrid Kusch-Brandt, Sonia Heaven and Elsayed Elbeshbishy

Received: 19 February 2021
Accepted: 6 April 2021
Published: 8 April 2021

Abstract: The aim of this paper is to describe a study of the anaerobic digestion of industrial citrus solid waste (ISCW) in both batch and semi-continuous modes for the production of bioenergy without the elimination of D-limonene. The study was conducted at the pilot plant level in an anaerobic reactor with a working volume of 220 L under mesophilic conditions of 35 ± 2 °C. Cattle manure (CM) was used as the inoculum. Three batches were studied. The first batch had a CM/ISCW ratio of 90/10, and Batches 2 and 3 had CM/ISCW ratios of 80/20 and 70/30, respectively. In the semi-continuous mode an OLR of approximately 8 g total chemical oxygen demand (COD)/Ld (4.43 gVS/Ld) was used. The results showed that 49%, 44%, and 60% of volatile solids were removed in the batch mode, and 35% was removed in the semi-continuous mode. In the batch mode, 0.322, 0.382, and 0.316 LCH_4 were obtained at $STP/gVS_{removed}$. A total of 24.4 L/d (34% methane) was measured in the semi-continuous mode. Bioenergy potentials of 3.97, 5.66, and 8.79 kWh were obtained for the respective batches, and 0.09 kWh was calculated in the semi-continuous mode. The citrus industry could produce 37 GWh per season. A ton of processed oranges has a bioenergy potential of 162 kWh, which is equivalent to 49 kWh of available electricity ($3.90).

Keywords: anaerobic digestion; industrial solid citrus waste; cattle manure; citrus industry; bioenergy potential

1. Introduction

Citrus belongs to the group of products with the highest world consumption, with oranges being the most consumed. According to the official data, currently around four million hectares are harvested [1], with the world's production of oranges being about 47.5 million tons [2]. In this context, Brazil (15.1%), China (7.3%), the European Union (5.8%), the United States (4.9%), and Mexico (4.4%) are the countries that produce the most oranges [2]. Approximately, 28.7 million tons (60%) are consumed as fresh oranges and 18.8 million tons (40%) are used in the citrus industry for the production of concentrated orange juice, essential oils, marmalades, jellies, potpourris, candied peel, jams, flavoring agents for beverages, health drinks, and essences that are used as food-grade products [3–5]. From the processing of oranges, 40–60% of the total tonnage of oranges is discarded as solid waste [6,7], so, worldwide, the citrus industry generates from 7.5 to 11.3 million tons of industrial solid citrus waste (ISCW) per year. The large amounts of ISCW that are produced and the peculiar characteristics of ISCW involve considerable constraints for their management due to both economic and environmental factors [8]. ISCW is characterized by a high water content (80%); acid pH (i.e., pH values in the range of 3–5) due to the presence of organic acids [9]; total solids (20.17%, wet basis); total mineral solids (0.87%, wet basis); volatile solids (19.31%, wet basis); chemical organic demand (1085 mgO_2/g, dry basis) [10]; and D-limonene, which is a toxic compound [11]. As a matter of fact, traditional citrus peel waste disposal strategies (e.g., incineration and landfilling) currently are insufficient and problematic in terms of environmental impacts and energy efficiency [9]. Some viable alternatives for treating this type of waste are anaerobic digestion and anaerobic co-digestion, which have strong potential benefits to contribute to both pollution control and energy recovery [12,13]. The main advantages of anaerobic digestion and anaerobic co-digestion are (i) environmental friendly solutions compared to other practices, (ii) organic waste with a low nutritional content can be degraded via co-digestion with different substrates in anaerobic reactors, (iii) improved methane yield because of the supply of additional nutrients from the co-digestates, (iv) more efficient use of equipment and cost-sharing by processing multiple waste streams in a single facility, and (v) the process produces biogas with a low cost, and this could be vitally important in meeting our energy needs in the future [14–16].

Therefore, a method to limit the inhibition of the process because it produces toxic compounds, such as D-limonene, consists of co-digesting citrus peel waste with other substrates to dilute the toxic compounds [5]. Anaerobic co-digestion, and also anaerobic digestion, produce a valuable biogas, mainly composed of methane (65–80%) and carbon dioxide (20–35%), and a wet residue (digestate) [17]. Methane has a heating power of 9.94 kWh/m^3 at standard temperature and pressure (STP) [18], but the heating power of biogas varies from 5.2 to 6.2 kWh/m^3 at STP [19]. Biogas has great potential for various applications, such as heating, combined heat and electricity [20], the improvement of the quality of transport fuel, and the replacement of natural gas for various uses.

In the last few years, various scientific articles have been published that focused on the anaerobic digestion and co-digestion of solid waste from oranges, but the studies reported in these articles have required either total or partial elimination of D-limonene to avoid its inhibitory effect during this biological process. Specifically, various substrates and co-substrates have been used for anaerobic digestion and co-digestion, such as biowaste, municipal waste, catering waste, and orange peel waste. As inoculums have been used, e.g., a mixture of sludge, co-digested municipal solid waste and melon residues digestate, and liquid digestate coming from a full-scale plant, among others. Some interesting papers in this area are: (1) Calabrò et al. [21], who indicated that orange peel waste can produce up to 370 $LCH_4/kgVS$ (under normal conditions) in mesophilic conditions and up to 300 $LCH_4/kgVS$ (under normal conditions) in thermophilic conditions, and the presence of increasingly high concentrations of essential oils temporarily inhibits methanogenesis. (2) Ruiz and Flotats [22] indicated that the biochemical methane potential values of the citrus waste that was tested (i.e., orange peel, mandarin peel, mandarin, pulp, and rotten fruit) were 354–398 $LCH_4/kgSV$ and that grinding the orange peel (2.5 g limonene/L) did not influence the potential value of biochemical methane. (3) Anjum et al. [23] studied the synergistic effect of co-digestion to enhance anaerobic degradation of catering waste and orange peel, and their findings indicated that the highest degradation of organic matter (49%) was achieved with co-digestion of catering waste and orange peel at a 50%/50% mixing ratio. (4) Calabró and Panzera [24] performed the anaerobic digestion of ensiled orange peel waste (OPW), and their findings indicated that the highest production was attained for samples of OPW ensiled for 37 days, with a value of 365 normal $mLCH_4/gVS$, and OPW ensiled for 7 days inoculated with sludge already adapted to the substrate yielded 513.7 normal $mLCH_4/gVS$ versus 187.2 normal $mLCH_4/gVS$ of the corresponding test using non-adapted inoculum.

Other works have proposed different strategies to minimize D-limonene's toxic effects on anaerobic digestion, i.e., the combination of recirculation and filtration can be a promising strategy for the anaerobic digestion of citrus waste at high OLRs [25]. Orange peel waste alkaline pretreatment after the addition of a moderate amount of granular activated carbon can render the anaerobic digestion of OPW sustainable as long as the organic loading does not exceed 2 gVS/L and the nutrients are supplemented [26]. Further, the addition and pre-treatment of zero valent iron/granular activated carbon enhance process stability up to a loading of 3 kgVS/m^3 d and increases the production of methane, even at a suboptimal pH [27]. Similarly, two-stage anaerobic digestion systems have replaced the use of other pretreatments and increased the concentration of methane (approximately 60% compared with about 50%) and volume (by 13%) relative to one-stage anaerobic digestion. The accumulated biogas yield was 0.79 L/gSVT and 0.49 L/gSVT for the methanogenic and control reactors, respectively [28].

Cattle manure is used extensively as an inoculum in many studies [29] or as a co-substrate due to its pH characteristics (7.25), chemical oxygen demand (COD) (24.85 mg/L), TS (189.8 mg/L), and VS (34 mg/L) [30]. In addition, cattle manure is used directly as fertilizer in agriculture, but this can cause environmental problems, such as foul odors and contamination of both soil and water [31].

Despite the efforts made by the scientific community in the search for new and better alternatives for the use of solid citrus waste, it remains a challenge to find new ways of treatments that can be implemented on an industrial scale. A practical and affordable way is to take advantage of its physicochemical properties of cattle manure by using it as an inoculum in the anaerobic digestion of these wastes. Thus, the objective of this paper is to report the results of a study of anaerobic digestion for the production of bioenergy in both the batch and semi-continuous modes of ISCW with CM, without the elimination of D-limonene.

2. Materials and Methods

The methods used in this work are explained in detail below.

2.1. Experimental Device

It was constructed an anaerobic reactor (AR) made of fiberglass with a wall thickness of 0.64 cm, a height of 1.04 m, and a total volume of 250 L. The AR had a working volume of 220 L and a 30-L biogas chamber. Fiberglass is known to be a good thermal insulator, to be inert to diverse substances (e.g., the volatile fatty acids produced during the anaerobic digestion), to have resistance to deformation, and to be stable at relatively high temperatures. The valves, tubes, and connectors were made of schedule 80 PVC material that had a diameter of 5.08 cm. As shown in Figure 1, the reactants were placed in the AR. In addition, during the operation of the AR, the mesophilic conditions (35 ± 2 °C) using a heating jacket with a capacity of 25 L capacity was kept with an automatic thermostat and a 600 L/h capacity recirculation submersible pump inside the water reservoir. The inoculum and the substrate were mixed through a recirculation system installed at the bottom of the AR. The recirculation system consisted of a reservoir tank and a Masterflex Cole-Parmer variable-velocity, peristaltic pump. The biogas was collected in an inverted water displacement system, and HCl was used to maintain the pH at approximately 5.5 in order to avoid the dissolution of CO_2 into the water.

2.2. Inoculum

CM was used as the source of inoculum for this study. The CM was obtained from a geomembrane reactor located at a cattle farm in the city of Orizaba in Veracruz, Mexico. After collection, the CM was filtered using a pore size of 2 mm to remove large particles. The CM was characterized according to the parameters shown in Table 1, after which it was used to inoculate the AR.

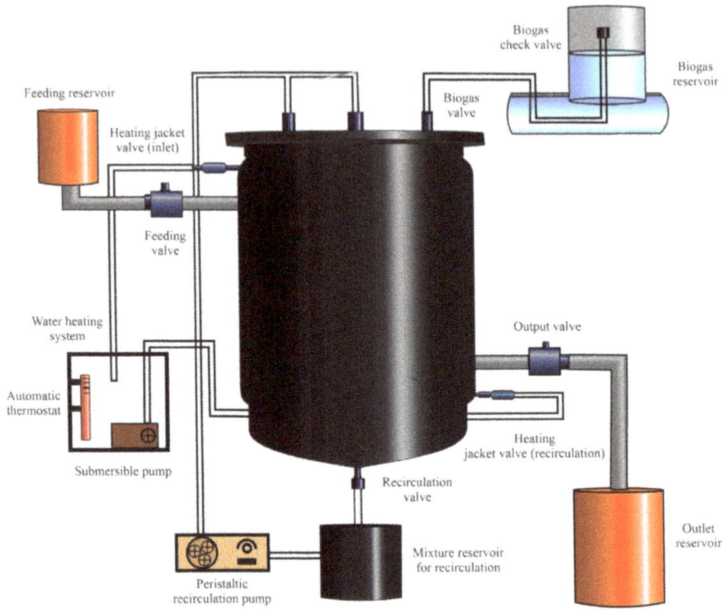

Figure 1. Anaerobic reactor for the digestion of industrial citrus solid waste (ISCW) with cattle manure (CM).

Table 1. Characterization of cattle manure and industrial solid citrus waste.

Parameter	CM		ISCW	
	Mean	**Standard Deviation**	**Mean**	**Standard Deviation**
Total COD (g/L)	7.83	1.45	54.11	1.25
Soluble COD (g/L)	4.22	1.95	32.96	1.27
TS (g/L)	4.32	1.40	61.22	1.33
VS (g/L)	3.59	2.30	57.85	1.24
pH	7.45	0.18	4.46	0.17

2.3. *Substrate*

ISCW from a citrus processing plant in the municipality of Martínez de la Torre, Veracruz, Mexico was used as the substrate. This ISCW was generated at different points in the process of obtaining various products, such as fresh juice, concentrated juice, essential oil, and dry peel. ISCW was mixed before its final disposal, as part of the operating practices into the citrus processing plant. After, the ISCW samples were obtained, they were preserved at 4 °C. Then, the ISCW was cut into small pieces that were approximately 1–1.5 cm in length, and they were grinding to reduce the particle size for later physicochemical characterization. Table 1 shows the main parameters of the characterization of CM and ISCW.

All samples were analyzed in triplicate.

2.4. *Methodology of the Experiment*

The experiment consisted of three stages, as described below:

Stage 1. Inoculation, start-up, and stabilization of the anaerobic reactor with CM

The AR was inoculated with 160 L of fresh CM, and it was operated in a short batch mode over a period of seven days. Then, the start-up was initiated by extracting 140 L of CM, which left 20 L of CM along with a 200 L supply of fresh excreta, which ensured the availability of a working volume of 220 L of CM, and the AR was operated for 30 days. The reactor was stabilized by using two batches, each one for 30 days, and removing 200 L from the 220 L CM and adding 200 L of fresh CM into the reactor. In Stage 1, it was not necessary to regulate the pH, because, as shown in Table 1, the CM presented an almost neutral pH, and it was recirculated at a constant rate under mesophilic conditions at 35 ± 2 °C for a duration of 97 days.

Stage 2. Digestion stage of the ISCW with CM

After the previous stage, the AR was fed with ISCW, thus starting the anaerobic digestion process of the ISCW using CM in batch mode. Due to the low pH values of the ISCW, it was necessary to use a 3 M NaOH solution to increase the values to approximately 7.5 so the anaerobic digestion process could occur. The ISCW were fed at three different ratios because the bacteria inside the reactor had not adapted to the new substrate. Each ratio was evaluated as a batch, which means that one batch one corresponded to the CM/ISCW ratio of 90/10 (*v*/*v*), as shown in Table 2. Each batch was operated for 30 days with a total volume of 220 L of substrates.

Table 2. Anaerobic digestion of CM/ISCW.

Batch	CM/ISCW (%*V*/*V*)	CM Remaining in the Reactor (L)	Fresh ISCW (L)	Batch Duration (d)
1	90 / 10	198	22	30
2	80 / 20	176	44	30
3	70 / 30	154	66	30

Stage 3. Digestion stage in semi-continuous mode

After evaluating the anaerobic digestion process of the ISCW using CM in the batch mode, the adaptation of the inoculum to other conditions was proven. For this, the AR was maintained in recirculation for 15 additional days with the same CM/ISCW mixture used in the last stage, i.e., 70/30. Later, 44 L of the mixture was replaced by 44 L of fresh ISCW, and the remaining 176 L was used as an inoculum since it was perfectly adapted to degrade ISCW. This process was repeated four times, the first time for 5 days and 10 days each for the three remaining times. Thus, this digestion process took place during short periods of time, i.e., at the same conditions as Batch 2 of Stage 2.

To evaluate the operation of the AR in a semi-continuous mode, the CW/ISCW ratio was modified again, i.e., 90/10 was used. This last ratio was equivalent to manage an OLR of 8 $gCOD_T$/Ld, thus the AR was operated during two short periods of 10 days each one.

2.5. Analytical Determinations

The pH was determined by an Orion Model 250 A potentiometer. The total COD (COD_T) and the soluble COD (COD_S) were monitored every 24 h in both operational batches and semi-continuous modes using the colorimetric method [32]. Due to its nature, ISCW cannot be analyzed directly [33], and, for that reason, the particle size was reduced to less than 1 mm in a conventional blender with the aim of homogenizing the samples in order to avoid the obstruction of the measuring instruments. In order to conduct the COD_S, 12 mL of the sample were centrifuged in an HERMLE Z 383 centrifuge at 3500 rpm for 10 min, after which the supernatant was analyzed. The CODs of the digested samples were measured with a HACH spectrophotometer at 620 nm. The total solids (TS) and volatile solids (VS) were determined by the gravimetric method [32].

The composition of the biogas was determined using a Buck 310 gas chromatograph equipped with an All-Tech CRT I capillary column that was 6 inches long and had a diameter of 1/4 inch. The gas chromatograph detected CH_4, CO_2, O_2, and N_2. The 2 mL doses were injected directly into the gas chromatograph, helium at 70 psi was used as the carrier gas, the temperature of the column was 36 °C, and the temperature of the detector was 121 °C.

2.6. Bioenergy Potential of ISCW with CM

Bioenergy potential was calculated from the methane yields obtained during the anaerobic digestion of the industrial solid citrus waste with cattle manure in both batch and semi-continuous mode. The following conditions were assumed:

- ISCW had no physical or chemical pre-treatment to eliminate the D-limonene contained in the waste. This was done to maintain the same conditions that exist with industrial waste.
- In batch mode, three CM/ISCW ratios were considered, i.e., 90/10, 80/20, and 70/30. Each batch had a duration of 30 days.
- In semi-continuous mode, 22 L/day of fresh ISCW were fed, so, OLR 8 g total COD/Ld (4.43 gVS/Ld) was used. The experiment was conducted for a period of 10 days.
- The anaerobic digestion process in both the batch and semi-continuous cases was conducted at the mesophilic condition of 35 ± 2 °C.

The volume of methane generated by the anaerobic digestion (CH_4 AD) was calculated using Equation (1):

$$CH_4\ AD = (V_{AR})(VS_I)(VS_{\%})(Y_{CH4})(1 \times 10^{-5}) \quad (1)$$

where:

V_{AR}, is the volume of the anaerobic reactor (in liters) of the mixture (CM/ISCW) that was used.

VS_I, is the feeding concentration of the mixture (initial VS), expressed in g/L.

$VS_{\%}$, is the removal efficiency of the VS, expressed in %.

Y_{CH4}, is the methane yield obtained from the anaerobic co-digestion process, in LCH_4 at STP/gVS_{rem}.

1×10^{-5} is a conversion factor to express the methane generated by AD in m^3 at standard temperature and pressure (STP). Bioenergy potential was estimated using Equation (2):

$$BEP = (CH_4\ AD)(HP_{CH4}) \quad (2)$$

where:

HP_{CH4}, is the heating power of methane at standard temperature and pressure, 9.94 kWh/m^3 at STP [18].

From the bioenergy potential, the amount of electricity that can be used was calculated using Equation (3) based on the energy conversion efficiency (η) of a commercial generator, i.e., 30% [34], and 1×10^{-2} is a conversion factor that was used to express electricity in kWh. The remaining 70% corresponds to thermal energy.

$$Electricity = (BEP)(\eta)(1 \times 10^{-2}) \quad (3)$$

The cost of electricity tariff (ET) in Mexico is approximately 0.08 USD/kWh [35], so Equation (4) can be used to calculate the cost of electricity provided by the anaerobic digestion process:

$$Cost = (Electricity)(ET) \quad (4)$$

3. Results and Discussion

The results presented in this section are explained according to the research methodology, i.e., Stage 1: inoculation, start-up, and stabilization of the anaerobic reactor with CM; Stage 2: digestion stage of the ISCW with CM; and Stage 3: the digestion stage in the semi-continuous mode. In addition, an estimate of the bioenergy potential from anaerobic digestion process is presented. The findings of each stage are described below:

3.1. Inoculation, Start-Up, and Stabilization of the Anaerobic Reactor with CM

In this stage, only cattle manure was used, i.e., for inoculation (7 days), start-up (30 days), and stabilization of the anaerobic reactor (two batches at 30 days per batch).

During the inoculation, the pH values remained stable, ranging between 6.95 and 7.81, while the total COD decreased from 7.5 to 3.6 g/L, and the soluble COD decreased from 4.7 to 2.5 g/L. The initial TS value of 5.14 g/L decreased to a final level of 3.7 g/L, and the values of VS were 3.7 at the beginning and 1.6 g/L at the final of the inoculation. Subsequently, during the start-up, the pH at the inlet was 7.69, and it was 6.71 at the outlet. The initial and final values of the total COD were 6.5 and 1.45 g/L, respectively. Likewise, there was a decrease in the soluble COD from 2.97 to 0.84 g/L, and the TS and VS values varied from 4.36 to 2.76 g/L and from 2.74 to 1.11 g/L, respectively. Finally, in the stabilization, the pH was equal to the value in the previous phases, i.e., very close to 7. The CM used to feed the AR maintained an average total COD of 7.35 g/L and an average soluble COD of 3.8 g/L. Each batch presented 79 and 88% of total COD removal, with a similar tendency for the removals of soluble COD, TS, and VS. The accumulated biogas was quantified as approximately 72.6, 659, 647, and 741 L in the inoculation, start-up, and stabilization phases (two batches), respectively. Baek et al. [36] found that the biochemical methane potential of the cattle manure though mono-digestion can produce around 109.2 L/kgVS. These same researchers concluded that co-digesting cattle manure with food waste and pig manure proved effective in accelerating the initiation of anaerobic digestion, which suggested that the co-digestion strategy could be applied to promote the start-up of a digester to treat cattle manure. Usually manure is considered to be an output product in livestock systems, which leads to the idea that it is simply residual; however, manure should be considered to be a valuable product because of its nutrients and biogas potential [37]. Alatriste-Mondragón et al. [13], mention that the main wastes most used in co-digestion processes are municipal wastewater sludge, the organic fraction of municipal solid waste and cattle manure. Cattle dung proved to be beneficial to achieve enhanced biogas production with supplementation of four residues: compost, landfill waste, paddy soil and kitchen waste [38]. Silva and Abud [39], evaluated the use of bovine manure as inoculum in the vinasse biodigestion process, using 0.5, 3.0 and 5.5% of manure.

3.2. Digestion Stage of the ISCW with CM

After conditioning the inoculum during Stage 1, the anaerobic reactor was operated in three batches at the conditions shown in Table 2, i.e., using the ISCW as the substrate and the CM as the inoculum.

Regarding the monitoring of the pH, the three batches were fed so as to regulate the pH of their respective CM/ISCW relationships with values close to 7.5. During the 30 days of operation, the pH remained relatively stable, finally reaching values below 7, as shown in Figure 2. However, in Batch 3 during the first six days, a decrease in pH was observed, and it reached 5, which may have been due to the bacterial medium's being affected by the increased ISCW feed. In order to avoid greater levels of acidification, an extra addition of NaOH solution was required, which successfully stabilized the pH, as had been the case in the previous batches. Continuing with the analysis of Batch 3, despite having suffered a significant initial change in pH, the bacteria regained their activity, and the anaerobic digestion process was not inhibited. In anaerobic digestion or co-digestion, pH is an important operating parameter, and, in this experiment, the pH oscillated between values that were close to 7, i.e., in a range of 6.80 to 7.58. Ward et al. [40], found that the ideal range of pH for anaerobic digestion was 6.8−7.2. Lee et al. [41] reported that methanogenesis occurs efficiently in an anaerobic reactor at pH values between 6.5 and 8.2. Marín-Peña et al. [42] also reported that pH values in the range of 7.0–7.5 favor the methanogenic stage of anaerobic digestion. In addition, during the anaerobic co-digestion of 50% orange peel with 50% catering waste, Anjum et al. [23] observed that the pH was in the range of 6.38–7.01 between days 39 and 42, which is the optimum range for methanogenesis.

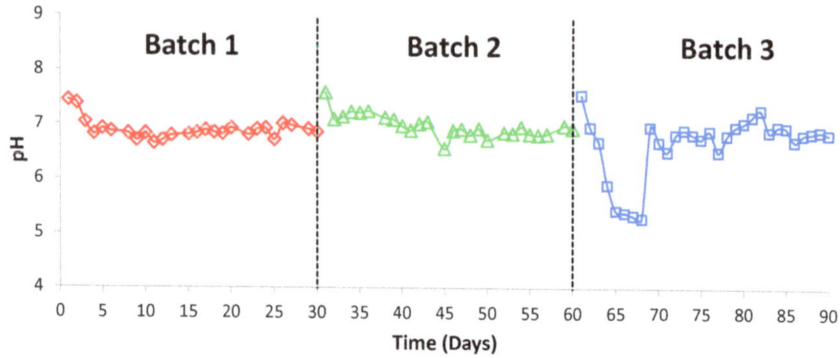

Figure 2. Values of pH during the digestion of SCW with CM.

Total COD and soluble COD varied from batch to batch depending on the amount of ISCW that was added. Thus, as shown in Figure 3, the total COD values at the beginning of the three batches were 19.78, 25.18, and 37.34 g/L, respectively. After 30 days of operation, the total COD removals for the three batches were 58%, 50%, and 62%, respectively. Similarly, 60%, 57%, and 35% of the soluble COD were removed from Batches 1, 2, and 3, respectively. Figure 4 shows that, as the CM/ISCW ratio increases, a wider relative difference exists between the total COD and the soluble COD. Anjum et al. [23] noted that, after 42 days of the digestion process, the highest decrease in insoluble COD, i.e., from 30,080 to 14,720 mg/L, was observed at the 50% orange peel ratio with catering waste. This means that 51% of insoluble COD was transformed into soluble COD. Comparing this findings with the results of the present work, the digestion process of ISCW using CM is adequate.

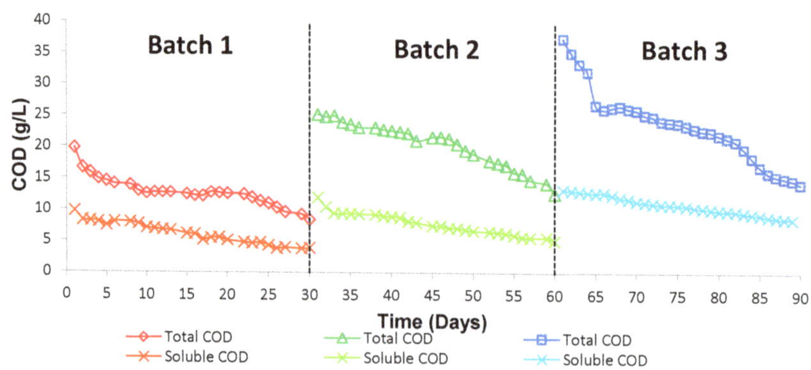

Figure 3. Behavior of the total chemical oxygen demand (COD) and soluble COD.

However, the addition of ISCW resulted in increases in the volatile solids concentrations and the total concentrations of each batch. Figure 4 shows that the TS values at the beginning of each batch were 15.17, 19.66, and 27.75 g/L, respectively, and they had removal levels of 53, 49, and 58%, respectively. As expected, the VS levels followed a trend similar to the trend of the TS levels, with removal percentages of 49, 44, and 60%, respectively. In considering other types of substrates, Li et al. [43] reported that the initial concentration of the substrate influences the mesophilic anaerobic digestion of solid, organic, municipal waste, as was the case for Batches 1, 2, and 3 in terms of both COD and solids. Thus, comparing these findings with the literature, Anjum et al. [23], found that the co-digestion of catering waste and orange peel at a 50% ratio presented 66% of the organic matter removal efficiency (volatile solids) and 55% of total solids, but the experiment was conducted only for a period of 80 days.

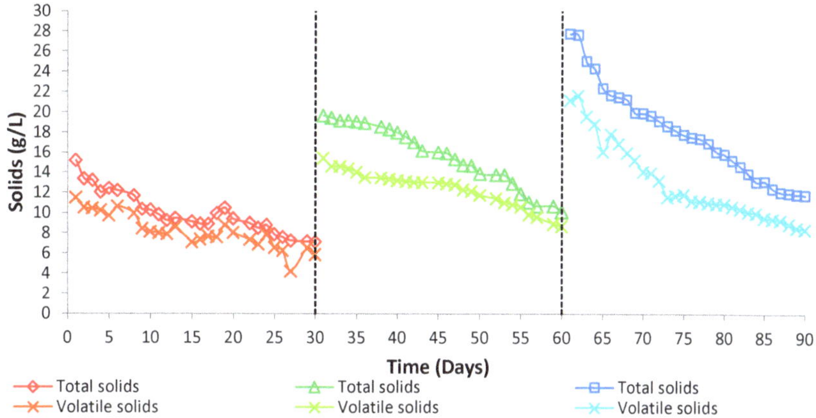

Figure 4. Degradation profiles for total solids (TS) and volatile solids (VS) during digestion in the batch mode.

While the bacteria were adapting to the ISCW that was added, Batch 1 showed low levels of biogas production during the initial days of the experiment. Then, an increase was observed until Day 8, and, later, 29 L was reached at 18 days. Batch 2 reacted similarly, but it had higher biogas production during the first days of operation than the previous batch, which was attributed to the increase of the concentration of ISCW. The highest production of 28 L was reached on day 43. However, Batch 3 was affected from 61 to 71 days due to the new increase in the concentration of ISCW. However, Figure 5 shows that, from day 73 forward, this batch showed a very clear recovery, and it reached its highest point of 28 L on day 80. The biogas that accumulated in the three batches was 550, 606, and 467 L, respectively. The highest quantity of biogas was accumulated in Batch 2, followed by Batches 1 and 3. The maximum methane yield for the three batches were 0.305, 0.337, and 0.331 LCH_4 at STP/g total COD_{rem}, respectively. These same methane yield expressed in terms of volatile solids for Batches 1, 2, and 3 presented averages of 0.322, 0.382, and 0.316 LCH_4 at STP/gVS_{rem}, respectively. Despite the elevated concentrations of organic material, the bacterial medium was capable of adapting and degrading this type of waste, thereby generating high methane yields. SRTs of 300, 150, 100 days for Batches 1, 2 and 3, respectively; were calculated.

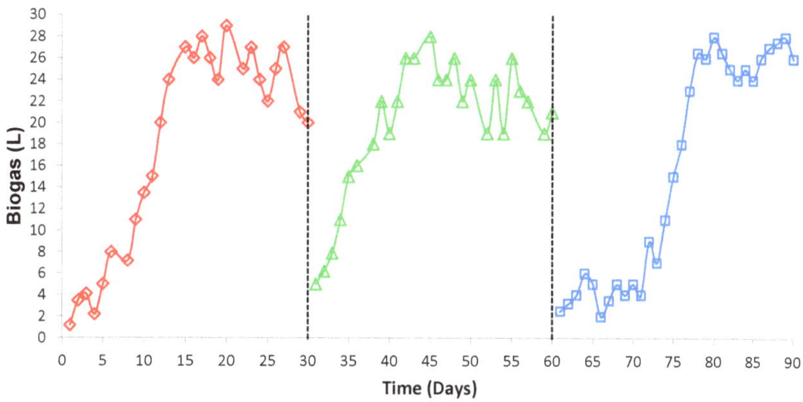

Figure 5. Biogas production in the digestion stage in batch mode.

Calabrò et al. [21] reported a methane yield of about 0.396 LCH_4/gVS (under normal conditions) in 30 days, utilizing a co-digestion process on orange peel waste (50%) with biowaste (50%), but they extracted a large amount of the D-limonene present in the fresh residue. This yield was higher than that obtained in this work, but the difference is due to the proportions of the citrus waste that were used and the fact that the remaining D limonene was eliminated during the pretreatment. Similarly, Ruiz and Flotats [22], found that the biochemical methane potential obtained for orange peel samples was an average 356 $LCH_4/kgVS$ over a period of 28 days. Finally, the concentration of essential oils, in orange peel waste, is 5.4 g/kg, being 90–98% D-limonene [44]. The concentrations of D-limonene above 200 mg/kg has an inhibitory effect on anaerobic digestion of citrus peel [22]. The inhibitory effect of essential oils (up to 2 g/L) on anaerobic digestion of orange peel waste under mesophilic conditions results in a methane yields up to 370 LCH_4 at STP/kgVS [21].

Table 3 presents a summary of the results of the anaerobic digestion process compared with common studies.

Table 3. Summary of the current findings with other common studies.

Parameter	Current Findings			Reference					
	Batch 1	Batch 2	Batch 3	[10]	[21]	[22]	[23]	[24]	[28]
Inoculum	Cattle manure	Cattle manure	Cattle manure	Granular sludge	Mixture of sludges	Digestate from cow manure digesters	Co-digested municipal solid waste and melon residues digestate	Liquid digestate	Mesophilic anaerobic sludge
Substrate	ISCW	ISCW	ISCW	Orange peel waste	Orange peel waste	Citrus waste of different origins	Catering waste	Industrial orange peel waste	Industrial orange peel waste
Co-substrate	-	-	-	-	Biowaste	-	Orange peel	-	-
Stages	Single-stage	Single-stage	Single-stage	Single-stage	Single-stage	Single-stage	Single-stage	Single-stage	Two-stage
Ratio	90/10	80/20	70/30	-	50/50	2.6	50/50	0.3	Orange peel/Inoculum/Water 35/26/39
Reactor volume (L)	250	250	250	3.5	0.5	2	0.5	1.1	4.3
Period (d)	30	30	30	5	30	28	42	37	25.8
Temperature (°C)	35 ± 2	35 ± 2	35 ± 2	37	35 ± 0.5	38	30 ± 1	35 ± 0.5	35
pH	6.86–7.45	6.98–7.58	6.84–7.50	6.70–8.60	-	-	6.38–7.01	7.58–7.65	7.00–8.00
COD_T removed (%)	58	50	62	84–90	-	75	49	-	-
COD_S removed (%)	60	57	35	-	-	77	51	-	-
TS removed (%)	53	49	58	-	-	-	55	-	-
VS removed (%)	49	44	60	-	-	11	66	-	-
Y_{CH4} (LCH_4/gVS)	0.322	0.382	0.316	0.230	0.396	0.354–0.398	1.06 L/dt_{subs}	0.365	0.79

3.3. Digestion Stage in Semi-Continuous Mode

After the completion of the digestion in batches over a period of 30 days, it was necessary to verify the performance of the inoculum and evaluate the digestion of the waste in a shorter operating period. The reactor was operated with an 80/20 ratio, but this was done according to the conditions established in Table 4.

Table 4 shows that the percentage of total COD and VS removed from short Batch 1 was low, and this was due to the short time the bacteria had to degrade the organic material. Thus, it was necessary to increase the degradation time for the subsequent short batches, in which there was a significant increase in the removal percentages, with similar behavior observed for the degradation in the other parameters, i.e., total COD and VS, in short Batches 2 and 3. However, removal efficiencies decreased drastically for short Batch 4, which probably was due to the fact that there was a short time period in which to digest the CM/SCW ratio of 80/20. At standard temperature and pressure, the maximum methane yields for each short batch were 0.103, 0.164, 0.117, and 0.134 LCH_4 per gram of total COD_{rem}, respectively. However, the average methane yields for each batch, according to the VS readings, were 0.052, 0.104, 0.074. and 0.297 LCH_4 at STP/gVS_{rem}, in which it was observed that the methane yields for the last batch were high compared to the low percentages for the removal of total COD and VS. STRs from 25 to 50 days were calculated.

Table 4. Evaluation of digestion over short periods of time.

Short Batch	Operation (Days)	ISCW Fed to the Anaerobic Reactor (L)	Initial Total COD (g/L)	Removed Total COD (%)	Initial VS (g/L)	Removed SV (%)
1	5	44	13.58	30	11.98	48
2	10	44	12.17	47	10.46	67
3	10	44	14.25	40	12.96	67
4	10	44	17.86	12	10.01	16

Once the evaluation of the short batches was completed, the operation of the reactor in semi-continuous feeding mode began with a daily 22-L dose of fresh ISCW in order to keep the organic loading rate (OLR) at approximately 8 g total COD/Ld (4.43 gVS/Ld). During the adaptation phase, an average of 8.03 g/L of total COD was used, while, during the stabilization phase 8.35 g/L of total COD was used.

Figure 6 shows a similar behavior in the removal of total COD and VS in the adaptation and stabilization phases, presenting average values of 29% for total COD in the adaptation phase and 24% in the stabilization phase. For the removal of VS, the adaptation phase presented 25%, while the stabilization phase presented 35%. The generation of biogas for the adaptation and stabilization phases were, on average, 21.6 L/d and 24.4 L/d, respectively, with about 34% methane.

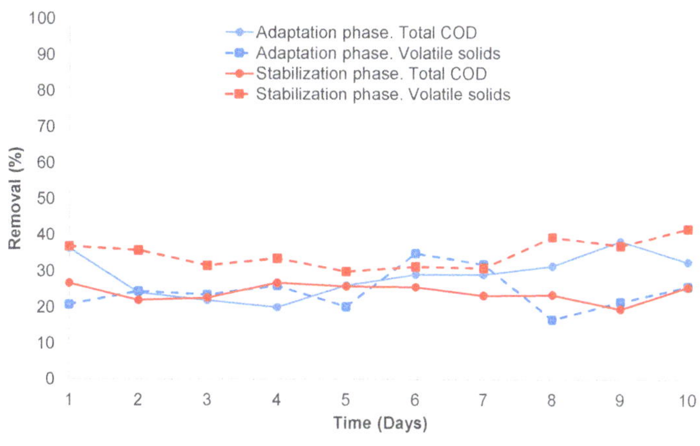

Figure 6. ISCW feeding in semi-continuous mode in both the adaptation and stabilization phases: percent of total COD and VS removed.

Operating in the semi-continuous mode, an SRT of 10 days was obtained, which is in accordance with those described by Martín et al. [10], who found an 84–90% level for the COD removal at an OLR of 1.20–3.67 gCOD/Ld. When they increased the load to 4 gCOD/Ld, they observed a strong inhibition of COD. However, the methane yield was 0.27–0.29 LCH_4 at STP/$gCOD_{add}$. In contrast with this study, Martín et al. [10] conducted their experiments at thermophilic conditions with an SRT of 25 days.

While the removal of total COD and VS, as well as the methane yield, were lower in the semi-continuous mode than in the batch mode, the anaerobic reactor can treat up to 220 L of ISCW in 10 days, whereas only 22 L were fed during 30 days in the batch mode.

This finding highlights the adaptation of a specialized inoculum for the treatment of greater volumes of ISCW in shorter periods.

3.4. Bioenergy Potential from Anaerobic Digestion

Equation (1) was used to calculate the volume of methane generated by anaerobic digestion. For Batches 1, 2, and 3, 220 L of the CM/ISCW mixture were fed with 90/10, 80/20, and 70/30 ratios, respectively. The initial concentrations of total VS for the three batches were 11.52, 15.41, and 21.19 g/L, respectively. The percentages of VS removed, which were obtained experimentally after 30 days of operation for each batch, were 49%, 44%, and 60%, and the maximum methane yields were 0.322, 0.382, and 0.316 LCH_4 at STP/gVS removed. Thus, the methane generated by AD for each batch was 0.40, 0.57, and 0.88 m^3 at STP, respectively. In the semi-continuous mode, an average of 8.35 g/L of total COD was fed, and it was equivalent to 4.63 g/L of VS, reaching 35% of VS removed, with 0.024 LCH_4 at STP/gVS removed, all during a period of 10 days. The methane generated by AD in semi-continuous mode was approximately 0.01 m^3 at STP.

Bioenergy potential was estimated using Equation (2) for Batches 1, 2, and 3, and the values of 3.97, 5.66, and 8.79 kWh were obtained, respectively, during 30 days for each batch. To conduct the experiments in batch mode, it was necessary to supply fresh ISCW, and, for Batches 1, 2, and 3, approximately 10, 20, and 30 kg of fresh ISCW, respectively, were supplied. Therefore, the specific bioenergy potential from a kg of fresh ISCW was 0.40, 0.28, and 0.29 kWh/kg fresh ISCW for each batch. The bioenergy potential in the semi-continuous mode was calculated as 0.09 kWh during a period of 10 days, as shown in Figure 7. It is important to mention that, although in semi-continuous mode a much lower bioenergy potential was obtained compared to batch mode, the retention time to treat 220 L of ISCW was 10 days, which is equivalent to 22 L of ISCW/day, and to treat 22 L of ISCW in batch mode (Batch 1), 30 days were required.

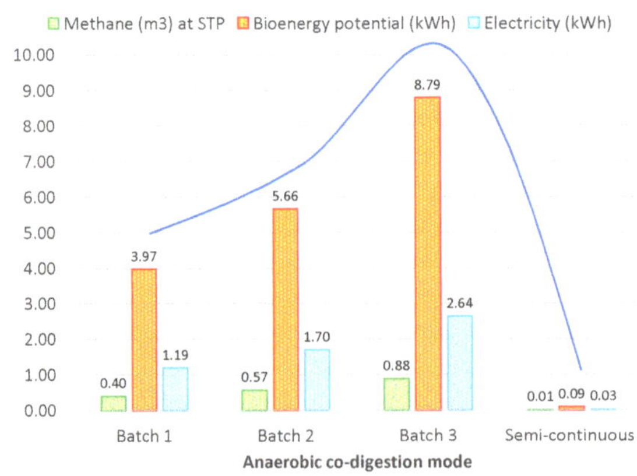

Figure 7. Bioenergy potential derived from anaerobic digestion ISCW with cattle manure (batch and semi-continuous modes).

From the bioenergy potential, using Equation (3) the amount of electricity that can be available for use was obtained, and it was 1.19, 1.70, and 2.64 kWh for Batches 1, 2, and 3, respectively, while 0.03 kWh was obtained for the semi-continuous mode. The thermal energy values derived from the bioenergy potentials were 2.78, 3.96, and 6.15 kWh for the aforementioned batches, and it was 0.06 kWh for the semi-continuous mode. The costs of consuming these amounts of electricity were estimated by Equation (4), and they were 0.10, 0.14, and 0.421 USD for each respective batch and 0.002 USD for the semi-continuous mode.

The citrus processing plant mentioned previously was used as a case study, and it had a processing capacity per season (124 days) of 230,000 tons of oranges, which is equivalent to about 1850 tons per day. From each ton of oranges processed, 0.5 to 0.6 tons of ISCW were generated, which is equivalent to about 925 tons of ISCW per day and 115,000 tons of ISCW per season. Using the results obtained from the bioenergy potential in batch mode, it can be assumed that the citrus industry could produce 46, 33, and 34 GWh per season using Batches 1, 2, and 3, respectively, resulting in an average of 37 GWh per season. In terms of available electricity, there would be 14, 10, and 10 GWh for each batch, respectively, and, on average, 11 GWh per season, which is equivalent to 1.1, 0.8, and 0.8 million USD per season, with an average of 0.9 million USD per season. As a consequence, one ton of ISCW has a bioenergy potential of approximately 324.5 kWh, corresponding to 97 kWh of electricity (7.8 USD). In other words, a ton of processed oranges has a bioenergy potential of 162 kWh and could provide 49 kWh of electricity equivalent to 3.9 USD. Koppar and Pullammanappallil [45] conducted an analysis of the bioenergy potential using 270 wet tons/day of orange peel waste, and they found that 106 GWh/year could be obtained, which would be equivalent to 78 GWh in a season of 270 days, as was analyzed in their case study. With a 25% conversion efficiency, 26 GWh/year of electricity could be obtained, i.e., 19.5 GWh in a season of 270 days.

For the calculation of the bioenergy potential on an industrial scale, it is possible to consider the use of a single reactor and the information obtained from the anaerobic digestion of the waste, however; this would be a hypothetical case, since the reactor would have to be of considerable volume. A practical alternative to solve this could be through the implementation of modular reactors, so that when the amount of solid waste produced by the citrus industry increases or decreases, modular units would be available to cover both treatment and energy demands. On the other hand, as previously mentioned, this work was focused on the treatment of solid orange waste, whose season is 124 days, i.e., from January to April. The rest of the year the citrus plant continues processing lemons (May to October), grapefruit (September to December) and tangerine (December). As can be seen, there are solid citrus wastes throughout the year, which have similar characteristics, so there would be raw material available for the anaerobic digestion process, in the case of an annual operation.

4. Conclusions

A study was conducted of the anaerobic digestion of citrus solid waste in both the batch mode and the semi-continuous mode without the elimination of D-limonene. Cattle manure was used as the inoculum, and once stabilized through two batches of 30 days each, it reached values in the range of 79–88% removal of COD, with similar values of other parameters, such as soluble COD, total solids, and volatile solids with the pH maintained at about 6.71. The accumulated biogas was quantified as being in the range of 647 to 741 L.

In batch mode, industrial solid citrus waste was treated by means of anaerobic digestion, with the previously conditioned inoculum (CM), using ratios CM/ISCW 90/10 (Batch 1), 80/20 (Batch 2), and 70/30 (Batch 3) with a retention time of 30 days for each batch. The values of pH were around 7, the total COD removals were 58%, 50%, and 62%, the soluble COD removals were 60%, 57%, and 35%, the total solids removals were 53%, 49%, and 58%, and the volatile solids removals were 49%, 44%, and 60%. The maximum methane yields were 0.305, 0.337, and 0.331 LCH_4 at STP/g of total COD removed, which was equivalent to 0.322, 0.382, and 0.316 LCH_4 at STP/gVS removed. All of these values are for each respective batch.

The same inoculum that was used in batch mode also was used in the semi-continuous mode to treat 22 L of ISCW per day, equivalent to an OLR of around 8 g total COD/Ld. The semi-continuous mode was proved over a 10-day period. The total COD removal was 24%, and the volatile solids removal was 35% on average, while maintaining a pH value around 7. The biogas generation that was obtained was 24.4 L/d, which was about 34% methane. The semi-continuous mode presented disadvantages compared to the batch mode, but the 22 L of ISCW that were digested in Batch 1 in semi-continuous mode during a 30-day period were treated in only one day.

Bioenergy potentials of 3.97, 5.66, and 8.79 kWh were obtained for Batches 1, 2, and 3, respectively, and 0.09 kWh was calculated in the semi-continuous mode. With these findings, it can be assumed that the citrus industry could produce 46, 33, and 34 GWh per season using Batches 1, 2, and 3, respectively, with an average of 37 GWh per season. This implies that, at an industrial level, it is feasible to operate the anaerobic digestion process, so, a ton of processed oranges has a bioenergy potential of 162 kWh and could provide 49 kWh of electricity equivalent to 3.9 USD.As a part of our future work, a general analysis of the anaerobic digestion process will be conducted using both solid and liquid waste (effluents) from the citrus industry, with attention to priority areas, such as energy and the environment. In addition, the biodegradation of the D-limonene concentration inside the reactor will be monitored in order to study in depth the inhibitory effects and its effect on biogas production.

Author Contributions: Conceptualization, E.S.R.-M. and A.A.-L.; methodology, E.S.R.-M., R.S.-C. and S.M.-H.; validation, A.A.-L.; formal analysis, R.S.-C. and S.M.-H.; investigation, E.S.R.-M. and A.A.-V.; resources, N.A.V.-C. and A.A.-V.; writing—original draft preparation, E.S.R.-M. and A.A.-V.; writing—review and editing, A.A.-L. and N.A.V.-C.; visualization, R.S.-C. and S.M.-H.; supervision, A.A.-L.; project administration, N.A.V.-C. All authors have read and agreed to the published version of the manuscript.

Funding: This research received no external funding.

Acknowledgments: Authors acknowledge the support provided by the Tecnológico Nacional de México (TecNM) and the Consejo Nacional de Ciencia y Tecnología (CONACYT).

Conflicts of Interest: The authors declare no conflict of interest.

Abbreviations

AD	Anaerobic digestion
AR	Anaerobic reactor
BEP	Bioenergy potential
CH_4 AD	Volume of methane generated by the anaerobic digestion
CM	Cattle manure
CO_2	Carbon dioxide
COD	Chemical oxygen demand
COD_{add}	Chemical oxygen demand added
COD_{rem}	Chemical oxygen demand removal
COD_S	Soluble chemical oxygen demand
COD_T	Total chemical oxygen demand
d	Day
ET	Electricity tariff
g	Grams
GWh	Gigawatt-hour
HCl	Hydrochloric acid
HP_{CH4}	Heating power of methane
ISCW	Industrial solid citrus waste
kg	Kilogram
kWh	Kilowatt-hour
L	Liter
LCH_4	Liters of methane
m^3	Cubic meter
η	Energy conversion efficiency
N_2	Molecular nitrogen
O_2	Molecular oxygen
OLR	Organic loading rate
OPW	Orange peel waste
pH	Potential of hydrogen
psi	Pound per square inch

PVC	Polyvinyl chloride
SRT	Solid retention time
STP	Standard temperature and pressure
TS	Total solids
USD ($)	United States dollars
V_{AR}	Volume of the anaerobic reactor
VS_I	Initial concentration of the VS
$VS_{\%}$	Removal efficiency of the VS
VS	Volatile solids
Y_{CH4}	Methane yield

References

1. FAOSTAT. Available online: http://www.fao.org/faostat/es/#compare (accessed on 11 January 2021).
2. USDA. Available online: https://apps.fas.usda.gov/psdonline/circulars/citrus.pdf (accessed on 12 January 2021).
3. Kimball, D.A. *Citrus Processing: A Complete Guide*, 2nd ed.; Aspen Publishers Inc.: Gaithersburg, MD, USA, 1999; ISBN 0834212587.
4. Sharma, K.; Mahato, N.; Cho, M.H. Converting citrus wastes into value-added products: Economic and environmently friendly approaches. *Nutrition* **2017**, *34*, 29–46. [CrossRef]
5. Sharma, K.; Mahato, N.; Lee, Y.R. Extraction, characterization and biological activity of citrus flavonoids. *Rev. Chem. Eng.* **2019**, *35*, 265–284. [CrossRef]
6. Rosas-Mendoza, E.S.; Vallejo-Cantú, N.A.; Méndez-Contreras, J.M.; Aguilar-Lasserre, A.A.; Alvarado-Lassman, A. Evaluation of bioenergy potential from citrus effluents through anaerobic digestion. *J. Clean. Prod.* **2020**, *254*, 120128. [CrossRef]
7. Wilkins, M.R.; Widmer, W.W.; Grohmann, K.; Cameron, R.G. Hydrolysis of grapefruit peel waste with cellulase and pectinase enzymes. *Bioresour. Technol.* **2007**, *98*, 1596–1601. [CrossRef]
8. Zema, D.A.; Calabrò, P.S.; Folino, A.; Tamburino, V.; Zappia, G.; Zimbone, S.M. Valorisation of citrus processing waste: A review. *Waste Manag.* **2018**, *80*, 252–273. [CrossRef] [PubMed]
9. Satari, B.; Karimi, K. Citrus processing wastes: Environmental impacts, recent advances, and future perspectives in total valorization. *Resour. Conserv. Recycl.* **2018**, *129*, 153–167. [CrossRef]
10. Martín, M.A.; Siles, J.A.; Chica, A.F.; Martín, A. Biomethanization of orange peel waste. *Bioresour. Technol.* **2010**, *101*, 8993–8999. [CrossRef]
11. García-Gonzalo, D.; Espina, L.; Gelaw, T.; De Lamo-Castellvi, S.; Pagán, R. Mechanism of bacterial inactivation by (+)-limonene and its potential use in food preservation combined processes. *PLoS ONE* **2013**, *8*, e56769. [CrossRef]
12. Chow, W.L.; Chong, S.; Lim, J.W.; Chan, Y.J.; Chong, M.F.; Tiong, T.J.; Chin, J.K.; Pan, G.-T. Anaerobic Co-Digestion of Wastewater Sludge: A Review of Potential Co-Substrates and Operating Factors for Improved Methane Yield. *Processes* **2020**, *8*, 39. [CrossRef]
13. De Gioannis, G.; Muntoni, A.; Polettini, A.; Pomi, R.; Spiga, D. Energy recovery from one- and two-stage anaerobic digestion of food waste. *Waste Manag.* **2017**, *68*, 595–602. [CrossRef] [PubMed]
14. Carstensen, F.V.; Graziano, M.; Vorotytseva, N.; Waite, W.E.; Parr, K.E. Economic Analysis of Bioenergy: An Integrated Multidisciplinary Approach. In *New and Future Developments in Catalysis*; Steven, L., Suib, S.L., Eds.; Elsevier: Amsterdam, The Netherlands, 2013; pp. 297–323. [CrossRef]
15. Alatriste-Mondragón, F.; Samar, P.; Cox, H.H.; Ahring, B.K.; Iranpour, R. Anaerobic codigestion of municipal, farm, and industrial organic wastes: A survey of recent literature. *Water Environ. Res.* **2006**, *78*, 607–636. [CrossRef]
16. Khalid, A.; Arshad, M.; Anjum, M.; Mahmood, T.; Dawson, L. The anaerobic digestion of solid organic waste. *Waste Manag.* **2011**, *31*, 1737–1744. [CrossRef] [PubMed]
17. Siciliano, A.; Limonti, C.; Curcio, G.M.; Calabrò, V. Biogas Generation through Anaerobic Digestion of Compost Leachate in Semi-Continuous Completely Stirred Tank Reactors. *Processes* **2019**, *7*, 635. [CrossRef]
18. Muylaert, M.S.; Sala, J.; Freitas, M.A.V. *Consumo de Energia e Aquecimento do Planeta: Analise do Mecanismo de Desenvolvimento Limpo (MDL) do Protocolo de Quioto*; Case Studies; Post-Graduate Engineering Programs Coordination (COPPE): Rio de Janeiro, Brazil, 2000.
19. Gerardi, M. *The Microbiology of Anaerobic Digesters*; John Wiley & Sons: Hoboken, NJ, USA, 2003; p. 51. ISBN 0-471-20693-8.
20. Prussi, M.; Padella, M.; Conton, M.; Postma, E.D.; Lonza, L. Review of technologies for biomethane production and assessment of Eu transport share in 2030. *J. Clean. Prod.* **2019**, *222*, 565–572. [CrossRef]
21. Calabrò, P.S.; Pontoni, L.; Porqueddu, I.; Greco, R.; Pirozzi, F.; Malpei, F. Effect of the concentration of essential oil on orange peel waste biomethanization: Preliminary batch results. *Waste Manag.* **2016**, *48*, 440–447. [CrossRef]
22. Ruiz, B.; Flotats, X. Effect of limonene on batch anaerobic digestion of citrus peel waste. *Biochem. Eng. J.* **2016**, *109*, 9–18. [CrossRef]
23. Anjum, M.; Khalid, A.; Qadeer, S.; Miandad, R. Synergistic effect of co-digestion to enhance anaerobic degradation of catering waste and orange peel for biogas production. *Waste Manag. Res.* **2017**, *35*, 967–977. [CrossRef]
24. Calabró, P.; Panzera, M. Anaerobic digestion of ensiled orange peel waste: Preliminary batch results. *Therm. Sci. Eng. Progr.* **2018**, *6*, 355–360. [CrossRef]
25. Lukitawesa; Wikandari, R.; Millati, R.; Taherzadeh, M.J.; Niklasson, C. Effect of effluent recirculation on biogas production using two-stage anaerobic digestion of citrus waste. *Molecules* **2018**, *23*, 3380. [CrossRef] [PubMed]

26. Calabrò, P.S.; Fazzino, F.; Folino, A.; Paone, E.; Komilis, D. Semi-Continuous Anaerobic Digestion of Orange Peel Waste: Effect of Activated Carbon Addition and Alkaline Pretreatment on the Process. *Sustainability* **2019**, *11*, 3386. [CrossRef]
27. Calabrò, P.S.; Fazzino, F.; Folino, A.; Scibetta, S.; Sidari, R. Improvement of semi-continuous anaerobic digestion of pre-treated orange peel waste by the combined use of zero valent iron and granular activated carbon. *Biomass Bioenergy* **2019**, *129*. [CrossRef]
28. Jiménez Castro, M.P.; Buller, L.S.; Zoffreo, A.; Timko, M.; Forster-Carneiro, T. Two-stage anaerobic digestion of orange peel without pre-treatment: Experimental evaluation and application to São Paulo state. *J. Environ. Chem. Eng.* **2020**, *8*. [CrossRef]
29. Yao, Y.; Wei, Y.; An, L.; Zhou, J. Effect of inoculum on anaerobic co-digestion of vegetable processing wastes and cattle manure at high solids concentration. *Waste Biomass Valori.* **2018**, *9*, 2091. [CrossRef]
30. Nasir, I.M.; Ghazi, T.I.M.; Omar, R.; Idris, A. Batch and semi-continuous biogas production from cattle manure. *Int. J. Eng. Technol.* **2013**, *10*, 16–21.
31. Ferreira, J.G.; Mendes de, E.; Oliveira de Araújo, V.; val, M.; Lopes, H.; Silva Melgaço, H.; Oliveira, Y.A.; Rosado Fernandes, A. Optimum Co-Digestion Ratio of Cattle Manure and Manipueira in a Single-Stage Anaerobic Digester for Biogas Production. *Clean Soil Air Water* **2020**, *48*, 2000096. [CrossRef]
32. APHA. *Standard Methods for the Examination of Water and Wastewater*, 21st ed.; American Public Health Association/American Water Works Association/Water Environment Federation: Washington, DC, USA, 2005.
33. Buffiere, P.; Frederic, S.; Marty, B.; Delgenes, J. A comprehensive method for organic matter characterization in solid wastes in view of assessing their anaerobic biodegradability. *Water Sci. Technol.* **2008**, *58*, 1783–1788. [CrossRef]
34. Rosas-Mendoza, E.S.; Méndez-Contreras, J.M.; Martínez-Sibaja, A.; Vallejo-Cantú, N.A.; Alvarado-Lassman, A. Anaerobic digestion of citrus industry effluents using an Anaerobic Hybrid Reactor. *Clean Technol. Environ.* **2018**, *20*, 1387–1397. [CrossRef]
35. CFE. Available online: https://app.cfe.mx/Aplicaciones/CCFE/Tarifas/Tarifas/Tarifas_industria.asp?Tarifa%C2%BCCMAMF&Anio%C2%BC2021 (accessed on 3 February 2021).
36. Baek, G.; Kim, D.; Kim, J.; Kim, H.; Lee, C. Treatment of Cattle Manure by Anaerobic Co-Digestion with Food Waste and Pig Manure: Methane Yield and Synergistic Effect. *Int. J. Environ. Res. Public Health* **2020**, *17*, 4737. [CrossRef]
37. Leip, A.; Ledgard, S.; Uwizeye, A.; Palhares, J.C.P.; Aller, M.F.; Amon, B.; Binder, M.; Cordovil, C.M.D.S.; De Camillis, C.; Dong, H.; et al. The value of manure—Manure as co-product in life cycle assessment. *J. Environ. Manag.* **2019**, *241*, 293–304. [CrossRef]
38. Mehta, S.; Malik, K.; Verma, N.; Anand, R. Effect of Microbial Inoculum on Biogas Production from Cattle Dung under Anaerobic Batch Digestion. *Int. J. Curr. Microbiol. App. Sci.* **2018**, *7*, 897–904. [CrossRef]
39. Silva, C.E.; Abud, A.K. Influence of manure concentration as inoculum in anaerobic digestion of vinasse. *Acta Sci. Biol. Sci.* **2017**, 173–180. [CrossRef]
40. Ward, A.J.; Hobbs, P.J.; Holliman, P.J.; Jones, D.L. Optimisation of the anaerobic digestion of agricultural resources. *Bioresour. Technol.* **2008**, *99*, 7928–7940. [CrossRef] [PubMed]
41. Lee, D.H.; Behera, S.K.; Kim, J.; Park, H.S. Methane production potential of leachate generated from Korean food waste recycling facilities: A lab scale study. *Waste Manag.* **2009**, *29*, 876–882. [CrossRef] [PubMed]
42. Marín-Peña, O.; Alvarado-Lassman, A.; Vallejo-Cantú, N.A.; Juárez-Barojas, I.; Rodríguez-Jarquín, J.P.; Martínez-Sibaja, A. Electrical Conductivity for Monitoring the Expansion of the Support Material in an Anaerobic Biofilm Reactor. *Processes* **2020**, *8*, 77. [CrossRef]
43. Li, Y.; Park, S.Y.; Zhu, J. Solid-state anaerobic digestion for methane production from organic waste. *Renew. Sustain. Energy Rev.* **2011**, *15*, 821–826. [CrossRef]
44. Fisher, K.; Phillips, C.A. The effect of lemon, orange and bergamot essential oils and their components on the survival of Campylobacter jejuni, Escherichia coli O157, Listeria monocytogenes, Bacillus cereus and Staphylococcus aureus in vitro and in food systems. *J. Appl. Microbiol.* **2006**, *101*, 1232–1240. [CrossRef]
45. Koppar, A.; Pullammanappallil, P. Anaerobic digestion of peel waste and wastewater for onsite energy generation in a citrus processing facility. *Energy* **2013**, *60*, 62–68. [CrossRef]

processes

Article

Hydrothermal Pretreatment of Wheat Straw—Evaluating the Effect of Substrate Disintegration on the Digestibility in Anaerobic Digestion

Timo Zerback [1,*], Britt Schumacher [1], Sören Weinrich [1], Benedikt Hülsemann [2] and Michael Nelles [1,3]

1 Department Biochemical Conversion, DBFZ Deutsches Biomasseforschungszentrum gemeinnützige GmbH, Torgauer Straße 116, D-04347 Leipzig, Germany; britt.schumacher@dbfz.de (B.S.); soeren.weinrich@dbfz.de (S.W.); michael.nelles@uni-rostock.de (M.N.)
2 State Institute of Agricultural Engineering and Bioenergy, University of Hohenheim, Garbenstraße 9, D-70599 Stuttgart, Germany; benedikt.huelsemann@uni-hohenheim.de
3 Department Waste and Resource Management, University of Rostock, Justus-von-Liebig Weg 6, D-18057 Rostock, Germany
* Correspondence: timo.rolf.zerback@dbfz.de; Tel.: +49-341-2434-383

Abstract: The increasing demand for renewable energy sources and demand-oriented electricity provision makes anaerobic digestion (AD) one of the most promising technologies. In addition to energy crops, the use of lignocellulosic residual and waste materials from agriculture is becoming increasingly important. However, AD of such feedstocks is often associated with difficulties due to the high content of lignocellulose and its microbial persistence. In the present work, the effect of hydrothermal pretreatment (HTP) on the digestibility of wheat straw is investigated and evaluated. Under different HTP temperatures (160–180 °C) and retention times (15–45 min), a significant increase in biomethane potential (BMP) can be observed in all cases. The highest BMP (309.64 mL CH4 g^{-1} volatile solid (VS) is achieved after pretreatment at 160 °C for 45 min, which corresponds to an increase of 19% of untreated wheat straw. The results of a multiple linear regression model show that the solubilization of organic materials is influenced by temperature and time. Furthermore, using two different first-order kinetic models, an enhancement of AD rate during hydrolysis due to pretreatment is observed. However, the increasing intensity of pretreatment conditions is accompanied by a decreasing trend in the conversion of intermediates to methane.

Keywords: anaerobic digestion; wheat straw; hydrothermal pretreatment; biomethane potential; batch test

Citation: Zerback, T.; Schumacher, B.; Weinrich, S.; Hülsemann, B.; Nelles, M. Hydrothermal Pretreatment of Wheat Straw—Evaluating the Effect of Substrate Disintegration on the Digestibility in Anaerobic Digestion. *Processes* **2022**, *10*, 1048. https://doi.org/10.3390/pr10061048

Academic Editors: Sonia Heaven, Sigrid Kusch-Brandt and Charles Banks

Received: 29 April 2022
Accepted: 20 May 2022
Published: 24 May 2022

1. Introduction

Given the change from fossil to renewable energy sources, biomass has a special significance in Germany. According to a recent study, around 26% of primary energy demand in Germany could be covered by biomass in 2050. The largest share comes from cultivated biomass (such as energy crops) and agricultural residues such as straw, slurry and manure. However, only one-third of the disposable potential has currently been exploited [1]. Anaerobic digestion (AD) is a universal technology that provides different options for waste treatment, provision of renewable energy (biogas) and production of an organic fertilizer. Thus, different types of organic materials, such as sewage sludge, manure, municipal organic waste or lignocellulosic agricultural waste can be used for biogas production. Among agricultural residues, cereal straw (e.g., wheat straw) is an interesting feedstock for biogas production due to its high available potential [2].

However, challenges in AD of lignocellulosic feedstocks are due to the inherent structure, conferring the resistances to hydrolysis and further conversion to biogas. The cause is primarily to be found in the crosslinked structure of the lignin polymer, which essentially consists of aromatic bonds, double bonds and phenolic groups. This form of structure gives lignin molecules hydrophobic properties that are difficult to access biochemically. Moreover, both characteristic polysaccharides cellulose and hemicellulose are linked to the lignin molecule by covalent bonds. Consequently, the strong integration of lignin within the lignocellulose complex leads to increased resistance of the two polysaccharides to the biochemical degradation processes [3].

In order to facilitate the AD of lignocellulosic biomasses, studies are focusing on different disintegration procedures. Among them, physical (extrusion, irradiation), chemical (acids, bases, ionic liquids) or biological (fungal, microbial, enzymatic) methods are regularly the subject of research. Moreover, thermal processes (e.g., steam explosion, liquid hot water) or combined process are regarded as a key technology for the rapid and largely complete conversion of lignocellulosic biomasses [4]. However, most technologies have rarely been transferred to large-scale implementations. The reasons can be found in their inherent drawbacks, such as high costs on energy and chemical consumption, partly corrosive properties or possible pollutions [5]. Because hydrothermal pretreatment (HTP) only uses lignocellulosic biomasses and water, it has been widely accepted as a green technology [6]. The process describes a disintegration method, where the recalcitrant structure breaks under elevated process conditions (150–300 °C, initial pressure 0–60 bar, 5–75 min) [7]. Under these conditions, water and organic acids (especially from hemicellulose) accelerate hemicellulose degradation into short chain products, rendering them as a soluble fraction in the AD process [7,8]. In recent years, numerous studies dealt with HTP of lignocellulosic feedstocks and its influence on biogas production. Here, most of these investigations examine the effect of pretreatment temperature [9,10]. Studies focusing on the influence of retention time as an additional process variable are rarely available. Furthermore, some experiments partly provide contradictory results. As an example, Chandra et al. [11] found that the hydrothermal disintegration of wheat straw (200 °C, 15 min) led to an increase of 20% in biomethane potential (BMP) compared to an untreated reference. This was in contrast to the findings of Wang et al. [9], who obtained that BMP was even reduced by 30% when pretreatment was conducted under nearly similar process conditions (210 °C, 15 min). More promising results were achieved at 180 °C, which was associated with an increase in the specific BMP of up to 3%, indicating a slight improvement in feedstock digestibility [9]. Similar findings were provided by Shang et al. [12], who investigated the effect of pretreatment temperature and duration on wheat straw within a range of 150–225 °C and 5–60 min, respectively. The maximum increase in methane yield (62.9%) was achieved, when pretreatment was conducted for 30 min at 175 °C. In contrast, substrate disintegration at 200 and 225 °C revealed a decreasing trend, which was particularly pronounced with increasing retention times. The authors explained this effect by means of dissolved lignin, which was deposited on the surface of the straw particles during the cooling process. The formation of the so-called pseudo-lignin resulted in a deterioration of the enzymatic accessibility and thus the anaerobic degradability. Furthermore, formation of by-products from lignocellulose that inhibit microbiological activity is suspected [12]. The most frequently mentioned inhibitors are furfurals and 5-Hydroxymethylfurfural (5-HMF), whose influence is assumed by the inhibition of cell growth, DNA damage and an enzymatic inhibition of the glycolysis pathway [13,14].

Moreover, HTP processes are often associated with the formation of phenolic compounds (e.g., vanillin or syringaldehyde), which originate from thermal lignin-degradation. In contrast to furanic compounds, microbial cell damages are assumed to be changes in membrane permeability, which are associated with the leakage of intracellular components and the inactivation of essential enzymatic systems [14–17]. In order to minimize the risk of microbiological inhibition, more attention was attributed to the AD of the solid phase but studies dealing with the total slurry are still limited.

However, such an approach is associated with the loss of easily degradable organic matter, such as monomeric sugars (e.g., glucose or xylose) or organic acids (e.g., acetic acid or formic acid), which consequently leads to a loss in biogas and methane production.

The present study examines the influence of different pretreatment temperatures and retention times on the performance of AD in terms of BMP. The focus is on the anaerobic conversion of the total slurry, which has been rarely investigated in previous studies. Furthermore, the influence of the most important pretreatment parameters (temperature and time) on the composition of processed water samples, with respect to easily degradable components and potential inhibitors, is statistically examined and evaluated in more detail. Such an approach represents an important aspect for determining the most appropriate pretreatment parameters but, to our knowledge, has not yet been investigated. In addition, reaction kinetics were analyzed by, among other things, an extended model structure that has not been used in the context of HTP. The model provides information on the formation and degradation of intermediates (such as sum VFA) and differs from the investigations of most other authors [9,10,12].

2. Materials and Methods

2.1. Origin of Wheat Straw and Sample Preparation

Wheat straw was collected from a farm near Leipzig, Germany (51°17′26.5″ N 12°46′09.7″ E) and chopped by a hammer mill "CHM 230" (Erich Netzsch GmbH & Co. KG, Selb, Germany) to a particle size of 10 mm. A first visual assessment of the samples showed an inhomogeneous distribution of fiber lengths. In order to minimize the influence of different particle sizes, mechanically chopped wheat straw samples were further sieved using a vibrating sieve machine "AS control" (Retsch GmbH, Haan, Germany) at an amplitude setting of 0.65 mm. Individual disintegration steps and BMP tests were carried out using the fiber fractions that settled on the screen with a mesh size between 3.15 and 2.00 mm.

2.2. Hydrothermal Pretreatment and Process Conditions

The HTP was performed in a 500 mL stainless, batch reactor "BR 500" (Berghof Products + Instruments GmbH, Eningen unter Achalm, Germany), which was heated with a synthetic thermal oil at the desired temperature range (160–180 °C) with a heating rate of 2 K min^{-1} for 15, 30 or 45 min. The individual pretreatment conditions were based on a partial factorial experimental design, which was created with "Design-Expert" (Version 12, Stat-Ease Inc., Minneapolis, MN, USA) (Table 1).

Table 1. Pretreatment conditions for hydrothermal wheat straw disintegration based on 2^k-factorial design.

Temperature [°C]	Retention Time [min]		
	15	30	45
160	HTP-160–15		HTP-160-45
170		HTP-170-30	
180	HTP-180-15		HTP-180-45

To estimate the variability of the results and to increase the accuracy of estimation, every pretreatment, with the exception of HTP-170-30, was conducted in duplicate. Hydrothermal process conditions according to HTP-170-30 represented the center point (CP) and had to be performed six times to increase the information content of the design space and estimate the repetition error.

In order to avoid combustion or coking reactions [18], a similar mass ratio of wheat straw and tap water (1:17) was chosen, as shown in Tian et al. [19]. For each pretreatment, 25 g wheat straw (on fresh matter (FM) basis) was filled into the reactor and mixed with 424 g tap water.

Once preparation was completed, the sealed reactor was positioned in the heating jacket and the process was initiated when target temperature was reached. After the pretreatment time was obtained, the reactor stopped immediately, and samples were automatically cooled down to a temperature of 30 °C.

The reactor was opened on the next day after 24 h and the liquid phase of each batch was separated from the solid phase by means of a stainless juice press and a commercially available filter fleece (pore size < 100 µm). In order to achieve consistent separation results, the juice press was attached to a laboratory scale and downforce was increased until the scale displayed a mass of 300 kg. Considering the piston diameter of 11.5 cm, a pressure of 2.83×10^5 Pa was applied for each separation. To cover the whole period from HTP to the BMP tests, HTP factions (liquid and solid) were immediately stored in a freezer (−20 °C) after masses of both fractions were determined. Furthermore, untreated wheat straw samples were stored under identical conditions. Thus, potential effects of freezing or thawing should be considered for all samples.

2.3. Biomethane Potential Test

BMP tests were performed using the Hohenheim Biogas Yield Test (HBT) according to VDI 4630. The system consists of a continuously rotating drum (1.2 rpm), placed in an incubator. The drum is loaded with syringes, each with a total volume of 100 mL [20]. The experimental setup is shown in Figure 1. The liquid and solid phases of the hydrothermally pretreated wheat straw samples were mixed and loaded to AD. For this purpose, syringes were filled with 30 g FM of inoculum and sample material. The inoculum was taken from a 400 L laboratory scale reactor, that was fed with maize silage, shredded wheat, soybean meal, rapeseed oil and digestate from biogas plants in Baden-Württemberg, Germany [19]. Here, both the inoculum to substrate ratio of 2.5 (g VS of substrate/g VS of inoculum) as well as the mass ratio between separated solid and liquid phase were considered to calculate the required amount of pretreated substrate materials. After completion of sample preparation, the syringes were transferred to rotating drum and AD process was conducted under mesophilic conditions (37 °C) for 35 days.

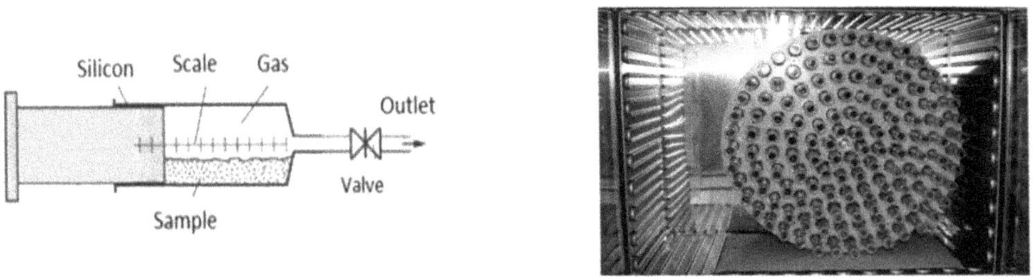

(a) (b)

Figure 1. Experimental setup of Hohenheim Biogas Yield Test (HBT) with syringe (**a**) and rotational drum (**b**) placed in the incubator [21,22].

Determination of methane content was based on dry gas measurement. For this purpose, the biogas volume was first measured by manually reading the scale on the syringes (Figure 1). As soon as measuring volume (20 mL) was reached, the valve was opened and biogas passed through a flexible pipe (not shown) containing an adsorbent (“SICAPENT®”, Merck, Darmstadt, Germany) for gas drying. The determination of methane concentration was subsequently carried out by means of infrared-spectrometric methane sensor “Advanced Gasmitter” (Pronova Analysetechnik, Berlin, Germany) and cumulated methane production of dry gas under standard conditions (273.15 K, 1013.25 hPa) was calculated according to guideline given in VDI 4630 (2016) [23].

BMP tests of every HTP approach were performed in the context of a six-fold determination. Moreover, two test series using cellulose and untreated wheat straw were conducted as positive control and reference, respectively. A blank group without any substrate addition was set to exclude biogas and methane produced by the inoculum in the net volume calculation of the samples.

2.4. Analytical Methods

The total and volatile solid content (TS and VS) of the solid and liquid fraction were analyzed according to the standard methods given under DIN EN 15934 [24] and 15935 [25], respectively. Moreover, the soluble chemical oxidation demand (COD), concentration of organic acids and carbohydrates and the concentration of potential inhibitory compounds (e.g., Furfural, 5-HMF, phenols) from liquid have also been determined.

The analysis of individual volatile fatty acids (VFA) was conducted according to Schumacher et al. [26]. The determination of the COD for liquid samples was based on the use of the cuvette quick test "LCK 014" (Hach Lange GmbH, Düsseldorf, Germany) and the analysis was performed according to the manufacturer's specifications.

The analysis of phenolic compounds was performed on a gas chromatograph with mass spectrometry coupling (GC-MS) "Trace 1310-ISQ LT Single Quadropul Mass Spectrometer" (Thermo Fisher Scientific Inc., Waltham, MA, USA). The injection mode of the gas chromatograph was split 10:1 and injection temperature was kept at 250 °C. For separation, a "ZB-5HT-Inferno" column (30 m × 0.25 mm × 0.25 μm) (Phenomenex, Torrance, CA, USA) and helium 5.0 with a flow rate of 1.2 mL min^{-1} were used as carrier gas. The column was operating with two linear ramps at programmed temperature-mode as follows: heating from an initial temperature (50 °C) with 5 °C min^{-1} till 225 °C; heating with 3 °C min^{-1} till 300 °C and hold for 5 min. The ionization took place by means of ionization module EI. The temperature of the ion source was 250 °C and the MS transfer line was 310 °C. For the determination of phenol compounds, 300 μL of an internal standard (4-tert-butylphenol c = 117 mg L^{-1}) and 400 μL toluene were added to 700 μL of the liquid aqueous sample in a 2 mL vial and extracted by intensive mixing. After 20 min, the organic phase was carefully transferred to a vial and analyzed by GC-MS.

The analysis of soluble sugars (C5-C6) and detection of furfural and 5-HMF were done on a high performance liquid chromatograph (HPLC) using a "Security Guard Carbo Pb" column (4 × 3.00 mm) (Phenomenex Torrance, CA, USA) for sample purification and a second column "MetaCarb 87P" (300 mm × 7.8 mm) (Aligent Technologies, Inc., Santa Clara, CA, USA) for separation with temperature setting at 75 °C. Ultra-pure water with an injection volume of 15 μL and a flow rate of 0.35 mL min^{-1} was used as eluent for every measurement where total run time was set at 80 min. The measurement of sugars (e.g., glucose, xylose, arabinose, mannose, fructose) was performed using a refractive index detector (RID) at a temperature of 40 °C. Depending on its concentrations, determination of furfural and 5-HMF was performed in two different ways. For concentrations ≥ 200 mg L^{-1}, the RID detector was used, whereas lower concentrations required the use of a diode array detector (DAD) measuring at 254 nm. For analysis, aqueous sample was centrifuged at room temperature and 3900 rpm for 5 min. Then, the supernatant was transferred to a filter vial (0.45 μm) and centrifuged at 10 °C and 10,000 rpm for 10 min. The supernatant was diluted 1:10 with ultrapure water and analyzed by the HPLC.

2.5. Kinetic Modelling

Based on characteristic first-order reactions summarized by Brulé et al. [27], different kinetic model structures were chosen to evaluate the effect of HTP on the anaerobic degradation kinetics in BMP tests. Because hydrolysis is often considered to be the rate-limiting step during AD of lignocellulosic biomass [28], data sets were analyzed by a simple first-order reaction model (first-order one-step). Furthermore, an extended model (first-order two-step) was used to evaluate the formation and degradation of additional intermediates (such as sum of VFA). Cumulated methane production over time of both model structures can be described according to Equations (1) and (2), respectively [29].

$$S(t) = S_{max} \times \left(1 - e^{-k \times t}\right) \tag{1}$$

$$S(t) = S_{max} \times \left(1 + \frac{\left(k_H \times e^{-k_{VFA} * t} - k_{VFA} \times e^{-k_H \times t}\right)}{(k_{VFA} - k_H)}\right) \tag{2}$$

where S_{max} (mL g^{-1} VS) is the maximum specific methane potential and k (d^{-1}) represents the first-order rate constant. Additionally, k_H (d^{-1}) is the first-order rate constant of substrate degradation into intermediates (first step), whereas k_{VFA} describes first-order kinetics of VFA degradation into methane (second step).

Individual equations and related procedures for parameter estimation were implemented in the software environment "Matlab" (The Mathworks, Inc., Natick, MA, USA). The non-linear trust region reflective algorithm implemented in the lsqcurvefit function was applied for constrained least-squares estimation of unknown model parameter [29,30].

2.6. Experimental Design and Statistical Analysis

For statistical analyses, a two level (2^k) factorial design arrangement was used to evaluate the effect of pretreatment temperature (X_1) and retention time (X_2) at two different levels each one (160 and 180 °C and 15 and 45 min, respectively) in 14 runs were randomly generated. The experimental data were analyzed by a linear model (Equation (3)) using the commercial software environment "Design-Expert" (Version 12, Stat-Ease, Inc., USA). The statistical significance of the regression terms was examined by analyses of variance (ANOVA) for each response, according to Equation (3).

$$y = \beta_0 + \beta_1 \times X_1 + \beta_2 \times X_2 + \beta_{12} \times X_1 \times X_2 \quad (3)$$

where y is the variable response, X_1 and X_2 are the independent variables and β_1, β_2 and β_{12} represent the regression coefficients of the model. Moreover, a direct comparison between untreated and pretreated wheat straw samples was carried out by the software "SPSS statistics" (Version 20, IBM, Chicago, IL, USA). In order to determine whether there was a significant difference between the BMP of untreated and pretreated samples, experimental data were evaluated using a Welch's ANOVA with a confidence level of 95%. If differences existed, a post-hoc test according to Games–Howell ($\alpha = 0.05$) was applied [31].

3. Results and Discussion

3.1. Effect of Hydrothermal Pretreatment on Process Water Composition

In order to evaluate the effect of pretreatment temperature and retention time on HTP of wheat straw, process water samples were analyzed with respect to the concentration of easily degradable components, such as monomeric sugars (glucose, xylose, arabinose) and organic acids (acetic acid, formic acid), which may be beneficial for AD. Furthermore, the influence of both pretreatment parameters on the content of potential inhibitors (mainly phenolic and furanic compounds) was evaluated. Table 2 shows mean concentrations of the most represented organic soluble fractions in the process water, depending on individual pretreatment conditions.

Table 2. Content of soluble organic fractions in process water under different hydrothermal pretreatment conditions considering acids, sugars and COD.

Experiment	Acetic Acid [a]	Formic Acid [a]	Glucose [a]	Xylose [a]	Arabinose [a]	COD [a]
	[mg L^{-1}]]	[mg L^{-1}]	[mg L^{-1}]	[mg L^{-1}]	[mg L^{-1}]	[g_{O2} L^{-1}]
HTP-160-15	627.4 ± 11.7	84.7 ± 1.2	132.5 ± 32.5	n. d. [b]	171.5 ± 29.5	8.6 ± 0.1
HTP-160-45	826.4 ± 22.8	113.9 ± 3.2	49.6 ± 6.6	n. d. [b]	343.0 ± 3.0	11.6 ± 0.1
HTP-170-30	1078.5 ± 17.6	161.3 ± 11.6	33.4 ± 7.7	n. d. [b]	372.6 ± 16.0	16.1 ± 0.8
HTP-180-15	1252.3 ± 20.9	210.4 ± 4.3	31.3	n. d. [b]	374.5 ± 12.5	18.6 ± 0.1
HTP-180-45	1626.8 ± 15.8	368.0 ± 3.2	41.3 ± 3.7	346.5 ± 7.5	543.0 ± 76.0	21.2 ± 0.3

[a] Data in table are exhibited in form of "Mean ± standard deviation", [b] Component not detected (n. d.).

The highest acetic acid concentration (1626.8 mg L^{-1}) was achieved at HTP-180-45, which resulted in an increase of 159.3% compared to the lowest severe pretreatment conditions given under HTP-160-15. Similar effects were observed for formic acid, where an increase in pretreatment temperature and/or time led to higher concentration of saturated carbonic acid. Acetic acid represents degradation products formed by hydrolysis of acetyl groups of hemicellulose, whose thermal decomposition usually starts at temperatures above 150 °C [7]. Formic acid refers to a by-product whose formation is mainly due to the thermal decomposition of furanic compounds (furfural and 5-HMF) and whose origin in turn is the dehydration of pentoses and hexoses [32]. Thus, increasing acid concentrations could indicate an accelerated hydrolysis of hemicelluloses depending on the pretreatment conditions [12,33]. For monosaccharides, glucose, arabinose and xylose were the main monosaccharides in the process water. Among them, an increase in concentration, depending on pretreatment conditions, was especially observed for arabinose. The highest concentration (543.0 mg L^{-1}) related to the longest pretreatment time and highest temperature (180 °C, 45 min). These findings were in line with results from Ilanidis et al. [34], who described an increasing trend when pretreating wheat straw within a temperature range from 160 °C to 190 °C [34]. Here, the tendency of lower arabinose concentrations could be attributed to a higher liquid-to-solid-ratio (LSR) during the HTP.

For example, Ilanidis et al. [34] (LSR = 7:1) found a concentration of 400 mg L^{-1} whereas the detected concentration in the present work (LSR = 17:1) was 171.5 mg L^{-1} despite the same pretreatment conditions (160 °C, 15 min) [34]. A similar trend was also reported by Chen et al. [35], who published an increasing concentration of hemicellulose derived arabinose with rising temperature up to 180 °C. More severe pretreatment temperatures (e.g., 200 °C) led to a decrease, which indicated thermal degradation of the monosaccharide compound [35]. In contrast, the results for glucose showed a downward trend throughout the whole range of HTP severities. The highest concentration of 132.5 mg L^{-1} was attributed to the HTP-160-15, whereas rising temperatures or longer pretreatment times led to a concentration of 41.3 mg L^{-1}. Because thermal decomposition of cellulose usually starts at a temperature above 200 °C [36,37], it was unlikely that glucose was originated from the polysaccharide under comparatively mild pretreatment conditions. According to similar results of Chen et al. [35], it was more likely that the monosaccharide compounds, found in the process water samples, were probably hemicellulose derived degradation products or β-glcucan [35,38].

Table 3 shows the resulting regression coefficients estimated by multiple linear regression. Individual coefficients are applied to identify the relative impact of influencing factors (temperature and time) when changing the input variables by one level.

Table 3. Coefficients estimated by multiple linear regression of the physicochemical characterization of process water samples of wheat straw considering acids, sugars and COD.

Fitted Model [2FI]	Responses				
Coefficients	**Acetic Acid**	**Formic Acid**	**Glucose**	**Arabinose**	**COD**
Intercept	1081.76 *	179.88 *	53.26 *	370.14 *	15.49 *
X_1	356.34 *	94.94 *	−29.46 *	100.75 *	4.89 *
X_2	143.37 *	46.69 *	−16.11	85.00 *	1.41 *
X_1X_2	43.87 *	32.09 *	25.31 *	−0.7500	−0.0850
R^2	0.9979	0.9992	0.6679	0.8564	0.9632
p of F model	<0.0001 *	<0.0001 *	0.0155 *	0.0002 *	<0.0001 *

X_1: Temperature, X_2: Time, X_1X_2: Interaction between temperature and time. * Indicates estimates of significant parameters ($p < 0.05$).

It is interesting to notice that the pretreatment temperature showed a significant influence ($p < 0.05$) on almost all output parameters, whereas the temporal influence was discernible but could only be statistically proven in the case of organic acids, arabinose and chemical oxygen demand (COD).

Based on estimated regression coefficients, it can be proven that changes in output variables are primarily influenced by pretreatment temperature. As an example, in the case of acetic acid, the effect of temperature (356.34) on its concentration was more than twice as high as the pretreatment time (140.37). Moreover, only a small but also significant interaction between temperature and time has been observed for organic acids and sugar compounds.

Besides organic acids, similar trends have been found for the number of potential inhibitors, shown in Table 4. Although the influence of increasing pretreatment severity was mainly reflected in the furfural concentration (368.3 ± 11.1 mg/L), an increase in concentration was observed for almost all components. Because the reaction mechanisms were not investigated in detail, the cause of the increase in furfural concentration can only be suspected. In general, furfural is a reaction product formed by the thermal pentose sugars (e.g., D-xylose or L-arabinose).

Table 4. Content of soluble organic fractions in process water under different hydrothermal pretreatment conditions considering furanic and phenolic compounds.

Experiment	Furfural [a]	5-HMF [a]	Guaiacol [a]	Syringol [a]	Vanillin [a]	Syringaldehyde [a]
	[mg L^{-1}]	[mg L^{-1}]	[mg L^{-1}]	[mg L^{-1}]	[mg L^{-1}]	[mg L^{-1}]
HTP-160-15	13.8 ± 2.0	n. d. [b]	2.6 ± 0.0	1.6 ± 0.0	8.6 ± 0.0	3.0 ± 0.0
HTP-160-45	41.2 ± 9.0	n. d. [b]	3.2 ± 0.2	2.0 ± 0.1	11.9 ± 0.0	4.2 ± 0.1
HTP-170-30	65.7 ± 5.7	n. d. [b]	3.5 ± 0.2	2.4 ± 0.1	15.2 ± 0.4	5.5 ± 0.2
HTP-180-15	88.1 ± 4.7	n. d. [b]	4.7 ± 1.2	3.3 ± 0.8	17.1 ± 0.1	6.3 ± 0.0
HTP-180-45	368.3 ± 11.1	12.6 ± 0.8	4.9 ± 0.3	4.5 ± 0.0	22.0 ± 1.2	9.1 ± 0.2

[a] Data in table are exhibited in form of "Mean ± standard deviation", [b] Component not detected (n. d.).

As xylose could only be detected at most severe experimental conditions (HTP-180-45), it is more likely that the amount of furfural under less severe conditions (HTP-160-15 to HTP-180-15) could be attributed to the thermal decomposition of arabinose [34]. The larger increase from HTP-180-15 to HTP-180-45 might be explained by the longer pretreatment time, which had a comparatively smaller but nevertheless significant influence on the furfural concentration (Table 5). Moreover, it is well known that the formation of acetic acid, as a by-product, is capable of facilitating the dehydration of hemicellulose to pentose sugars [34,39]. Because there was a sharp increase in acetic acid concentration between these pretreatment conditions, the higher furfural concentration may also be attributed to an acid catalyzed degradation effect. In comparison to furfural, only a minor amount of 5-HMF (12.6 mg L^{-1}), originated from the dehydration of hexose (e.g., glucose and mannose), was detected in the liquid phase obtained at HTP 5. These findings were in line with other authors [32,35], who also found minor amounts, especially at high temperatures or time ranges.

Table 5. Coefficients estimated by multiple linear regression of the physicochemical characterization of process water samples of wheat straw considering furanic and phenolic compounds.

Fitted Model (2FI)	Responses				
Coefficients	Furfural	Guaiacol	Syringol	Vanilin	Syringaldehyde
Intercept	101.53 *	3.69 *	2.69 *	15.12 *	5.48 *
X_1	100.35 *	0.9387 *	1.04 *	4.65 *	2.04 *
X_2	76.93 *	0.1787 *	0.4075 *	2.07 *	1.01 *
X_1X_2	63.20 *	−0.0937	0.1700 *	0.4000	0.4038 *
R^2	0.9216	0.6480	0.8388	0.9708	0.9617
p of F model	<0.0001 *	<0.0123	<0.0003	<0.0001	<0.0001

X_1: Temperature, X_2: Time, X_1X_2: Interaction between temperature and time. * Indicates estimates of significant parameters ($p < 0.05$).

The concentration of phenolics in total ranged from 3.18 to 40.42 mg L^{-1} and also showed an upward trend with rising pretreatment severities. For each compound, a significant effect of both pretreatment time and temperature was found, with the latter having a greater effect on the change in concentration (Table 5). Among these, vanillin and syringaldehyde were the most represented fractions in process water. Both compounds are typical by-products, whose origin can be explained by the depolymerization of guaiacyl and syringyl units of lignin polymers [40].

3.2. Evaluation of Biochemical Methane Potential (BMP)

The influence of pretreatment temperature and time on the measured BMP was examined based on a linear regression model (Equation (3)). The statistical analysis showed very poor accuracy of fit, which was confirmed by most model-relevant parameters. The F-value of 1.14 implied that the model was not significant relative to the noise. In addition, the low adjusted R^2 (0.0361) and negative predicted R^2 (−0.7263) values also indicated insignificance and the overall mean might be a better predictor for the BMP than the current model.

Table 6 illustrates the different temperature and time combinations and the measured BMP for untreated and pretreated wheat straw samples after 35 days. Moreover, both the statistical results of the Games–Howell post-hoc test and the percentage changes due to the pretreatment are shown.

Table 6. Finale methane yields of untreated and hydrothermally pretreated (HTP) wheat straw samples, including increase in BMP, significance (Games–Howell post-hoc) and confidence level of 95%.

Experiment	BMP			Confidence Level of 95%	
		Increase	Significance	Downer Limit	Upper Limit
	[mL g^{-1} VS]	[%]	[-]	[-]	[-]
Untreated	261 ± 15	-	-	-	-
HTP-160-15	302 * ± 17	16	0.013	−79.40	−7.64
HTP-160-45	310 * ± 14	19	0.004	−72.50	−12.72
HTP-170-30	299 * ± 14	15	0.002	−74.96	−18.56
HTP-180-15	298 * ± 9	14	0.013	−66.95	−6.94
HTP-180-45	289 * ± 9	11	0.043	−58.40	−0.80

* Indicates significant differences ($p < 0.05$) in BMP between untreated and pretreated wheat straw samples (confidence level 95%).

AD of untreated wheat straw resulted in a specific BMP of 261 mL g^{-1} VS, which was in approximate agreement with the investigation of Ferreira et al. [41]. Other publications report specific methane yields ranging from 125 to 276 mL g^{-1} VS [10,12,42]. The difference in yield can be attributed to sample fractionation (mesh size 2 mm) prior to AD. An increase in biogas yield can usually be observed with decreasing fiber or particle size due to the larger surface and improved microbial accessibility [43,44]. The statistical evaluation of the BMP, using Games–Howell post hoc test, revealed a significant difference ($p < 0.05$) between the untreated sample and every single temperature–time combination. However, there were no significant differences between the individual parameter combinations (HTP-160-15–HTP-180-45).

Thus, an upward trend was observed especially under mild pretreatment conditions, where the highest increase (19%) in BMP was attributed to the parameter combination HTP-160-45.

In contrast, more intense conditions (HTP-170-30, HTP-180-15, HTP-180-45) seemed to result in a decline in methane production. A decreasing BMP is often attributed to the formation of inhibitory substances [12]. However, based on low concentrations (Table 4), inhibition of furfural, 5-HMF or phenolic compounds can probably be ruled out in the current experimental findings.

Sugar derived degradation products, especially furfural at low concentrations (1 g L^{-1}), was often found to have a beneficial effect on the biogas process [40,45]. However, an inhibitory effect of methanogenic microorganisms due to furfural or 5-HMF had been observed only at higher concentrations (2 g L^{-1}) [46]. Regarding lignin-derived degradations products, an inhibition due phenolic compounds can occur even at low thresholds [47]. Nevertheless, recent literature reports anaerobic degradation or adaption to a certain degree [40,48,49].

Barakat et al. [40] added degradation products such vanillin and syringaldehyde at concentrations of 1 g L^{-1} to AD of C5-sugar and found no reduction in final BMP. Chapleur et al. [48] investigated the influence of different phenol concentrations on AD of cellulose, whereby inhibitory effects only occurred at maximum values of 1.5 g L^{-1} and 2.0 g L^{-1}, respectively. Wirth [49] found an adaption of anaerobic microbial community and complete degradation by using phenol (2 g L^{-1}) as the sole carbon source. A complete inhibition was only detected at a concentration of 5 g L^{-1}.

3.3. Evaluation of Kinetic Parameters

In order to evaluate the influence of HTP on anaerobic degradation kinetics, two different model structures based on first-order reactions (one- and two-step) were applied to depict experimental data. Figure 2 shows the progression of individual measurements and both model results of cumulative methane production in individual BMP tests.

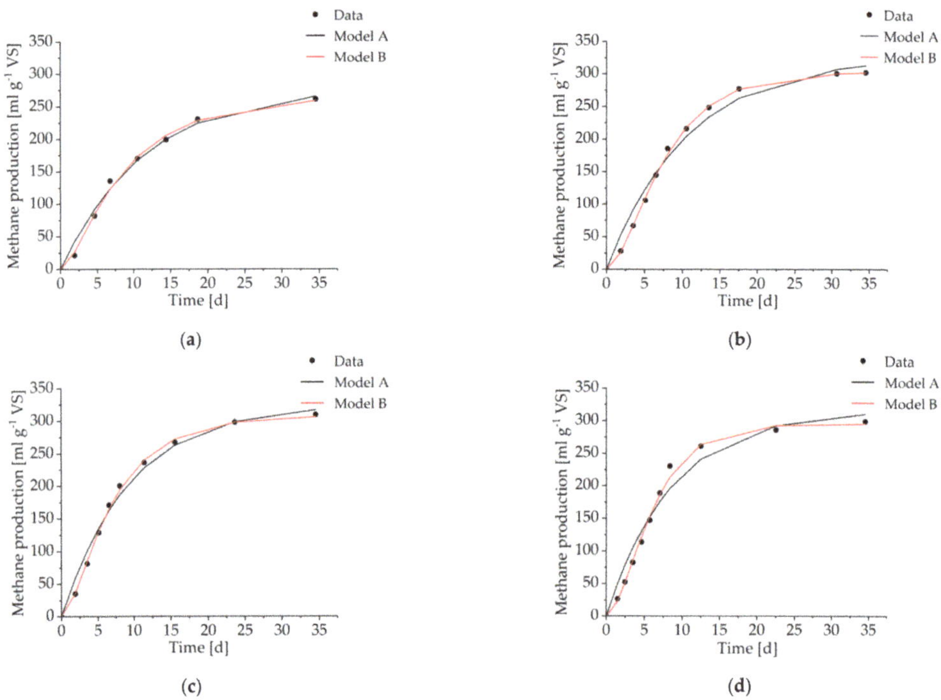

Figure 2. *Cont.*

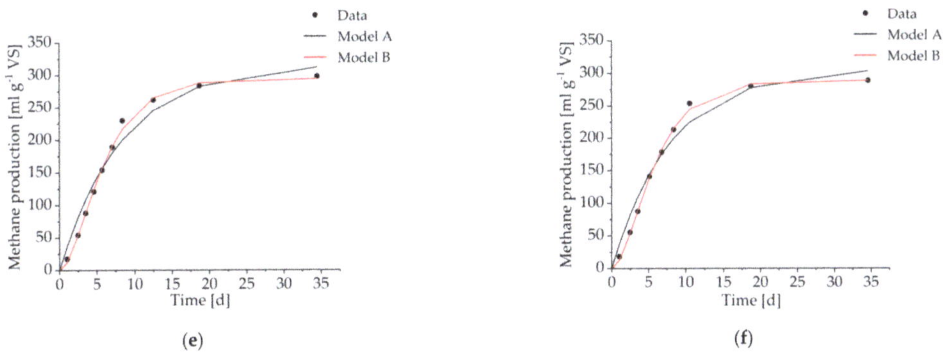

Figure 2. First order model fits illustrated by experimental data (black points) and predicted values by the first-order model (black line) and the first-order two-step model (red line) for untreated wheat straw (**a**) and hydrothermally pretreated wheat straw samples (**b**) HTP-160-15; (**c**) HTP-160-45; (**d**) HTP-170-30; (**e**) HTP-180-15; (**f**) HTP-180-45.

It can be observed that both models were capable of predicting the experimental measurements with high precision. Furthermore, determination coefficients (R2) of 0.98 and 1.00 indicate that the first-order one-step (Model A) and the first-order two-step model (Model B) were able to explain between 98% and 100% of the observed data, respectively.

Based on the single first-order rate constant of Model A (Table 7), it can clearly be observed that HTP had a positive effect on the hydrolysis rate. With increasing pretreatment severity, an increase in the first-order rate constant was observed, which could be explained by an increasing release of organic content and the improvement of substrate accessibility due to an increase in pretreatment severity.

Table 7. Results of kinetic parameter determination.

	Model A			Model B			
Experiment	S_{max}	k	R^2	S_{max}	k_H	k_{VFA}	R^2
	[mL g^{-1} VS]	[d^{-1}]	[-]	[mL g^{-1} VS]	[d^{-1}]	[d^{-1}]	[-]
Untreated	279 ± 10.27	0.08 ± 0.01	0.99	266 ± 11.43	0.10 ± 0.01	1.02 ± 0.23	1.00
HTP-160-15	324 ± 31.13	0.10 ± 0.00	0.98	306 ± 27.30	0.15 ± 0.02	0.65 ± 0.23	1.00
HTP-160-45	317 ± 12.90	0.11 ± 0.00	0.98	301 ± 13.84	0.17 ± 0.01	0.62 ± 0.07	1.00
HTP-170-30	318 ± 22.68	0.12 ± 0.00	0.98	300 ± 19.50	0.23 ± 0.04	0.49 ± 0.10	1.00
HTP-180-15	311 ± 24.01	0.13 ± 0.01	0.98	293 ± 19.98	0.29 ± 0.03	0.39 ± 0.07	1.00
HTP-180-45	308 ± 16.49	0.13 ± 0.00	0.97	291 ± 15.80	0.32 ± 0.01	0.32 ± 0.01	1.00

The highest values were assigned to the highest temperature (180 °C), which seems to be a typical observation, especially for the considered temperature range [12,41]. Because the degradation constant of untreated wheat straw was comparably low ($k = 0.08\ d^{-1}$), its increase may also indicate that HTP helps to facilitate the rate limiting step of the biogas process.

The positive effect of substrate disintegration on degradation kinetics is partly reflected in Model B. In the first stage, an increase in the degradation constant (k_H) was also observed, which indicates a faster formation of intermediates and consequently an acceleration in hydrolysis. In contrast, the second stage shows a deterioration in the degradation kinetics. As shown in Table 7, k_{VFA} decreases from 0.65 d^{-1} to 0.32 d^{-1}, when pretreatment severity increases from HTP-160-15 to HTP-180-45. A decreasing trend of k_{VFA} indicates a slowed degradation of intermediates (e.g., VFA) originated from the first step and consequently a negative influence on methane formation. This observation is supported by declining BMP predicted by the model.

Because Model B is not widely used in the context of substrate disintegration, a direct comparison with other sources is rather difficult. Most authors refer to the modified Gompertz model, in which negative effects on methane or biogas formation due to more severe pretreatment conditions are often shown by prolonging lag phases or reduction in the maximum biogas production rates [10,12,50,51]. As already mentioned in Section 3.2, the low concentrations of the most commonly known compounds (Table 4) did not suggest an inhibitory effect on microbiology. Thus, no clear lag phase was observed in the current experiments and the application of the modified Gompertz model does not yield additional information [29].

However, the negative effect on degradation kinetics and/or methane production may strongly be affected by the source and adaptability of the applied inoculum. For example, in direct comparison of two different inocula, anaerobically digested waste activated sludge (ADWAS) and anaerobically digested cattle manure (ADCM), Phuttaro et al. [52] found significantly better biodegradability of hydrothermally pretreated Napier grass (200 °C, 15 min) by ADWAS, suggesting that ADCM was more sensitive to inhibitory compounds. Because the ADWAS inoculum was frequently exposed to toxic substances (e.g., household chemicals), the authors suspected a better adaption of the microbial consortium to inhibitory compounds [46,52–54].

As the inoculum used in our study was taken from a 400 L laboratory reactor, only fed with maize silage, shredded wheat, soybean meal, rapeseed oil and digestate from biogas plants, the slowed degradation kinetic (k_{VFA}) as well as the decreasing BMP might be explained by an insufficient adaption to furanic and phenolic compounds.

Furthermore, it is well known that hydrothermal pretreatment may lead to reactions between proteins and carbohydrates, which starts to occur at temperatures higher than 150 °C or longer pretreatment times [55]. The so-called Maillard reaction—often indicated by a light to dark brown substrate discoloration—is accompanied by the formation of melanoidines which are less biodegradable due to their complex structure [55,56]. A lesser biodegradability in turn means lower biogas production, which also could explain the decrease in specific BMP (Table 6). In addition, a few studies revealed a negative effect on acidogenesis [56] which resulted in a delayed production of VFA and consequently a reduced BMP [57]. Wang et al. [57] suspected a certain degree of toxicity and a competitive effect for electrons within the microbiological community [57]. With regard to the results illustrated in Figure 3, the darkening of the wheat straw samples clearly increases with increasing pretreatment intensity from HTP-160-15 to HTP-180-45.

(a) (b) (c)

Figure 3. *Cont.*

(d) (e) (f)

Figure 3. Photographs of untreated wheat straw (**a**) and hydrothermally pretreated wheat straw samples with different pretreatment severity (**b**) HTP-160-15; (**c**) HTP-160-45; (**d**) HTP-170-30; (**e**) HTP-180-15; (**f**) HTP-180-45.

An increasing change in color from light to dark brown may indicate an increasing concentration of maillard products (melanoidines), which could explain the decreasing trend in substrate degradation (k_{VFA}), as illustrated by first-order two-step kinetic model.

According to the two different model approaches, all pretreatment conditions led to an increase in degradation constants (k, k_H), indicating that hydrothermal substrate disintegration was capable of accelerating hydrolysis as a rate-limiting step and that it helps to overcome the recalcitrant structure of lignocellulosic biogas substrates (such as wheat straw). However, more severe pretreatment conditions seem to have a negative impact on the degradation kinetics of consecutive process phases (k_{VFA}). Thus, higher temperatures (>180 °C) and longer pretreatment times should be avoided.

4. Conclusions and Outlook

In this study, the effect of HTP on the digestibility of wheat straw was investigated. The results showed that the solubilization of organic matter was a consequence of both pretreatment parameters (temperature and time), where the effect was primary influenced by temperature.

Here, the effect was primarily manifested by an increasing concentration of VFA (e.g., acetic and formic acid) and individual monosaccharides (e.g., arabinose) indicating that HTP was capable of accelerating hydrolysis of hemicellulose, depending on its pretreatment intensity. Furthermore, an increase in furanic and phenolic compounds (e.g., furfural, vanillin, syringaldehyde) was observed, but an inhibitory effect was not to be assumed due to comparatively low concentrations.

Taking the BMP results into account, a significant increase in specific BMP was found for all parameter combinations compared to untreated wheat straw, with the highest increase of 16 to 19% at the pretreatment temperature of 160 °C, where the optimum appeared after 45 min. Moreover, first-order models showed that HTP at all conditions increased the degradation rate, indicating that the disintegrations method helps to facilitate the rate-limiting step of the biogas process. However, because the first-order two-step model revealed a negative influence on consecutive conversion steps (AD of VFA), as indicated by a decreasing rate constant of the second step (k_{VFA}), more severe pretreatment conditions—temperatures (>180 °C) and longer pretreatment times (>45 min)—should be avoided.

A specific application aspect of HTP might be the ReBi (ReBi—dispatchable biogas plant) configuration [58] for dynamic biogas production and demand-oriented electricity provision [58,59]. The ReBi configuration is based on a multistage system, where hydrolysis and acidogenesis as well as acetogenesis and methanogenesis proceed in two separated steps. The effluent formed in the first stage is separated into a liquid and solid fraction by a screw press. The solid fraction is fed into a conventional stirred tank reactor (CSTR), whereas the liquid fraction, containing rapidly degradable substrates, is stored in a tank and fed to a fixed-bed reactor for demand-oriented biogas production [58]. To cover time-dependent biogas demand, rapidly degradable substrates with high BMP and suitable reactors systems are required [59,60]. For a specific application in the ReBi concept, the HTP-reactor might replace the first step. Thus, process water, with its high content of easily digestible organics, might be used to feed a fixed-bed reactor, whereas the fermentation of hydrothermally pretreated wheat straw occurs in a conventional stirred tank reactor.

However, because BMP results only serve a first assessment of substrate digestibility, further research should focus on the AD of hydrothermally pretreated wheat straw under continuous conditions. This allows a long-term evaluation of the effects of substrate disintegration and an assessment of its feasibility in practical applications. In addition, such test procedures can be used to determine the optimal conditions, such as the organic loading rate or the hydraulic retention time, in the biogas process. Furthermore, a techno-economic assessment of the lab-scale results is mandatory to evaluate the applicability of hydrothermal substrate disintegration under full-scale conditions [61,62].

Author Contributions: Conceptualization, T.Z. and B.S.; methodology, T.Z., B.S. and S.W.; software, T.Z. and S.W.; validation, T.Z., B.S. and S.W.; formal analysis, T.Z.; investigation, T.Z. and B.H.; writing—original draft preparation, T.Z.; writing—review and editing, B.S., S.W. and B.H.; visualization, T.Z.; supervision, M.N. All authors have read and agreed to the published version of the manuscript.

Funding: The authors gratefully acknowledge the German Federal Ministry for Digital and Transport for financial support for the project "Pilot-SBG-Bioresources and hydrogen as raw materials for methane as a fuel—conceptual design and realization of a pilot plant" and the German Federal Ministry of Food and Agriculture for the institutional funding of DBFZ.

Data Availability Statement: The data presented in this study are part of a market project and are consequently not available.

Acknowledgments: The authors express their appreciation to colleagues from DBFZ-laboratory as well as the State Institute of Agricultural Engineering and Bioenergy for technical support and conducting the Hohenheim Biogas Yield Test. The authors would also like to thank the company Uwe & Olaf Kuper GbR for providing the wheat straw free of charge.

Conflicts of Interest: The authors declare no conflict of interest. The funders had no role in the design of the study; in the collection, analyses, or interpretation of data; in the writing of the manuscript; or in the decision to publish the results.

References

1. FNR. *Bioenergy in Germany—Facts and Figures 2020*; Agency for Renewable Resources: Gülzow, Germany, 2019.
2. Brosowski, A.; Bill, R.; Thrän, D. Temporal and spatial availability of cereal straw in Germany—Case study: Biomethane for the transport sector. *Energy Sustain. Soc.* **2020**, *10*, 42. [CrossRef]
3. Zoghlami, A.; Paës, G. Lignocellulosic Biomass: Understanding Recalcitrance and Predicting Hydrolysis. *Front. Chem.* **2019**, *7*, 874. [CrossRef] [PubMed]
4. Mahmood, H.; Moniruzzaman, M.; Iqbal, T.; Khan, M.J. Recent advances in the pretreatment of lignocellulosic biomass for biofuels and value-added products. *Curr. Opin. Green Sustain. Chem.* **2019**, *20*, 18–24. [CrossRef]
5. Hendriks, A.T.W.M.; Zeeman, G. Pretreatments to enhance the digestibility of lignocellulosic biomass. *Bioresour. Technol.* **2009**, *100*, 10–18. [CrossRef]
6. Saha, B.C.; Yoshida, T.; Cotta, M.A.; Sonomoto, K. Hydrothermal pretreatment and enzymatic saccharification of corn stover for efficient ethanol production. *Ind. Crops Prod.* **2013**, *44*, 367–372. [CrossRef]
7. Ahmed, B.; Aboudi, K.; Tyagi, V.K.; Álvarez-Gallego, C.J.; Fernández-Güelfo, L.A.; Romero-García, L.I.; Kazmi, A.A. Improvement of Anaerobic Digestion of Lignocellulosic Biomass by Hydrothermal Pretreatment. *Appl. Sci.* **2019**, *9*, 3853. [CrossRef]
8. Rezania, S.; Oryani, B.; Cho, J.; Talaiekhozani, A.; Sabbagh, F.; Hashemi, B.; Rupani, P.F.; Mohammadi, A.A. Different pretreatment technologies of lignocellulosic biomass for bioethanol production: An overview. *Energy* **2020**, *199*, 117457. [CrossRef]
9. Wang, D.; Shen, F.; Yang, G.; Zhang, Y.; Deng, S.; Zhang, J.; Zeng, Y.; Luo, T.; Mei, Z. Can hydrothermal pretreatment improve anaerobic digestion for biogas from lignocellulosic biomass? *Bioresour. Technol.* **2018**, *249*, 117–124. [CrossRef]
10. Rajput, A.A.; Ayub Zeshan, S.; Visvanathan, C. Effect of thermal pretreatment on chemical composition, physical structure and biogas production kinetics of wheat straw. *J. Environ. Manag.* **2018**, *221*, 45–52. [CrossRef]

11. Chandra, R.; Takeuchi, H.; Hasegawa, T.; Kumar, R. Improving biodegradability and biogas production of wheat straw substrates using sodium hydroxide and hydrothermal pretreatments. *Energy* **2012**, *43*, 273–283. [CrossRef]
12. Shang, G.; Zhang, C.; Wang, F.; Qiu, L.; Guo, X.; Xu, F. Liquid hot water pretreatment to enhance the anaerobic digestion of wheat straw—Effects of temperature and retention time. *Environ. Sci. Pollut. Res.* **2019**, *26*, 29424–29434. [CrossRef] [PubMed]
13. Almeida, J.R.; Bertilsson, M.; Gorwa-Grauslund, M.F.; Gorsich, S.; Liden, G. Metabolic effects of furaldehydes and impacts on biotechnological processes. *Appl. Microbiol. Biotechnol.* **2009**, *82*, 625–638. [CrossRef] [PubMed]
14. Palmqvist, E.A.; Hahn-Hägerdal, B. Fermentation of lignocellulosic hydrolysates. II: Inhibitors and mechanisms of inhibition. *Bioresour. Technol.* **2000**, *74*, 25–33. [CrossRef]
15. Campos, F.M.; Couto, J.A.; Figueiredo, A.R.; Toth, I.V.; Rangel, A.O.S.S.; Hogg, T.A. Cell membrane damage induced by phenolic acids on wine lactic acid bacteria. *Int. J. Food Microbiol.* **2009**, *135*, 144–151. [CrossRef]
16. Heipieper, H.J.; Weber, F.; Sikkema, J.; Keweloh, H.; de Bont, J.A.M. Mechanisms of resistance of whole cells to toxic organic solvents. *Trends Biotechnol.* **1994**, *12*, 409–415. [CrossRef]
17. Hierholtzer, A.; Chatellard, L.; Kierans, M.; Akunna, J.C. The impact and mode of action of phenolic compounds extracted from brown seaweed on mixed anaerobic microbial cultures. *J. Appl. Microbiol.* **2012**, *114*, 964–973. [CrossRef]
18. Ibrahim, B. Investigation of the Applicability of Wetland Biomass for Producing Hydrochar by Hydrothermal Carbonization and Its Effectiveness for the Adsorption of Ammonia. Ph.D. Thesis, University of Rostock, Rostock, Germany, 2018.
19. Tian, W.; Chen, Y.; Shen, Y.; Zhong, C.; Gao, M.; Shi, D.; He, Q.; Gu, L. Effects of hydrothermal pretreatment on the mono- and co-digestion of waste activated sludge and wheat straw. *Sci. Total Environ.* **2020**, *732*, 139312. [CrossRef]
20. Helffrich, D.; Oechsner, H. The Hohenheim biogas yield test. *Landtechnik* **2003**, *58*, 148–149.
21. Hülsemann, B.; Zhou, L.; Merkle, W.; Hassa, J.; Müller, J.; Oechsner, H. Biomethane Potential Test: Influence of Inoculum and the Digestion System. *Appl. Sci.* **2020**, *10*, 2589. [CrossRef]
22. Bavarian State Research Center for Agriculture (LfL). Available online: https://www.lfl.bayern.de/mam/cms07/ipz/dateien/aggf_2008_nussbaum_3.pdf (accessed on 20 December 2021).
23. VDI 4630. *Fermentation of Organic Materials—Characterisation of the Substrate, Sampling, Collection of Material Data, Fermentation Tests*; Beuth Verlag GmbH: Berlin, Germany, 2016.
24. DIN. EN 15934; *Characterization of Sludges—Determination of Dry Residue and Water Content*; DIN Deutsches Institut für Normung e.V.: Berlin, Germany, 2012.
25. DIN. EN 15935; *Characterization of Sludges—Determination of the Loss on Ignition of Dry Masses*; DIN Deutsches Institut für Normung e.V.: Berlin, Germany, 2012.
26. Schumacher, B.; Wedwitschka, H.; Weinrich, S.; Mühlenberg, J.; Gallegos, D.; Oehmichen, K.; Liebetrau, J. The influence of pressure swing conditioning pre-treatment of chicken manure on nitrogen content and methane yield. *Renew. Energy* **2019**, *143*, 1554–1565. [CrossRef]
27. Brulé, M.; Oechsner, H.; Jungbluth, T. Exponential model describing methane production kinetics in batch anaerobic digestion: A tool for evaluation of biochemical methane potential assays. *Bioprocess Biosyst. Eng.* **2014**, *37*, 1759–1770. [CrossRef] [PubMed]
28. Noike, T.; Endo, G.; Chang, J.-E.; Yaguchi, J.-I.; Matsumoto, J.-I. Characteristics of carbohydrate degradation and the rate-limiting step in anaerobic digestion. *Biotechnol. Bioeng.* **1985**, *27*, 1482–1489. [CrossRef] [PubMed]
29. Weinrich, S.; Astals, S.; Hafner, S.D.; Koch, K. Kinetic modelling of anaerobic batch tests. In *Collection of Methods for Biogas—Methods to Determine Parameters for Analysis Purposes and Parameters that Describe Processes in the Biogas Sector*, 2nd ed.; Liebetrau, J., Pfeiffer, D., Eds.; DBFZ: Leipzig, Germany, 2020; Volume 7, pp. 349–369.
30. Hofmann, J.; Müller, L.; Weinrich, S.; Deeber, S.; Schumacher, B.; Velghe, F.; Liebetrau, J. Assessing the Effects of Substrate Disintegration on Methane Yield. *Chem. Eng. Technol.* **2020**, *43*, 47–58. [CrossRef]
31. Field, A. *Discovering Statistics Using SPSS*, 3rd ed.; SAGE Publications: London, UK, 2009; p. 821.
32. Nitsos, C.K.; Matis, K.A.; Triantafyllidis, K.S. Optimization of Hydrothermal Pretreatment of Lignocellulosic Biomass in the Bioethanol Production Process. *ChemSusChem* **2013**, *6*, 110–122. [CrossRef] [PubMed]
33. Gao, Y.; Wang, H.; Guo, J.; Peng, P.; Zhai, M.; She, D. Hydrothermal degradation of hemicelluloses from triploid poplar in hot compressed water at 180–340 °C. *Polym. Degrad. Stab.* **2016**, *126*, 179–187. [CrossRef]
34. Ilanidis, D.; Stagge, S.; Jönsson, L.J.; Medina, C.M. Hydrothermal Pretreatment of Wheat Straw: Effects of Temperature and Acidity on Byproduct Formation and Inhibition of Enzymatic Hydrolysis and Ethanolic Fermentation. *Agronomy* **2021**, *11*, 487. [CrossRef]
35. Chen, X.; Li, H.; Sun, S.; Cao, X.; Sun, R. Co-production of oligosaccharides and fermentable sugar from wheat straw by hydrothermal pretreatment combined with alkaline ethanol extraction. *Ind. Crops Prod.* **2018**, *111*, 78–85. [CrossRef]
36. Sakaki, T.; Shibata, M.; Yasuda, S. Saccharification of cellulose using hot-compressed water-flowreactor. *Ind. Eng. Chem. Res.* **2002**, *41*, 661–665. [CrossRef]
37. Jin, F.M.; Zhou, Z.Y.; Enomoto, H.; Moriya, T.; Higashijima, H. Conversion mechanism of cellulosic biomass to lactic acid in subcritical water and acidbase catalytic effect of subcritical water. *Chem. Lett.* **2004**, *33*, 126–127. [CrossRef]
38. Cara, C.; Ruiz, E.; Carvalheiro, F.; Moura, P.; Ballesteros, I.; Castro, E.; Girio, F. Production, purification and characterisation of oligosaccharides from olive tree pruning autohydrolysis. *Ind. Crops Prod.* **2012**, *40*, 225–231. [CrossRef]
39. Delbecq, F.; Wang, Y.; Muralidhara, A.; Ouardi, K.E.; Marlair, G.; Len, C. Hydrolysis of Hemicellulose and Derivates—A Review of Recent Advances in the Production of Furfural. *Front. Chem.* **2018**, *6*, 146. [CrossRef] [PubMed]

40. Barakat, A.; Monlau, F.; Steyer, J.-P.; Carrere, H. Effect of lignin-derived and furan compounds found in lignocellulosic hydrolysates on biomethane production. *Bioresour. Technol.* **2012**, *104*, 90–99. [CrossRef] [PubMed]
41. Ferreira, L.C.; Donoso-Bravo, A.; Nilsen, P.J.; Fdz-Polanco, F.; Pérez-Elvira, S.I. Influence of thermal pretreatment on the biochemical methane potential of wheat straw. *Bioresour. Technol.* **2013**, *143*, 251–257. [CrossRef] [PubMed]
42. Theuretzbacher, F.; Lizasoain, J.; Lefever, C.; Saylor, M.K.; Enguidanos, R.; Weran, N.; Gronauer, A.; Bauer, A. Steam explosion pretreatment of wheat straw to improve methane yields: Investigation of the degradation kinetics of structural compounds during anaerobic digestion. *Bioresour. Technol.* **2015**, *179*, 299–305. [CrossRef]
43. Menardo, S.; Airoldi, G.; Balsari, P. The effect of particle size and thermal pre-treatment on the methane yield of four agricultural by-products. *Bioresour. Technol.* **2012**, *104*, 708–714. [CrossRef]
44. Surendra, K.C.; Khanal, S.K. Effects of crop maturity and size reduction on digestibility and methane yield of dedicated energy crops. *Bioresour. Technol.* **2015**, *178*, 178–193. [CrossRef]
45. Pekařová, S.; Dvořáčková, M.; Stloukal, P.; Ingr, M.; Šerá, J.; Koutny, M. Quantitation of the Inhibition Effect of Model Compounds Representing Plant Biomass Degradation Products on Methane Production. *Bioresources* **2017**, *12*, 2421–2432. [CrossRef]
46. Ghasimi, D.S.M.; Aboudi, K.; de Kreuk, M.; Zandvoort, M.H.; van Lier, J.B. Impact of lignocellulosic-waste intermediates on hydrolysis and methanogenesis under thermophilic and mesophilic conditions. *Chem. Eng. J.* **2016**, *295*, 181–191. [CrossRef]
47. Chen, Y.; Cheng, J.J.; Creamer, K.S. Inhibition of anaerobic digestion process: A review. *Bioresour. Technol.* **2008**, *99*, 4044–4064. [CrossRef]
48. Chapleur, O.; Civade, R.; Hoyos-Hernandez, C.F.; Mazéas, L.; Bouchez, T. Growing concentrations of phenols increasingly modify microbial communities' dynamics and perfomances' stability of anerobic digesters. In Proceedings of the 13th World Congress on Anaerobic Digestion: Recovering (bio) Resources for the World, Santiago de Compostella, Spain, 25–28 June 2013. Available online: https://hal.archives-ouvertes.fr/hal-01118701 (accessed on 20 April 2022).
49. Wirth, B. Anaerobic Treatment of Liquid By-Products from Hydrothermal Carbonization of Biomass. Ph.D. Thesis, Technische Universität Berlin, Berlin, UK, 2020.
50. Myszograj, S. Biogas and Methane Potential of Pre-Thermally Disintegrated Bio-Waste. *Energies* **2019**, *12*, 3880. [CrossRef]
51. Şenol, H.; Açikel, Ü.; Demir, S.; Oda, V. Anaerobic digestion of cattle manure, corn silage and sugar beet pulp mixtures after thermal pretreatment and kinetic modeling study. *Fuel* **2020**, *263*, 116651. [CrossRef]
52. Phuttaro, C.; Sawatdeenarunat, C.; Surendra, K.C.; Boonsawang, P.; Chaiprapat, S.; Khanal, S.K. Anaerobic digestion of hydrothermally-pretreated lignocellulosic biomass: Influence of pretreatment temperatures, inhibitors and soluble organics on methane yield. *Bioresour. Technol.* **2019**, *284*, 128–138. [CrossRef] [PubMed]
53. Boopathy, R. Anaerobic biotransformation of furfural to furfuryl alcohol by a methanogenic archaebacterium. *Int. Biodeterior. Biodegrad.* **2009**, *63*, 1070–1072. [CrossRef]
54. Meyer, T.; Edwards, E.A. Anaerobic digestion of pulp and paper mill wastewater and sludge. *Water Res.* **2014**, *65*, 321–349. [CrossRef] [PubMed]
55. McVoitte, W.P.A.; Clark, O.G. The effects of temperature and duration of thermal pretreatment on the solid-state anaerobic digestion of dairy cow manure. *Heliyon* **2019**, *5*, e02140. [CrossRef]
56. Yin, J.; Liu, J.; Chen, T.; Long, Y.; Shen, D. Influence of melanoidins on acidogenic fermentation of food waste to produce volatility fatty acids. *Bioresour. Technol.* **2019**, *284*, 121–127. [CrossRef]
57. Wang, S.; Hu, Z.Y.; Geng, Z.Q.; Tian, Y.C.; Ji, W.X.; Li, W.T.; Dai, K.; Zeng, R.J.; Zhang, F. Elucidating the production and inhibition of melanoidins products on anaerobic digestion after thermal-alkaline pretreatment. *J. Hazard. Mater.* **2021**, *424*, 127377. [CrossRef]
58. Hahn, H.; Krautkremer, B.; Hartmann, K.; Wachendorf, M. Review of concepts for a demand-driven biogas supply for flexible power generation. *Renew. Sustain. Energy Rev.* **2014**, *29*, 383–393. [CrossRef]
59. Szarka, N.; Scholwin, F.; Trommler, M.; Jacobi, H.F.; Eichhorn, M.; Ortwein, A.; Thrän, D. A novel role for bioenergy: A flexible, demand-oriented power supply. *Energy* **2013**, *61*, 18–26. [CrossRef]
60. Peters, L.; Uhlenhut, F.; Biernacki, P.; Steinigeweg, S. Status of demand-driven biogas concepts to cover residual load rises. *ChemBioEng Rev.* **2018**, *5*, 163–172. [CrossRef]
61. Scherzinger, M.; Kaltschmitt, M. Techno-economic assessment of common reed and sewage suldge co-fermentation improved by vapothermal pre-treatment. *Clean Technol. Environ. Policy* **2021**, *23*, 2741–2755. [CrossRef]
62. Mainardis, M.; Buttazzoni, M.; Gievers, F.; Vance, C.; Magnolo, F.; Murphy, F.; Goi, D. Life cycle assessment of sewage sludge pretreatment for biogas production: From laboratory test to full-scale applicability. *J. Clean. Prod.* **2021**, *322*, 129056. [CrossRef]

processes

Article

Anaerobic Co-Digestion of Sewage Sludge and Trade Wastes: Beneficial and Inhibitory Effects of Individual Constituents

Olivia Berzal de Frutos, Martin Götze, Marc Pidou and Yadira Bajón Fernández *

Cranfield Water Science Institute (CWSI), School of Water, Energy and Environment (SWEE), Cranfield University, Bedford MK43 0AL, UK

* Correspondence: y.bajonfernandez@cranfield.ac.uk

Abstract: Anaerobic digestion (AD) of sewage sludge can be optimised by adding trade wastes (TWs) because of their nutrient content and boost in biogas formation if non-inhibitory. However, some components in TWs might have an inhibitory impact, such as nitrogen compounds, sulphate, heavy metals, metalloids, halogens and organic pollutants (e.g., phenol). This study aimed to understand the impact of TWs on the co-digestion with sewage sludge to identify appropriate TW loads for sustainable AD operation. The composition of 160 TWs was evaluated and the constituents with potential to cause inhibition or toxicity were tested in bio-methane potential (BMP) tests. The compounds studied in BMP tests included ammonia, zinc, copper, aluminium, mercury, arsenic, chloride, sulphate and nitrate. An improvement was observed at concentrations 2–746 mg Zn/L, 1066–2821 mg Cl/L as zinc sulphate and sodium chloride in biogas production, and 2–746 mg Zn/L, 162 mg SO_4/L, 25 mg Hg/L as zinc sulphate, sodium chloride and mercury sulphate in methane production, respectively. Considering the TWs characterised and the results of the BMP tests, a volumetric ratio of 10/90 of TWs and sewage sludge is proposed as a suitable feedstock for co-digestion.

Keywords: inhibition; salts; counter-ions; biogas; kinetics; methane

Citation: Berzal de Frutos, O.; Götze, M.; Pidou, M.; Bajón Fernández, Y. Anaerobic Co-Digestion of Sewage Sludge and Trade Wastes: Beneficial and Inhibitory Effects of Individual Constituents. *Processes* **2023**, *11*, 519. https://doi.org/10.3390/pr11020519

Academic Editors: Sonia Heaven, Sigrid Kusch-Brandt and Charles Banks

Received: 18 December 2022
Revised: 27 January 2023
Accepted: 28 January 2023
Published: 8 February 2023

1. Introduction

Sewage sludge is widely recognised as a valuable resource because of its potential to generate renewable energy and organic fertiliser when treated by anaerobic digestion (AD), with additional benefits of reducing its pathogen content, odour nuisance and volumes [1,2]. Despite these benefits, only 30 to 50% of dry organic solids from sewage sludge are converted to methane during AD, which affects the efficiency of the process and its economic profitability [3]. Optimum biogas production from sewage sludge digestion is also frequently limited by the organic loading rate (OLR) at which sludge digesters are operated (1.6 to 4.8 kg VSS/m^3/d), with OLR generally limited by the risk of accumulating inhibitors such as free ammonia or fatty acids [4]. Furthermore, the carbon-to-nitrogen ratio associated with sewage sludge digestion is typically low, between 6 to 16, compared to the optimum values of 20 to 30 [5], and there is frequently a deficiency of heavy metals that act as micronutrients [6]. Anaerobic co-digestion (Aco-D) can be used to solve these bottlenecks and increase process efficiency.

Aco-D consists of treating several substrates in the same digester to achieve a synergistic benefit that balances their disadvantages and maximises process yield. A range of substrates has been used to increase the performance of sewage sludge AD; for example, organic fraction of municipal solid waste (OFMSW) [7], grass [2], fats, oils and greases (FOG) [5], food waste [6], manure [8] and trade wastes (TWs). TWs typically contain nutrients favourable for the AD process, but also high levels of toxic constituents and insufficient buffer capacity to be digested as single substrates [3]. Co-digestion of sewage sludge and TWs has often been considered to digest organic wastes that could not be treated alone. To illustrate, Hagelqvist [9] co-digested 50% volatile solids (VS) of paper-mill sludge, a substrate with low biodegradability, and 50% VS of sewage sludge. This resulted in an enhancement of the methane yield from 53 mL/g VS for paper-mill sludge mono-digestion up to 84 mL/g VS and an improvement of VS degradation from 5 to more than 10% [9].

Previous studies have highlighted the benefit of biogas production of sewage sludge Aco-D with industrial wastewaters. Berenjkar et al. recorded that the use of 10% volume of landfill leachate as co-digestate with sewage sludge increased methane production from 0.75 to 1.01 CH_4 m^3/kg VS [1]. The addition of 50% weight of petrochemical effluent to sludge improved bio-methane production by 18–32% [10]. In another study, 5% v/v of olive mill wastewater, cheese whey and glycerol addition to sewage sludge increased biogas production by up to 220%, 350% and 86%, respectively [3].

Although the performance benefits of co-digesting TWs with sewage sludge have been widely reported, careful consideration of TW loads is paramount as the AD process can be inhibited by the high concentrations of inhibitors expected in TWs (see Supplementary Materials Table S1). To illustrate, TWs from food processing industries can contain up to 3800 mg/L of total Kjeldahl nitrogen (TKN), which could degrade to total ammonia nitrogen (TAN) in the digester [11]. Other TWs, such as coal-gasification wastewater, can contain 3000 to 9000 mg/L of TAN, resulting in potential AD inhibition [12]. Free ammonia nitrogen (FAN) in crab industry wastewater was 50% inhibitory to methane production when it reached values of 85 mg/L in a toxicity test [13]. Other nitrogen compounds that could affect AD stability are nitrate (NO_3) and cyanide (CN), which have been found at levels potentially inhibitory for AD in TWs from pectin factories, cellophane and explosives production, metal finishing, fertiliser and distilleries [14,15] and from the casava industry [16], among others. Sulphate (SO_4) can also be found in TWs from the chemical and saline food industries and it has been reported to generate competition between the sulphate reducing microorganisms and methanogens [17,18]. Moreover, these sulphate reducing organisms transform sulphate to sulphide, which in the form of hydrogen sulphide gas is known to inhibit AD processes [17]. Heavy metals, which are commonly found in TWs, have a positive effect as nutrients for the microorganisms and a negative effect over a certain level [4]. Some examples of metals in TWs are iron (Fe), zinc (Zn), cadmium (Cd), nickel (Ni), chromium (Cr), copper (Cu), mercury (Hg), aluminium (Al), lead (Pb), arsenic (As) and selenium (Se). Halogens can also be inhibitory to AD, such as fluorine (F) and chloride (Cl); organic pollutants common in chemical industries, such as phenol, could also pose a risk of inhibition (see Supplementary Materials Table S1).

Previous studies on Aco-D of TWs and sludge only covered one specific type of TWs at a time. This is useful for the producer industry, but not for waste collection companies such as water utilities receiving hundreds of types of TWs daily. In the UK, water utilities are currently only allowed to treat these wastes with the municipal wastewater in the main treatment train, but there is potential to use some of the TWs received by direct addition into the AD process, potentially increasing renewable energy generation. There is then a necessity for a more systematic and holistic approach to control the addition of those TWs and avoid any inhibitory events; therefore, it is critical to understand the impact that TWs' characteristics have on the AD of sewage sludge. A significant number of studies are available in the literature on inhibitory constituents for AD that are commonly found in TWs (see Supplementary Materials Table S1). However, these studies use different reactor types and experimental conditions, including inoculum, feed and temperature, making it difficult to compare the results obtained and draw general conclusions.

The aim of this study was then to understand the beneficial and inhibitory effect of key components of TWs on the AD process during co-digestion with sewage sludge. The associated objectives were to identify the key constituents of TWs that might affect AD, determine their impact in Aco-D with sewage sludge, and establish the loads of each constituent that can be added to sludge digesters without hindering methane and biogas production. Through characterisation of 160 real TWs commonly sent to wastewater treatment sites and evaluation of the impact that typical TWs constituents have on biomethane potential (BMP) tests, this work addresses the gaps in knowledge identified above and provides insights into selection of TWs for Aco-D with sludge.

2. Materials and Methods

2.1. Feedstock and Inoculum

The sewage sludge used as a substrate for the AD trials consisted of a mixture of primary sludge and secondary waste activated sludge from two wastewater treatment plants in the West Midlands, England. The inoculum was the digestate from the same sites. The substrate and inoculum were stored at 4 °C on the day of sampling and were allowed to reach room temperature for 24 h before the start of the experiments. The average organic content in the feed was about 50 g COD/L with 34% being soluble (Table 1). Storage at 4 °C was chosen to temporarily suppress biological activity and hence avoid organic matter degradation until AD trials were started, as it has previously been reported to keep methanogenic activity close to that of fresh inoculums [19].

Table 1. Characteristics of the feed and inoculum used for the BMP experiments.

	Feed	Inoculum
pH	6.9 ± 0.7	7.5 ± 0.4
COD total (mg COD/L)	49,657 ± 16,717	17,125 ± 6029
COD soluble (mg COD/L)	6346 ± 4438	480 ± 266
TAN (mg-N/L)	1376 ± 1216	825 ± 415
FAN (mg-N/L)	17 ± 36	44 ± 50
Total alkalinity (mg $CaCO_3$/L)	2785 ± 989	5389 ± 1699
Partial alkalinity (mg $CaCO_3$/L)	199 ± 165	3611 ± 754
Intermediate alkalinity (mg $CaCO_3$/L)	2105 ± 814	1504 ± 949
Total solids (TS) (% over sample)	4.4 ± 1.3	2.4 ± 0.7
TS (g/kg over sample)	44 ± 13	24 ± 7
VS (% over TS)	77.5 ± 2.4	64.89 ± 2.2
VS (g/kg over TS)	775 ± 24	648.9 ± 22
Cl (mg/L)	441 ± 375	179 ± 229
NO_3 (mg/L)	134 ± 230	4 ± 6
SO_4 (mg/L)	49 ± 47	4 ± 8
Na (mg/L)	2337 ± 3415	1119 ± 1345
Al (mg/L)	9 ± 16	3 ± 4
Cu (mg/L)	2 ± 5	4 ± 5
Zn (mg/L)	31 ± 60	8 ± 6
As (mg/L)	0.3 ± 0.6	0.1 ± 0.1
Hg (mg/L)	1.1 ± 1.7	0.3 ± 0.5

2.2. Trade Wastes Data Analysis

A dataset containing the characterisation of TWs that were requested to be treated in a treatment plant between 2010 and 2017 was obtained from a water utility. A total of 1122 TWs were included, with a highly variable composition observed for individual TWs. In order to account for variability in their composition, only those with a minimum of 5 samples characterised in the dataset were selected for this study to provide some insight on it. This resulted in a total of 160 TWs characterised for 25 parameters: pH, chemical oxygen demand (COD), suspended solids, ammonia, chromium, copper, lead, nickel, zinc, sulphate, phenol, cyanide, fluoride, sulphide, chloride, silver, arsenic, mercury, iron, aluminium, cadmium, antimony, bromide, nitrate and nitrite. These parameters were selected based on previous experience to represent the typical pollutants found on the type of TWs analysed. The average values for each parameter of each TW were used to evaluate and categorise TWs.

2.3. Batch Tests

BMP tests were used to understand the impact of TW addition in sewage sludge AD. To elucidate the impact of individual constituents of TWs, reactors were spiked with targeted chemicals informed by the dataset characterisation, as opposed to real TWs where multiple inhibitory components can be present simultaneously. The BMP tests were carried out in 125 mL serum glass bottles consisting of a blank with only inoculum, a control with inoculum and sewage sludge as feed; and test reactors containing inoculum and a feedstock mixture of 90% sewage sludge on mass basis and 10% of a solution spiked with a compound of interest. All the control and test digesters were operated with an inoculum to feedstock VS ratio of 1:1, assuming the compound of interest added no VS. The constituents studied included ammonia, zinc, copper, aluminium, mercury, arsenic, chloride, sulphate and nitrate, which were added using salts (Table 2). Ammonia, zinc, copper and aluminium were studied using both chloride and sulphate salts to understand the effect of the counter-ion dosed. The pH change resulting from aluminium salts dosing was further studied by controlling the pH with sodium hydroxide (NaOH) and through the evaluation of the impact of pH alone by reducing it with hydrochloric acid (HCl). Arsenic, chloride, sulphate and nitrate were all used with sodium as counter-ions, assumed to be inert (Table 2). The impact of each constituent was assessed by testing for at least 5 different concentrations and each reactor was operated in triplicates. The total concentration of each studied constituent is reported as the background concentration in the control of the experiment plus the concentration added with the salt. The temperature of the reactors was kept at 38 °C using water baths and incubators with continuous shaking. The tests were ended when the daily biogas production variation was less than 1% [7,20], with the tests lasting between 32 and 92 days.

Table 2. Salts added to the biological methane potential tests.

Parameter Studied	Chemical Added
Ammonia	Ammonium chloride Ammonium sulphate
Zinc	Zinc chloride Zinc sulphate
Copper	Copper chloride Copper sulphate
Aluminium	Aluminium chloride Aluminium sulphate Aluminium sulphate + NaOH
pH	HCl
Mercury	Mercury sulphate
Arsenic	Sodium arsenate
Chloride	Sodium chloride
Sulphate	Sodium sulphate
Nitrate	Sodium nitrate

The BMP experiments were carried out over a year; therefore, some seasonality was observed in the characteristics of the sewage sludge used as feedstock and inoculum (Table 1). Performance of all test reactors was normalised to their respective blanks to compensate for those variations and allow relevant comparison of the results obtained.

Biogas (Y_{Bg}, mL/g) and methane (Y_M, mL/g) yields were calculated from their accumulated production in mL ($AccV_S$) subtracted by the accumulated biogas production from the blank in mL ($AccV_B$) and normalising over the mass of VS from sewage sludge fed to the reactor (g VS_f). Yield ratio (Y_i/Y_c) was calculated dividing the sample yield over the control (Equations (1) and (2)).

$$Y_{iBg}/Y_{cBg} = \frac{AccV_S - AccV_B}{g\ VS_f}/Y_C \quad (1)$$

$$Y_{iM}/Y_{cM} = \frac{AccV_S - AccV_B}{g\ VS_f}/Y_C \tag{2}$$

Biogas production kinetics were estimated using first-order kinetics and Gompertz modified models [7], as represented by Equations (3) and (4), respectively.

$$\text{1st order kinetics } P(t) = P_{max}\cdot\left(1 - e^{(-k_h\cdot t)}\right) \tag{3}$$

$$\text{Gompertz modified model } P(t) = P_{max}\cdot e^{\left(-e^{\left(\frac{R_m\cdot(\lambda - t)}{P_{max}}+1\right)}\right)} \tag{4}$$

where P(t) is the production of biogas at time t (d); P_{max} is the maximum biogas production or biogas yield (mL biogas/g VS); k_h is the hydrolysis constant (d^{-1}); R_m is the specific rate constant (mL biogas/g VS/d); λ is the lag phase time constant (d).

The differences between both models and the real data were calculated using normalised root mean square error ($RMSE_n$) and r^2 (Equations (5) and (6)) [21].

$$RMSE_n = \left(\frac{1}{n}\sum_{i=1}^{n}\left(\frac{X_i - Y_i}{Y_i}\right)^2\right)^{1/2} \tag{5}$$

$$r^2 = \left(\frac{m(\sum_{i=1}^{n} X_iY_i) - (\sum_{i=1}^{n} X_i)(\sum_{i=1}^{n} Y_i)}{\left(\left[m(\sum_{i=1}^{n} X_i^2) - (\sum_{i=1}^{n} X_i)^2\right]\left[m(\sum_{i=1}^{n} Y_i^2) - (\sum_{i=1}^{n} Y_i)^2\right]\right)^{1/2}}\right)^2 \tag{6}$$

where X_i is the predicted value of biogas (mL biogas/g VS), and Y_i is the measured value of biogas (mL biogas/g VS).

The impact of each salt and its concentration tested on biogas and methane yield and formation kinetics was classified as beneficial, non-effect, inhibitory and/or toxic. Beneficial was defined as the samples that had a biogas or methane yield above the value of the control sample [22]. Non-effect samples had values close to the control [23]. Inhibition was divided into two types, which were yield inhibition if there was a reduction on the maximum yield of biogas or methane [24] and kinetic inhibition if a condition resulted in a lag phase [25]. Toxic refers to samples that had less biogas production than the blank and hence resulted in negative yields achieving complete inhibition [23].

2.4. Analytical Methods

Sludge samples were analysed before and after the BMP tests. Alkalinity, TS and VS were measured using standard methods [20,26]. The measurement of pH was carried out with a Mettler Toledo MA235 pH/ion analyser and a portable Hanna HI 991003 pH meter. Total and soluble COD, TAN and total nitrogen (TN) were analysed using cell test kits from Millipore (VWR, UK) and a spectrometer Spectroquant NOVA 60 from the same company. Samples for analyses of soluble COD, TAN, volatile fatty acid (VFA), ions and metals were obtained after centrifuging at 5000 rpm for 8 h and filtering the supernatant through 0.45 µm Millipore syringe filters. FAN was calculated from the TAN concentration and pH value using Equation (7) [27].

$$[FAN] = \frac{[TAN]}{1+\frac{10^{-pH}}{K_a}} \tag{7}$$

where [FAN] is the FAN concentration (mg/L), [TAN] is TAN concentration (mg/L) and K_a is the dissociation constant with a value of 1.29×10^{-9} for 37 °C [27].

A Shimadzu prominence high-performance liquid chromatographer (HPLC) was used to analyse VFA with a similar methodology as [26]. A Dionex ICS-1600 ion chromatographer (IC) was used for the determination of nitrate, nitrite, sulphate, phosphate, fluoride, chloride and bromide with a mobile phase of 9 mM sodium carbonate and an AS 9HC column.

Finally, a NexION 350D inductively coupled plasma (ICP) mass spectrometer was used to measure cations (calcium, magnesium, potassium, sodium, manganese, zinc, cobalt, molybdenum, nickel, copper, iron, selenium, chromium and tungsten, lithium, aluminium, silica, titanium, arsenic, cadmium, lead, mercury and antinomy). ICP samples were acidified with 100 μL of >68% nitric acid in 10 mL of diluted soluble sample.

Gas samples were characterised for biogas and methane. Biogas was measured using a 1 M hydrochloric acid trap with an inverted cylinder to measure liquid displacement. The biogas was collected in a 10 mL syringe from the measuring cylinder after filtering it through a 0.45 μm Millipore syringe filter to avoid humidity intrusion for methane analysis with a Servomex 1440 gas analyser (Surrey, UK).

2.5. Trade Wastes Load Calculation

The last part of the study aimed to establish the loads of each TW that could be positively co-digested in sewage sludge ADs. To estimate the impact that TWs could have on the Aco-D process, the characterisation data of TWs previously analysed was compared against the different effect ranges for individual constituents found in the BMP tests. For some of the ions of interest, two counter-ions (salts) were studied. The range selected when two ranges were obtained in BMP tests was the one that showed higher inhibition, to account for the worst-case scenario. Scenarios corresponding to TWs/sewage sludge volume ratios of 0/100, 10/90, 30/70, 50/50, 70/30, 90/10 and 100/0 were considered to calculate the total concentration of the ions of interest in the digester. The percentage of the 160 TWs investigated that constituted an inhibitory risk for those scenarios was then calculated.

2.6. Statistics

ANOVA analysis was applied to determine if observed differences were statistically significant. ANOVA uses the F-test to analyse the similarity between groups by comparing the variances between the groups. The similarity between control and test reactors occurred when the F-value (F) was below the F-critical value (Fcrit). In contrast, when F was above Fcrit, the control and test had statistically different yields (see Supplementary Materials Table S2).

3. Results and Discussion

3.1. Trade Wastes Characterisation

The most common categorisation of TWs is by their source [28]. In the current research, data were divided into landfill leachates, glycol/alcohol/beverage industry, food industry, chemical industry, oil-containing waste, soap-containing wastes, sludge and other (Figure 1). Landfill leachates and sludges are not actual industrial wastewaters but, as they are often tankered to water utilities for treatment in the same way as TWs, they were considered as TWs in this work. There was a great variation of COD in the glycol, beverage and alcohol, food and chemical groups with values as high as 700 g/L of COD down to 1 g/L of COD (Figure 1). These three groups were the most predominant ones after landfill leachate with 20 to 30 different TWs from a total of 160 TWs evaluated. Oily and soapy wastewaters that did not fit in the other sections showed a lower variation from 400 to 1 g/L of COD (Figure 1). The lowest variation and the lowest maximum COD level were observed in the groups of landfill leachate, sludge and other TWs that were below 100 g/L of COD (Figure 1). Only landfill leachates could be distinguished with a direct correlation when plotting COD and TAN concentrations against each other (Figure 2). Hence, grouping the TWs by source type did not show any correlation with the characteristics of those TWs, due to the variability in their characteristics within each group. It was therefore decided to evaluate the TWs based on their composition.

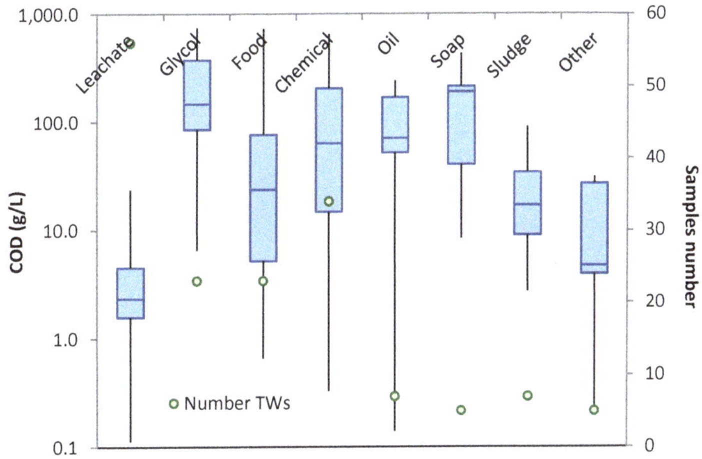

Figure 1. TWs divided by the type of industry where they were generated with their typical COD content.

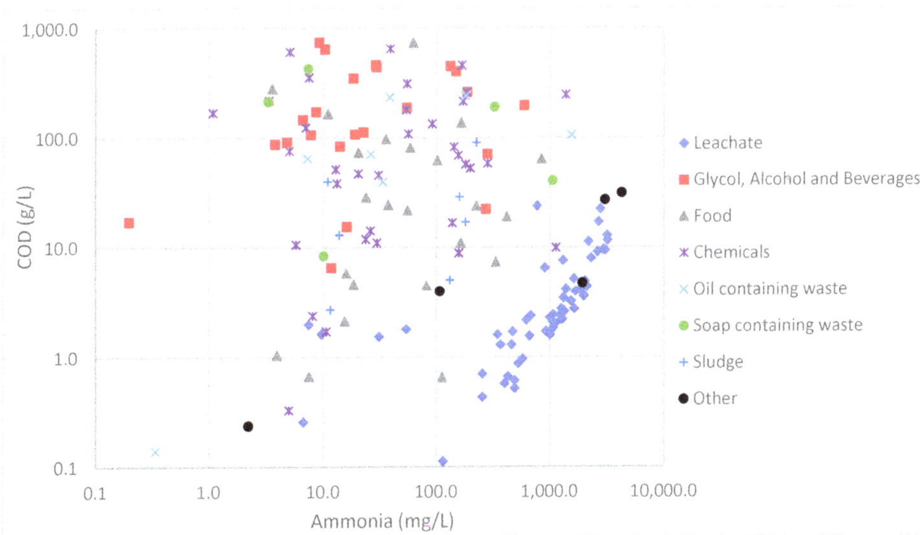

Figure 2. COD versus ammonia concentrations of TWs grouped by their source.

According to previous practice on the full-scale site where the TWs dataset was obtained, the TWs were first divided based on their COD content. Low-strength TWs with a COD below 50 g/L were typically fed at the head of wastewater treatment plants, undertaking full treatment and serving as an additional carbon source in biological nutrient removal (BNR) processes [14]. TWs with a COD between 50 and 250 g/L could go directly to the AD, and high-strength wastes with COD above 250 g/L would go to a high-strength tank (HST) from where they were dosed slowly into the AD. The TWs divided into those groups were then further categorised according to concentrations of all parameters available and compared to inhibitory values reported in the literature (Figure 3). TWs showed high variability in all the compounds and elements considered. The group that was most constant in its composition was the TWs that would go to the HST, for which COD ranged between 600 and 750 g/L (Figure 3). This type of TWs was found to have lower chloride, ammonia, phenol and aluminium levels than the lowest inhibitory values to AD reported in the literature (see Supplementary Materials Table S1). For the wastes that would be directly dosed into the AD, several were found to be above the lowest inhibitory value for all constituents evaluated, demonstrating a risk for the digestion process. The highest concentrations observed for chloride, ammonia, phenol, aluminium, nitrate, zinc, sulphate, copper and arsenic were all above the lowest inhibitory value for AD reported in previous literature (see Supplementary Materials Table S1 and red line Figure 3). Therefore, there was a clear risk that those TWs would negatively affect the sludge digestion if their loads to the reactor were not carefully selected. The TWs going to BNR were low in phenol, aluminium and nitrate, although several constituents of concern were observed, such as chloride, ammonia, zinc, sulphate, copper and arsenic.

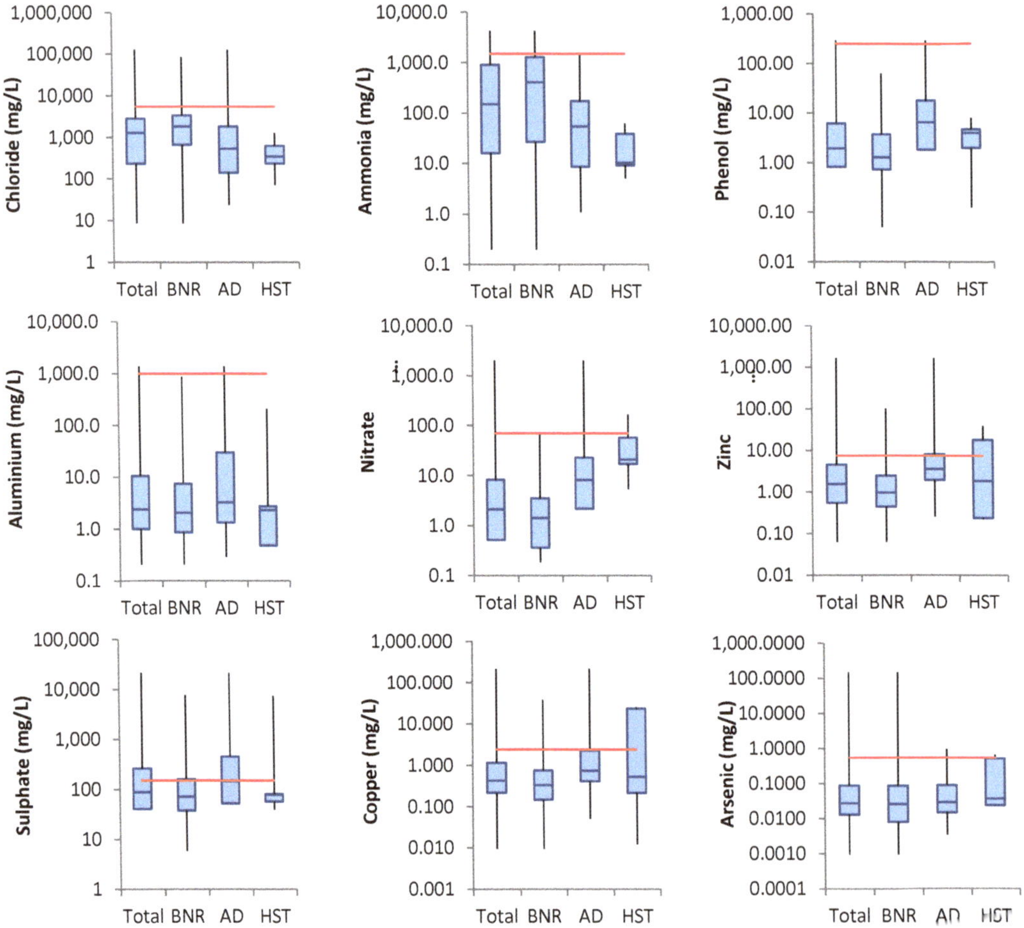

Figure 3. Box and whiskers plot values representing TWs' composition depending on the treatment route generally chosen by water utilities. BNR represents TWs with COD below 50 g/L that are typically fed at the head of wastewater treatment plants, AD represents TWs with a COD between 50 and 250 g/L that go directly to the AD, and HST represents TWs with COD above 250 g/L that normally go to high-strength tanks. The lowest inhibitory value for AD reported in the literature is represented by the red lines (Supplementary Materials Table S1). The lowest point in the whisker represents the minimum value of the data, the highest point on the whisker represents the maximum value, the bottom of the box is the 25 percentile, the top of the box is the 75 percentile and the line dividing the box is the median.

3.2. Biological Methane Potential Tests

3.2.1. Yield

Ammonia

Similar trends and values in biogas and methane formation were observed when dosing ammonium chloride and ammonium sulphate in the digesters (Figure 4a). The biogas and methane yields decreased from 631 to 521 mL/g VS with a 0.8 yield ratio in relation to the control and 415 to 354 mL/g VS with a 0.8 yield ratio over the control, respectively, when the TAN concentration was increased between 635 and 3405 mg/L with ammonium sulphate (Figure 4a). Furthermore, addition of TAN between 586 and 5162 mg/L with ammonium chloride decreased biogas and methane yields from 624 to 431 mL/g VS with a 0.7 yield ratio in relation to the control and 377 to 257 mL/g VS with a 0.7 yield ratio over the control, respectively (Figure 4a). This demonstrates increased inhibition as the TAN concentration increased and suggests that the chloride and sulphate counter-ions had similar or no effects on the results. Nevertheless, the statistical analysis suggested that the yields of test ADs dosed with ammonium chloride were not statistically different from the control reactors except for that of 5162 mg TAN/L, but the performance of ammonium sulphate dosed units was statistically different from that of the control for all the concentrations tested (see Supplementary Materials Table S2). A similar concentration of TAN, 5000 mg/L, was reported as 25% inhibitory for methane yield by Borja et al. in cattle manure digesters [29].

As it has been previously reported, FAN is the ammonia form most toxic to the AD process as it permeates through the cell membrane more easily than ammonium, inhibiting enzymes or leading to proton imbalance [23]. In the current study, there was an increase from 7 to 255 mg/L of FAN using ammonium chloride that decreased the methane yield by 32% (Figure 4b). Additionally, a decrease of 15% in methane formation was observed at 691 mg/L of FAN using ammonium sulphate (Figure 4b). Ammonium chloride addition led to a similar reduction of the methane yield against the control (11%) to the level reported by Bujoczek et al. [30] with a FAN concentration of 173 mg/L when food waste and sewage sludge were co-digested in a ratio 40%/60%. There have been other studies that showed inhibition of methanogens at similar FAN ranges to those tested. Angelidaki and Ahring reported a reduced growth rate of methanogens by 50% at FAN concentrations of 200 and 500 mg N/L [31] and a negative effect in a continuous AD operated at 55 °C when FAN reached 600 and 800 mg N/L [32]. An even lower range of 280 and 520 mg N/L of FAN was found to reduce by 50% the growth in acetate consumers and hydrogen consumer methanogens, respectively, by Borja et al. [29].

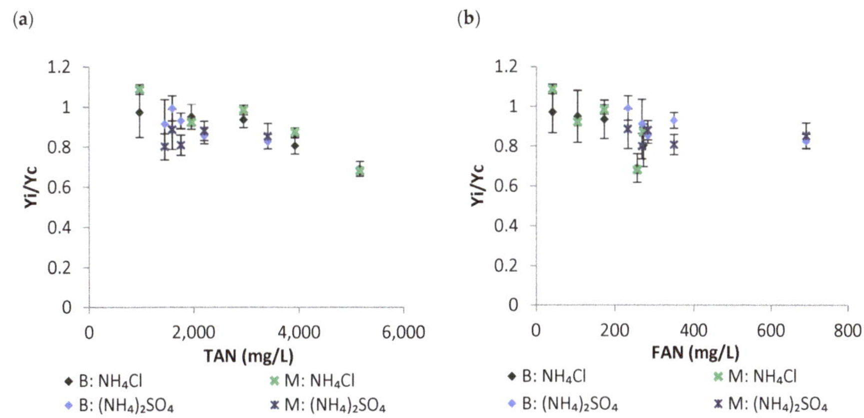

Figure 4. *Cont.*

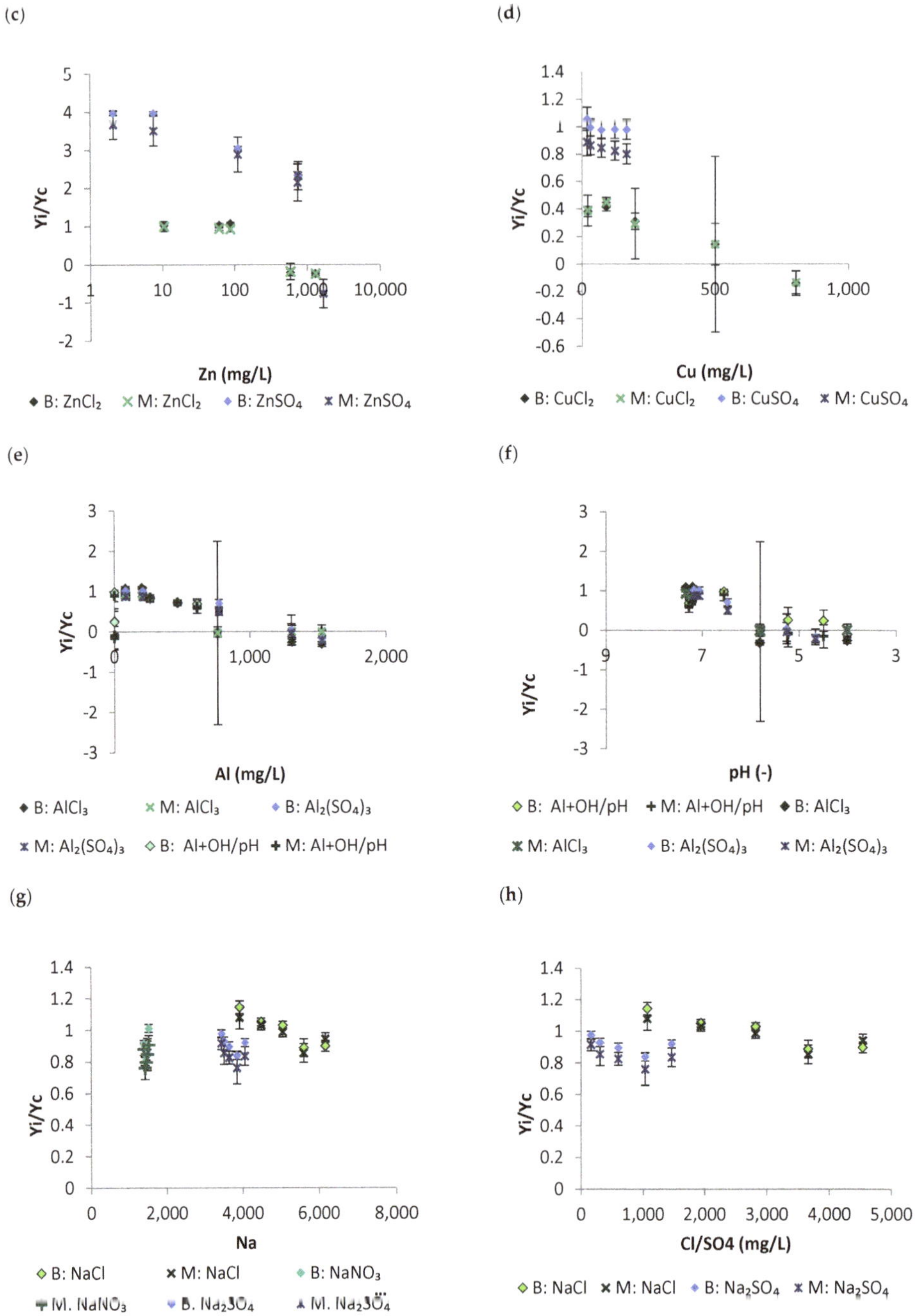

Figure 4. *Cont.*

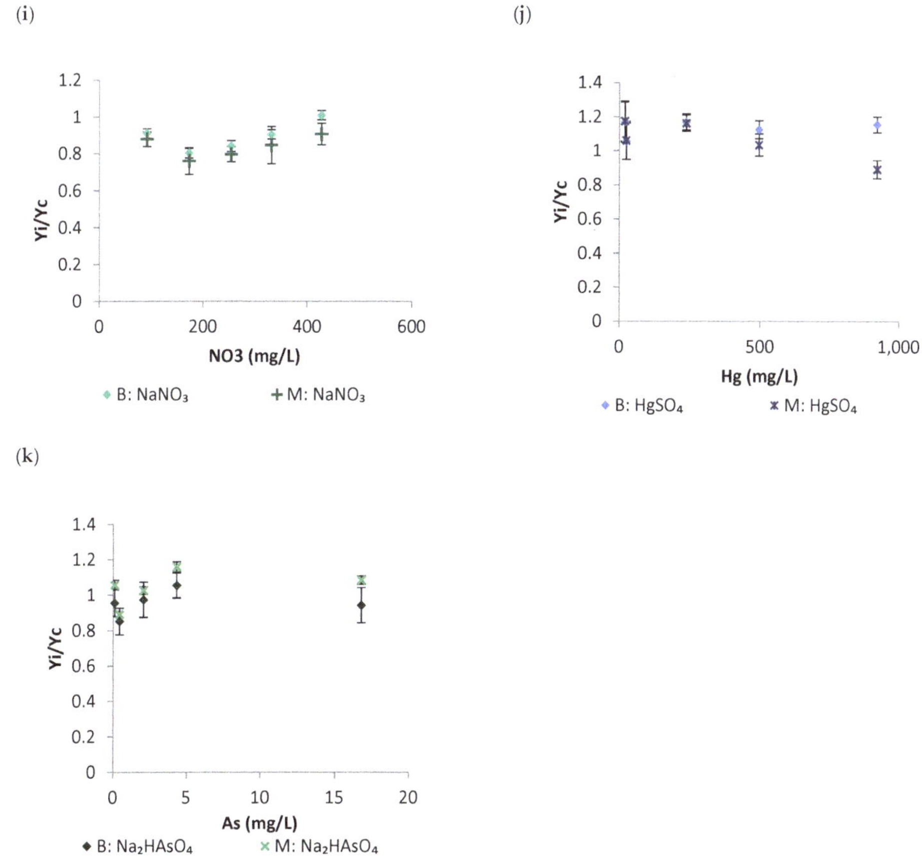

Figure 4. Biogas (B) and methane (M) yield ratios over the control when TAN (**a**) and FAN (**b**) were added using ammonium sulphate and ammonium chloride, zinc was added as zinc sulphate and zinc chloride (**c**), copper was added as copper sulphate and copper chloride (**d**), aluminium (**e**) was added and pH (**f**) was changed using aluminium sulphate, aluminium chloride, aluminium sulphate plus sodium hydroxide and hydrochloric acid, and sodium (**g**), sulphate and chloride (**h**) were added as sodium sulphate and sodium chloride, nitrate was added as sodium nitrate (**i**), mercury was added as mercury sulphate (**j**), and arsenic was added as sodium arsenate (**k**).

Zinc

An increment of Zn from 0.2 mg to 2 mg Zn/L improved the biogas and methane yields, reaching a ratio of 4 and a ratio of 3.7 over the control yield, respectively, for the experiments using zinc sulphate (Figure 4c). This could be associated with zinc stimulating the growth of methanogens as it is needed in RNA polymerase and other biosynthetic enzymes [33]. From 2 to 760 mg Zn/L there was a steady decrease of approximately 40% in both biogas (42%) and methane (37%) yields (Figure 4c). Zinc sulphate in a range between 7 and 112 mg Zn/L still showed an increase of performance over the control, similarly to the increase in yield observed with a zinc concentration of 8 mg/L in the study by Altaş [22]. At 1676 mg Zn/L, there was a toxic effect in the reactors as the biogas and methane productions recorded were lower than the ones of the blank units (i.e., negative yield value) (Figure 4c). Using zinc chloride, the biogas and methane yields were statistically similar to that of the control for the concentrations between 2.8 and 87 mg Zn/L, indicating stable conditions as the biogas production only varied between 630 and 664 mL/g VS. Complete inhibition of biogas and methane production was reached at a concentration of zinc of 592 mg/L using zinc chloride (Figure 4c). A concentration of 760 mg Zn/L increased iso-butyric concentration up to 147 mg/L, compared with the control concentration of 55 mg/L. A further increase up to 1676 mg Zn/L led to an increase in concentrations of all VFAs except valeric and iso-valeric above those of the control. The VFA concentrations at 1676 mg Zn/L were 1085 mg/L of acetic, 2109 mg/L of propionate, 160 mg/L of iso-butyric, and 301 mg/L of butyric, while levels in the control were under the detection limit except for iso-butyric previously mentioned. Zinc has been shown to have an inhibitory effect interfering with enzymes required for methanogenesis [33]. The accumulation of VFA in the current study at 760 and 1676 mg Zn/L agrees with an effect of zinc in the methanogenesis phase of the digestion process. Further trials on continuously operated ADs would be recommended to link dynamics of VFA accumulation to specific pathways of inhibition, which could inform developing a control strategy for TW dosed reactors.

Copper

Biogas and methane yields remained in a similar range with 604–559 mL biogas/g VS and 431–388 mL methane/g VS when the concentration of copper sulphate was increased from 6 to 167 mg Cu/L (Figure 4d). The consistency in biogas and methane yields for the range studied differed from that of previous studies that found inhibition between 2.2 and 50 mg Cu/L (see Supplementary Materials Table S1). This difference could be linked to the zinc concentration in the system, as it has been reported to reduce copper inhibition by competition to enter the microorganism cell [33]. Nevertheless, the background zinc concentration in the current study was between 17 and 46 mg Zn/L in comparison to 65 mg Zn/L, for example, in the work reported by Abdel Azim et al. [33]. Therefore, other mechanisms such as copper polysulphide precipitation, which were reported to be the most common form in sewage sludge at pH 7–7.5 at 37 °C, might be responsible for the reduction of soluble copper [34]. This decrease of soluble copper would have reduced the effect of copper on the current experiments. Attachment to the organic matter could also reduce the available copper concentration for anaerobic microorganisms [35].

Alternatively, the use of copper chloride reduced the biogas and methane yields by up to 62 and 66%, respectively, with an increase in Cu concentration from 0.15 to 22 mg/L (Figure 4d). This value is similar to that of a study with thermophile methanogen *Methanobacterium thermoautotrophicum* that was 50% inhibited at 14 mg Cu/L [36]. The biogas and methane yields further decreased with increasing copper chloride doses, resulting in 86% inhibition and toxic effect at 501 and 804 mg Cu/L, respectively (Figure 4d). An accumulation of acetate at concentrations of 613 mg acetate/L was observed at the end of the BMP test when copper reached 811 mg Cu/L, compared with the control test that had 4 mg acetate/L. Therefore, despite the methanogens being affected by the high concentration of copper, acetogens were able to grow, which agrees with previous studies on copper inhibition that have reported methanogens to be more affected than acidogens [37].

Aluminium

With aluminium sulphate, the biogas and methane yields remained constant at 574 mL biogas/g VS and 454 mL methane/g VS between 3 and 210 mg Al/L, with a small 13% reduction of methane at 210 mg Al/L. When increasing the aluminium concentration to 771 mg Al/L, the biogas and methane yields were reduced by 30 and 51%, respectively. The pH at 771 mg Al/L remained at 6.5, which is within the limit for AD operation, hence suggesting that the impact on process yield was attributable to the aluminium toxicity. Toxic effects for both biogas and methane productions were reached at 1350 mg/L Al, although the low pH value of 5.3 recorded (Figure 4e), prevents a clear understanding of whether inhibition was related to aluminium toxicity, acidification of the reactors or a combination of both.

In the case of aluminium chloride, there was a 9% increase in biogas yield and a 9% reduction of methane yield with concentrations between 1 and 203 mg Al/L. A toxic effect was then observed for both biogas and methane production as the concentration was further increased to 780 mg Al/L and pH reached 5.8 (Figure 4e,f). For both salts, aluminium chloride and aluminium sulphate, dosed reactors toxic effects occurred at a pH of 5.8 or below (Figure 4f). To dissociate the impacts of the pH and aluminium on the system, an experiment with pH control for ADs dosed with aluminium sulphate between 262 and 608 mg Al/L was performed, and other reactors where pH was reduced from 6.55 to 4.5 without aluminium addition were also operated.

The addition of aluminium even with pH control reduced biogas production by 18, 29 and 42% at concentrations of 262, 464 and 608 mg Al/L, confirming that at least part of the toxicity observed was due to aluminium. Aluminium can negatively affect microorganisms by disrupting glucose degradation by attaching to ATP and substitution of divalent ions, as it forms stronger complexes [38]. Additionally, aluminium cationic form (Al^{3+}) can compete with the hydrogen cationic form (H^+) at low pH [38]. An accumulation of VFA was also observed with an increase in acetate concentration from 833 mg/L in the control to 1193 mg/L at 608 mg Al/L addition. VFA accumulation due to aluminium addition has been previously reported [39]. Nevertheless, the experiment with pH control resulted in 42% less biogas at 608 mg Al/L than that of the control unit, suggesting a combined inhibition due to aluminium and acidification in ADs dosed with aluminium sulphate and without pH correction. This synergistic effect could be confirmed when looking at the 84 and 86% reduction in biogas yield in ADs operated at pH 5 and 4.5 without aluminium addition, as its impact is lower than what was observed in aluminium dosed units with pH of 5.3, where toxicity was complete and resulted in negative biogas yields.

Nitrate

Increasing levels of nitrate by dosing sodium nitrate resulted in an initial decrease in both biogas and methane performance, followed by an increase in yields reaching similar production to that of the control reactors at the maximum concentration studied of 427 mg NO_3/L (Figure 4i). The 15% methane inhibition observed at 332 mg NO_3/L was similar to the 14% at that concentration reported by Tugtas and Pavlostathis [15]. The lowest point was at 174 mg NO_3/L with 20% inhibition in biogas yield with 606 mL/g VS and 24% inhibition of methane yield with 411 mL/g VS (Figure 4i). The change of yield performance with nitrate addition might be related to the accumulation of butyrate at the end of the experiment: >0.1 mg butyrate/L in the control against 113 mg butyrate/L at 254 mg NO_3/L, 118 mg butyrate/L at 332 mg NO_3/L and 135 mg butyrate/L at 427 mg NO_3/L. The accumulation of butyrate might result from the accumulation of hydrogen caused by the inhibition of hydrogen consumers by nitrate, as hydrogen has been seen to inhibit butyrate degradation [40,41]. Kluber and Conrad [40] found a 50% inhibition at 186 mg NO_3/L (3 mM) using *Methanosarcina barkeri,* which is double the inhibition at 174 mg NO_3/L observed in the current study, and at 1550 mg NO_3/L (25 mM) using *Methanobacterium bryantii* fed by hydrogen.

Arsenic and Mercury

With mercury, the biogas yield was mostly similar to that of the control units except for an increase of 17 and 13% in biogas yield above the control value of 754 mL/g VS at 239 and 495 mg Hg/L. However, the methane yield was constant at about 570 mL/g VS between 0.027 and 920 mg Hg/L (Figure 4j). Comparison with previous literature shows that 500 mg/L of mercury decreased by 90% the biogas production when a continuously fed system was spiked once with mercury [42]. The lower inhibition observed in the current research as compared with values reported in the literature could result from the low solubility of mercury sulphate, which could precipitate in the reactor or adsorb on the sludge, reducing its bio-availability [43] as its concentration was reduced below 0.1 mg Hg/L at the end of the current experiment for all the concentrations studied.

Arsenate experiments had similar biogas and methane yields throughout the experiments from 0.02 to 16 mg As/L by adding sodium arsenate (Figure 4k). Field et al. observed a competition for the organics between arsenate reducers and methanogens at a concentration lower than 150 mg As/L of arsenate [44]. A mechanism that might reduce the toxicity of arsenate is precipitation with sulphur compounds [43,44]. Nonetheless, in previous experiments, a value of 960 mg SO_4/L (10 mM SO_4), which is higher than the concentration in this experiment of 31 mg SO_4/L, did not show precipitation with sulphate when pH was close to neutral [44].

Counter-Ion Effect

As commented in the previous sections, there are differences between the experiments using chloride or sulphate as counter-ions. The TAN experiments showed a higher methane inhibition, 32%, using ammonium sulphate at 3405 mg N/L than for ammonium chloride at 5162 mg N/L, 15%. In contrast, the FAN levels for those experiments showed a similar inhibition with ammonium sulphate and ammonium chloride at 255 mg N/L, decreasing methane yield around 12% (Figure 4a,b). Zinc chloride presented a constant production of biogas and methane between 3 and 87 mg/L, in contrast with the decreasing trend resulting from zinc sulphate from 2 to 112 mg Zn/L. Nonetheless, the ratios of yields for biogas and methane were lower in the zinc chloride, close to 1, than the ratio using zinc sulphate, which decreased from 4 to 2 (Figure 4c). The copper sulphate ratios of biogas and methane yields were close to 1, exhibiting a better performance than copper chloride experiments that showed a decrease of performance to less than half that of the control (Figure 4d). This difference could be because copper sulphate's solubility, 22.2% in weight, is lower than copper chloride's solubility, 44.8% in weight at 40 °C [45], precipitating a portion of copper. Additionally, the higher content of zinc in the experiment using copper sulphate could have competed with copper for transport into the cells, reducing its toxicity [33]. Aluminium chloride addition decreased pH more than similar concentrations using aluminium sulphate because of the formation of hydrochloric acid when aluminium chloride is mixed in water [46] (Figure 4e,f). At 771 mg Al/L, aluminium sulphate showed a better performance of biogas and methane yields; 402 mL biogas/g VS with a ratio of 0.7 over the control and 241 mL methane/g VS with a ratio of 0.5 over the control; while aluminium chloride at 760 mg Al/L showed a toxic effect (Figure 4e,f). For sodium salts concentrations ranged between 3360 and 4054 mg Na/L, there was an increase in biogas yield of 14% and similar methane yield than that of the control at 3915 mg Na/L using sodium chloride, while sodium sulphate decreased the biogas and methane yields by 8 and 16%, respectively, at 4054 mg Na/L. In the current study, there was a decrease of 16% in methane yield at 1466 mg SO_4^{2-}/L. Jeong et al. observed 33% reduction in methane production at 1500 mg SO_4^{2-}/L [47]. Addition of chloride up to 4535 mg Cl/L decreased methane yield by only 6%. Vijayaraghavan et al. found a decreasing trend from the optimum when chloride was added above 5500 mg Cl/L [48].

Therefore, there was an effect of the counter-ion in the experiments. Ammonia and sodium had a synergistic effect with sulphate, while zinc, copper and aluminium had a synergistic effect with chloride. Previous studies have found a different effect for chloride salts than sulphate salts in soil bacteria because of different inhibitory mechanisms of chloride and sulphate. Chloride can affect some enzymes, such as phosphoenolpyruvate carboxylase, glutathione reductase and ribulose biphosphate carboxylase, and it can inhibit protein synthesis by preventing the binding of ribosomes to mRNA [49,50]. Meanwhile, sulphate can be metabolised by some bacteria, competing with the methanogens for the substrate [17]. Sulphate addition has an effect on sulphate-reducing bacteria that can be inhibited by some of the elements studied. Therefore, the negative impact on sulphate reducing species will have a beneficial effect on methanogens, because they are competitors. For these reasons, the use of similar counter-ions is important when various cations are compared, and the concentration of chloride and sulphate might be an important parameter in the limit set for a cation in AD reactors.

3.2.2. Kinetics

A standard first-order model and the Gompertz model were utilised to analyse the kinetic data of biogas formation in the BMPs previously described. The first-order model gives a good approximation for substrates that contain easily hydrolysable organic content without inhibition in the form of a lag phase [51]. Instead, Gompertz accounts for any lag phase, leading to a better fit of the data if samples are inhibitory for the process or present a delayed degradation [51,52]. In general, the Gompertz model showed a better fit for the data in this study than first-order kinetics, as confirmed by r^2 and $RMSE_n$ values (see Figure 5 and Supplementary Materials Table S3).

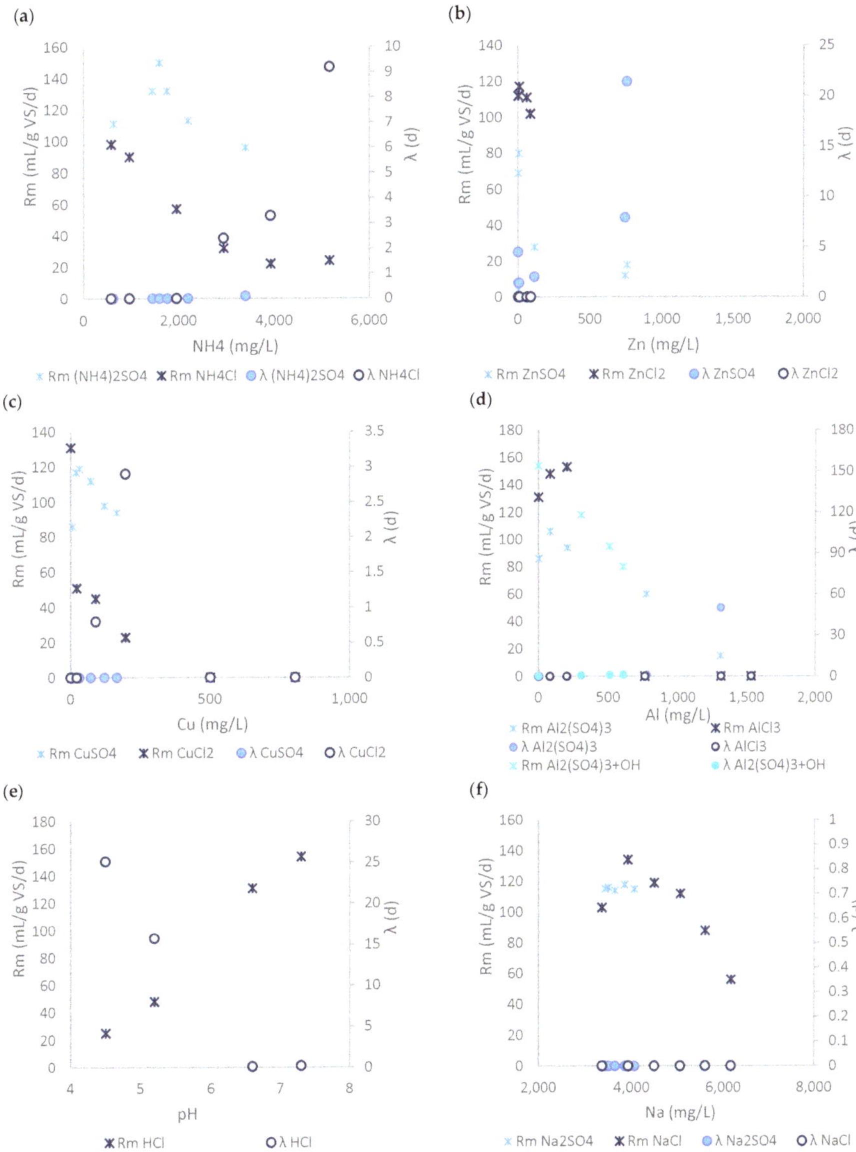

Figure 5. *Cont.*

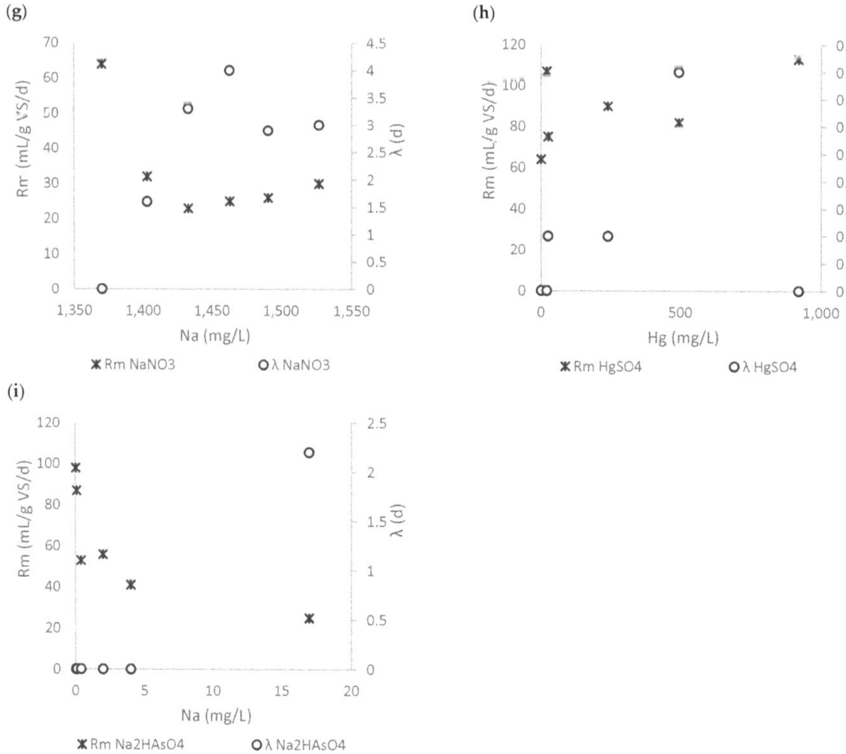

Figure 5. Results from Gompertz model for the biogas formation kinetics on the BMP experiments where TAN was increased adding ammonium sulphate and ammonium chloride (**a**), zinc was added as zinc sulphate and zinc chloride (**b**), copper was added as copper sulphate and copper chloride (**c**), aluminium was added with aluminium sulphate, aluminium chloride, and aluminium sulphate plus sodium hydroxide (**d**), pH was modified using hydrochloric acid (**e**), sodium was added as sodium sulphate, sodium chloride (**f**), and sodium nitrate (**g**), mercury was added as mercury sulphate (**h**), and arsenic was added as sodium arsenate (**i**).

The faster biogas formation kinetics of ammonium sulphate dosed reactors compared to ammonium chloride dosed units is notable in the Gompertz model results. The lag phase was found to increase from 2 up to 9 d when TAN was between 2946 and 5162 mg N/L using ammonium chloride, while ammonium sulphate showed a much smaller lag phase, 0.1 d, at 3405 mg N/L of TAN. A higher maximum R_m was also obtained at 1602 mg N/L of TAN of ammonium sulphate, 150 mL biogas/g VS/d, than in the ammonium chloride experiment, 98 mL biogas/g VS/d, at 586 mg/L of TAN (Figure 5). Therefore, despite the biogas yield being similar to that of the control for most of the concentrations of ammonium chloride, except for 5162 mg N/L, there was a kinetic inhibition in the shape of a delayed biogas formation between 2946 and 5162 mg N/L (Figure 5). In contrast, ammonium sulphate resulted in a decrease in biogas yield from a concentration of 1764 mg N/L, with kinetic inhibition resulting in a lag phase noticeable only at 3405 mg N/L (Figure 5). Therefore, the inhibition of kinetic parameters for ammonium starts at values close to or above 3000 mg N/L for both salts. These data showed that ammonia might not be important in kinetic inhibition, but it was the counter-ion which was responsible for the observed lag phase or a synergistic effect between ammonium and chloride.

Zinc exerted a different behaviour with different counter-ions that differed from what was observed with the ammonium salts. The lag phase in the Gompertz model observed in the control reactors, 4.5 d, decreased to 1.4 d and 2 d when zinc sulphate was added between 2 and 112 mg Zn/L, respectively, with a later increase reaching a value of 21.4 d at a concentration of 760 mg Zn/L (Figure 5). This contrasts with the results reported by Atlaş [22] that showed an increase of the length of the lag phase when zinc was increased from 4 to 62 mg Zn/L, while in this study the zinc had to be increased up to 476 mg Zn/L to observe a longer lag phase than the control value of 7.9 d. The R_m value from the Gompertz model increased to 80 mL biogas/g VS/d when zinc sulphate was added up to 7 mg Zn/L (Figure 5). Zinc chloride did not result in any lag phase between 10 and 87 mg Zn/L. Therefore, zinc kinetic inhibition is enhanced by sulphate. In contrast with the yield results, kinetics were faster in the experiments using zinc chloride rather than for zinc sulphate. Moreover, the effect of chloride and sulphate was reversed to what was found for ammonia. Therefore, the cause of the inhibition might not be one ion or the other, but the synergistic effect of both ions or other components in the sludge on the kinetic parameters.

Copper sulphate at 32 mg Cu/L improved the substrate degradation rate, as evidenced by a 38% increase in R_m from the Gompertz model (Figure 5). No lag phase was observed within the concentrations studied which, combined with the lack of impact on biogas and methane yields, suggest that copper within 7 and 167 mg Cu/L using sulphate is not inhibitory for the digestion of sewage sludge and can indeed accelerate gas formation. The control from the experiment was faster than when copper chloride was dosed as R_m from the Gompertz model, which produced the highest values in the experiment, 131 mL biogas/g VS/d (Figure 5). Additionally, copper chloride led to a lag phase equal to 3 d at 198 mg Cu/L, while copper sulphate did not show any lag phase at a similar concentration. It was hence concluded that any inhibition in copper chloride reactors was attributable to the chloride counter-ion and not the copper itself.

Aluminium sulphate dosing resulted in a positive impact at a dose of 85 mg Al/L, as the Gompertz model showed a 23% increase on R_m up to 106 mL biogas/g VS/d (Figure 5). A further increase of aluminium sulphate, up to 771 mg Al/L, increased the lag phase to 0.7 d, impacting substrate degradation rate. The aluminium chloride maximum k_h from the first kinetic model was 0.49 d^{-1} at 82 mg Al/L with an increase of 19% from the control, and a maximum R_m from Gompertz model at 204 mg Al/L, increasing by 17% from the control (see Figure 5 and Supplementary Materials Table S3). Aluminium addition using aluminium sulphate with sodium hydroxide increased λ from 0.2 d in the control to 1.2 d at 607 mg Al/L (Figure 5). A decrease in pH alone increased λ from 0.2 d to 25.1 d, evidencing the need for the microbial communities to recover from the initial acidic conditions (Figure 5). Both salts alone triggered a decrease in pH, but aluminium chloride did not show any λ despite the lag phase observed when decreasing the pH alone, suggesting that pH alone did not affect the lag phase increase. The similarities between the lag phase data at 771 mg Al/L in the experiments of aluminium sulphate without pH control, and 607 mg Al/L in the experiment of aluminium sulphate with adjusted pH, suggest a higher effect of aluminium in those experiments. It can be concluded that the difference in synergistic and antagonistic effects between the salt ions and the pH decreases because the salt addition is the cause of the different behaviour between aluminium sulphate and aluminium chloride and the reason for the lag phase.

Sodium nitrate impacted process kinetics when dosed, resulting in a λ of 4 d at 1463 mg Na/L and 254 mg NO_3/L (Figure 5), which also matched a decrease of 15% in biogas yield (Figure 4i). In contrast to what was observed in the yield, the lag phase did not reduce to the values of the control, reaching 3 d at 1527 mg Na/L and 427 mg NO_3/L. Therefore, nitrate concentrations above 92 mg NO_3/L require time to acclimatise, which could be associated with a shift in the microbial population as commented in the previous section.

Sodium experiments using chloride and sulphate as counter-ions achieved a faster biogas production at 3915 and 3846 mg Na/L, and 1066 mg Cl/L and 1032 mg SO_4/L, respectively. First-order kinetics for sodium chloride and sodium sulphate showed an increase of k_h of 24% and 48% from the control, respectively (see Supplementary Materials

Table S3). This suggested better substrate degradation when sulphate was used as a counter-ion. In contrast, the Gompertz model parameter R_m increased by 30 and 14% respectively from the control (Figure 5). Sodium sulphate data fitted better on the first-order kinetics between 3499 and 4054 mg Na/L than in the Gompertz model, suggesting a lack of lag phase, while sodium chloride fitted better in the Gompertz model. Despite the small differences, there were no clear differences between the use of chloride and sulphate as counter-ions and using sodium.

Mercury addition improved the speed of biogas conversion when it was added up to 920 mg/L, increasing R_m from the Gompertz model by 77% from the control. An increase of λ occurred at 495 mg/L up to 0.8 d, showing a slight inhibition. Sodium arsenate affected the kinetic parameters more, resulting in a λ of 2 d when arsenic was added up to 17 mg/L, in contrast to what was observed in the yield, which remained similar to that of the control for every concentration tested (Figure 5). This could mean that at the concentration of 17 mg As/L there was a slight inhibition and an acclimation of the microorganisms, or a competition between the arsenate reducers and the methanogens for the organic substrate [44]. Therefore, despite no inhibitory effect on the yield, there might be an inhibitory effect on the kinetics of AD due to the competition between arsenate reducers and methanogens [43,44].

In general, speed of biogas formation using chloride differed from that using sulphate, which differs from the results obtained in the previous section for yields. Ammonium chloride and sodium arsenate exhibited a lag phase despite having similar yields to the control. Zinc sulphate and sodium nitrate showed a slower biogas formation than the control for concentrations that resulted in similar or greater yields than the control reactors. Aluminium sulphate, copper sulphate and ammonium sulphate showed decreasing trends in the yield and kinetics. There is a difference between the impact on gas formation kinetics of the ion studied depending on the counter-ion used, as was observed with sulphate and chloride. The difference in kinetics parameters might be caused by an interaction of the two ions in the salt rather than the separate effect of one of them.

3.3. Potential for Co-Digestion of Trade Wastes

One of the objectives was to establish the loads of each type of TW that can be added to sludge anaerobic digesters without hindering methane production. A preliminary understanding of the maximum TW loads that can be co-digested optimally in sewage sludge ADs can be obtained with results from the inhibition tests in BMPs, summarised in Table 3, and information on the characterisation of TWs obtained from the database. To understand the impact of adding TWs in different loads, scenarios considering different volume ratios of TWs to sewage sludge were evaluated: 0/100, 10/90, 30/70, 50/50, 70/30, 90/10 and 100/0. The concentrations of individual constituents calculated for those scenarios were then compared with the BMP inhibition and toxicity values to obtain Figure 6.

Ammonia, aluminium, zinc, copper, chloride, nitrate, sulphate and arsenic showed a high number of TWs with a potentially negative effect on sewage sludge AD, with the number of potentially inhibitory TWs increasing when TW loads to the reactor were increased. In a scenario with only TWs fed to the ADs (TWs/sludge volume ratio of 100/0) 2, 4, 1, 7, 16, 4, 32 and 1% of the TWs studied were regarded as potentially inhibitory when considering calculated concentrations of ammonia, aluminium, zinc, copper, chloride, nitrate, sulphate and arsenic, respectively (Figure 6). The only element that did not show any effect at any volume ratio was mercury (Figure 6).

Sulphate and chloride constituted the largest risk of inhibition, with 7 and 4% of the TWs resulting in inhibitory calculated concentrations in a 10/90 scenario and 32 and 16% in a 100/0 scenario for sulphate and chloride, respectively (Figure 6). These two constituents showed potential to inhibit methane yield by up to 25% at the studied concentrations (Figure 2). Ammonia, aluminium, zinc and copper all led to a less pronounced increase in the proportion of TWs with potential to have a negative effect by those constituents as the TW/sludge volume ratio was increased (Figure 6). They also showed a higher methane yield inhibition at the concentrations studied with potential for 32, 39, 100 and 100% methane reduction, respectively (Figure 2).

Table 3. Summary of concentration ranges impacting AD of sewage sludge obtained from BMP tests.

	Beneficial	Non-Effect	Yield Inhibition	Kinetic Inhibition	Toxic	Unknown
Chloride (mg/L)	-	<3666	3666–4535	-	-	>4535
Ammonia (mg/L)	635–1460	<635 1460–1602	1602–3405	1602–3405	-	>3405
Aluminium (mg/L)	-	<309	309–607	309–607	-	>607
Nitrate (mg/L)	-	<92 332–427	92–332	92–427	-	>427
Zinc (mg/L)	-	<87	87–498	-	>498	-
Sulphate (mg/L)	-	<162	162–1466	-	-	>1466
Copper (mg/L)	-	<7	7–501	91–198	501–804	-
Arsenic (mg/L)	-	<17	-	17	-	>17
Mercury (mg/L)	-	<25 495–920	25–495	25–495	-	>920

Note: As specified in Section 2.3, "beneficial" was defined as the samples with a biogas yield above the control yield; "non-effect" samples had yields similar to the control; "yield inhibition" showed a reduction in the maximum yield of biogas in respect to the control one; "kinetic inhibition" showed an increase of lag phase; "toxic" refers to yield values below the blank production; and, "unknown" refers to conditions above the maximum concentration studied in the BMP tests, where no toxicity was observed. When two salts were tested for the same inhibitor, the range presented is the one that showed higher yield inhibition at the smallest concentration to account for the worst-case scenario, with the exception of chloride and sulphate salts, which had sodium as a counter-ion.

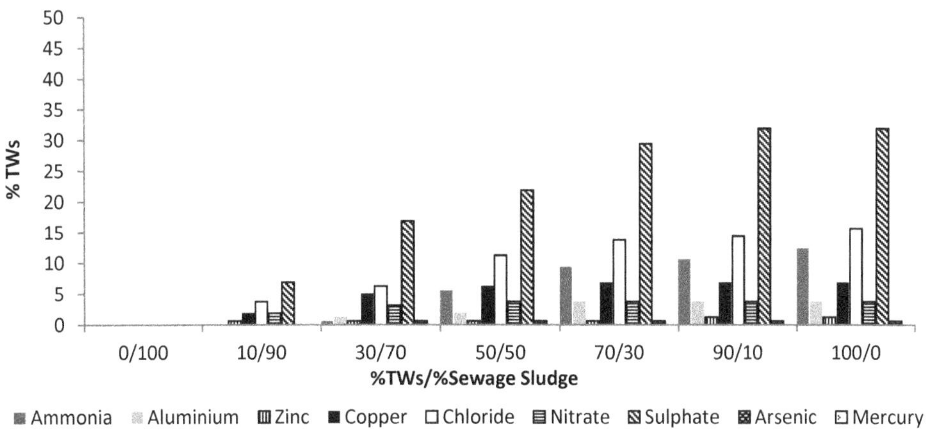

Figure 6. Percentage of TWs considered from the database (total of 160) that could result in a negative effect (inhibition or toxicity) if co-digested with sewage sludge, depending on the TWs/sewage sludge volume ratio.

Overall, a volume ratio of 10/90 TWs and sewage sludge (calculated as per Section 2.5) was estimated to be the best volume ratio for co-digestion considering all the ions studied and typical concentrations encountered in real TWs (Figure 6). Most of the constituents in TWs remain below the inhibitory levels obtained in BMP tests at this volume ratio, with only 13.7% of all the TWs studied presenting a potential for inhibition. Chemical TWs showed inhibition in 7 cases for sulphate, 3 for chloride, 3 for copper and 2 for nitrate.

Glycerol/beverage/alcohol wastes showed inhibition from sulphate, chloride and nitrate for 7, 1 and 1 of them, respectively. An oil containing TW showed inhibition for zinc and sulphate. Chloride inhibition was also found in one of the food TWs. Finally, chloride and sulphate inhibitory effects were observed in 1 and 2 landfill leachates, respectively. Moreover, there were constituents such as ammonia, aluminium, arsenic and mercury that did not show any inhibitory effect at that volume ratio. The average spare capacity in wastewater treatment plants digesters has been reported as 30%, which could be an opportunity for Aco-D with TWs up to this volume ratio [53]. An increase in TW in the feed to ADs from a volume ratio of 10 to 30% can, however, significantly impact the process. To illustrate, for a scenario with 10% of TWs in the feedstock, 1% of the TWs investigated constitute a risk of zinc inhibition, but a scenario with 30% of TWs showed 1% of the TWs considered could be completely toxic for the digestion process (Supplementary Materials Figure S1a,b). Therefore, further investigation in continuous experiments of the impact of the preferred TW/sludge volume ratio (10/90) and the volume ratio representing site spare capacity (30/70) is paramount to understand whether acclimation of microbial communities not observable in batch ADs could occur.

4. Conclusions

TW characterisation and BMP tests give an insight into the opportunities for Aco-D of sludge and TWs. Within the 160 TWs studied, the key constituents identified as potentially inhibitory (yield or lag phase) or toxic were ammonia, chloride, nitrate, sulphate, aluminium, copper, zinc, arsenic and mercury. Co-digestion with TWs as 10% of the feedstock volume in sewage sludge digesters can result in significant biogas uplifts, although inhibition would be expected for 13.7% of the TWs investigated, which requires further investigation of potential for microbial acclimation in continuous experiments. Those inhibitory TWs were 15, 9, 3, 1 and 1 TWs from chemical, glycerol/beverage/alcohol, landfill leachate, oil-containing, and food groups, respectively. Further work is recommended to investigate variability in composition of individual TWs and their impact on consistently sustaining biogas uplifts.

5. Disclaimer

This study was completed in cooperation with Severn Trent to gain a more in-depth understanding of co-digestion at a scientific and technical level. At the date of releasing the paper, Severn Trent did not co-digest materials with sewage sludge.

Supplementary Materials: The following supporting information can be downloaded at: https://www.mdpi.com/article/10.3390/pr11020519/s1, Table S1. Concentrations and impact of AD inhibitors reported in the literature; Table S2. Statistics of the biogas and methane yield using ANOVA compared with the control test; Table S3. Results from first-order and Gompertz model for the biogas formation kinetics of the BMP experiments; Figure S1. Percentage of TWs within the 160 studied that will have beneficial, non-effect, yield inhibition, kinetic inhibition, toxic or unknown impacts if dosed to sewage sludge anaerobic digesters as 10% of the feedstock (a) or 30% of the feedstock (%) (b). References [54–72] are listed in Supplementary Materials.

Author Contributions: Conceptualisation, O.B.d.F., M.P. and Y.B.F.; methodology, O.B.d.F., M.P. and Y.B.F.; formal analysis, O.B.d.F. and M.G.; writing—original draft preparation, O.B.d.F.; writing—review and editing, O.B.d.F., M.P. and Y.B.F.; supervision, M.P. and Y.B.F.; project administration, M.P. and Y.B.F.; funding acquisition, M.P. and Y.B.F. All authors have read and agreed to the published version of the manuscript.

Funding: This work was undertaken during O. Berzal de Frutos's Ph.D. research at Cranfield University, funded by Severn Trent.

Data Availability Statement: All relevant data are provided within the manuscript or as Supplementary Materials.

Conflicts of Interest: The authors declare no conflict of interest.

References

1. Berenjkar, P.; Islam, M.; Yuan, Q. Co-treatment of sewage sludge and mature landfill leachate by anaerobic digestion. *Int. J. Environ. Sci. Technol.* **2019**, *16*, 2465–2474. [CrossRef]
2. Cardona, L.; Levrard, C.; Guenne, A.; Chapleur, O.; Mazéas, L. Co-digestion of wastewater sludge: Choosing the optimal blend. *Waste Manag.* **2019**, *87*, 772–781. [CrossRef] [PubMed]
3. Maragkaki, A.E.; Fountoulakis, M.; Gypakis, A.; Kyriakou, A.; Lasaridi, K.; Manios, T. Pilot-scale anaerobic co-digestion of sewage sludge with agro-industrial by-products for increased biogas production of existing digesters at wastewater treatment plants. *Waste Manag.* **2017**, *59*, 362–370. [CrossRef] [PubMed]
4. Tchobanoglous, G.; Burton, F.L.; Stensel, H.D. *Wastewater Engineering Treatment and Reuse*, 4th ed.; Metcalf & Eddy, Inc.: McGraw Hill, NY, USA, 1991.
5. Silvestre, G.; Rodríguez-Abalde, A.; Fernández, B.; Flotats, X.; Bonmatí, A. Biomass adaptation over anaerobic co-digestion of sewage sludge and trapped grease waste. *Bioresour. Technol.* **2011**, *102*, 6830–6836. [CrossRef]
6. Wang, N.; Zheng, T.; Ma, Y. New insights into the co-locating concept on synergistic co-digestion of sewage sludge and food waste towards energy self-sufficient in future WWTPs. *Bioresour. Technol. Rep.* **2020**, *10*, 100351. [CrossRef]
7. Nielfa, A.; Cano, R.; Fdz-Polanco, M. Theoretical methane production generated by the co-digestion of organic fraction municipal solid waste and biological sludge. *Biotechnol. Rep.* **2015**, *5* (Suppl. C), 14–21. [CrossRef]
8. Bai, X.; Chen, Y.-C. Synergistic effect and supernatant nitrogen reduction from anaerobic co-digestion of sewage sludge and pig manure. *Bioresour. Technol. Rep.* **2020**, *10*, 100424. [CrossRef]
9. Hagelqvist, A. Batchwise mesophilic anaerobic co-digestion of secondary sludge from pulp and paper industry and municipal sewage sludge. *Waste Manag.* **2013**, *33*, 820–824. [CrossRef]
10. Siddique, M.N.I.; Munaim, M.S.A.; Wahid, Z.B.A. The combined effect of ultrasonic and microwave pre-treatment on bio-methane generation from co-digestion of petrochemical wastewater. *J. Clean. Prod.* **2017**, *145*, 303–309. [CrossRef]
11. Koster, I.W.; Lettinga, G. Anaerobic digestion at extreme ammonia concentrations. *Biol. Wastes* **1988**, *25*, 51–59. [CrossRef]
12. Gai, H.; Jiang, Y.; Qian, Y.; Kraslawski, A. Conceptual design and retrofitting of the coal-gasification wastewater treatment process. *Chem. Eng. J.* **2008**, *138*, 84–94. [CrossRef]
13. Boardman, G.D.; McVeigh, P.J. Use of UASB technology to treat crab processing wastewaters. *J. Environ. Eng.* **1997**, *123*, 776. [CrossRef]
14. Czerwionka, K.; Makinia, J.; Kaszubowska, M.; Majtacz, J.; Angowski, M. Distillery wastes as external carbon sources for denitrification in municipal wastewater treatment plants. *Water Sci. Technol.* **2012**, *65*, 1583–1590. [CrossRef]
15. Tugtas, A.E.; Pavlostathis, S.G. Inhibitory effects of nitrogen oxides on a mixed methanogenic culture. *Biotechnol. Bioeng.* **2007**, *96*, 444–455. [CrossRef]
16. Glanpracha, N.; Annachhatre, A.P. Anaerobic co-digestion of cyanide containing cassava pulp with pig manure. *Bioresour. Technol.* **2016**, *214*, 112–121. [CrossRef]
17. Hu, Y.; Jing, Z.; Sudo, Y.; Niu, Q.; Du, J.; Wu, J.; Li, Y.-Y. Effect of influent COD/SO42− ratios on UASB treatment of a synthetic sulfate-containing wastewater. *Chemosphere* **2015**, *130*, 24–33. [CrossRef]
18. Soto, M.; Méndez, R.; Lema, J.M. Sodium inhibition and sulphate reduction in the anaerobic treatment of mussel processing wastewaters. *J. Chem. Technol. Biotechnol.* **1993**, *58*, 1–7. [CrossRef]
19. Astals, S.; Koch, K.; Weinrich, S.; Hafner, S.D.; Tait, S.; Peces, M. Impact of storage conditions on the methanogenic activity of anaerobic digestion inocula. *Water* **2020**, *12*, 1321. [CrossRef]
20. Holliger, C.; Fruteau de Laclos, H.; Hafner, S.D.; Koch, K.; Weinrich, S.; Astals, S.; Alves, M.; Andrade, D.; Angelidaki, I.; Appels, L.; et al. Requirements for Measurement and Validation of Biochemical Methane Potential (BMP): Standard BMP Methods Document 100, Version 1.4. 2020. Available online: https://www.dbfz.de/en/BMP (accessed on 19 April 2020).
21. Bhattarai, S.; Oh, J.-H.; Euh, S.-H.; Krishna Kafle, G.; Hyun Kim, D. Simulation and model validation of sheet and tube type photovoltaic thermal solar system and conventional solar collecting system in transient states. *Sol. Energy Mater. Sol. Cells* **2012**, *103*, 184–193. [CrossRef]
22. Altaş, L. Inhibitory effect of heavy metals on methane-producing anaerobic granular sludge. *J. Hazard. Mater.* **2009**, *162*, 1551–1556. [CrossRef]
23. Rajagopal, R.; Massé, D.I.; Singh, G. A critical review on inhibition of anaerobic digestion process by excess ammonia. *Bioresour. Technol.* **2013**, *143*, 632–641. [CrossRef] [PubMed]
24. Fang, C.; Boe, K.; Angelidaki, I. Anaerobic co-digestion of desugared molasses with cow manure; focusing on sodium and potassium inhibition. *Bioresour. Technol.* **2011**, *102*, 1005–1011. [CrossRef] [PubMed]
25. van Houten, R.T.; Lettinga, G. Biological sulphate reduction with synthesis gas: Microbiology and technology. In *Progress in Biotechnology*; Wijffels, R.H., Buitelaar, R., Bucke, C., Tramper, J., Eds.; Elsevier: Amsterdam, The Netherlands, 1996; Volume 11, pp. 793–799. [CrossRef]
26. Bajón Fernández, Y.; Soares, A.; Vale, P.; Koch, K.; Masse, A.L.; Cartmell, E. Enhancing the anaerobic digestion process through carbon dioxide enrichment: Initial insights into mechanisms of utilization. *Environ. Technol.* **2019**, *40*, 1744–1755. [CrossRef]
27. Fotidis, I.A.; Karakashev, D.; Kotsopoulos, T.A.; Martzopoulos, G.G.; Angelidaki, I. Effect of ammonium and acetate on methanogenic pathway and methanogenic community composition. *FEMS Microbiol. Ecol.* **2013**, *83*, 38–48. [CrossRef]
28. Ng, W.J. *Industrial Wastewater Treatment*; Imperial College Press: London, UK, 2006.

29. Borja, R.; Sánchez, E.; Weiland, P. Influence of ammonia concentration on thermophilic anaerobic digestion of cattle manure in upflow anaerobic sludge blanket (UASB) reactors. *Process. Biochem.* **1996**, *31*, 477–483. [CrossRef]
30. Bujoczek, G.; Oleszkiewicz, J.; Sparling, R.; Cenkowski, S. High Solid Anaerobic Digestion of Chicken Manure. *J. Agric. Eng. Res.* **2000**, *76*, 51–60. [CrossRef]
31. Angelidaki, I.; Ahring, B.K. Thermophilic anaerobic digestion of livestock waste: The effect of ammonia. *Appl. Microbiol. Biotechnol.* **1993**, *38*, 560–564. [CrossRef]
32. Angelidaki, I.; Ahring, B.K. Anaerobic thermophilic digestion of manure at different ammonia loads: Effect of temperature. *Water Res.* **1994**, *28*, 727–731. [CrossRef]
33. Abdel Azim, A.; Rittmann, S.K.-M.R.; Fino, D.; Bochmann, G. The physiological effect of heavy metals and volatile fatty acids on Methanococcus maripaludis S2. *Biotechnol. Biofuels* **2018**, *11*, 301. [CrossRef]
34. Shakeri Yekta, S.; Svensson, B.H.; Björn, A.; Skyllberg, U. Thermodynamic modeling of iron and trace metal solubility and speciation under sulfidic and ferruginous conditions in full scale continuous stirred tank biogas reactors. *Appl. Geochem.* **2014**, *47*, 61–73. [CrossRef]
35. Gonzalez-Silva, B.M.; Briones-Gallardo, R.; Razo-Flores, E.; Celis, L.B. Inhibition of sulfate reduction by iron, cadmium and sulfide in granular sludge. *J. Hazard. Mater.* **2009**, *172*, 400–407. [CrossRef]
36. Ahring, B.K.; Westermann, P. Sensitivity of thermophilic methanogenic bacteria to heavy metals. *Curr. Microbiol.* **1985**, *12*, 273–276. [CrossRef]
37. Yue, Z.-B.; Yu, H.-Q.; Wang, Z.-L. Anaerobic digestion of cattail with rumen culture in the presence of heavy metals. *Bioresour. Technol.* **2007**, *98*, 781–786. [CrossRef]
38. Amonette, J.E.; Russell, C.K.; Carosino, K.A.; Robinson, N.L.; Ho, J.T. Toxicity of Al to Desulfovibrio desulfuricans. *Appl. Environ. Microbiol.* **2003**, *69*, 4057–4066. [CrossRef]
39. Cabirol, N.; Barragán, E.J.; Durán, A.; Noyola, A. Effect of aluminium and sulphate on anaerobic digestion of sludge from wastewater enhanced primary treatment. *Water Sci. Technol.* **2003**, *48*, 235–240. [CrossRef]
40. Klüber, H.D.; Conrad, R. Inhibitory effects of nitrate, nitrite, NO and N2O on methanogenesis by Methanosarcina barkeri and Methanobacterium bryantii. *FEMS Microbiol. Ecol.* **1998**, *25*, 331–339. [CrossRef]
41. Schmidt, J.E.; Ahring, B.K. Effects of hydrogen and formate on the degradation of propionate and butyrate in thermophilic granules from an upflow anaerobic sludge blanket reactor. *Appl. Environ. Microbiol.* **1993**, *59*, 2546–2551. [CrossRef]
42. Abdel-Shafy, H.I.; Mansour, M.S.M. Biogas production as affected by heavy metals in the anaerobic digestion of sludge. *Egypt. J. Pet.* **2014**, *23*, 409–417. [CrossRef]
43. Stasinakis, A.S.; Thomaidis, N.S. Fate and Biotransformation of Metal and Metalloid Species in Biological Wastewater Treatment Processes. *Crit. Rev. Environ. Sci. Technol.* **2010**, *40*, 307–364. [CrossRef]
44. Field, J.A.; Sierra-Alvarez, R.; Cortinas, I.; Feijoo, G.; Moreira, M.T.; Kopplin, M.; Gandolfi, A.J. Facile Reduction of Arsenate in Methanogenic Sludge. *Biodegradation* **2004**, *15*, 185–196. [CrossRef]
45. Haynes, W.M.; Lide, D.R.; Bruno, T.J. *CRC Handbook of Chemistry and Physics*; CRC Press: Boca Raton, FL, USA, 2017.
46. Bratby, J. *Coagulation and Flocculation in Water and Wastewater Treatment*, 3rd ed.; IWA Publishing: London, UK, 2016. Available online: https://iwaponline.com/ebooks/book/286/Coagulation-and-Flocculation-in-Water-and (accessed on 15 September 2020).
47. Jeong, T.-Y.; Chung, H.-K.; Yeom, S.H.; Choi, S.S. Analysis of methane production inhibition for treatment of sewage sludge containing sulfate using an anaerobic continuous degradation process. *Korean, J. Chem. Eng.* **2009**, *26*, 1319–1322. [CrossRef]
48. Vijayaraghavan, K.; Ramanujam, T.K. Effect of chloride and condensable tannin in anaerobic degradation of tannery wastewaters. *Bioprocess. Eng.* **1999**, *20*, 499. [CrossRef]
49. Rath, K.M.; Maheshwari, A.; Bengtson, P.; Rousk, J. Comparative Toxicities of Salts on Microbial Processes in Soil. *Appl. Environ. Microbiol.* **2016**, *82*, 2012–2020. [CrossRef] [PubMed]
50. Serrano, R. *Salt Tolerance in Plants and Microorganisms: Toxicity Targets and Defense Responses*; Jeon, C., Ed.; Academic Press: Cambridge, MA, USA, 1996; Volume 165, pp. 1–52. [CrossRef]
51. Ware, A.; Power, N. Modelling methane production kinetics of complex poultry slaughterhouse wastes using sigmoidal growth functions. *Renew. Energy* **2017**, *104*, 50–59. [CrossRef]
52. Rocamora, I.; Wagland, S.T.; Villa, R.; Simpson, E.W.; Fernández, O.; Bajón-Fernández, Y. Use of Inoculum, Water and Percolate as Strategy to Avoid Inhibition on Dry-Batch Anaerobic Digestion of Organic Fraction of Municipal Solid Waste. *Waste Biomass Valorization* **2021**, *13*, 227–239. [CrossRef]
53. Maragkaki, A.E.; Fountoulakis, M.; Kyriakou, A.; Lasaridi, K.; Manios, T. Boosting biogas production from sewage sludge by adding small amount of agro-industrial by-products and food waste residues. *Waste Manag.* **2018**, *71*, 605–611. [CrossRef]
54. McCarty, P.L. Anaerobic waste treatment fundamentals. *Public Work* **1964**, *95*, 107–112.
55. Nielsen, H.B.; Angelidaki, I. Strategies for optimizing recovery of the biogas process following ammonia inhibition. *Bioresour. Technol.* **2008**, *99*, 7995–8001. [CrossRef]
56. Bhattacharya, S.K.; Parkin, G.F. The Effect of Ammonia on Methane Fermentation Processes. *J. Water Pollut. Control. Fed.* **1989**, *61*, 55–59.
57. Akunna, J.C.; Bizeau, C.; Moletta, R. Nitrate reduction by anaerobic sludge using glucose at various nitrate concentrations: Ammonification, denitrification and methanogenic activities. *Environ. Technol.* **1994**, *15*, 41–49. [CrossRef]

58. Yi, X.-H.; Wan, J.; Ma, Y.; Wang, Y.; Guan, Z.; Jing, D.-D. Structure and Succession of Bacterial Communities of the Granular Sludge during the Initial Stage of the Simultaneous Denitrification and Methanogenesis Process. *Water Air Soil Pollut.* **2017**, *228*, 1–21. [CrossRef]
59. Clarens, M.; Bernet, N.; Delgenès, J.-P.; Moletta, R. Effects of nitrogen oxides and denitrification by *Pseudomonas stutzeri* on acetotrophic methanogenesis by *Methanosarcina mazei*. *FEMS Microbiol. Ecol.* **1998**, *25*, 271–276. [CrossRef]
60. Borges, L.I.; López-Vazquez, C.M.; García, H.; van Lier, J.B. Nitrite reduction and methanogenesis in a single-stage UASB reactor. *Water Sci. Technol.* **2015**, *72*, 2236–2242. [CrossRef]
61. Yan, G.; Wang, J.; Guo, S. Anaerobic Biochemical Treatment of Wastewater Containing Highly Concentrated Organic Cyanogen. *Energy Sources Part A Recovery Util. Environ. Eff.* **2007**, *29*, 529–535. [CrossRef]
62. Gijzen, H.J.; Bernal, E.; Ferrer, H. Cyanide toxicity and cyanide degradation in anaerobic wastewater treatment. *Water Res.* **2000**, *34*, 2447–2454. [CrossRef]
63. Li, C.; Fang, H.H.P. Inhibition of heavy metals on fermentative hydrogen production by granular sludge. *Chemosphere* **2007**, *67*, 668–673. [CrossRef]
64. Yu, B.; Lou, Z.; Zhang, D.; Shan, A.; Yuan, H.; Zhu, N.; Zhang, K. Variations of organic matters and microbial community in thermophilic anaerobic digestion of waste activated sludge with the addition of ferric salts. *Bioresour. Technol.* **2015**, *179*, 291–298. [CrossRef]
65. Van Bodegom, P.M.; Scholten, J.C.M.; Stams, A.J.M. Direct inhibition of methanogenesis by ferric iron. *FEMS Microbiol. Ecol.* **2004**, *49*, 261–268. [CrossRef]
66. Colussi, I.; Cortesi, A.; Vedova LDella Gallo, V.; Robles, F.K.C. Start-up procedures and analysis of heavy metals inhibition on methanogenic activity in EGSB reactor. *Bioresour. Technol.* **2009**, *100*, 6290–6294. [CrossRef]
67. Lin, C.-Y.; Shei, S.-H. Heavy metal effects on fermentative hydrogen production using natural mixed microflora. *Int. J. Hydrogen Energy* **2008**, *33*, 587–593. [CrossRef]
68. O'Connor, O.A.; Young, L.Y. Toxicity and anaerobic biodegradability of substituted phenols under methanogenic conditions. *Environ. Toxicol. Chem.* **1989**, *8*, 853–862. [CrossRef]
69. Hansen, K.H.; Angelidaki, I.; Ahring, B.K. Anaerobic digestion of swine manure: Inhibition by ammonia. *Water Res.* **1998**, *32*, 5–12. [CrossRef]
70. Sierra-Alvarez, R.; Cortinas, I.; Yenal, U.; Field, J.A. Methanogenic Inhibition by Arsenic Compounds. *Appl. Environ. Microbiol.* **2004**, *70*, 5688–5691. [CrossRef] [PubMed]
71. Lenz, M.; Janzen, N.; Lens, P.N.L. Selenium oxyanion inhibition of hydrogenotrophic and acetoclastic methanogenesis. *Chemosphere* **2008**, *73*, 383–388. [CrossRef]
72. Ochoa-Herrera, V.; Banihani, Q.; León, G.; Khatri, C.; Field, J.A.; Sierra-Alvarez, R. Toxicity of fluoride to microorganisms in biological wastewater treatment systems. *Water Res.* **2009**, *43*, 3177–3186. [CrossRef]

processes

MDPI

Article

Operation of Submerged Anaerobic Membrane Bioreactors at 20 °C: Effect of Solids Retention Time on Flux, Mixed Liquor Characteristics and Performance

Santiago Pacheco-Ruiz [1,2,*], Sonia Heaven [2] and Charles J. Banks [2]

[1] Biothane, Veolia Water Technologies Techno Center Netherlands BV, Tanthofdreef 21, 2623 EW Delft, The Netherlands

[2] Faculty of Engineering and Physical Sciences, University of Southampton, Southampton SO17 1BJ, UK; S.Heaven@soton.ac.uk (S.H.); C.J.Banks@soton.ac.uk (C.J.B.)

* Correspondence: santiago-pacheco.ruiz@veolia.com; Tel.: +31-643-836-991

Abstract: Four flat-sheet submerged anaerobic membrane bioreactors ran for 242 days on a simulated domestic wastewater with low Chemical Oxygen Demand (COD) and high suspended solids. Organic loading was maintained around 1.0 g COD L^{-1} day^{-1}, while solids retention time (SRT) was varied from 20–90 days. This was achieved at a constant membrane flux, maintained by adjusting transmembrane pressure (TMP) in the range 1.8–9.8 kPa. Membrane fouling was assessed based on the required TMP, with mixed liquors characterised using capillary suction time, frozen image centrifugation and quantification of extracellular polymeric substances (EPS). SRT had a significant effect on these parameters: fouling was least at an SRT of 30 days and highest at 60 days, with some reduction as this extended to 90 days. Operation at SRT < 30 days showed no further benefits. Although operation at a short SRT was optimal for membrane performance it led to lower specific methane productivity, higher biomass yields and higher effluent COD. Short SRT may also have accelerated the loss of essential trace elements, leading to reduced performance under these conditions. A COD-based mass balance was conducted, including both biomass and methane dissolved in the effluent.

Keywords: anaerobic membrane bioreactors; ambient temperature; membrane fouling; mean cell residence time; wastewater treatment

Citation: Pacheco-Ruiz, S.; Heaven, S.; Banks, C.J. Operation of Submerged Anaerobic Membrane Bioreactors at 20 °C: Effect of Solids Retention Time on Flux, Mixed Liquor Characteristics and Performance. *Processes* **2021**, *9*, 1525. https://doi.org/10.3390/pr9091525

Academic Editor: Bipro R. Dhar

Received: 19 July 2021
Accepted: 24 August 2021
Published: 29 August 2021

1. Introduction

Anaerobic technologies for wastewater treatment may offer advantages over aerobic systems, as they produce methane-rich biogas and have much lower sludge yields [1,2]. Anaerobic processes are normally operated at around 35 °C, which is known to be an optimum for maintaining a high metabolic activity [3]. It is, however, rarely economical to work at this temperature when treating low-strength municipal wastewaters, as the energy yield may be lower than the parasitic energy demand for heating [4].

Although it is well-known that lower temperatures reduce the rate of biological reaction, there is increasing awareness that effective operation is possible using retained and acclimated biomass [5]. Lower temperature operation does, however, raise other issues such as the increased solubility of methane in the effluent stream; potentially lower removal efficiencies for Chemical Oxygen Demand (COD); and an increase in water viscosity that can reduce the membrane flux and change the settling characteristics of biological solids [3,6–9]. The development of the anaerobic membrane bioreactor (AnMBR) has made it possible to produce high-quality effluents while operating at ambient temperatures and at a reasonably short hydraulic retention time (HRT). A number of reviews [10–12] have demonstrated a growing interest in the application of AnMBRs to a variety of wastewater types. The successful treatment of real and simulated municipal wastewaters with biogas production in AnMBR at operating temperatures as low as 15 °C has been demonstrated [13,14].

Yet, there still remain considerable knowledge gaps and technical challenges before the technology can become more widely adopted at full scale [10,15].

The retention of biomass in an AnMBR allows mixed liquor suspended solids (MLSS) concentrations to be maintained without carry-over of solids into the effluent; it also gives the potential for selecting the MLSS concentration for optimum organic matter degradation and membrane performance [7,15]. The decoupling of solids retention time (SRT) from HRT provides a way to control the mean cell residence time (MCRT) in the reactor [8]. This approach to process control has been used extensively in aerobic wastewater treatment, as MCRT is the reciprocal of biomass growth rate and, therefore, directly influences both metabolic activity and sludge yield [16]. The simplest way to control MCRT, and, therefore, gain kinetic control over the treatment process, is through proportional biomass wastage [17]. This important kinetic control parameter has not, however, been widely used in anaerobic systems, even though MCRT has been shown to influence the production of extracellular polymeric substances (EPS) and soluble microbial products [3,18]. These are particularly important in submerged AnMBR (SAnMBR), where the membrane is directly immersed in the mixed liquor, and both EPS and soluble microbial products are known to affect membrane fouling [19,20]. In aerobic treatment systems, it is thought that a long MCRT reduces the concentration of EPS and soluble microbial products and because these are considered to be more critical in inducing membrane fouling than is the MLSS concentration, a long MCRT is usually chosen for operation [3,20,21].

Relatively few AnMBR studies to date have used MCRT or solids retention time (SRT) as a control parameter, while many have operated at very extended or near-infinite solids retention times, with the SRT in some cases determined only by the need to remove small volumes of MLSS for analysis. Studies in which SRT has been varied, however, have not always run for long enough to reach steady-state conditions: this is often defined as operation for at least 3 SRT. Without this, it is unlikely that MLSS properties will be representative in each case. Baek et al. [22] operated a 10-L AnMBR with a sidestream membrane filter on settled municipal wastewater over a range of SRT from 19–217 days but only achieved 3 SRT at the lowest value of 19 days. Yeo and Lee [23] tested SRT of 20 and 40 days in a 5-L SAnMBR operated at 23 °C on a feed of glucose at 5 g COD L^{-1}. It is likely that only 1 SRT was achieved in each case, although the duration of this experiment was not explicitly stated. Dong et al. [24,25] used the bench and pilot-scale AnMBRs with external hollow fibre membrane (HFM) units fed on municipal wastewater at 23 °C. They tested SRT of 100, 70 and 40 days, but none of these appear to have run for 3 SRT. Thanh et al. [26] reduced the SRT from 100 to 75, 50 and then 25 days over a 60-day period in a flat-sheet SAnMBR at 35 °C fed on dilute synthetic wastewater. Yurtsever et al. [27] treated high-salinity synthetic textile wastewater in aerobic and anaerobic flat-sheet MBRs at 60-days, 30-days and near-infinite SRT, but in each case for much less than 3 SRT. Ji et al. [28] altered the SRT from 53 days to near-infinite in response to changes in MLSS caused by varying load: this is a familiar alternative method for process control in aerobic systems, but they did not attempt to use SRT itself as a control parameter.

A small number of studies using SRT as the main control parameter have run for long enough to approach steady-state operation. In some cases, these have used high-strength effluents under mesophilic conditions. Dereli et al. [29,30] operated two mesophilic cross-flow AnMBR on high-strength corn stillage for 3 months at SRT of 20 and 30 days, respectively, then ran for a further 3 months at a 50-day SRT. Szabo-Corbacho et al. [31] investigated the effect of SRT in a crossflow AnMBR treating high-strength synthetic dairy wastewater at 35 °C and completed ~3 SRT at 20 and 40-day SRT. Pacheco-Ruiz et al. [32] used a low-to-intermediate strength synthetic wastewater at 36 °C and showed that a short SRT gave enhanced membrane performance but resulted in lower specific methane production and higher waste sludge yield. Very few such studies have used low-strength wastewaters at lower temperatures, however, and the limited results to date have been conflicting. When treating synthetic low-strength wastewater in a flat-sheet membrane at 25–30 °C, Huang et al. [33] noted that a longer SRT led to greater fouling when using the same AnMBR to treat municipal wastewater, however, they also found higher fouling at a short SRT [34]. While more studies are coming through using real municipal wastewaters, at ambient temperatures, and/or in larger-scale systems [1,6,15,24,35], data interpretation and performance prediction can be challenging due to the number of influencing variables. This is especially the case where multiple factors can alter simultaneously [3,4,7], and there is thus a clear need for studies focusing on the effects of individual parameters such as SRT.

The current research aimed to assess the effect of changes in SRT on the performance of a SAnMBR treating low-strength wastewater at 20 °C, and, in particular, the influence on membrane fouling, sludge yield, COD removal efficiency, and physico-chemical properties of the MLSS once steady-state operation has been achieved. The results may thus contribute towards resolving some conflicting reports in the literature and could help to establish MCRT or SRT as a principal control parameter for anaerobic wastewater treatment systems, in much the same way as it is considered the control parameter of choice in aerobic systems.

2. Materials and Methods

2.1. SAnMBR Design and Operation

Four SAnMBRs (S1, S2, S3 and S4) were used, in which the transmembrane pressure (TMP) was applied via a gravity head [28]. Each SAnMBR had a working volume of 9.6 L, with 2.1 L of headspace (Figure 1). Each was fitted with a chlorinated polyethylene flat-plate membrane cartridge (Type 203, Kubota Corporation, Osaka, Japan) with nominal pore size 0.4 μm and effective surface area 0.113 m^2, at a membrane packing density of 0.012 m^2 L^{-1}. The SAnMBRs were fed continuously with the chilled substrate at a controlled flow achieved by maintaining a constant head differential with the outlet from the membrane cartridge lumen. The driving force for the passage of effluent through the membrane was thus gravity-induced, and the TMP could be altered within a range of 1.8–9.8 kPa by changing the differential head. The SAnMBRs could thus be operated at a constant membrane flux, allowing control of the organic loading rate to be maintained as the membranes became progressively fouled. Flow was measured on a weighing scale with 32 kg capacity and readable to 1.0 g (Model CBK 32, Adam Equipment Co Ltd., Milton Keynes, UK), with the weight of collected permeate logged automatically at 5 min intervals. The internal headspace pressure was maintained at around 0.3 kPa, with biogas released via a fermentation gas lock into a gas-impermeable collection bag.

In situ cleaning of the membrane was achieved by recirculating headspace biogas at approximately 0.57 L L^{-1} of reactor min^{-1} (corresponding to 48.5 L min^{-1} m^{-2} of membrane surface area) using a stainless-steel sparger and a diaphragm pump (AIRPO, Ningbo forever Electronic Appliances Co Ltd., Ningbo, Zheijiang, China). This gas flow also maintained the mixed liquor in suspension. SAnMBR temperature during the experimental period was maintained at 20.5 ± 0.5 °C by a stainless-steel internal heat exchange coil coupled with a thermocirculator (FC15 and FH15V, Grant Instruments Europe BV, Amsterdam, Netherlands). Feed and SAnMBR temperature were continuously recorded by solid-state IC temperature probes (LM35DZ, Texas Instruments, Dallas, TX, USA) and a datalogger (Model U3-LV, Labjack Corporation, Lakewood, CO, USA).

2.2. Inoculum and Substrate

The substrate used was a synthetic wastewater [36], which was prepared fresh each day as a concentrate and diluted to the required working strength. It was designed to give a C:N:P ratio of around 100:20:4 and solids contents similar to those of typical municipal wastewaters [37,38].

The four AnMBRs were seeded from a mesophilic anaerobic digester at a municipal wastewater treatment plant in Southampton, UK, and the headspace was purged with nitrogen. To allow temperature acclimatisation they were initially fed with the concentrated substrate at an organic loading rate (OLR) of 0.5 g COD L^{-1} day^{-1} for 48 days (not counted as part of the trial). A total of 50% of the MLSS was then removed and replaced by tap water to reduce the solids content. From this point onwards (taken as day 0 of the experimental period) the SAnMBRs were fed with substrate prepared by diluting the concentrate to the required COD strength. The SAnMBRs were then run at different SRTs in 3 experimental phases (EP), giving a total operational period of 242 days. Details of the applied SRT, TMP and other operational parameters are given in Table A1.

Two trace element stock solutions were used containing (g L^{-1}): Al 0.1, B 0.1, Co 1.0, Cu 0.1, Fe 10.0, Mn 1.0, Ni 1.0, Zn 1.0; and Se 0.1, Mo 0.1, W 0.1.

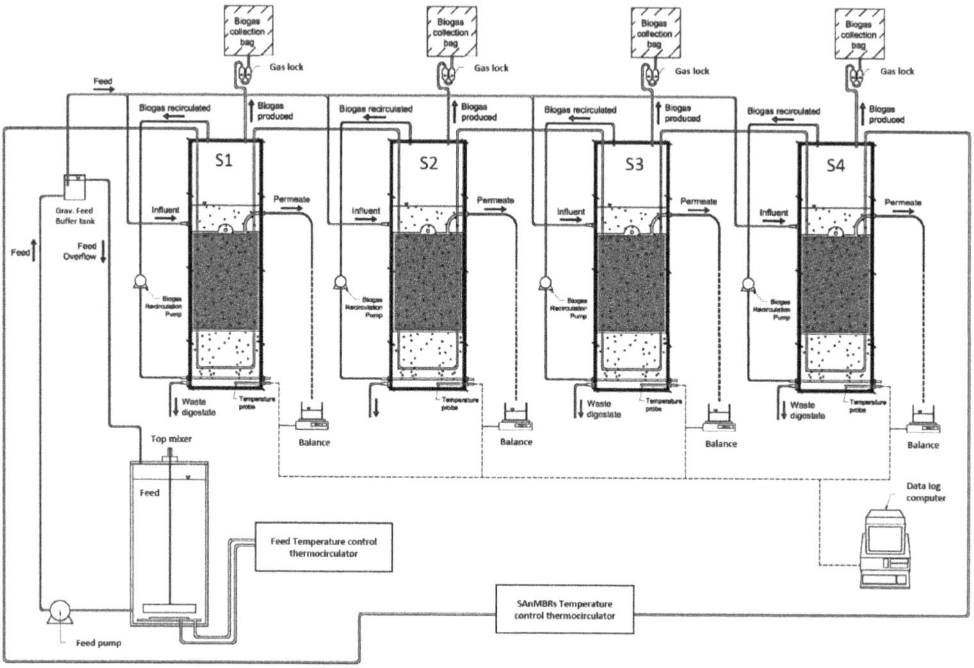

Figure 1. Schematic of experimental set-up for SAnMBRs.

2.3. Performance and Stability

Membrane performance was evaluated on the basis of the flux achieved at constant TMP, calculated as described in Pacheco-Ruiz et al. [36]. SAnMBR performance and operational stability were assessed on the basis of COD percentage conversion, specific methane production (SMP) per g of COD removed, MLSS and mixed liquor volatile suspended solids (MLVSS) concentrations, mixed liquor pH, EPS content and composition, capillary suction time (CST) (Triton Electronics Ltd., Dunmow, Cambridge, UK), and frozen image centrifugation (FIC) (Triton Electronics, UK). FIC uses a technique in which a 'frozen image' of the sample is generated by matching the frequency of a stroboscopic light to the centrifuge rotor speed, allowing measurement of the height of the solid-liquid interface in real time without interrupting the test. The centrifugation speed was fixed at 660 ± 10 rpm with observations made every minute up to 8 min. COD of fresh feed, feed after chilled storage for 24 h, and AnMBR effluent was measured using a closed-tube reflux method with titrometric end-point determination [39]. Biogas composition was analysed by gas chromatography (GP-3400, Varian Inc., Palo Alto, CA, USA) using 36% CO_2 with 64% CH_4 (v/v) (BOC, Guildford, Surrey, UK) as a standard gas. Biogas volumes were determined by a weight-type gasometer [40] and are reported at standard temperature and pressure (STP, 0 °C and 101.3 kPa).

Organic loading rate (OLR) and SMP were calculated using the average of the COD values for the fresh and stored feed. The reported SMP value includes both methane in gas collected from the reactor headspace and dissolved methane in the effluent. This allows comparison of SAnMBR performance under different sconditions since the proportion of methane that leaves the system in the effluent will differ at different flux rates. Dissolved methane content was estimated based on Henry's Law using 20 °C saturation concentration. The resulting average value 29.0 mL CH_4 L^{-1} was confirmed by empirical measurement, using the method in Walsh and McLaughlan [41].

MLSS and MLVSS concentrations were quantified using Standard Method 2540-D [42]. pH measurements were made using a pH meter (3310, Jenway Ltd., London, UK) calibrated in pH 4, 7 and 9.2 buffers (Fisher Scientific UK Ltd., Loughborough, Leicestershire, UK). For the COD mass balance, the only input was influent COD while outputs were taken as effluent COD, COD in gaseous or dissolved methane and COD in biomass. The last of these was estimated from daily MLSS removal (g MLVSS day^{-1}) divided by the average ratio of COD/VSS in the mixed liquor (g COD g^{-1} MLVSS). COD balances did not consider changes in stored biomass, as small errors in MLSS and MLVSS measurement could introduce large variations in the overall result. Thus the COD balances were only valid for steady-state conditions or other periods with stable solids contents. The COD value of methane was taken as 2.855 g COD L^{-1} CH_4 based on stoichiometric considerations.

EPS was extracted using the formaldehyde plus NaOH procedure [43], modified in accordance with Domínguez et al. [44] and Liang et al. [45] to enable identification of bound and soluble components. Soluble EPS was extracted by centrifugation of mixed liquor, as suggested by Chabaliná et al. [46]. EPS composition was quantified by measuring the concentration of carbohydrate and protein using colorimetric methods. Carbohydrate was determined by the phenol-sulphuric acid method [47], using a glucose standard. Protein contents were analysed according to the modified Lowry Folin–Ciocalteu method suggested by Frølund et al. [48], with bovine serum albumin as a standard.

3. Results and Discussion

3.1. Operational Performance

Figure 2 presents graphical data on the operation and performance of the four AnMBRs during the 242-day experimental period, with results summarised in Tables A1 and A2. Discussion of key parameters in relation to operating conditions during each phase of the experimental period is provided below.

3.1.1. Membrane Flux, TMP, MLSS, HRT and OLR

Start-up (days 0–59). After the 48-day temperature acclimatisation phase and following dilution, the MLSS in each SAnMBR was initially 16.2 g L^{-1}. No sludge was wasted for the next 10 days, after which time proportional wasting was introduced with the aim of applying a SRT of 90 days (Figure 2a). In the following period, a number of changes were made to the TMP and feed concentration in order to find a combination of conditions that allowed acclimatisation to an increase in OLR. The TMP was initially set at 7 kPa, giving an initial flux of 16.1 L m^{-2} h^{-1}, which after 5 days had decreased to 10.2–11.0 L m^{-2} h^{-1} (Figure 2b). This high flux produced short HRTs with low COD removal rates of between 54–64%, and on day 9 TMP was reduced from 7.0 to 2.5 kPa with the aim of reducing the flux (Figure 2c). The flux remained high, however, at around 9.9–10.6 L m^{-2} h^{-1}. Thus to avoid organic overloading, the COD of the feed was reduced to around 160 mg L^{-1}. The pH immediately rose to 6.9–7.0, and the COD removal rate increased to 74–80%. The OLR was subsequently increased to 0.9, 1.0, 1.1 and 1.5 g COD L^{-1} day^{-1} stepwise at 3-day intervals by increasing the feed concentration. This led to changes in gas composition, with a higher methane content indicating greater biogas production, but the last increment in OLR resulted in a fall in pH to 6.6 and a decline in %COD removal. The OLR was, therefore, reduced on day 22 to 0.75 g COD L^{-1} day^{-1} by reducing the feed concentration, whilst the flux was maintained at 8.0–8.8 L m^{-2} h^{-1}. To improve COD removal, on day 28, the flux was then further reduced to around 6 L m^{-2} h^{-1} by reducing the TMP to 2.2 kPa (Figure 2c), and from this time onward, SAnMBR performance improved to a point where between day 40 and 52 the OLR was successfully raised to 1.0 g COD L^{-1} day^{-1}. By day 60, stability had been achieved in all four SAnMBRs, with flux rates between 6.3–6.7 L m^{-2} h^{-1}, and MLSS contents from 12.4–12.7 g L^{-1} (Figure 2d), COD removal rates between 88–93% (Figure 2e), and stable biogas composition and production. At this point, the start-up phase was considered complete.

Experimental phase 1 (EP1, days 60–111). On day 60, the SRT was reduced to 30, 45 and 60 days in S1, S2 and S3, respectively, by increasing the volume of mixed liquor removed each day, while the SRT in S4 remained unchanged at 90 days. At the start of this phase, the TMP in all four SAnMBR was 2.5 kPa, while flux rates were 6.7, 6.4, 6.3 and 6.4 L m^{-2} h^{-1} in S1, S2, S4 and S4, respectively (Figure 2b,c). As the flux gradually declined, TMP was adjusted in each SAnMBR individually, with the aim of stabilising at a value of around 5 L m^{-2} h^{-1}. Feed concentrations were also modified slightly with respect to the achieved flux, with the aim of providing a consistent OLR of around 1 g COD L^{-1} day^{-1}. The amount of adjustment applied during this phase was, however, much less than that required in the start-up period.

By day 67, the effect of the applied changes in SRT was already evident in S3, where the onset of membrane fouling was indicated by the need for substantial increases in TMP to maintain a flux of 5 L m^{-2} h^{-1}. Similar behaviour soon followed in S2 and S4, which from day 80 also needed higher TMPs to achieve the target flux value, while in S1 it became necessary to reduce the TMP as the flux was increasing. At this stage, it was concluded that a flux of 5 L m^{-2} h^{-1} was too high for these experimental conditions. On day 100, the target was, therefore, reduced to 4 L m^{-2} h^{-1}, achieved by applying reduced TMPs of 1.8, 2.9, 4.3 and 2.2 kPa in S1, S2, S3 and S4, respectively (Figure 2c). This gave HRTs of approximately 20 h and a OLR of around 1.0 g COD L^{-1} day^{-1}, at an average COD concentration in the feed of 860 ± 23 mg COD L^{-1}. Minor variations in feed concentration here were caused by slight day-to-day variations in batch preparation and in analytical results, with no further deliberate adjustments being made to influent strength. In S1, S2 and S4 a flux of 4 L m^{-2} h^{-1} was maintained at constant TMP until the end of EP-1. In contrast, the flux in S3 continued falling irrespective of increases in the TMP, which at the end of this phase had reached 6.3 kPa (Figure 2c), giving a final OLR of 0.9 g COD L^{-1} day^{-1} at an HRT of 22.5 h in this reactor. The changes in SRT were also reflected in MLSS concentrations, with values between 12.4–12.7 g L^{-1} in all SAnMBR at the start of the phase diverging to 6.2, 7.3, 8.7 and 11.7 g L^{-1} in S1, S2, S3 and S4, respectively (Figure 2d).

Experimental phase 2 (EP-2, days 112–160). In response to the changes observed in EP-1, at the beginning of EP-2 the SRT in S3 was reduced in one step from 60 to 30 days, at an initial TMP of 6.3 kPa. S1, S2 and S4 remained at their previous SRT of 30, 45 and 90 days with corresponding initial TMPs of 1.7, 2.9 and 2.2. While flux in S1 and S2 remained steady at 4 L m^{-2} h^{-1} without any adjustment of TMP, flux in S3 continued falling, despite continuing rises in TMP (Figure 2c). On day 143, the TMP in S3 reached the limiting value of 9.8 kPa, at a flux rate of 3.6 L m^{-2} h^{-1}. The flux in S4 remained steady until day 115 when it began to decline, necessitating increases in the TMP. By day 160, S4 had also reached the maximum TMP at a flux of 3.2 L m^{-2} h^{-1} (Figure 2b). In S1 and S2, the constant flux rates resulted in stable HRT of around 20.5 h and an OLR of 1.0 g COD L^{-1} day^{-1} for both SAnMBRs. In contrast the continuous decline in flux in S3 and S4 led to higher HRTs of 28.5 and 26.6 h with OLR of 0.7 and 0.8 g COD L^{-1} day^{-1}, respectively, at the end of EP-2.

MLSS concentrations in this phase continued to reflect the different SRT, with values in S1 and S2 stabilising at around 4.4 and 5.8 g L^{-1}, respectively (Figure 2d). The fall in MLSS in S3 reflected the change in SRT to 30 days at the start of the phase, and the reduction in applied OLR due to lower flux, which resulted in a MLSS concentration of 4.3 g L^{-1} in S3 at the end of EP-2. MLSS concentrations in S4 initially stabilised at 10.8 g L^{-1} (Figure 2d) but by the end of the phase had fallen to 9.5 g L^{-1} as a result of the decrease in flux rate and consequently in OLR.

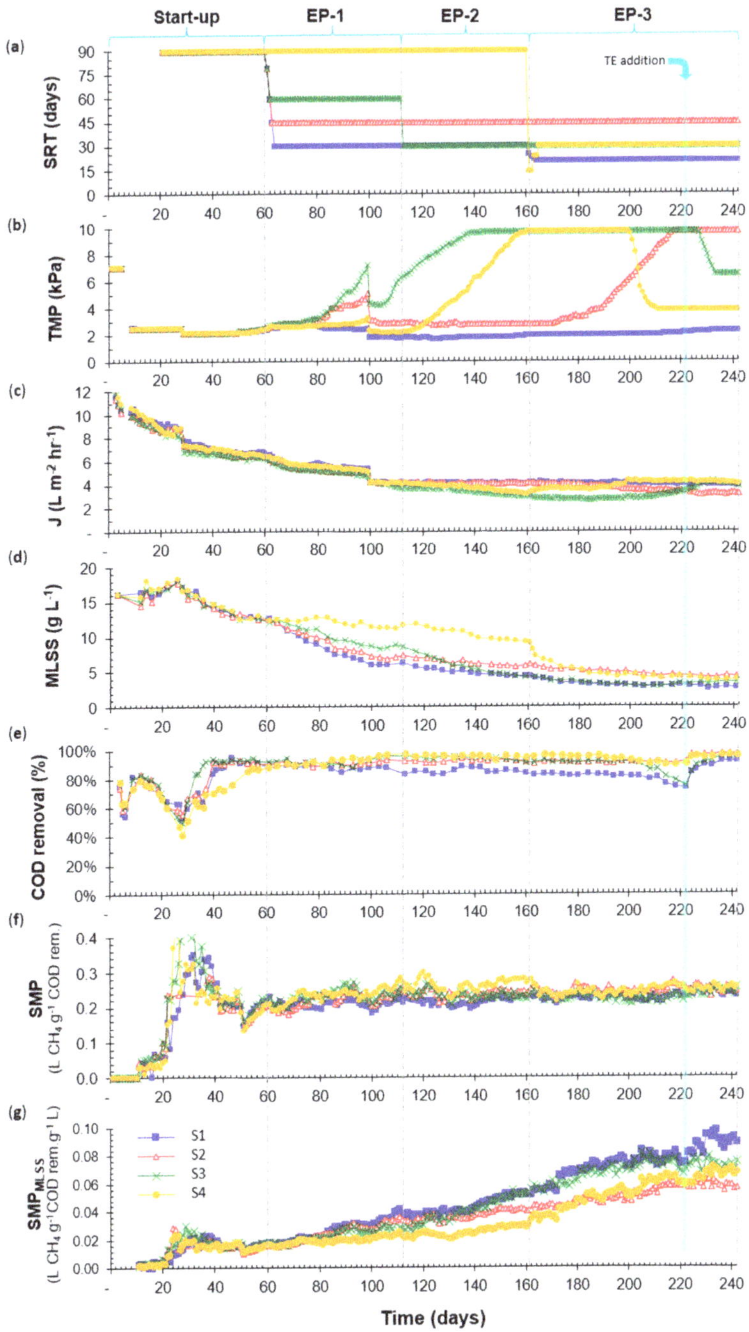

Figure 2. Operational parameters in SAnMBRs during the whole experimental period: (**a**) SRT; (**b**) TMP; (**c**) flux (daily average); (**d**) MLSS; (**e**) COD removal; (**f**) SMP; (**g**) specific SMP (MLSS basis). Vertical dotted lines indicate start of experimental phases.

At the end of EP-2, S1 had completed 101 days (equivalent to 3.4 solid retention times) at a 30-day SRT and was, therefore, regarded as having achieved steady-state operation in these conditions.

Experimental phase 3 (EP-3, days 161–242). At the start of this phase, SRT was reduced to 20 days in S1, maintained at 45 days in S2 and at 30 days in S3, and reduced to 30 days in S4. At this point, S4 had completed only 160 days at a 90-day SRT, equivalent to 1.78 solid retention times or approximately 83% washout of the MLSS originally present at the start of EP-1, but it was clear that operation was no longer sustainable within the limits of the target flux and TMP. As an alternative to the slow reduction in MLSS content occurring in S3, however, MLSS in S4 was reduced abruptly from 9.1 to 7.2 mg L^{-1}, targeting a value close to that expected for a 30-day SRT.

In the case of S4, this reduction was achieved by removing a known volume of mixed liquor and replacing it with influent, to maintain a constant working volume in the SAnMBR. The process was repeated on four consecutive days, until the MLSS concentration approached the target value. The daily volume of mixed liquor replaced was 700 mL on days 160 and 161, and 400 mL on days 162 and 163.

In S1 at a 20-day SRT, the flux during this phase was successfully maintained at 4 L m^{-2} h^{-1}, with only a slight TMP increase to 2.2 kPa. In contrast, flux in S2 at a 45-day SRT remained steady at 4 L m^{-2} h^{-1} until day 180 when it began to fall, necessitating repeated increases in TMP (Figure 2c). By day 216, TMP in S2 had reached the limiting value of 9.8 kPa, and from then on flux continued gradually to decline, stabilising at around 3.2 L m^{-2} h^{-1} by the end of EP-3. Flux in S3 at a 30-day SRT continued to fall (Figure 2b), reaching its lowest value of 2.7 L m^{-2} h^{-1} on day 183. After this it slowly recovered, up to the 4 L m^{-2} h^{-1} target by day 228 (Figure 2b). Over the next 5 days, TMP in S3 was reduced to 6.5 kPa, while flux remained steady at 4.1 L m^{-2} h^{-1} to the end of the experiment.

The sharp reduction in MLSS in S4 produced an immediate increase in flux, reaching 4 L m^{-2} h^{-1} by day 200 (Figure 2b). Over the next 10 days, the TMP was reduced to 3.8 kPa, while a steady flux of 4.1 L m^{-2} h^{-1} was maintained until the end of EP-3.

In S1 at a 20-day SRT, the HRT and OLR remained steady at around 21.0 h and 1.0 g COD L^{-1} day^{-1}. Similar values were also achieved in S3 and S4 at a 30-day SRT once they approached the same sustainable rate of 4 L m^{-2} h^{-1}. In S2 at a 45-day SRT a continuing slow decline in flux increased the HRT to 26.3 h with a corresponding fall in OLR to 0.8 g COD L^{-1} day^{-1} by the end of the phase.

At the end of EP-3, reactors S1, S2 and S3 had, respectively, completed 82, 183 and 131 days at 20-, 45- and 30-day SRT (equivalent to 4.1, 4.1 and 4.4 solid retention times in each case), making the results representative of steady-state operation in these conditions. S4 had completed 82 days at a 30-day SRT, corresponding to 2.73 solid retention times with removal of around 93.5% of the MLSS present at the start of EP-3. It was, therefore, regarded as approaching steady-state operation.

The above results indicate the complexity of interactions between SRT, required TMP, flux, MLSS, HRT and OLR, and factors associated with this are discussed in the following sections.

3.1.2. COD Removal Rates and TE Requirements

During the start-up period COD removal initially showed quite a high variability, which stabilised towards the end of the period. COD removal rates were similar in all four SAnMBRs for the first 20 days of EP-1 at 89–92% (Figure 2e). Differences in SRT began to affect COD removal during EP-1, however, with S1, S2, S3 and S4 at 30-, 45-, 60- and 90-day SRTs reaching the end of the phase with removal rates of 88%, 92%, 96% and 98%, respectively. In EP-2, COD removal rates in S1, S2 and S4 stabilised at 85 ± 2%, 93 ± 1% and 96 ± 1%, respectively. In contrast, the removal rate for S3, in which the SRT had been reduced from 60 to 30 days, fell steadily from 95% to 91%.

In the first 40 days of EP-3, there was a slight decrease in COD removal in all four SAnMBRs (Figure 2e), followed by significant falls of 8% and 20% in S1 and S3, respectively, between days 204 and 222. It was hypothesised that this was due to a lack of trace elements (TE), since the shorter SRT in these reactors meant they had the highest overall biomass turnover, and thus the greatest risk of depletion of essential micro-nutrients. Based on this, for a 3-day period from day 223, trace elements were added to the feed for all four SAnMBRS at a dosage of 0.1 mL of each TE solution per litre of dilute influent. This led to an immediate rise in COD removal in all SAnMBRs, which stabilised at 92%, 97%, 95% and 96% in S1, S2, S3 and S4, respectively, by the end of EP-3. It was, therefore, likely that some of the decline in COD removal during this period was due to TE deficiencies, but removal rates after the TE supplementation were still slightly lower at the shorter SRTs. Other authors [30,31] have also reported slightly higher COD removal rates at longer SRT, though these studies involved high-strength substrates. In contrast, Huang et al. [33] observed no significant differences in COD removal rates for low-strength synthetic wastewater at 30-day, 60-day and near-infinite SRT.

Determination of TE requirements in AnMBR is likely to be especially challenging due to the combined effects of uncoupled liquid and solid retention times with uncertainties on bioavailability and partitioning. A number of studies have reported the addition of trace elements to feed without providing detailed justifications of the concentration or dosages used [14,19,23,33,49]. Yu et al. [50] investigated the effects of individual and combined TE supplementation on samples from the methanogenic reactor in a 2-phase AnMBR treating industrial starch wastewater at 37 °C.

They reported little effect from low doses, but with an SRT of 200 days and inoculum from a municipal sewage treatment plant, it is possible that washout of some TE to critical levels had not yet occurred. Sierra et al. [51] looked at the partitioning of trace elements B, Co, Cu, Mg, Mn, Mo, Ni, Se and W and at the effect of additional supplementation with Co and W in an AnMBR treating a highly saline phenolic wastewater. Additional Co had little effect, but W was beneficial. Doubling the overall TE dose also improved performance, and they concluded that bioavailability and partitioning were affected by high salinity. Thanh et al. [26] examined the effects of HRT, SRT and pH on bioavailability and speciation of Co, Fe, Mn, Mo, Ni and Zn. When reducing the SRT from 100 to 75, 50 and then 25 days over a 60-day period, they found varying depletion rates for different metals depending on affinity and previous accumulation. While the total trace metal content fell with the reduction in MLSS, this was partly countered by a change in speciation towards more bioavailable forms for all metals except Mn and Ni. While these studies have shown the importance of TE in ANMBR, relationships between TE, SRT and other operating parameters and optimum dosing strategies for these systems are likely to be key areas for future work [51,52].

3.1.3. COD Balances and Dissolved Methane

COD balances (shown in Figure 3) indicated that a substantial fraction of the COD was converted to methane, including a considerable proportion dissolved in the effluent. The response to the addition of TE on day 223 can also be clearly seen. COD balances for steady-state periods closed at between 92–96% in this work.

Small errors in COD balances are typical, and studies reporting better closure tend to operate with higher strength substrates and/or temperatures or at a larger scale [31]. Closures of 93–94% were obtained for a 20-L SAnMBR treating municipal wastewater at 25 °C at HRT from 4–12 h [28], while a slight excess of around 101% was reported for a 5 m^3 SAnMBR operating at the same temperature on a similar municipal effluent [1]. The missing fraction in the COD balance could be attributable in part to small quantities of biogas leaving the SAnMBRs through the permeate line in gaseous form as bubbles [36] and by any H_2S fraction in the biogas. H_2S was not taken into account in the COD balance as concentrations in the biogas were considered to be low, but the inclusion of this component may be essential in the treatment of protein- or sulphate-rich wastewaters [53]. Using Henry's Law to estimate the amount of CH_4 in the effluent may also lead to underestimation, as the calculation assumed a saturation concentration, while the elevated

TMP could cause additional dissolution and apparent methane supersaturation in the effluent [8]. In such cases, the actual volume of dissolved methane lost would be greater than the estimated value, especially at lower operating temperatures where saturation concentrations are higher. Differences between observed and simulated mass flows of methane in pilot and bench-scale sidestream AnMBRs treating screened municipal wastewater at ambient temperature were attributed to oversaturation [24], but were insufficient to account for errors of ~20% in COD balances, which may have been linked to the absence of steady-state conditions as well as to the reduction of sulphates and ferric iron. Methane oversaturation was also reported in a pilot-scale gas-sparged sidestream AnMBR operating on screened municipal wastewater at temperatures of 18.8–31.5 °C [12]. Yeo and Lee [23] noted variations in dissolved methane content with SRT in SAnMBR at 23 °C fed on glucose at 5.2 g COD L^{-1} with an HRT of 10 days. They reported oversaturation of methane and a lower biogas methane percentage at a 20-day SRT and presented COD-based balances closing at 3% and 7% for the 20 and 40-day SRT, respectively.

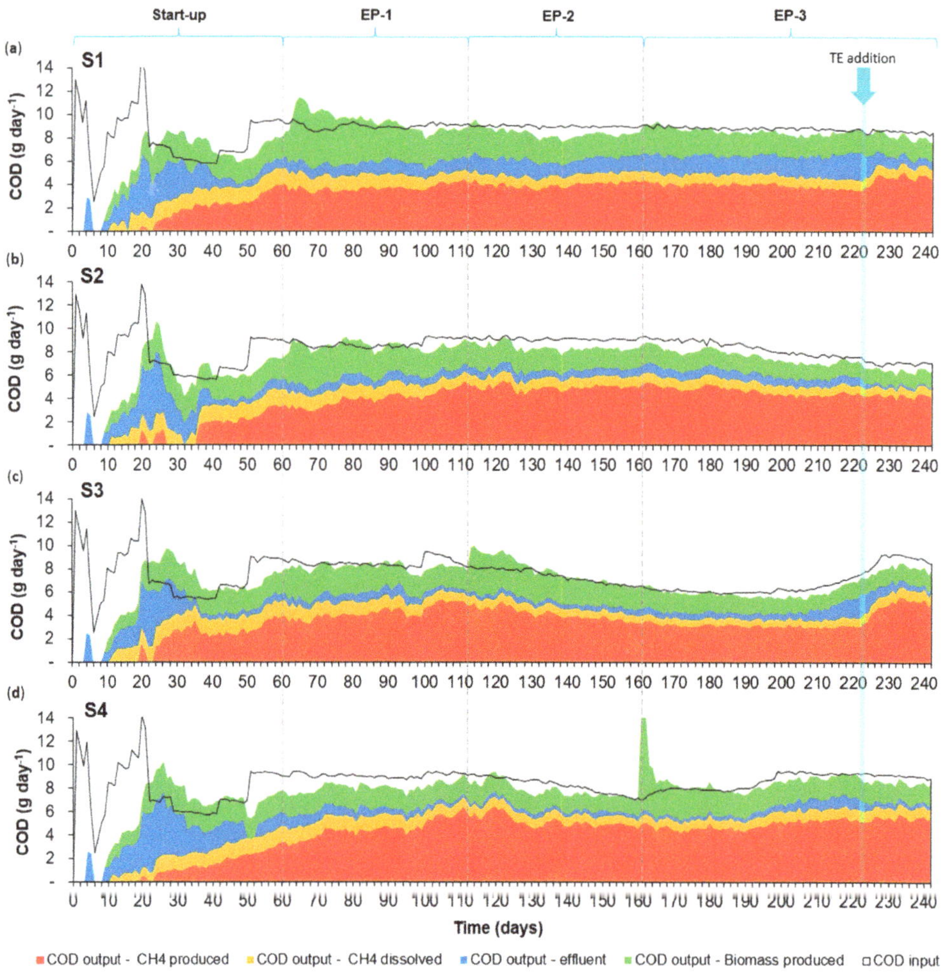

Figure 3. COD balances in SAnMBRs during the whole experimental period: (**a**) S1, (**b**) S2, (**c**) S3 and (**d**) S4. Vertical dotted lines indicate the start of experimental phases. Values for control parameters (SRT and TMP) in each phase are shown in Figure 2 and Table A2.

3.1.4. Specific Methane Productivity

Biogas composition for all SAnMBRs remained stable throughout the three experimental phases at 78% CH_4, 8% CO_2 and 14% nitrogen, the latter from headspace equilibration with atmospheric gases dissolved in the influent. SMP in all SAnMBRs showed the same trend during start-up, beginning at zero, increasing sharply to between 0.29–0.42 L CH_4 g^{-1} COD as residual feed and intermediate products were consumed, then stabilising at 0.20–0.22 L CH_4 g^{-1} COD. As can be seen from Figure 2f and Table A1, during the stable operating periods of each phase, the SMP in L CH_4 g^{-1} COD removed (COD_{rem}) was around 0.23 ± 0.01 for operation at a 20-day SRT in S1; 0.22 ± 0.01, 0.23 ± 0.01 and 0.24 ± 0.01 at a 30-day SRT in S1, S3 and S4, respectively; 0.25 ± 0.01 at a 45-day SRT in S2; and 0.26 ± 0.02 at a 90-day SRT in S4. SMP at a 60-day SRT in S3 was around 0.23 ± 0.02 L CH_4 g^{-1} COD_{rem}, although full steady-state operation under these conditions was not achieved. These results again demonstrate a decline in SMP at the shorter SRT, most likely due to the incorporation of a greater proportion of carbon into microbial biomass at higher growth rates.

This outcome was consistent with the results of earlier research at 36 °C [32], with the lower SMP values most probably due to reduced rates of reaction at a lower operating temperature. Huang et al. [33,34] also found that SMP rose with increases in SRT in the treatment of low-strength wastewaters. They reported values of 0.13, 0.20 and 0.22 L CH_4 g^{-1} COD_{rem} at SRT of 30, 60 and infinite days respectively in treatment of synthetic wastewater at OLR from 1.1–1.67 g COD L^{-1} day^{-1} [33]; and around 0.03, 0.08 and 0.08 L CH_4 g^{-1} COD_{rem} at SRT of 30, 60 and 90 days, respectively, when treating municipal wastewater at an OLR of around 1 g COD L^{-1} day^{-1} [34]. Their suggested explanation was that the longer SRT would provide better conditions for methanogenesis, allowing higher biogas productivity. No reason was given, however, for the very low values of SMP per g of COD removed when treating real municipal wastewater [34]. SMP values obtained at 20 °C in the current work are higher than those found by Huang et al. [33,34] at 25–30 °C. This may be partly due to the loss of dissolved methane in the effluent, which was not considered in their first study [33]. The pattern of lower SMP values at shorter SRTs is nevertheless evident in both of these studies and in this and earlier work at 36 °C [32]. Dong et al. [24] also reported a decline in SMP from 0.13 to 0.10 and then 0.08 L CH_4 g^{-1} COD_{rem} as SRT was stepped down from 100 to 70 and then 40 days over a 400-day experimental period when treating municipal wastewater at 23 °C.

Consideration of the SMP normalised to MLSS (SMP_{MLSS}) indicated that the biomass methane conversion efficiency was higher at shorter SRT (Figure 2g), as also seen in prior work at 36 °C [32]. SMP values obtained were similar to those at 36 °C, with a maximum of 0.095 L CH_4 g^{-1} COD_{rem} L g^{-1} MLSS for operation at a 20-day SRT in S1; and around 0.2 L CH_4 g^{-1} COD_{rem} L g^{-1} MLSS during a stable performance at a 90-day SRT between days 110–130 in S4. Data in Huang et al. [33] indicated a higher SMP_{MLSS} value at longer SRTs, in contradiction to the outcomes both of this study at 20 °C and of the earlier research at 36 °C [32]. While the values reported by Huang et al. [33] did not include dissolved CH_4, the same trend was repeated for all trials at a given HRT, whereas in theory, the effluent methane content in the effluent should be the same in each case. The absence of steady-state conditions under which stable MLSS concentrations can be achieved makes it difficult to identify reliable values for normalised SMP from other similar studies. For higher-strength wastewaters in mesophilic conditions, however, the same trends can be seen as in the current work. During AnMBR treatment of thin corn stillage, the SMP_{MLSS} of 0.016 L CH_4 g^{-1} COD_{rem} L g^{-1} MLSS at day SRT of 20 and 30 days was higher than that of 0.010 L CH_4 g^{-1} COD_{rem} L g^{-1} MLSS for a 50-day SRT [29], while for dairy wastewater 20 and 40-day SRT gave SMP_{MLSS} values of 0.046 and 0.026 L CH_4 g^{-1} COD_{rem} L g^{-1} MLSS, respectively [31].

3.2. Membrane Performance and Fouling Phenomena

This work at 20 °C showed that shorter SRT resulted in better performance with respect to flux but indicated that there was no clear advantage in operating at SRT of <30 days (Figure 2b). In EP-1, however, a more rapid increase in TMP was needed to maintain the flux at a 45-day SRT in S2 and a 60-day SRT in S3 (Figure 2c), compared to that required in S4 at a 90-day SRT, despite considerably higher MLSS concentration in the latter (Figure 2d). This suggests that the onset of membrane fouling was slower at SRT of 30 and 90 days than at 45 or 60 days. In EP-2, the required TMP at a 45-day SRT in S2 was above that for a 90-day SRT in S4 until day 120, while the TMP in S3 began rising sharply from day 106, despite the introduction of a 30-day SRT on day 100, indicating that performance was better at the longer 90-day SRT. TMP in S4 with a 90-day SRT started to increase slightly from day 80, although it was only after day 120 that it began to rise at a rate similar to that seen previously in S3 with a 60-day SRT (Figure 2c). When SRT in S3 and S4 was reduced to 30 days, this led to a complete recovery in flux to the target value of 4 L m^{-2} h^{-1} (Figure 2b), together with reductions in required TMP. Faster recovery was seen in S4, however, most likely because of the abrupt drop in MLSS compared to the slower transition in S3. It should also be noted that the SAnMBR were operated with in situ gas cleaning only through the experimental period, with no external or chemical cleaning. These responses thus not only demonstrate the considerable effect of SRT on membrane fouling but also show that fouling of this type can be at least partially reversed if an optimal SRT is applied.

The above results confirmed that the effect of SRT on membrane fouling is not simply due to the corresponding changes in MLSS concentration. Other research shows that fouling is also affected by components such as EPS, the production of which is strongly related to microbial growth and hence to SRT [3,19]. Research into the effects of SRT on fouling and overall performance in AnMBR is still scarce, however, and the interactions between multiple different factors are often unclear. Work by Huang et al. [33] with synthetic wastewater as well as prior work at 36 °C [32], found that membrane fouling was more severe at longer SRTs. When treating real sewage, however, Huang et al. [34] obtained the highest flux rate at a 60-day SRT, with more severe fouling at both longer and shorter SRTs.

3.3. Mixed Liquor Characteristics

3.3.1. Capillary Suction Time

Results of CST measurements are given in Figure 4a and Table A1. CST values ranged between 94–569 s, and SRT had a strong effect on the mixed liquor's ability to retain moisture: samples taken from SAnMBR operating at shorter SRT released liquid much more readily than those at longer SRT. During the start-up period, when all four SAnMBR were operating as replicates at a 90-day SRT, the increase in CST was similar in all cases. Differences in CST began to appear in EP-1, with values generally higher at longer SRT. By the end of EP-1, however, CST values in S3 and S4 were similar at 438 and 430 s, respectively, despite a 25% difference in MLSS concentrations. This result indicates that the 60-day SRT in S3 represented the least favourable conditions for moisture removal and is reflected in the fall in membrane flux during this period despite continuing increases in TMP. After SRT in S3 was reduced to 30 days at the start of EP-2, the CST began to fall. The value of CST per unit of MLSS (CST_{MLSS}, Figure 4b) continued to rise, however, showing that normalised mixed liquor filterability was still dropping. At the end of EP-2, the CST_{MLSS} value in S3 was twice that in S4, although S3 was operating at one third of the SRT in S4.

At the start of EP-3, SRT was reduced to 20 days in S1 and to 30 days in S4, in conjunction with a sharp reduction in MLSS content in the case of the latter. While CST values in S1 fell gradually, reaching around 100 s by the end of the phase, the sharp MLSS reduction in S4 produced a much steeper decline in CST, which fell from over 550 to 259 s by day 207 then stabilised at around 250 s. CST values continued to fall in S3, while in contrast, in S2 the CST rose significantly, stabilising at around 480 s, in parallel with a major fall in flux rate.

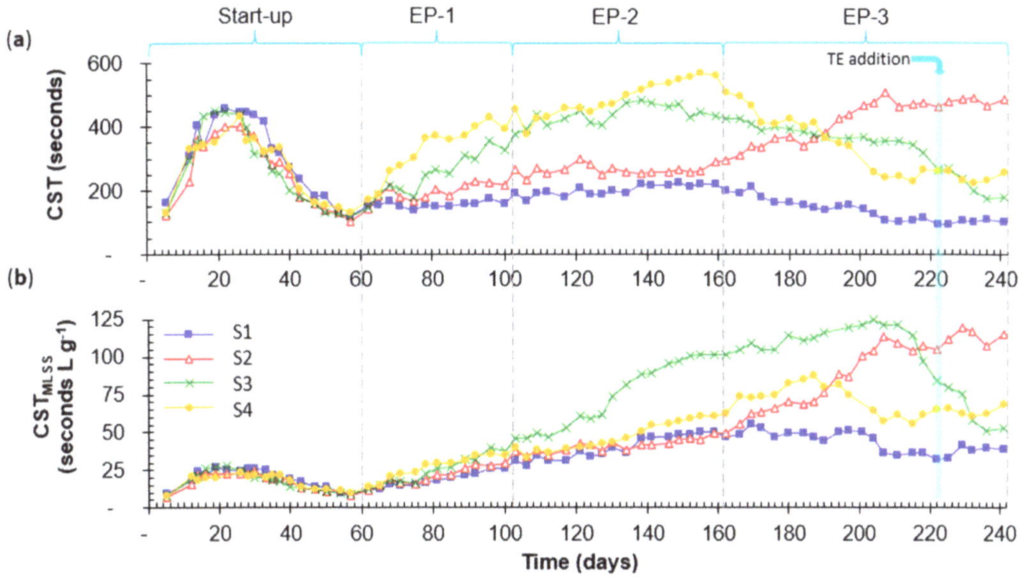

Figure 4. Mixed liquor filterability in SAnMBRs during the whole experimental period: (**a**) CST; (**b**) CST_{MLSS}. Vertical dotted lines indicate the start of experimental phases. Values for control parameters (SRT and TMP) in each phase are shown in Figure 2 and Table A2.

The sharp MLSS reduction in S4 at the start of EP-3 resulted in an almost-immediate rise in CST_{MLSS} values, indicating an increase in normalised filterability of the mixed liquor. This was reflected in the observed recovery in flux, which allowed significant reductions in the required TMP (Figure 2). In contrast, although CST values in S3 levelled off and then began to decrease after SRT was reduced to 30 days in EP-2, it was a further 100 days before the CST_{MLSS} started to fall. A reduction in SRT evidently produces only a gradual change in MLSS characteristics, as time is required for any 'fouling substances' present or microbial species particularly responsible for their production to be eliminated from the reactor. Conversely, if the MLSS concentration is decreased abruptly at the same time as SRT is reduced (e.g., by replacing a proportion of the mixed liquor as in S4), MLSS filterability can improve almost instantaneously as fouling materials are removed or disrupted, and the enhancements in CST and flux occur much more rapidly.

The steady-state CST values of around 100, 480 and 190 s in S1, S2 and S3 operating at 20, 45 and 30-day SRTs, respectively, at the end of EP-3 was higher than those generally reported for aerobic membrane bioreactors (MBR) treating municipal wastewater, which are typically on the order of 10 s [54,55]. CST_{MLSS} values in S1, S2 and S3 ranged from 39 to 113 s L g^{-1} MLSS, again higher than the normalised values of around 1–2 s L g^{-1} MLSS for aerobic MBR. CST and CST_{MLSS} values for combined or co-mingled primary and secondary sewage sludges show more variability but typically range from 50–200 s and 2–10 s L g^{-1}, with digestion sometimes leading to an increase in one or both parameters [56–58]. CST values of around 70 s were reported for MLSS from a full-scale SAnMBR treating source-separated blackwater in Spain [59], with the corresponding CST_{MLSS} of around 15–17 L g^{-1} MLSS, both similar in scale to those found in the current work. Dong et al. [25] operated a pilot-scale AnMBR on screened municipal wastewater at 23 °C at SRT of 40, 70 and 100 days, and reported a reduction in CST at the shorter SRT, although the system did not reach steady-state in each case. A similar pattern of reduction in CST at shorter SRT was also noted in a cross-flow AnMBR operating at 37 °C on thin corn stillage [29]. The authors reported that CST_{MLSS} values were similar at a 30 or 50-day SRT, and lower at 20 days, but again did not complete a full three SRT in all conditions.

3.3.2. Frozen Image Centrifugation

Frozen Image Centrifugation was developed by the UK's Water Research Centre to allow assessment of sludge dewaterability and thickening characteristics [60]. Although it has never been widely adopted in the water industry, it provides the basis for approaches to the design and operation of dewatering facilities [61,62].

Results from FIC testing are shown in Figure 5a, with the centrifuged sludge volume expressed as a percentage of the original MLSS sample at one-minute intervals during the FIC test. Values normalised against the original MLSS concentration are shown in Figure A2 in Appendix A. Similar to the CST values, these FIC results show clearly that SRT had a significant effect on solid-liquid separation, with MLSS samples taken at short SRT much more readily separable than those at longer SRT. As can be seen from Figure 5a, the rate of separation in the first few minutes of centrifugation was also considerably more rapid in SAnMBRs at shorter SRT, and the final sludge volumes were achieved in a shorter time. At the end of EP-3 the final sludge volumes for FIC tests were 13%, 18% and 29% at 20, 30 and 45-day SRT in S1, S3 and S2, respectively, while the corresponding final centrifuged solids concentrations were 20.1, 19.6 and 14.9 mg L^{-1} and separation rates in the first minute were 5.25, 3.75 and 0.50 mm min^{-1}. After one minute of centrifugation, S1 and S3 were close to their final volumes, while S2 was still compacting at the end of the test. Similar values at a 30-day SRT were also found in S1 at the end of EP-2 and S4 at end of EP-3, indicating some replicability of results.

One benefit of the FIC test is its ability to provide information on several aspects of dewaterability, all of which are potentially important with respect to equipment design and operating costs. The potential for a high ultimate solids concentration, for example, may be less significant if the time required to achieve it is inordinately long [62]. FIC can also reveal details and nuances not readily observable in other tests. In the current work, for example, from day 154 onwards, three separate phases were seen in many of the test samples. These consisted of a clear supernatant, then a distinct cloudy white layer overlying the darker layer of centrifuged sludge solids (Figure 5b). Times at which these layers were present are indicated by solid bar-colouration in Figures 5a and A1, and it can be seen that they remained visible for longer in samples taken from the SAnMBRs operating at longer SRT. The proportion of the test run during which three phases could be distinguished increased in S2 at 45-day SRT, but fell in S3 where the SRT had been reduced from 60 to 30 days, corresponding to improvements in TMP and CST. The appearance of the third phase of this type was also noted in FIC tests on digestate from sugar beet pulp [63]: it was referred to as the light fraction and constituted a high molecular weight material most probably composed of EPS and/or soluble microbial products. Its presence was more evident during periods of foaming, which, such as poor dewaterability, is associated with high sludge viscosity and EPS content. After the change from 60- to 30-day SRT in S3 at the end of EP-1, the CST remained high while final sludge volumes in the FIC test fell. Normalised values of both parameters continued to rise for almost 100 days, however, before dropping quite sharply around day 211. In S4 where the change to 30-day SRT at the end of EP-2 was achieved by MLSS dilution, a sharp fall in CST_{MLSS} after day 189 corresponding to a rapid increase in the separation rate at the start of the FIC test, matched by a reduction in time needed to reach the final volume. The reasons for these delayed but sudden shifts are unknown: they may be related to the passing of a threshold concentration for some component in the MLSS, or even to quorum behaviour of MLSS and biofilm organisms, a topic of growing interest for AnMBR performance and fouling [10,64].

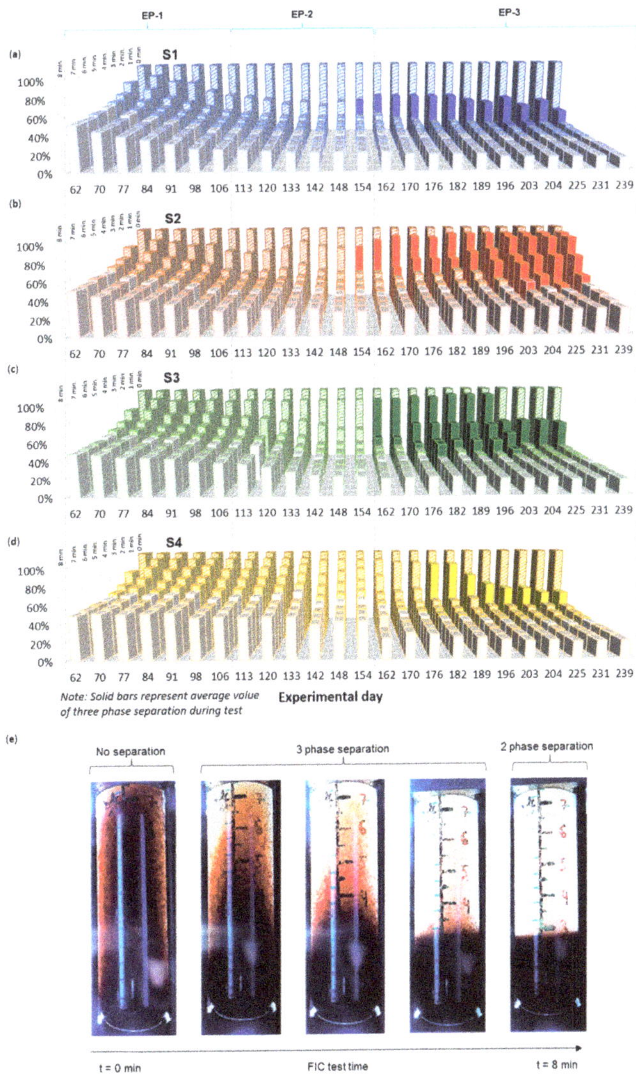

Figure 5. Mixed liquor centrifugation during experimental period: FIC test—Separated sludge volumes as % of original sample volume in (**a**) S1, (**b**) S2, (**c**) S3 and (**d**) S4, (**e**) separation phases during FIC test in centrifuged mixed liquor from S2 on day 204. Vertical dotted lines indicate start of experimental phases.

Several researchers have investigated relationships between parameters such as CST or specific resistance to filtration (SRF) and MLSS dewaterability, membrane flux and fouling in both aerobic and anaerobic MBR [25,29,59,65]. CST is generally considered to be one of the most useful parameters because it shows a reasonably good correlation with membrane performance and is relatively easy to evaluate on-site [27,35]. In the current work, regression analysis was carried out to investigate any relationship between applied TMP (as an indicator of filterability and membrane fouling) with CST and FIC. However, in both cases, interpretation was made difficult by periods at maximum TMP. Although neither showed a very strong correlation, the relationship between TMP and CST was generally stronger than for TMP and FIC. This is as might be expected as the water removal mechanism in a CST test where liquid is drawn through the MLSS is more similar to that in membrane bioreactor than in FIC. Several authors have noted the importance of using a type of test, which reflects the dewatering technology [66,67], and FIC testing has greater similarity to both sludge thickening and centrifugation. The value of expanding the range of tests currently utilised for sludge characterisation was emphasised by Spinosa and Doshi [68], and further work on FIC and other tests is needed to provide an enhanced understanding of different parameters and the significance of the relationships between them.

3.3.3. Extracellular Polymeric Substances

Under steady-state conditions at the end of EP-3, bound EPS concentrations at 20, 30 and 45-day SRT were 378, 483 and 511 mg L^{-1} in S1, S3 and S2, respectively. Specific values normalised against MLSS were around 165 mg g^{-1} VSS for the SAnMBR at 20 and 30-day SRT, and slightly lower at 134 mg g^{-1} VSS at a 45-day SRT. Other authors have also reported higher EPS content at shorter SRT: examples include treatment of municipal wastewater in a sidestream AnMBR at SRT ranging from 19–217 days, although steady-state operation was achieved only at the 19-day SRT [22]; and of synthetic wastewater at 31 °C in a AnMBR with a ceramic membrane at 100, 50 and 25-day SRT [69], although with full steady-state operation for the 25-day SRT only.

As can be seen in Figure 6a,b, for much of the experimental period, the specific concentrations of bound protein and carbohydrate were highest in S1 operating at a 30-day and 20-day SRT. When SRT in S3 was reduced from 60 to 30 days at the start of EP-2, however, specific concentrations increased in this reactor, with the carbohydrate content rising first. Specific protein and carbohydrate concentrations also increased in S4 after the SRT was reduced to 30 days at the start of EP-3. At the end of the experimental period, both the specific bound protein and the bound carbohydrate concentrations were lowest in S4 at a 45-day SRT, and higher in the SAnMBR at shorter SRT. Huang et al. [33,34] found higher specific protein contents in EPS at a 30-day SRT than at longer SRT, and suggested that this might produce larger flocs with a lower membrane fouling propensity. They also noted that long SRTs were associated with smaller median particle size, as the lower EPS content reduces flocculation, and hence promotes more rapid fouling. These suggestions appear to be consistent with the outcomes of the current work.

There was also evidence of a relationship between soluble EPS content and membrane performance at longer SRTs. As can be seen in Figure 6c,d, the specific carbohydrate and protein concentrations of soluble EPS in S3, which had shown the most rapid fouling of all at a 60-day SRT, were generally higher than in the other SAnMBRs. After SRT in S3 was reduced to 30 days from the start of EP-2, the specific soluble protein and carbohydrate contents continued rising until around day 205 when they finally reduced. At that point, the CST value also fell, the rate of separation in the FIC test increased, and the flux began to recover. Conversely, towards the end of EP-3, the specific soluble protein and carbohydrate contents in S2, operating at a 45-day SRT, increased relative to the other SAnMBRs, which at that point were all running at shorter SRTs. This corresponded to the period in which CST values in S2 began to rise and flux to reduce, despite an increase in TMP, to 9.8 kPa. This reactor reached the end of the experiment with the highest soluble protein and carbohydrate contents (specific and absolute), highest TMP, highest CST and FIC values, lowest separation rate in the FIC test and lowest flux.

Huang et al. [33] found that the concentration of soluble microbial products was inversely related to SRT, while higher SMP carbohydrate and protein contents at longer SRT resulted in higher fouling rates. In contrast, when real municipal wastewater feed was used, the minimum specific and absolute values for SMP carbohydrate and proteins were found at a 60-day SRT [34]. This was attributed to incomplete substrate degradation at the 30-day SRT, and an increased concentration of residual cell materials at the 90-day SRT. Although EPS and MLSS concentrations were lowest at the shortest SRT, this was insufficient to counteract the fouling effect of the higher SMP content. Soluble microbial products were not measured in the current study, but this explanation may be partly supported by the results of this experiment and of the previous work at 36 °C [32], which showed no further enhancement in membrane performance at SRT of <30 and <25 days, respectively. Laspidou and Rittmann [70] proposed that under some conditions soluble EPS and SMP may be similar, although other researchers were unable to confirm this [71]. Trends in bound and soluble EPS observed in this work are consistent with those in other studies carried out under steady-state conditions, where the consensus is that bound EPS tends to be high at shorter SRTs while microbially-induced SMP concentrations tend to decrease [33,34]. Given that soluble EPS also decreased at shorter SRTs in the current work, it can be concluded that even if it is not identical to SMP, its relationship to SRT and its effects on membrane fouling are closely similar.

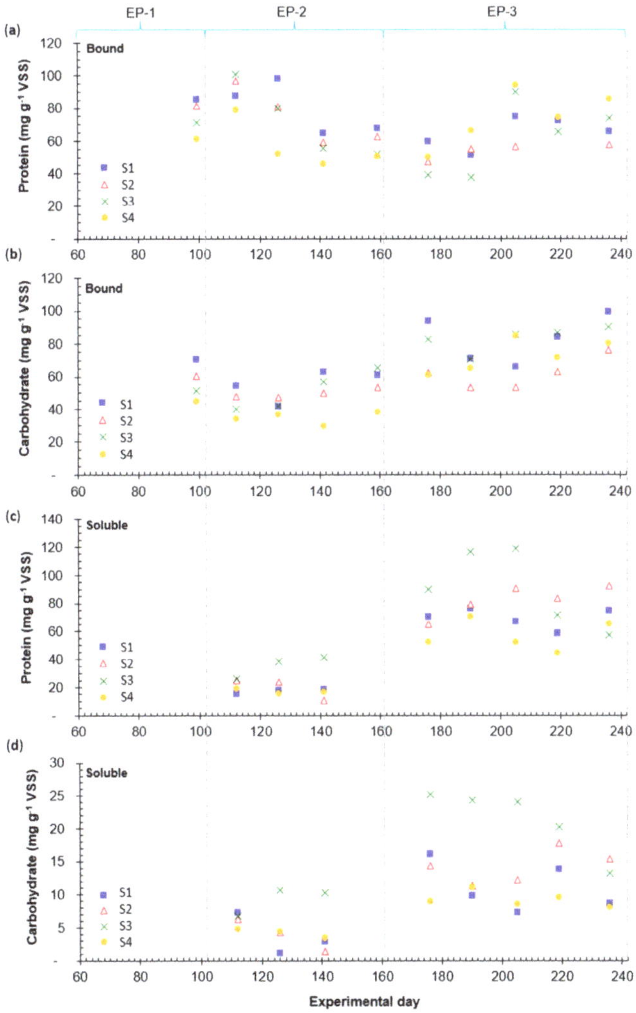

Figure 6. EPS composition during experimental period: (**a**) bound protein, (**b**) bound carbohydrate; (**c**) soluble protein, (**d**) soluble carbohydrate. Vertical dotted lines indicate start of experimental phases. Values for control parameters (SRT and TMP) in each phase are shown in Figure 2 and Table A2.

EPS content was not measured in the previous work with the same wastewater at 36 °C [32], and thus the effect of temperature in the current study is unknown. EPS and soluble microbial products normalised to MLSS content were reported to increase with decreasing temperature in a flat-sheet SAnMBR fed on a synthetic municipal wastewater at 25, 15 and 10 °C [14]. The ratio of protein to carbohydrate also rose with decreasing temperature, leading to higher rates of fouling at low temperatures. The applied SRT was not reported, however, thus it is likely that this was high and determined only by the withdrawal of samples for analysis. Similar trends in EPS and soluble microbial product concentrations were observed in two hollow fibre SAnMBR fed on synthetic wastewater at 25 and 35 °C with an SRT of 370 days [72], and in a hollow fibre SAnMBR fed on municipal wastewater at 25, 20 and 15 °C at SRT of 93.9, 40.3 and 20.7 days, respectively [73], although the operating periods under each set of conditions were less than 3 SRT. Kong et al. [1] operated a 5 m^3 pilot-scale SAnMBR with a HFM unit on municipal wastewater for over 200 days at 25 °C. No change in EPS concentration was observed when HRT was varied between 6–24 h, but SRT was not reported.

EPS and soluble microbial products were characterised in flat-sheet SAnMBR treating low-strength synthetic wastewater at 25 °C and operated at various OLR [19]. Properties were linked to membrane fouling mechanisms, but the SRT of the system was near-infinite as no MLSS was discharged other than for sampling.

While some differences in EPS production and composition were seen in the current study, these were not sufficiently marked enough to show a clear correlation between SRT and observed fouling behaviour. Nonetheless, the findings are consistent with those reported elsewhere. While there may be a specific range of SRT for each system in which membrane performance can be optimised, however, the effect of EPS concentrations and compositions on fouling in AnMBR is still not well understood, as a function of SRT or other operating parameters. For this reason, further work may be needed to assess the interaction between operational variables, MLSS characteristics and membrane fouling.

3.3.4. Biomass Growth and Kinetics

As well as having a significant impact on mixed liquor characteristics, the MCRT or SRT controls microbial growth rates, and thus potential growth yield. As previously seen for the work carried out at 36 °C [32], in the current experiment, the biomass yield appeared to show a sharp increase at each reduction in SRT. This reflected the fact that, as noted in the Methods section, changes in stored biomass were not taken into account to eliminate major variations. The apparent increases in yield were followed by a gradual decrease as conditions stabilised. The stable value was taken as the representative yield for each corresponding SRT and was equal to 0.131 ± 0.008 g VSS g^{-1} COD_{rem} at a 20-day SRT, 0.124 ± 0.012 g VSS g^{-1} COD_{rem} at a 30-day SRT, and 0.114 ± 0.004 g VSS g^{-1} COD_{rem} at a 45-day SRT. Biomass yields for the 60- and 90-day SRTs are not reported here as steady-state operation of the SAnMBRS was not achieved for these conditions, and thus representative values were not available.

The above results confirmed that shorter SRT gave higher biomass yields, in agreement with prior work at 36 °C [32]. These outcomes provide further support for the idea that at shorter SRT a higher proportion of the available carbon is directed into biomass growth rather than methane production, thus reducing the substrate SMP. Conversely, at longer SRT, lower growth rates and increased hydrolysis of MLSS leads to reduced biomass yields and higher SMP values [24]. In a full-scale system, a shorter SRT with a higher biomass yield would be linked to larger volumes of waste sludge for disposal and potentially to a greater requirement for TE supplementation, both leading to higher operating costs.

These low yields are typical of those reported elsewhere for similar systems, including 0.11 g VSS g^{-1} COD for low-strength synthetic wastewater treatment at 25 °C and near-infinite SRT [74]. Ji et al. [73] found sludge yields of 0.12, 0.19 and 0.388 g VSS g^{-1} COD_{rem} with real municipal wastewater at SRT of 93.9, 40.3 and 20.7 days; but the effect was confounded by accompanying changes in temperature from 25 to 20 and 15 °C and the system did not operate for 3 SRT in each set of conditions. Sludge yields of 0.07–0.11 g VSS g^{-1} COD_{rem} were reported using the same wastewater at 25 °C for SRT from 65 days to near-infinite with HRT from 6–12 h and OLR between 0.7–1.5 g COD L^{-1} day^{-1} [28].

The current study showed changes occurring over different timescales, which could be categorised as follows: (i) those which happen slowly, such as stabilisation of MLSS concentrations after a change in SRT; these changes cannot easily be accelerated as they are both growth-mediated and affected by washout rate, although interventions such as partial removal of MLSS may reduce the time needed to approach stable values. (ii) Those which can happen more rapidly, such as a response to TE addition or to other factors affecting the microbial population; as these are metabolically mediated, they can trigger an immediate response through stimulation or inhibition. (iii) Those which may happen rapidly but after a delay, such as the observed sharp changes in CST and FIC values following reductions in SRT, which could be metabolically or physico-chemically determined in relation to threshold concentrations. The factors that cause these changes may also be interdependent, and hence to allow full evaluation of their effects on system performance and mixed liquor characteristics long-term operation is advisable, wherever possible for at least 3 SRT under a given set of operating conditions.

4. Conclusions

Extended operation of four SAnMBRs at 20 °C on a low-to-intermediate strength substrate with a high suspended solids content was conducted at different SRTs. This enabled accurate determination of flux rates at specific SRT, accompanied by evaluation of COD removal efficiencies; estimation of biomass yields; a COD-based mass balance; physical characterisation of mixed liquors using CST and FIC tests; and analysis of EPS concentration, type and composition. The results showed that SRT had a considerable effect on flux rates, with shorter SRT allowing enhanced membrane performance and improved mixed liquor filterability, at a higher bound EPS content but with lower soluble EPS concentrations in the MLSS. Operation at shorter SRT led to a reduction in specific methane productivity and in COD removal rates, accompanied by higher biomass yields. Whilst no further enhancement of membrane performance was found at SRT of <30 days, operation at a 60-day SRT resulted in more rapid onset of membrane fouling and declining performance than at a 90-day SRT. Reduced COD removal rates at shorter SRTs were probably due to the increased washout of essential trace elements caused by the higher biomass turnover: this was supported by a rapid recovery in COD removal efficiency after TE supplementation was carried out. COD removal efficiencies remained slightly lower at shorter SRTs, however, suggesting that the lower available biomass concentration may also have affected this parameter. Overall COD removal efficiencies achieved after TE addition were very close to those seen in an earlier study using the same substrate at 36 °C, and the effect of the lower operating temperature on this parameter was, therefore, considered to be negligible. CST values gave some indication of changes in membrane performance, while frozen image centrifugation provided additional insights into MLSS properties and fouling behaviour, with three separate phases clearly visible at longer SRT. The ability of the FIC test to identify several parameters such as final solids concentration and rates of solid separation may also make it especially appropriate for assessing sludge dewaterability in gravity thickening or centrifugation, and further exploration of this approach is recommended. Operation over the full experimental period, without chemical or external cleaning, not only demonstrated the effects of SRT on performance parameters but also indicated that membrane fouling could be at least partially reversed if an optimal SRT is applied. These results indicated that responses to a change in SRT may be significantly delayed, probably as a result of the different timescales on which growth-related, metabolic and physico-chemical and/or the various interactions between them. Thus confirming the importance of long-term operation to allow full evaluation of system performance and mixed liquor characteristics under steady-state conditions. Together with earlier findings from operation at 36 °C, this work confirmed that there are potential trade-offs to be made between membrane performance, specific methane productivity and sludge yields when selecting a suitable SRT for AnMBR systems of this type.

Author Contributions: Conceptualization and methodology, S.P.-R., S.H., C.J.B.; experimental work, S.P.-R.; data analysis and interpretation, S.P.-R., S.H., C.J.B.; writing—original draft preparation, S.P.-R.; writing—review and editing, S.H., C.J.B.; supervision, C.J.B., S.H.; funding acquisition, C.J.B. All authors have read and agreed to the published version of the manuscript.

Funding: This research was supported by the Mexican National Council on Science and Technology (CONACYT), the Faculty of Engineering and Physical Sciences at the University of Southampton, and the BBSRC ERA-Net AmbiGAS project BB/L000024/1.

Institutional Review Board Statement: Not applicable.

Informed Consent Statement: Not applicable.

Data Availability Statement: The data presented in this study are openly available on DOI: 10.5258/SOTON/D1950.

Conflicts of Interest: The authors declare no conflict of interest. The funders had no role in the design of the study; in the collection, analyses, or interpretation of data; in the writing of the manuscript, or in the decision to publish the results.

Appendix A

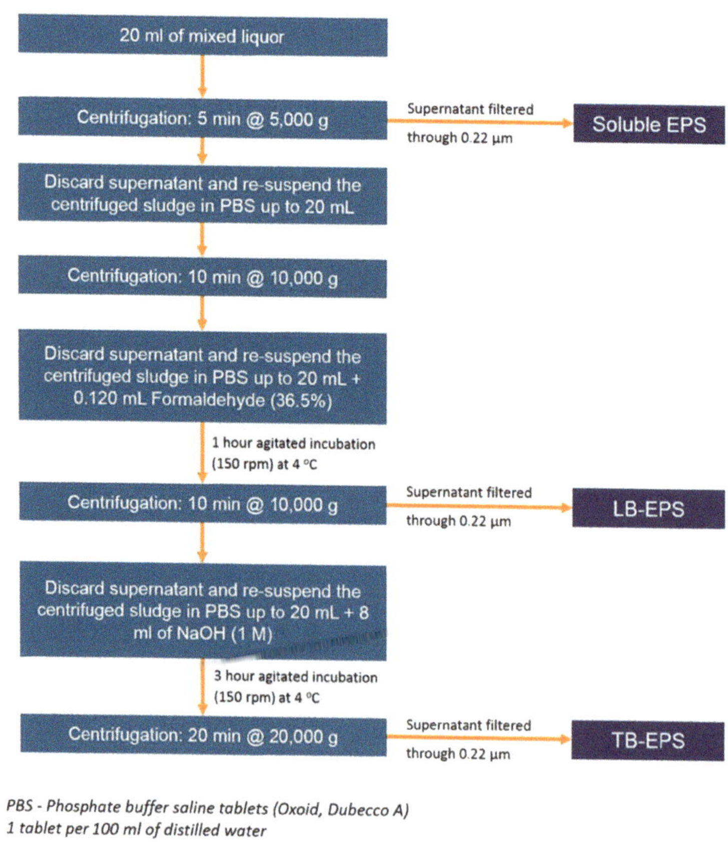

Figure A1. Flowchart for EPS extraction procedure.

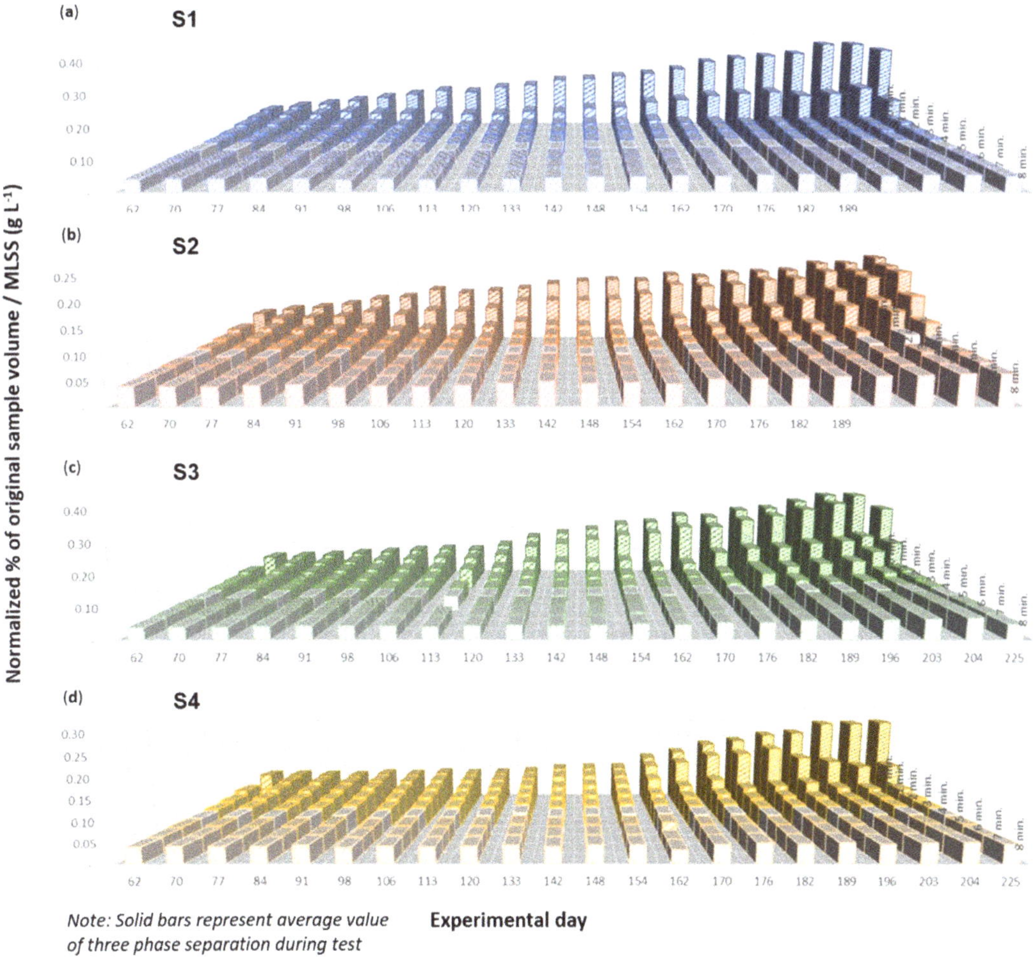

Figure A2. FIC test sludge volumes results normalised to MLSS content in (**a**) S1, (**b**) S2, (**c**) S2 and (**d**) S4.

Table A1. Experimental results; SRT effect on: membrane flux, OLR, HRT, feed COD, effluent COD and COD removal rates.

Parameter Phase	Reactor	SRT (days)	TMP (kPa)	Membrane Flux (J) * ($L\ m^{-2}\ h^{-1}$)	OLR ($g\ COD\ L^{-1}\ day^{-1}$)	HRT (hours)	Feed COD ($mg\ L^{-1}$)	Effluent COD ($mg\ L^{-1}$)	COD Removal Rate (%)
Start-up (day 0–59, total 60 days)	A	90	7.1 → 2.1 → 2.5	16.1 → 6.7	0.50–1.45	5.8 → 12.5	305 → 507	99 → 200 → 59	74% → 53% → 89%
	B	90	7.1 → 2.1 → 2.5	16.1 → 6.4	0.50–1.47	5.8 → 13.1	305 → 507	100 → 222 → 51	74% → 55% → 90%
	C	90	7.1 → 2.1 → 2.5	16.1 → 6.3	0.50–1.49	5.8 → 13.3	305 → 507	83 → 213 → 39	79% → 49% → 93%
	D	90	7.1 → 2.1 → 2.5	16.1 → 6.4	0.50–1.46	5.8 → 13.1	305 → 507	83 → 224 → 62	72% → 42% → 88%
EP-1 (day 60–111, total 52 days)	A	30	2.5 → 3.2 → 1.8	6.7 → 5.3 → 4.1	0.98 ± 0.03	12.5 → 15.8 → 20.2	567 → 860	59 → 144	88% ± 1%
	B	45	2.5 → 5.1 → 2.9	6.4 → 5.1 → 4.1	0.94 ± 0.04	13.1 → 16.5 → 20.3	567 → 860	51 → 65	91% ± 1%
	C	60	2.5 → 6.7 → 4.3 → 6.3	6.3 → 4.2 → 3.7	1.00 → 0.90	13.3 → 17.7 → 22.5	567 → 860	41 → 36	92% ± 2%
	D	90	2.5 → 3.2 → 2.2	6.4 → 5.3 → 4.1	0.97 ± 0.02	13.1 →16.2 → 20.4	567 → 860	61 → 21	88% → 97%
EP-2 (day 112–160, total 48 days)	A	30	1.8 ± 0.5	4.1 ± 0.0	0.98 ± 0.01	20.6 ± 0.2	843 ± 26	144 → 125	85% ± 2%
	B	45	2.8 ± 0.6	4.1 ± 0.0	0.99 ± 0.01	20.5 ± 0.2	843 ± 26	65 → 70	93% ± 1%
	C	30	6.3 → 9.8	3.7 → 3.0	0.90 → 0.72	22.5 → 28.5	843 ± 26	36 → 77	95% → 91%
	D	90	2.2 → 9.8	4.1 → 3.2	0.97 → 0.76	20.4 → 26.7	843 ± 26	21 → 44	96% ± 1%
EP-3 (day 161–242, total 82 days)	A	20	1.9 → 2.2	4.0 ± 0.1	0.98 ± 0.02	21.0 ± 0.4	853 ± 21	136 → 228 → 68	84% → 74% → 92%
	B	45	2.8 → 9.8	4.1 → 3.2	0.99 → 0.78	20.5 → 26.3	853 ± 21	70→ 84 → 22	92% → 92% → 97%
	C	30	9.8 → 3.8	3.0 → 2.7 → 4.0	0.72 → 1.02	28.5 → 31.5 → 21.9	853 ± 21	77 → 217 → 39	91% → 75% → 96%
	D	30	9.8 → 6.5	3.2 → 4.1	0.76 → 0.99	26.7 → 21.1	853 ± 21	44 → 91 → 34	95% → 89%→ 96%

(→) Variable trend: initial → middle → final; (±) Stable performance: One standard deviation to show spread of data from the average value under stable performance; (*) Daily average.

Table A2. Experimental results; SRT effect on: CH_4 content in biogas, SMP, MLSS, MLVSS, CST and pH.

Parameter Phase	Reactor	SRT (days)	CH_4 in Biogas * (%)	SMP ** (L CH_4 g^{-1} COD Removed)	MLSS (g L^{-1})	MLVSS (g L^{-1})	CST (seconds)	pH
Start-up (day 0–59, 60 days)	A	90	88% → 93%	0.00 → 0.33 → 0.21	16.2 → 12.7	12.5 → 10.0	162 → 461 → 117	7.0 → 6.2 → 6.0
	B	90	81% → 92%	0.00 → 0.29 → 0.20	16.2 → 12.5	12.6 → 10.2	122 → 402 → 101	7.0 → 6.2 → 6.8
	C	90	76% → 93%	0.00 → 0.42 → 0.22	16.2 → 12.4	13.3 → 10.0	130 → 449 → 117	7.0 → 6.2 → 6.8
	D	90	87% → 93%	0.00 → 0.42 → 0.21	16.2 → 12.7	12.7 → 10.4	132 → 435 → 131	7.0 → 6.2 → 6.9
EP-1 (day 60–111, 52 days)	A	30	92% ± 1%	0.21 ± 0.01	12.7 → 6.2	10.3 → 5.4	117 → 198	6.8 ± 0.0
	B	45	91% ± 1%	0.22 ± 0.02	12.5 → 7.3	10.2 → 6.2	101 → 252	6.8 ± 0.0
	C	60	92% ± 2%	0.23 ± 0.02	12.4 → 8.7	10.0 → 7.4	117 → 438	6.8 ± 0.0
	D	90	92% ± 2%	0.23 ± 0.02	12.7 → 11.7	10.4 → 9.8	131 → 430	6.8 ± 0.0
EP-2 (day 112–160, 48 days)	A	30	90% ± 1%	0.22 ± 0.01	6.2 → 4.4	5.4 → 3.8	198 →215	6.7 ± 0.0
	B	45	90% ± 1%	0.24 ± 0.01	7.3 → 6.0	6.2 → 5.2	252 →291	6.7 ± 0.1
	C	30	90% ± 1%	0.23 ± 0.01	8.7 → 4.3	7.4 → 3.7	407 → 482 → 433	6.7 ± 0.0
	D	90	90% ± 1%	0.26 ± 0.02	11.7 → 9.3	9.8 → 8.1	430 → 569	6.8 ± 0.0
EP-3 (day 161–242, 82 days)	A	20	90% ± 1%	0.23 ± 0.01	4.4 → 2.6	3.8 → 2.2	222→ 101	6.7 ± 0.0
	B	45	90% ± 1%	0.25 ± 0.01	6.0 → 4.2	5.2 → 3.7	291 → 488	6.8 ± 0.0
	C	30	90% ± 1%	0.23 ± 0.01	4.3 → 2.8 → 3.4	3.7 → 2.9	433 → 182	6.8 ± 0.0
	D	30	90% ± 1%	0.24 ± 0.01	9.3 → 6.4 → 3.8	8.1 → 3.3	569 → 240	6.8 ± 0.0

(→) Variable trend: initial → middle → final; (±) Stable performance: One standard deviation to show spread of data from the average value under stable performance; (*) Normalised to total biogas content in sample (i.e., neglecting air introduced dissolved through feed); (**) Takes into account the methane dissolved in the effluent.

References

1. Kong, Z.; Wu, J.; Rong, C.; Wang, T.; Li, L.; Luo, Z.; Ji, J.; Hanaoka, T.; Sakemi, S.; Ito, M. Large pilot-scale submerged anaerobic membrane bioreactor for the treatment of municipal wastewater and biogas production at 25 °C. *Bioresour. Technol.* **2021**, *319*, 124123. [CrossRef]
2. Smith, A.L.; Stadler, L.B.; Cao, L.; Love, N.G.; Raskin, L.; Skerlos, S.J. Navigating wastewater energy recovery strategies: A life cycle comparison of anaerobic membrane bioreactor and conventional treatment systems with anaerobic digestion. *Environ. Sci. Technol.* **2014**, *48*, 5972–5981. [CrossRef]
3. Stuckey, D.C. Recent developments in anaerobic membrane reactors. *Bioresour. Technol.* **2012**, *122*, 137–148. [CrossRef]
4. Lei, Z.; Yang, S.; Li, Y.-y.; Wen, W.; Wang, X.C.; Chen, R. Application of anaerobic membrane bioreactors to municipal wastewater treatment at ambient temperature: A review of achievements, challenges, and perspectives. *Bioresour. Technol.* **2018**, *267*, 756–768. [CrossRef]
5. McKeown, R.M.; Hughes, D.; Collins, G.; Mahony, T.; O'Flaherty, V. Low-temperature anaerobic digestion for wastewater treatment. *Curr. Opin. Biotechnol.* **2012**, *23*, 444–451. [CrossRef]
6. Lim, K.; Evans, P.J.; Parameswaran, P. Long-term performance of a pilot-scale gas-sparged anaerobic membrane bioreactor under ambient temperatures for holistic wastewater treatment. *Environ. Sci. Technol.* **2019**, *53*, 7347–7354. [CrossRef] [PubMed]
7. Ozgun, H.; Dereli, R.K.; Ersahin, M.E.; Kinaci, C.; Spanjers, H.; van Lier, J.B. A review of anaerobic membrane bioreactors for municipal wastewater treatment: Integration options, limitations and expectations. *Sep. Purif. Technol.* **2013**, *118*, 89–104. [CrossRef]
8. Smith, A.L.; Stadler, L.B.; Love, N.G.; Skerlos, S.J.; Raskin, L. Perspectives on anaerobic membrane bioreactor treatment of domestic wastewater: A critical review. *Bioresour. Technol.* **2012**, *122*, 149–159. [CrossRef] [PubMed]
9. Vinardell, S.; Astals, S.; Peces, M.; Cardete, M.; Fernández, I.; Mata-Alvarez, J.; Dosta, J. Advances in anaerobic membrane bioreactor technology for municipal wastewater treatment: A 2020 updated review. *Renew. Sustain. Energy Rev.* **2020**, *130*, 109936. [CrossRef]
10. Anjum, F.; Khan, I.M.; Kim, J.; Aslam, M.; Blandin, G.; Heran, M.; Lesage, G. Trends and progress in AnMBR for domestic wastewater treatment and their impacts on process efficiency and membrane fouling. *Environ. Technol. Innov.* **2020**, 101204. [CrossRef]
11. Dvořák, L.; Gómez, M.; Dolina, J.; Černín, A. Anaerobic membrane bioreactors—a mini review with emphasis on industrial wastewater treatment: Applications, limitations and perspectives. *Desalination Water Treat.* **2016**, *57*, 19062–19076. [CrossRef]
12. Lin, H.; Peng, W.; Zhang, M.; Chen, J.; Hong, H.; Zhang, Y. A review on anaerobic membrane bioreactors: Applications, membrane fouling and future perspectives. *Desalination* **2013**, *314*, 169–188. [CrossRef]
13. Smith, A.L.; Skerlos, S.J.; Raskin, L. Psychrophilic anaerobic membrane bioreactor treatment of domestic wastewater. *Water Res.* **2013**, *47*, 1655–1665. [CrossRef] [PubMed]
14. Watanabe, R.; Nie, Y.; Wakahara, S.; Komori, D.; Li, Y.-Y. Investigation on the response of anaerobic membrane bioreactor to temperature decrease from 25 °C to 10 °C in sewage treatment. *Bioresour. Technol.* **2017**, *243*, 747–754. [CrossRef] [PubMed]
15. Shin, C.; Bae, J. Current status of the pilot-scale anaerobic membrane bioreactor treatments of domestic wastewaters: A critical review. *Bioresour. Technol.* **2018**, *247*, 1038–1046. [CrossRef]
16. Tan, T.W.; Ng, H.Y.; Ong, S.L. Effect of mean cell residence time on the performance and microbial diversity of pre-denitrification submerged membrane bioreactors. *Chemosphere* **2008**, *70*, 387–396. [CrossRef]

17. Rittmann, B.E.; McCarty, P.L. *Environmental Biotechnology: Principles and Applications*; McGraw-Hill Education: New York, NY, USA, 2001.
18. WEF. *WEF Manual and Practice No. 36*; McGraw-Hill Education UK: London, UK, 2011.
19. Chen, R.; Nie, Y.; Hu, Y.; Miao, R.; Utashiro, T.; Li, Q.; Xu, M.; Li, Y.-Y. Fouling behaviour of soluble microbial products and extracellular polymeric substances in a submerged anaerobic membrane bioreactor treating low-strength wastewater at room temperature. *J. Membr. Sci.* **2017**, *531*, 1–9. [CrossRef]
20. Ng, H.Y.; Tan, T.W.; Ong, S.L. Membrane fouling of submerged membrane bioreactors: Impact of mean cell residence time and the contributing factors. *Environ. Sci. Technol.* **2006**, *40*, 2706–2713. [CrossRef]
21. Jinsong, Z.; Chuan, C.H.; Jiti, Z.; Fane, A. Effect of sludge retention time on membrane bio-fouling intensity in a submerged membrane bioreactor. *Sep. Sci. Technol.* **2006**, *41*, 1313–1329. [CrossRef]
22. Baek, S.H.; Pagilla, K.R.; Kim, H.-J. Lab-scale study of an anaerobic membrane bioreactor (AnMBR) for dilute municipal wastewater treatment. *Biotechnol. Bioprocess Eng.* **2010**, *15*, 704–708. [CrossRef]
23. Yeo, H.; Lee, H.-S. The effect of solids retention time on dissolved methane concentration in anaerobic membrane bioreactors. *Environ. Technol.* **2013**, *34*, 2105–2112. [CrossRef]
24. Dong, Q.; Parker, W.; Dagnew, M. Influence of SRT and HRT on bioprocess performance in anaerobic membrane bioreactors treating municipal wastewater. *Water Environ. Res.* **2016**, *88*, 158–167. [CrossRef] [PubMed]
25. Dong, Q.; Parker, W.; Dagnew, M. Long term performance of membranes in an anaerobic membrane bioreactor treating municipal wastewater. *Chemosphere* **2016**, *144*, 249–256. [CrossRef]
26. Thanh, P.M.; Ketheesan, B.; Zhou, Y.; Stuckey, D.C. Effect of operating conditions on speciation and bioavailability of trace metals in submerged anaerobic membrane bioreactors. *Bioresour. Technol.* **2017**, *243*, 810–819. [CrossRef] [PubMed]
27. Yurtsever, A.; Calimlioglu, B.; Sahinkaya, E. Impact of SRT on the efficiency and microbial community of sequential anaerobic and aerobic membrane bioreactors for the treatment of textile industry wastewater. *Chem. Eng. J.* **2017**, *314*, 378–387. [CrossRef]
28. Ji, J.; Chen, Y.; Hu, Y.; Ohtsu, A.; Ni, J.; Li, Y.; Sakuma, S.; Hojo, T.; Chen, R.; Li, Y.-Y. One-year operation of a 20-L submerged anaerobic membrane bioreactor for real domestic wastewater treatment at room temperature: Pursuing the optimal HRT and sustainable flux. *Sci. Total Environ.* **2021**, *775*, 145799. [CrossRef] [PubMed]
29. Dereli, R.K.; Grelot, A.; Heffernan, B.; van der Zee, F.P.; van Lier, J.B. Implications of changes in solids retention time on long term evolution of sludge filterability in anaerobic membrane bioreactors treating high strength industrial wastewater. *Water Res.* **2014**, *59*, 11–22. [CrossRef]
30. Dereli, R.K.; van der Zee, F.P.; Heffernan, B.; Grelot, A.; van Lier, J.B. Effect of sludge retention time on the biological performance of anaerobic membrane bioreactors treating corn-to-ethanol thin stillage with high lipid content. *Water Res.* **2014**, *49*, 453–464. [CrossRef]
31. Szabo-Corbacho, M.A.; Pacheco-Ruiz, S.; Míguez, D.; Hooijmans, C.M.; García, H.A.; Brdjanovic, D.; van Lier, J.B. Impact of solids retention time on the biological performance of an AnMBR treating lipid-rich synthetic dairy wastewater. *Environ. Technol.* **2021**, *42*, 597–608. [CrossRef]
32. Pacheco-Ruiz, S.; Heaven, S.; Banks, C.J. Effect of mean cell residence time on transmembrane flux, mixed-liquor characteristics and overall performance of a submerged anaerobic membrane bioreactor. *Environ. Technol.* **2017**, *38*, 1263–1274. [CrossRef]
33. Huang, Z.; Ong, S.L.; Ng, H.Y. Submerged anaerobic membrane bioreactor for low-strength wastewater treatment: Effect of HRT and SRT on treatment performance and membrane fouling. *Water Res.* **2011**, *45*, 705–713. [CrossRef]
34. Huang, Z.; Ong, S.L.; Ng, H.Y. Performance of submerged anaerobic membrane bioreactor at different SRTs for domestic wastewater treatment. *J. Biotechnol.* **2013**, *164*, 82–90. [CrossRef]
35. Baudry, M.; Zhou, T.; Van Gaelen, P.; Smets, I.; Pacheco-Ruiz, S. Protocol to evaluate and correlate membrane performance and mixed-liquor characteristics of full-scale and pilot-scale AnMBRs. In Proceedings of the 16th IWA World Conference on Anaerobic Digestion, Delft, The Netherlands, 23–27 June 2019.
36. Pacheco-Ruiz, S.; Heaven, S.; Banks, C.J. Development and testing of a fully gravitational submerged anaerobic membrane bioreactor for wastewater treatment. *Environ. Technol.* **2015**, *36*, 2328–2339. [CrossRef]
37. Henze, M.; Comeau, Y. Wastewater Characterization. In *Biological Wastewater Treatment: Principles, Modelling and Design*; Henze, M., Loosdrecht, M.C.M.V., Ekama, G.A., Brdjanovic, D., Eds.; IWA Publishing: London, UK, 2011.
38. Tchobanoglous, G.; Burton, F.L.; Stensel, H.D. *Wastewater Engineering—Treatment and Reuse*, 4th ed.; McGraw-Hill: New York, NY, USA, 2004.
39. SCA. *The Determination of Chemical Oxygen Demand in Waters and Effluents*; Standing Committee of Analysts; Environment Agency: Bristol, UK, 2007.
40. Walker, M.; Zhang, Y.; Heaven, S.; Banks, C. Potential errors in the quantitative evaluation of biogas production in anaerobic digestion processes. *Bioresour. Technol.* **2009**, *100*, 6339–6346. [CrossRef]
41. Walsh, K.P.; McLaughlan, R.G. Bubble extraction of dissolved gases from groundwater samples. *Water Air Soil Pollut.* **1999**, *115*, 525–534. [CrossRef]
42. APHA. *Standard Methods for the Examination of Water and Wastewater*; American Public Health Association: Washington, DC, USA, 2012.
43. Liu, H.; Fang, H.H. Extraction of extracellular polymeric substances (EPS) of sludges. *J. Biotechnol.* **2002**, *95*, 249–256. [CrossRef]

44. Domínguez, L.; Rodríguez, M.; Prats, D. Effect of different extraction methods on bound EPS from MBR sludges. Part I: Influence of extraction methods over three-dimensional EEM fluorescence spectroscopy fingerprint. *Desalination* **2010**, *261*, 19–26. [CrossRef]
45. Liang, Z.; Li, W.; Yang, S.; Du, P. Extraction and structural characteristics of extracellular polymeric substances (EPS), pellets in autotrophic nitrifying biofilm and activated sludge. *Chemosphere* **2010**, *81*, 626–632. [CrossRef] [PubMed]
46. Chabaliná, L.D.; Pastor, M.R.; Rico, D.P. Characterization of soluble and bound EPS obtained from 2 submerged membrane bioreactors by 3D-EEM and HPSEC. *Talanta* **2013**, *115*, 706–712. [CrossRef]
47. Dubois, M.; Gilles, K.A.; Hamilton, J.K.; Rebers, P.t.; Smith, F. Colorimetric method for determination of sugars and related substances. *Anal. Chem.* **1956**, *28*, 350–356. [CrossRef]
48. Frølund, B.; Griebe, T.; Nielsen, P. Enzymatic activity in the activated-sludge floc matrix. *Appl. Microbiol. Biotechnol.* **1995**, *43*, 755–761. [CrossRef]
49. Balcıoğlu, G.; Yilmaz, G.; Goender, Z.B. Evaluation of anaerobic membrane bioreactor (AnMBR) treating confectionery wastewater at long-term operation under different organic loading rates: Performance and membrane fouling. *Chem. Eng. J.* **2021**, *404*, 126261. [CrossRef]
50. Yu, D.; Li, C.; Wang, L.; Zhang, J.; Liu, J.; Wei, Y. Multiple effects of trace elements on methanogenesis in a two-phase anaerobic membrane bioreactor treating starch wastewater. *Appl. Microbiol. Biotechnol.* **2016**, *100*, 6631–6642. [CrossRef] [PubMed]
51. Sierra, J.D.M.; Lafita, C.; Gabaldón, C.; Spanjers, H.; van Lier, J.B. Trace metals supplementation in anaerobic membrane bioreactors treating highly saline phenolic wastewater. *Bioresour. Technol.* **2017**, *234*, 106–114. [CrossRef]
52. Thanh, P.M.; Ketheesan, B.; Yan, Z.; Stuckey, D. Trace metal speciation and bioavailability in anaerobic digestion: A review. *Biotechnol. Adv.* **2016**, *34*, 122–136. [CrossRef] [PubMed]
53. Giménez, J.; Carretero, L.; Gatti, M.; Martí, N.; Borrás, L.; Ribes, J.; Seco, A. Reliable method for assessing the COD mass balance of a submerged anaerobic membrane bioreactor (SAMBR) treating sulphate-rich municipal wastewater. *Water Sci. Technol.* **2012**, *66*, 494–502. [CrossRef]
54. Jin, B.; Wilén, B.-M.; Lant, P. Impacts of morphological, physical and chemical properties of sludge flocs on dewaterability of activated sludge. *Chem. Eng. J.* **2004**, *98*, 115–126. [CrossRef]
55. Pollice, A.; Laera, G.; Saturno, D.; Giordano, C. Effects of sludge retention time on the performance of a membrane bioreactor treating municipal sewage. *J. Membr. Sci.* **2008**, *317*, 65–70. [CrossRef]
56. Astals, S.; Esteban-Gutiérrez, M.; Fernández-Arévalo, T.; Aymerich, E.; García-Heras, J.; Mata-Alvarez, J. Anaerobic digestion of seven different sewage sludges: A biodegradability and modelling study. *Water Res.* **2013**, *47*, 6033–6043. [CrossRef]
57. Astals, S.; Venegas, C.; Peces, M.; Jofre, J.; Lucena, F.; Mata-Alvarez, J. Balancing hygienization and anaerobic digestion of raw sewage sludge. *Water Res.* **2012**, *46*, 6218–6227. [CrossRef]
58. Takashima, M.; Tanaka, Y. Acidic thermal post-treatment for enhancing anaerobic digestion of sewage sludge. *J. Environ. Chem. Eng.* **2014**, *2*, 773–779. [CrossRef]
59. Odriozola, M.; Lousada-Ferreira, M.; Spanjers, H.; van Lier, J.B. Effect of sludge characteristics on optimal required dosage of flux enhancer in anaerobic membrane bioreactors. *J. Membr. Sci.* **2021**, *619*, 118776. [CrossRef]
60. Lockyear, C.; White, M. *The WRc Thickenability Test Using a Low-Speed Centrifuge*; WRc Medmenham Laboratory: Stevenage, UK, 1979.
61. Hoyland, G.; Dee, A.; Day, M. Optimum design of sewage sludge consolidation tanks. *Water Environ. J.* **1989**, *3*, 505–516. [CrossRef]
62. Lockyear, C. Predicting Full-Scale Batch Thickener Performance Using the Frozen-Image Centrifuge. *Effl. Water Treat. J.* **1981**, *21*, 560–563.
63. Suhartini, S.; Heaven, S.; Banks, C.J. Comparison of mesophilic and thermophilic anaerobic digestion of sugar beet pulp: Performance, dewaterability and foam control. *Bioresour. Technol.* **2014**, *152*, 202–211. [CrossRef] [PubMed]
64. Tabraiz, S.; Petropoulos, E.; Shamurad, B.; Quintela-Baluja, M.; Mohapatra, S.; Acharya, K.; Charlton, A.; Davenport, R.J.; Dolfing, J.; Sallis, P.J. Temperature and immigration effects on quorum sensing in the biofilms of anaerobic membrane bioreactors. *J. Environ. Manag.* **2021**, *293*, 112947. [CrossRef]
65. Wang, Z.; Wu, Z.; Yu, G.; Liu, J.; Zhou, Z. Relationship between sludge characteristics and membrane flux determination in submerged membrane bioreactors. *J. Membr. Sci.* **2006**, *284*, 87–94. [CrossRef]
66. Kouzeli-Katsiri, A. Characterization of wastewater sludges end user's view. In *Monitoring of Water Quality*; Elsevier Science: Amsterdam, The Netherlands, 1998; pp. 75–88.
67. Spinosa, L. Technological characterization of sewage sludge. *Waste Manag. Res.* **1985**, *3*, 389–398. [CrossRef]
68. Spinosa, L.; Doshi, P. Re-conceptualizing Sludge Management: Focusing on Institutional and Governance Aspects. *J. Environ. Sci. Eng.* **2020**, *9*, 98–107.
69. Nilusha, R.T.; Yu, D.; Zhang, J.; Wei, Y. Effects of Solids Retention Time on the Anaerobic Membrane Bioreactor with Yttria-Based Ceramic Membrane Treating Domestic Wastewater at Ambient Temperature. *Membranes* **2020**, *10*, 196. [CrossRef]
70. Laspidou, C.S.; Rittmann, B.E. A unified theory for extracellular polymeric substances, soluble microbial products, and active and inert biomass. *Water Res.* **2002**, *36*, 2711–2720. [CrossRef]
71. Ramesh, A.; Lee, D.-J.; Hong, S. Soluble microbial products (SMP) and soluble extracellular polymeric substances (EPS) from wastewater sludge. *Appl. Microbiol. Biotechnol.* **2006**, *73*, 219–225. [CrossRef] [PubMed]

72. Ding, Y.; Guo, Z.; Liang, Z.; Hou, X.; Li, Z.; Mu, D.; Ge, C.; Zhang, C.; Jin, C. Long-term investigation into the membrane fouling behavior in anaerobic membrane bioreactors for municipal wastewater treatment operated at two different temperatures. *Membranes* **2020**, *10*, 231. [CrossRef]
73. Ji, J.; Ni, J.; Ohtsu, A.; Isozumi, N.; Hu, Y.; Du, R.; Chen, Y.; Qin, Y.; Kubota, K.; Li, Y.-Y. Important effects of temperature on treating real municipal wastewater by a submerged anaerobic membrane bioreactor: Removal efficiency, biogas, and microbial community. *Bioresour. Technol.* **2021**, *336*, 125306. [CrossRef] [PubMed]
74. Chen, R.; Nie, Y.; Ji, J.; Utashiro, T.; Li, Q.; Komori, D.; Li, Y.-Y. Submerged anaerobic membrane bioreactor (SAnMBR) performance on sewage treatment: Removal efficiencies, biogas production and membrane fouling. *Water Sci. Technol.* **2017**, *76*, 1308–1317. [CrossRef]

processes

MDPI

Article

Exploring Farm Anaerobic Digester Economic Viability in a Time of Policy Change in the UK

Angela Bywater [1,*] and Sigrid Kusch-Brandt [1,2,*]

1 Faculty of Engineering and Physical Sciences, University of Southampton, Southampton SO17 1BJ, UK
2 Faculty of Mathematics, Natural Sciences and Management, University of Applied Sciences Ulm, 89075 Ulm, Germany
* Correspondence: a.m.bywater@soton.ac.uk (A.B.); mail@sigrid-kusch.eu (S.K.-B.)

Abstract: The combination of a post-Brexit agricultural policy, the Global Methane Pledge announced during the last United Nations Climate Change Conference in Glasgow (COP26), and urgency of meeting climate goals means the UK has a unique opportunity to create an exemplar through recognition of the benefits of small-scale farm anaerobic digesters that valorise on-site wastes for renewable electricity and heat, cushioning agri-businesses against energy perturbations. To explore economic viability of farm-based biogas production, combinations of support levels, energy prices, capital cost, internal rate of return (IRR), and digestate value were analysed, employing a 550-cow dairy farm with access to other agricultural wastes. A 145 kWe system utilising 100% of CHP electricity (grid value: £0.1361 per kWh) and 70% of the heat (heating oil value: £0.055 per kWh) could achieve an IRR above 15.5% with a median electricity tariff of £0.1104 per kWh at a heat tariff from £0.0309 to £0.0873 per kWh thermal. Under a subsidy-free regime, the same system could achieve a 10% IRR with electricity prices in the range £0.149 to £0.261 per kWh. High fertiliser prices could increase digestate value, further improving viability. With late-2021 high energy prices, the technology approaches subsidy-free viability, but uptake is unlikely unless wider environmental and societal benefits of on-farm systems can be explicitly valued.

Keywords: agricultural wastes; biogas production; anaerobic digestion costs; economic viability; UK policy; Brexit; feed-in tariff; renewable heat incentive

Citation: Bywater, A.; Kusch-Brandt, S. Exploring Farm Anaerobic Digester Economic Viability in a Time of Policy Change in the UK. *Processes* **2022**, *10*, 212. https://doi.org/10.3390/pr10020212

Academic Editor: Elsayed Elbeshbishy

Received: 31 December 2021
Accepted: 17 January 2022
Published: 24 January 2022

1. Introduction

In the last decade, financial incentives based on energy production have created significant growth in anaerobic digestion (AD) installations in the United Kingdom (UK), with the number of agricultural plants increasing from 25 in 2010 to 344 in 2020 [1]. AD can capture uncontrolled greenhouse gas (GHG) emissions from the biodegradation of organic wastes and from farm management activities [2,3], making a useful contribution to overall GHG reductions (potentially by 6% for the UK, according to the Anaerobic Digestion and Bioresources Association [4]), thus helping the UK to meet its Paris Agreement and COP26 Global Methane Pledge commitments. Farm AD, particularly using on-site/local wastes for biogas generation and recovery of digestate, has myriad benefits beyond renewable energy generation alone: it can reduce GHG emissions [3,5–7], improve soil organic matter [8], facilitate improved nutrient management [9–11] thereby reducing the need for artificial fertiliser [12,13], kill pathogens and weed seeds if appropriately applied [14–16], provide opportunities for skilled rural employment [17,18], and create additional revenues in rural areas [19].

Clearly, a breakthrough in deployment of anaerobic digestion does not only depend on technical aspects and is strongly moderated by incentives provided [20]. Growth in the UK's on-farm AD industry started with the introduction of the Pollution Control Grants in the late 1980s, then stagnated when these were withdrawn in the mid-1990s [21,22]. The digesters were relatively small in size, usually under 350 m^3, and the biogas was primarily used in boilers for heating [21,23]. During the eight years prior to the introduction of the 2010 Feed-In Tariff (FIT), only seven projects [24] running on farm feedstocks had been commissioned under the 2002 Renewables Obligation system, which was designed for production of renewable electricity.

The AD FIT incentivised electricity production from biogas via combined heat and power (CHP) units. It was followed by the introduction in 2011 of the Renewable Heat Incentive (RHI), which was designed to encourage heat use and biomethane production. Since CHP has relatively low electrical efficiencies of approximately 30% for smaller units and up to 45% for larger ones, the RHI introduced the possibility of improved AD economics if most of the CHP heat produced could be beneficially utilised at a relatively low cost. The policy, however, primarily encouraged the introduction of biomethane-to-grid plants, with 108 biomethane plants operational when it closed in March 2021, compared to seven heat-only plants [1].

Incentives that recognise only the energy contributions of AD have encouraged the construction of larger plants with a greater proportion of crop inputs. Agricultural CHP plants over 250 kWe comprise 68% of the UK total at an average of 1 MWe_{eq}, with agricultural biomethane plants averaging 789 m^3 biomethane per hour or approximately 3.1 MWe_{eq} [1]. Concern over using land to provide feedstock for crop-only digesters led to the requirement for sustainability criteria to encourage the use of 'waste' AD feedstocks and to limit the amount of crop material fed to digesters [25], which can result in indirect land use change when land for biofuel crops displaces that used for food or feed. The sustainability criteria have been built into the Green Gas Support Scheme (GGSS), which was introduced in 2021, funded by a gas consumer levy and specifically designed for AD biomethane grid injection. The policy aims to support larger biomethane plants, with the highest out of three tariffs being paid to installations injecting up to 60,000 MWh $year^{-1}$ of biomethane into the gas grid (approximately 750 m^3 biomethane per hour).

Most farmers are unwilling or unable to find sufficient feedstocks or to raise the significant investment required for such a large plant [26], due to factors such as uncertainty about the future, where best to target new investments and concerns over the underlying profitability of their farm businesses. Thus, there remains an AD policy gap for UK farmers who wish to valorise their on-site feedstocks by installing smaller, less capital-intensive AD systems. Additionally, because on-farm organic wastes such as slurries and manures contain less energy than an equivalent tonnage of crop biomass, the challenge is to introduce a mechanism that can make such systems a sufficiently attractive economic proposition for farm businesses to invest in.

In their life cycle assessments on AD systems, Mesa-Dominguez et al. [27] and Styles et al. [28] noted that the greenhouse gas balances of AD plants improved by maximising waste and minimising crop inputs, utilising any CHP heat produced and covering digestate stores. Mesa-Dominguez et al. [27] concluded that energy-based incentives do not create the most sustainable deployment of AD. To illustrate this context, it is worth highlighting that a cubic metre of digester space costs the same whether it is fed on 1 m^3 of cow slurry at 23 m^3 biogas $tonne^{-1}$ of fresh feedstock or 1 m^3 of maize at 220 m^3 biogas $tonne^{-1}$ of fresh feedstock, but the waste slurry produces only one-tenth of the energy of the crop. The slurry digester, however, has better overall environmental credentials, mitigating 1449 kg CO_{2eq} $tonnne^{-1}$ of dry matter [27], because it utilises material that would otherwise create greenhouse gas emissions through its production, storage, and handling. Mesa-Dominguez et al. [27] concluded that public FIT/RHI funding should integrate consequential life cycle assessment (CLCA) and eco-systems services criteria into sustainability criteria for the most effective climate protection. In line with this, a CLCA conducted by Beausang et al. [29] for co-digestion of grass silage and cattle slurry found lower proportions of grass silage to be more sustainable, thus highlighting the environmental benefits of slurry-based farm AD plants.

Sustainability criteria were designed into the UK's non-domestic RHI and GGSS from the outset, and subsequently added to the accreditation process for the FIT from April 2017. With the 2020 Brexit withdrawal of the UK from the European Union (EU), spelling the end of EU agricultural support mechanisms in the UK, an opportunity arises to include anaerobic digestion in post-Brexit environmental schemes on UK livestock farms, particularly dairy farms, where it can make significant positive impacts [30]. The policy dilemma, therefore, is to address how on-farm AD, which utilises local or on-site wastes can be encouraged in a cost-effective way, particularly to mitigate the GHG emissions associated with livestock farming, thus valuing the technology for its wider ecosystem services deliverables (such as nutrient recycling, pollution mitigation, GHG reduction), rather than simply the production of energy [31]. While this is particularly acute in the UK due to current changes in regulation and support, similar challenges exist in many countries trying to encourage the implementation of AD by using financial support mechanisms that directly influence the digester size [32,33].

A unique opportunity to advance on-farm AD in the UK is currently given for policymakers. This opportunity is characterised by the following: a prevailing policy gap for small farm-scale AD to valorise on-site materials such as slurries and manures; the need to introduce a completely new agricultural policy that values the environment; increased impetus to meet climate targets; and high current fossil energy prices, which also means elevated prices for artificial fertilisers. Such prices favour the on-site production of renewable electricity and heat, including through AD.

In the light of COP26 aspirations, the introduction of new post-Brexit agricultural support mechanisms and a current UK policy vacuum for small farm AD that utilises on-site/local waste feedstocks, the aim of this study is to determine what level of policy support might be required against the high energy prices such as those that characterize the end of 2021. The economics of implementing an on-farm AD installation under current and potential future UK policy regimes are explored, using the Leckford Estate as an example. An economic model has been developed for this study, underpinned by AD process data from the University of Southampton's mass and energy balance model ADAT (Anaerobic Digestion Assessment Tool), available at http://borrg.soton.ac.uk/resources/adat (accessed on 18 January 2022). Combinations of heat and electricity tariff support levels, capital cost, and digestate value that could make farm AD projects viable are explored.

2. Materials and Methods

2.1. The Leckford Estate as Site under Study

To ensure that the situation in practice is adequately captured and the farm perspective appropriately considered, this study is based on analysing the Leckford Estate by way of an example. The estate, known as the 'Waitrose Farm', covers approximately 1600 hectares and is part of the John Lewis Partnership (JLP). In line with the ethos of Waitrose and the JLP, the estate is 'passionate about sustainable farming' [34] and the production of quality food. With an engaged, environmentally aware, and supportive management, availability of on-site wastes primarily from livestock operations and further feedstocks from diversification activities, AD is a technology that should be accessible to this and similar farm businesses.

At the Leckford Estate, an AD case study (not published) was originally carried out in 2017 by the first author of this paper. For the purpose of this current research, the situation at the Leckford Estate has been revisited because of the unique opportunity that the combined conditions under the current post-Brexit policy changes present.

2.2. Feedstock Pre-Assessment and Selection

To enable an assessment of the feasibility of AD at the Waitrose Farm, the various agricultural operations were examined to identify a range of potential digester feedstocks, their volumes, and seasonal availability.

The primary feedstock for the proposed AD installation is slurry and farmyard manure from the 550-cow Holstein/Friesian dairy herd. This herd size is larger than the UK average of 155 animals [35], the EU average of 45 [36], and the US average of 297 animals [37], although it should be noted that such average figures hide large variations in herd sizes, particularly in the US and UK.

The cows are housed for part of the year (1 November–1 April), with access to grazing for at least 120 days a year. Approximately 50 of these cattle will not be in milking, so will be housed on straw yards, producing 75% farmyard manure (FYM) and 25% slurry. The remaining 450 will be producing slurry. A further 50 cows will be in the 'dry cow yard' and will also be producing FYM during the winter housing season. In the summer, half of these will be out to grass. Of the 500 cows normally in the main accommodation shed, on average, 300 cows will spend 10 h per day at grass through this period, so 10/24 of the manure from these animals will not be collected.

Other farm operations generating waste feedstocks include free-range chickens; apple and pear orchards; grape vines; beef cattle and sheep, which are permanently out to pasture; a mushroom growing operation; and a cold-pressed rapeseed oil production facility. Further enterprises that produce small amounts of organic material include a golf course, a nursery, a farm shop, a lodge/campsite, a guest house, and a café.

In order to provide a conservative biogas production figure, control capital expenditure and minimise potential process problems, a number of potential feedstocks were excluded from the final calculation. These included feedstocks that were:

- intermittently produced after long-term in-situ storage and therefore likely to have little biogas potential;
- in quantities too small to justify the regulatory and equipment costs associated with its processing; or
- highly lignocellulosic, so less suited to AD, particularly without further processing.

Although there is an arable part of the business and energy crops could be a potential choice to underpin biogas production at agricultural AD installations [38], the John Lewis Partnership did not wish to introduce purpose-grown crops as an AD feedstock, thus maintaining their key focus on food production. Maize silage was, however, considered for inclusion as an option to improve the economic case for the digestion plant and/or level out seasonality.

2.3. AD Plant Site Selection and Request for Supplier Quotes

Before contacting potential AD plant suppliers active in the UK market, a suitable site for the plant installation was identified, based on the following main criteria: a large open area with suitable road access and proximity to the main feedstocks, existing feedstock storage, a water supply, and an electricity grid connection.

Three types of AD technology supplier were contacted to provide 'budget' quotes based on the digester location and feedstock types and volumes. These were: a supplier who designs simple cost-effective farm-based digesters (CAPL), one whose mid-range farm digester offering could provide automated de-gritting (CAPM), and a third who provides mainly industrial digesters (CAPH). As all suppliers provided quotes in British pound sterling (GBP) (£), monetary figures are shown in this currency.

Best efforts were made to ensure that the three quotes encompassed a similar scope of supply. It is important to note that, at this stage, suppliers use their own feedstock biogas production values for CHP and digester sizing. Whilst main components such as the tank and pipework, CHP container, feed system, control system, and any associated pumps will be included in a quote, many site-specific items may be approximate estimates only or may be considered out of scope at the budget quotation stage. These items could include planning, professional fees, road access, security, hardstanding, water supply, drainage, feedstock storage (e.g., silage clamps), rainwater/effluent catchment, data cabling, electricity supply, cost of electricity grid access, operator amenities, modifications to buildings or slurry/materials handling systems, permitting, compliance (e.g., bunding), commissioning (including cost of initial heating), digestate storage, and separation equipment. Therefore, quotes received were evaluated against literature data to check for reliability and whether they were realistic.

2.4. Digester System Modelling Using the ADAT Tool

The University of Southampton's ADAT mass and energy balance modelling tool [39] was used for the digester modelling. While several biogas production models are available online and offline, the ADAT tool, developed under the leadership of the University of Southampton in the context of various academic projects and support from the IEA (International Energy Agency) Task 37 UK, has a number of advantages, such as transparency about data assumptions and calculations, inclusion of fertiliser values in the underlying database, the option to easily add feedstocks as required by the specific user case, and availability of a comprehensive manual. The ADAT tool has been used in several studies [40–43].

Feedstock volumes were entered into the ADAT model. Where specific feedstocks were not available within the model, values for total solids, volatile solids, and methane production were taken from a range of other sources and added to the model as a 'user-defined' feedstock.

Two scenarios for energy use were considered:

1. Digester with CHP electrical production only (referred to as CHP-E scenario)
2. Digester with CHP electricity production and heat use at the mushroom farm (referred to as CHP-EH scenario)

Table 1 shows the parameters used in the ADAT model for a steel construction digester with integral gas storage operating at a mesophilic temperature of 37 °C located in the Southampton climatic region. Default ADAT values [39] were used for all parameters unless otherwise indicated.

Table 1. Main parameters for the farm anaerobic digestion system from the ADAT modelling tool.

Parameter	Value	Unit
Digester organic loading rate	4.0	kg VS m^{-3} day^{-1}
CHP electrical efficiency	35%	
CHP heat efficiency	56%	
Load factor	8000	h $year^{-1}$
Heating energy source replaced	Heating oil	L $year^{-1}$
Digester operational lifespan	20	years

The ADAT modelling tool was used to calculate the total feedstock volume, the digester size based on loading rate, the retention time, biomethane production, digestate output, CHP size, CHP electricity production, digester parasitic electrical energy, and net heat output after allocating required digester heating. The ADAT model can also calculate embodied energy [39], but as the aim of the study was to establish the potential economic viability of an on-farm digestion system rather than its carbon footprint, this was not considered.

2.5. Economic Modelling

Based on the ADAT model output and the three supplier quotes, a purpose-built economic model was created in Excel to explore the economic outcomes of the two energy options (CHP-E scenario, CHP-EH scenario). The economic model integrates the following three key parameters: (1) capital expenditure; (2) energy production expectations; (income and savings); and (3) operation and maintenance costs as described below.

2.5.1. Capital Expenditure

Budget digester cost from the suppliers is the main cost element considered here. It is also possible to include other costs such as planning, grid connection, consultants, digestate storage, groundworks, water supply, commissioning boiler fuel, permitting, and where applicable, heat meters, heat pipe, and trenching. In order to maintain a similar scope of supply and to minimise any skewing of budget costs, none of these other costs were specifically included in the economic viability calculations in the current study.

For the CHP-EH scenario, the following costs were added: heat pipe at £100 m^{-1}, pipe trenching at £8 m^{-1} (based on Estate costs), heat meters at £1000, plus a contingency margin of 3.7% of these total costs.

2.5.2. Energy Production

Figures derived from the ADAT model were included here: gross energy production, CHP size, CHP electrical efficiency, load factor, and electricity production based on the load factor.

For the CHP-EH scenario, the following additional parameters were considered: CHP heat efficiency and heat production based on the load factor.

2.5.3. Income and Savings

In order to ascertain what savings might be achieved through electricity and heat generation, net heat and electrical energy values are calculated after deduction of ADAT-derived digester parasitic heat and electricity values, respectively. The model allows a percentage of the net electricity to be used on site to displace bought-in electricity, with the remainder exported at £0.0557 kWh^{-1} [44]. Where excess CHP heat can be beneficially utilised on site, the model allows a percentage of heat to displace fossil fuels such as heating oil, liquefied petroleum gas (LPG), or natural gas. In this case study, heating oil was displaced.

Although the UK no longer has incentives designed specifically for heat or electricity production from biogas, historical tariff values exist, and provide an indication of the support levels that government deems practicable in terms of both budget and levels of deployment. Therefore, the model utilised selected tariff levels from the Feed-in Tariff (FIT) for electricity production [44] and the Renewable Heat Incentive (RHI) for heat production [45]. FIT levels were banded in order to provide smaller systems with greater tariff support. Between the FIT introduction in April 2010 and September 2011, these systems would have fallen into a sub-500 kWe CHP band; thereafter, they would be included in the sub-250 kWe CHP band.

The FIT tariff levels currently available [44] are higher than they were at the inception of the scheme, since they are index-linked and therefore adjusted annually in line with the UK Retail Prices Index (RPI). AD FITs for the relevant CHP capacity band were extracted and analysed for the highest (FITH) and lowest (FITL) values. To facilitate fiscal management of this fixed budget, the FIT scheme also included a degression mechanism that reduced tariffs as deployment increased, meaning that there was a significant difference between these two figures, so a median figure was also calculated (FITM).

A similar exercise was carried out to ascertain the minimum and maximum RHI tariff levels, RHIL and RHIH, respectively. Although the RHI scheme also included a degression mechanism, the differential between these two figures was not large, so a median figure was not calculated. Both tariff levels are given in Table 2.

Nutrients in digestate have a potential value because they can reduce or eliminate the need for fossil fertilisers [46]. The model calculates potential savings in fertiliser costs associated with digestate utilisation, but inclusion of these is optional since this value has historically been hard to realise.

Fertiliser values in terms of nutrients contained can be based either on ADAT figures [39] or on the RB209 farm nutrient management guide standard values for farm sourced digestate [47]. For the purposes of the case study, RB209 figures of 3.6 kg $tonne^{-1}$ (total N), 1.7 kg $tonne^{-1}$ (P_2O_5), and 4.4 kg $tonne^{-1}$ (K_2O) [47] were used since reliable nutrient data for the user-defined feedstocks in the ADAT model could not be ascertained.

Current market prices for equivalent fossil fertilisers were obtained [48] in order to provide a value for each kilogram of each nutrient. These fertilisers were ammonium nitrogen at 34.5% N content, triple super phosphate (TSP) at 46% P_2O_5 content, and muriate of potash at 60% K_2O content. The sum of these individual nutrient prices represents the displaced cost of synthetic fertiliser.

Table 2. Economic model parameter values.

Parameter	Value	Unit	Notes
FIT (feed-in tariff) rate	£0.0465 (FITL) £0.1104 (FITM) £0.1814 (FITH)	£ kWh^{-1}	2021 rate, indexed. From low of £0.0465 kWh^{-1} (start 1 January 2019) to high of £0.1814 kWh^{-1} (start 30 September 2011), as well as median of £0.1104 kWh^{-1} [44]
Cost of electricity replaced by on-site production	£0.1361	£ kWh^{-1}	Electricity average prices purchased by non-domestic consumers in UK [49]
RHI (renewable heat incentive) rate	£0.0309 (RHIL) £0.0873 (RHIH)	£ kWh$_{th}$$^{-1}$	2021 rate, indexed. From low of £0.0309 Wh$_{th}$$^{-1}$ (1 July 2017 tariff) to high of £0.0873 Wh$_{th}$$^{-1}$ (before 1 April 2016) [45]
Cost of heating fuel replaced (diesel/heating oil)	£0.055	£ kWh^{-1}	Converted from price per litre using net calorific value of 35.73 MJ per litre [50]
Digester operation and maintenance (O&M) costs	2.5%	Percentage of capital value	2–3% indicated in literature [51]
CHP O&M costs	£0.017	£ kWh^{-1} produced	Literature data available: £0.0075 to £0.014 kWh^{-1} [51], £0.03 kWh^{-1} [52]
Labour	£12.50	£ h^{-1}	Approximately 2 h daily [53]

2.5.4. Operation and Maintenance (O&M) Costs

As outlined in Table 2, these include AD operator wages, as well as annual permitting, insurance, and rates. Digester and CHP maintenance costs are also included here.

2.5.5. Economic Viability Calculations

An internal rate of return (IRR) is calculated based on the capital cost, total income/savings, total operational/maintenance costs, and a Retail Price Index (RPI) of 102%. It is assumed that short-lived equipment will be replaced every 5 years at 4% of the original capital expenditure and that, additionally, 20% of the original capital expenditure will be spent every 10 years [23]. The IRR is calculated over 20 years.

The internal IRR target for JLP in 2017 was 16%, although, the company recognised the wider environmental benefits of projects such as an AD installation, and so was prepared to consider lower returns.

Based on the ADAT model output and the three supplier quotes, the economic model was then used to explore the economic outcomes of the two energy options (CHP-E scenario, CHP-EH scenario) with the aim of answering the following questions:

- What internal rate of return (IRR) might be achievable by adjusting subsidy levels to the historical highest (FITH), median (FITM), and lowest (FITL) FIT tariff?
- Using the highest/median/lowest FIT/RHI subsidy levels as appropriate, what capital cost is required for the digester to achieve a 16% IRR?
- What level of policy support might be required against a background of high energy prices as witnessed in 2021?

The capital cost of digesters has not historically decreased, particularly for large systems that are complex and expensive civil engineering projects. Nevertheless, precedent exists for small digesters, which are modular and off-the-shelf, with minimal expensive site work, thus reducing capital costs [21].

3. Results and Discussion

3.1. Methane Generation and Digester Design

Table 3 shows the feedstocks as entered into the ADAT model and the resultant methane production estimates; where values were not available within the model, methane yields were taken from the sources shown in the notes. Feedstocks themselves can be variable within a single farm or even within a silage clamp, for example, so some suppliers recommend lab tests be carried out to ascertain site-specific values. Feedstock characteristics for mushroom waste were estimated based on typical energetic nutritional values for raw mushrooms with a 50% reduction for spoilage to give a conservative estimate of methane yields. Thus, this waste accounted for approximately 1.4% of the total methane production. For regulatory purposes, all feedstocks are likely to be considered as wastes, thereby meeting incentive scheme sustainability criteria, although interpretation across different schemes varies in practice.

Table 3. Potential AD feedstocks and corresponding methane production.

Feedstock	FM (tonnes $year^{-1}$)	Seasonal/ Variable?	Notes	TS (in % of FM)	VS (in % of TS)	CH_4 (m^3 $tonne^{-1}$ VS)	CH_4 (m^3 $year^{-1}$)
Dairy cattle slurry [1]	8017	Yes	Cattle out in April to October; dairy and heifer units	9%	83%	185	110,792
Farmyard manure (FYM) [1]	4432	No	High straw content; batch mucking out; dairy and heifer units. Chopped straw recommended.	25%	80%	190	168,412
Mushroom stalks and waste mushrooms	568	Yes	Supply and demand variable	8%	85%	135	5217
Rapeseed press cake and filter waste [1]	157	No	Some could be used as cattle feed or potentially sold off as feed	90.3%	94.7%	430	57,777
Waste milk	109	Yes	Variable depending upon calving and cattle illness; standard literature value [54] [2]	13.5%	94.7%	564.6	7903
Waste silage (feed) [1]	104	Yes	Cattle out in summer	30%	94%	350	10,292
Orchard Fruit (mainly apples, some pears)	30	Yes	Seasonal, currently left on trees and/or left to fall; figures based on Dubrovskis and Plume [55]	14.53%	96.75%	451	1902
Grass cuttings from golf course and grounds	14	Yes	KTBL [56]	50%	85%	300	1785
Maize silage	Optional	No	Possible addition to levelise seasonality	30%	94%	350	variable
TOTAL [3]	13,432						364,080

[1] Default ADAT values used; [2] Figure refers to whole milk (as reported by LfL [54] for "Vollmilch, Kuh, Frisch", English translation: whole milk, cow, fresh); [3] Without maize silage.

Cows are bedded on 1.5 tonnes day^{-1} of a paper pulp/sawdust product, which has a dry matter content of 95%. Whilst both paper [57] and paper sludge [58] can produce biogas, production from sawdust in negligible [59], so a conservative assumption was made that no additional biogas was produced from the bedding.

Dairy farm substrates (slurry, FYM, waste milk and waste feed) from the 550-cow herd contribute to 82% of the yearly methane production. In the UK, approximately 1.5% of the UK's holdings house 12.3% of the nation's dairy cows in herds greater than 499 cows [35]. Initial targeting of larger dairy farms (such as Leckford Estate) to improve slurry management by the introduction of AD makes logistical and financial sense, particularly where they are also able to utilise local substrates such as chicken manure in case that the farm may not have a land base sufficient to spread those nutrients.

Volumes of the following feedstocks vary throughout the year at the Leckford Estate:

- Mushroom stalks and waste mushrooms—vary due to supply and demand;
- Fruit—apples and pears. Depending upon variety, harvesting can be from August through to November;
- FYM—slightly seasonal in that more is produced in the winter housing period than in summer when cows are primarily living outdoors;
- Waste silage—less silage is available in summer when the cows are grazing in fields;
- Grass cuttings from the grounds and gardens are produced primarily between April and October and are dependent upon weather, grass type (i.e., lawn/rough meadow grass), and mowing frequency.

The seasonality of these feedstocks does not affect the biomethane production calculations, as these are on a yearly basis. However, when planning such a scheme, the estate may need to consider the costs and benefits of strategies such as storage or ensiling in order to include them in a way that maintains the relatively consistent gas production that a CHP requires.

For completeness, it is noted that some potentially available biomass types were not included in the calculation due to small quantities/unreliable availability, poor quality of material (i.e., too old), or low suitability for AD (e.g., strongly lignocellulosic biomass), thus relying on conservative figures to avoid any overestimation of the methane generation to be expected. The following potential feedstocks were not included:

- Spent mushroom compost (SMC) (86 tonnes year^{-1})—highly lignocellulosic material, and trials indicate some biogas production [60,61]. However, following a conservative approach, it was decided not to include this material due to its high straw content.
- Poultry litter (400 tonnes year^{-1})—litter is removed from the sheds on an annual basis, so it is likely to be significantly degraded. This is from egg laying hens, so it also contains an appreciable amount of grit, which can cause silting in many types of digesters.
- Tree/bush prunings (20 tonnes year^{-1}) and leaf litter (50 tonnes year^{-1})—woody materials and therefore not ideally suited to AD [62].
- Food waste—local food waste from a restaurant, several houses, a café, and a campsite could be included, and generally is a suitable AD co-substrate [63]. However, a collection round would need to be established and such materials would need to be shredded, pasteurised, and likely sieved to remove plastics [64]. This would add considerable cost and the volumes did not currently justify this.

Although these materials were not included in the following calculations, it is worth noting that with an operational digester, the first two feedstocks could also be trialled in order to level seasonal feedstock fluctuations.

To ensure proximity to the bulk of the feedstocks, the AD site selected was on the dairy unit, adjacent to silage clamps, two digestate storage lagoons, a holding tank, a separator, and with good road access. The site was approximately 2 km from the mushroom farm where CHP heat could potentially be utilised year-round.

Based on the above quantities of feedstocks included in the assessment, Table 4 summarises the results from the ADAT tool that were utilised in the economic modelling.

The on-site baseload electricity usage, as reported by Leckford Estate (figures not provided as commercially confidential), comfortably exceeds that of the potential CHP electrical production so all electricity could be used on site continuously. It was also confirmed that the mushroom farm could utilise 100% of the heat produced year-round: this is unusual, but many farms have a heat requirement for agricultural or diversification activities [65,66]. This is discussed further in the CHP-EH scenario below.

Table 4. Resultant ADAT model data (to be further utilised for the economic modelling).

Parameter	Value	Unit
Daily feedstock input	36.8	tonnes day^{-1}
Digester size	1524	m^3
Retention time	31.9	days
Biogas production	609,796	m^3 year^{-1}
Methane production	364,083	m^3 year^{-1}
Volatile solids (VS) destroyed	744	tonnes year^{-1}
Digestate	12,688	tonnes year^{-1}
CHP electrical capacity	145	kWe
CHP electricity production (gross)	1157.9	MWh year^{-1}
Digester parasitic electrical energy consumption	77.4	MWh year^{-1}
CHP heat (net) available after digester heating	1353.7	MWh year^{-1}
Digester parasitic heat energy consumption	499	MWh year^{-1}

3.2. Capital Cost

Supplier quotes received to install the AD plant at the site under study were £900,000 (CAPL), £1,400,000 (CAPM), and £1,600,000 (CAPH). It was assumed that the farm would provide 100% of the funding without applying for bank loans. If finance through bank loans were required, as is likely to be the case for many typical UK farm businesses, the business case would be further detrimentally affected, and thus the assumption of no bank loan is in line with this study's approach to analyse whether economic viability of AD is at all possible.

Despite the fact that suppliers were contacted with a detailed set of data describing the specific case, it is important to note that they will use their own procedures to frame the AD plant sizing and to estimate the site-specific requirements, and these assumptions might be rather complete or not, depending on the level of detail and the degree of complexity a supplier is ready to include. In general, the cost of such site-specific items means that the actual capital cost of an AD plant installation varies greatly because the assumed complexity of the scope of the project varies: for example, a project with a complex planning application where a road, extensive bunding, and several silage clamps are included as part of the AD plant finance figure will be considerably more expensive than one of the same size where these are not required. Additionally, data on costs and their breakdown are often not readily available, as they are considered to be commercially sensitive information.

Because of this uncertainty around digester costs, quotes were cross-checked against two sources that provide UK digester historical cost data, albeit based on relatively small sample sizes. The first was a tool developed by the World Biogas Association, based on actual historical costs for 64 UK plants [67]. Quotes were also checked against the relevant CHP capacity band for low, central, and high digester cost cases analysed by Parsons Brinckerhoff in a 2015 report to the UK government [68].

The costs indicated by the obtained supplier quotes for a system based on a 145 kWe appeared broadly in line with the central and high Parsons Brinckerhoff historical costs shown Table 5. In this context, it is interesting to note that the digester cost model developed by the World Biogas Association [67] suggests that a plant of 145 kWe is likely to have a cost of £449,500; but the relatively small number of digesters analysed at this scale may account for the lack of alignment with the supplier quotes.

Thus, the three supplier quotes (capital costs of £900,000, £1,400,000, £1,600,000) obtained for the site under study are considered realistic, indicating a lower-cost solution for a simple cost-effective farm-type digester (CAPL), a medium-cost solution (CAPM), and a solution at industrial standard with elevated costs (CAPH). These figures are therefore used in the following as a basis for the economic viability calculations.

In addition, for the scenario with heat valorisation, with the mushroom farm located 2 km from the proposed digester site, the capital cost of providing heat to the farm was calculated to be £225,000 (assumptions are documented in Section 2).

Table 5. Literature-based data indicating capital costs for a range of digester sizes (derived from figures available in [68]) and, thus, calculated capital cost for a 145 kWe plant (PBL: Parsons Brinckerhoff Low Case figure; PBM: Parsons Brinckerhoff Central Case (i.e., median) figure; PBH: Parsons Brinckerhoff High Case figure).

Capacity Band	Low Case	Central Case	High Case
<250 kWe	£3780 kWe^{-1}	£5953 kWe^{-1}	£8126 kWe^{-1}
145 kWe plant	£548,100 (PBL)	£887,206 (PBM)	£1,211,060 (PBH)
250–500 kWe	£3685 kWe^{-1}	£5804 kWe^{-1}	£7922 kWe^{-1}
>500 kWe	£2835 kWe^{-1}	£4465 kWe^{-1}	£6095 kWe^{-1}

3.3. *Economic Viability of the AD Plant with CHP Electrical Production Only (CHP-E Scenario)*

Table 6 shows the IRR based on the three capital expenditure levels given by suppliers. The £900,000 digester (CAPL) would easily reach the 16% IRR under levels FITM (median feed-in-tariff applicable for UK situation) and FITH (highest feed-in-tariff applicable for a UK situation).

Table 6. Internal rate of return (IRR) based on three quoted capital expenditure levels (supplier quotes obtained for the site under study), electricity production only (CHP-E scenario).

	IRR with FITL [1]	IRR with FITM [2]	IRR with FITH [3]
CAPL	13.57%	22.76%	32.55%
CAPM	4.45%	12.14%	19.36%
CAPH	1.66%	9.35%	16.13%

[1] FITL: £0.0465 kWh^{-1} (lowest electricity feed-in-tariff applicable for UK situation); [2] FITM: £0.1104 kWh^{-1} (median electricity feed-in-tariff applicable for UK situation); [3] FITH: £0.1814 kWh^{-1} (highest electricity feed-in-tariff applicable for UK situation).

Because of their broad alignment and inclusion in UK Government AD viability calculations, the calculations were repeated using the Parsons-Brinckerhoff cases [68], shown in Table 7. Due to the slightly lower capital costs, the model showed an increased IRR in all use cases.

Table 7. Internal rate of return based on historical Parsons Brinckerhoff figures (as available in [68]), electricity production only (CHP-E scenario).

	IRR with FITL [1]	IRR with FITM [2]	IRR with FITH [3]
PBL	25.4%	39.63%	54.92%
PBM	13.32%	23.15%	33.07%
PBH	6.85%	15.33%	23.20%

[1] FITL: £0.0465 kWh^{-1} (lowest electricity feed-in-tariff applicable for UK situation); [2] FITM: £0.1104 kWh^{-1} (median electricity feed-in-tariff applicable for UK situation); [3] FITH: £0.1814 kWh^{-1} (highest electricity feed-in-tariff applicable for UK situation).

These figures provide an indication of where a positive business case may lie for such systems. The capital cost of £1.2 million (PBH) to £1.4 million (CAPM) combined with the mid-range FIT of £0.1104 kWh^{-1} appears to provide IRRs large enough to warrant investment, but not so large as to create a 'boom' in construction activities.

Aligning historical deployment data with FIT levels at small-scale could shed further light on where an economic FIT level might lie, but this analysis is complicated by several factors:

- There is typically a lag time between the announcement of a FIT level and a digester being commissioned. Due to the long lead-in times (typically a year or more) required for financing, planning, and constructing a digester, the UK FIT scheme had a preliminary accreditation mechanism that allowed applicants to 'lock in' at a given tariff level in order to provide certainty of income, and therefore make a project financially

attractive to lenders. Thus, a digester could have locked into a tariff of £0.16 kWh^{-1} on a certain date, but by the time the digester was actually built and included in deployment figures a year later, the tariff on that date could have been much lower.
- The scheme had several breaks [69], which included a 'fast track' tariff review and the temporary removal of preliminary accreditation, which caused breaks in digesters being commissioned.
- It is difficult to identify these smaller systems in the aggregated total deployment of all system sizes.

The average FIT level has undergone considerable changes during its lifetime [69]. While the average FIT for an AD plant of the scale 146 kWe was above £0.10 kWh^{-1} in the years 2010 to 2015, it fell to lower levels in later years, and was below £0.06 kWh^{-1} in the years 2018 and 2019. In any case, when the FIT had dropped to less than £0.05 kWh^{-1}, the UK's Anaerobic Digestion and Bioresources Association (ADBA) noted in its April 2017 policy report [70] that the smaller-scale end of the market 'has been decimated', adding in the November 2017 report [71] that only 13 plants under 250 kWe had been commissioned in 2016. This coincided with the timing of the original case study carried out for JLP in 2017 (report not published), which concluded that the desired IRR could not be achieved at the more realistic CAPM and CAPH levels using these parameters: a FIT rate of £0.0499 kWh^{-1}, an RHI rate of £0.0226 kWh_{th}^{-1}, and energy prices of £0.10 kWh^{-1} and £0.42 kWh^{-1} for electricity and heating oil, respectively.

The 1 April 2015 FIT was set at £0.1013 kWh^{-1}, which when modelled using current energy prices, provides an IRR in double figures for all cases except CAPH; this therefore provides a further clear indication of the level of support below, for which few plants of this size are likely to be built.

3.4. Economic Viability of the AD Plant with CHP and Heat Use (CHP-EH Scenario)

In the case of the site under study in this research, the mushroom farm is able to use all of the CHP heat produced year-round, and thus full valorisation of the available heat is assumed. This represents a best-case setting with view to heat usage. The improved IRRs under this setting, shown in Table 8, reflect the financial benefits of utilising both heat and electricity.

Table 8. Internal rate of return based on three quoted capital expenditure levels, electricity production and 100% heat use (CHP-EH scenario).

	IRR with FITL [1]		IRR with FITM [2]		IRR with FITH [3]	
	RHIL	RHIH	RHIL	RHIH	RHIL	RHIH
CAPL	21.83%	29.20%	28.61%	35.73%	36.27%	43.25%
CAPM	13.00%	18.75%	18.30%	23.62%	24.02%	29.08%
CAPH	10.54%	15.97%	15.55%	20.48%	20.84%	25.47%

[1] FITL: £0.0465 kWh^{-1} (lowest electricity feed-in-tariff applicable for UK situation); [2] FITM: £0.1104 kWh^{-1} (median electricity feed-in-tariff applicable for UK situation); [3] FITH: £0.1814 kWh^{-1} (highest electricity feed-in-tariff applicable for UK situation).

Overall CHP efficiency is increased if a beneficial heat use can be found. It is easier to use the bulk of the heat produced by a smaller CHP, thus maximising the overall system efficiency. Leckford Estate is relatively unusual in its ability to utilise heat year-round. Nevertheless, a dairy farm could use CHP heat for hot water in buildings [65,66], for dairy washing (which is a highly relevant cost factor for dairy farms, typically amounting to nearly one third of electricity costs of a dairy unit [72]), space heating, crop drying, and to improve milk yields by warming cattle drinking water [73,74]. A number of increasingly common farm business diversification activities also use heat, e.g., greenhouses, campsites (space and water heating), and local food production operations.

A FIT tariff of £0.1013 kWh^{-1} has historically resulted in modest digester deployment. In a scenario where 70% of the heat could be used, it was decided to calculate what level of

RHI support would be required in order to achieve an IRR of 16% at CAPM and CAPH. This resulted in an RHI rate of £0.0459 kWh^{-1} and £0.0859 kWh^{-1} at CAPM and CAPH, respectively. This result is broadly in line with the RHIL of £0.0309 $kWh_{th}{}^{-1}$ and RHIH of £0.0873 $kWh_{th}{}^{-1}$, a level of support which has historically been considered reasonable.

With the end of the FIT in April 2019, electricity generation from biogas has fallen from regulatory favour, not least because the bulk of the energy produced from even the most electrically efficient CHP is in the form of heat. In the UK and countries with similar climates, it is relatively rare to find a year-round heat use, nevertheless, system economics can be improved by utilising a greater proportion of the CHP heat production. Particularly in colder climates where there is a much wider variation in the heat demand, not least for digester heating, the challenge is to ensure that there is sufficient heat produced to meet the demand. This means that in seasons of low demand, there is a greater excess of heat. Strategies to address this might include finding a further seasonal heat use, for example, a diversification activity such as hot water for campers. Conversely, any seasonal shortfall could be met through other renewables: biomass, solar PV, solar thermal, battery storage [75].

3.5. *Analysis of the Impact of Energy Prices*

In view of the recent sharp increases in energy prices across the UK and worldwide, it was decided to explore the potential for subsidy free support for the production of electricity while utilising 70% of the heat. Table 9 illustrates what electricity prices (required non-domestic prices) would have to be in order to achieve economic viability of an AD plant at the studied site, and includes a less ambitious IRR of 10% in a subsidy-free scenario where CHP heat is replacing heating oil (at a cost of £0.55 per litre [50]).

Table 9. Minimum economic electricity prices at varying capex and IRR under a tariff-free regime with 70% heat use.

	Electricity Price Required (Pence kWh^{-1} Electricity Sold) to Achieve Economic Viability of the AD Installation			
	Electricity Usage Only		Electricity and Heat Valorisation [1]	
	At 10% IRR	At 16% IRR	At 10% IRR	At 16% IRR
CAPL	16.76	20.57	14.92	19.62
CAPM	23.39	29.37	21.59	28.49
CAPH	26.1	32.89	24.26	32.01

[1] Replacing heating oil with an average price of £0.055 kWh^{-1} [50].

With an average non-domestic electricity price for very small non-domestic users of £0.1734 kWh^{-1} over the period from the fourth quarter (Q4) of 2019 to the second quarter (Q2) of 2021 [49], future electricity prices in the region of approximately £0.20 kWh^{-1} seem increasingly likely, which would make some scenarios an economic proposition.

These figures [49] show a gradual increase in non-domestic electricity prices from an average of £0.0416 kWh^{-1} in 2004 to £0.1361 kWh^{-1} in Q2 of 2021, an increase of 327%, with natural gas rising from £0.01254 kWh^{-1} to £0.0259 kWh^{-1}, an increase of 207% over the same period. These figures do not reflect the steep rise in wholesale gas prices experienced in the last quarter of 2021 [76]. Due to worldwide increases and volatility in the wholesale gas price market, UK prices reached as high as £0.042 kWh^{-1} [77], affecting both gas and electricity prices (as 35.7% of electricity is generated using gas [78]).

Support at the lowest FIT rate of £0.1013 kWh^{-1}, coupled with the lowest RHI rate at £0.0309 kWh^{-1}, an electricity price of £0.1492 kWh^{-1} and £0.1840 kWh^{-1} at CAPM and CAPH, respectively, would provide an IRR of 16%. These values are greater than the ten-year (2010 to 2019) non-domestic electricity price average of £0.1030 kWh^{-1}. However, the value of £0.1492 kWh^{-1}, although higher than the current average electricity price modelled of £0.1361 kWh^{-1}, is certainly within the range of 3rd quarter 2021 prices for small/medium users (£0.1502 kWh^{-1} for 500–1999 MWh annual usage) and less than that for small users (£0.1534 kWh^{-1} for 20–499 MWh annual usage) and very small users (£0.1818 kW^{-1} for 0–20 MWh annual usage) [49]. This provides a further indication of the support required if the general trend for average non-domestic electricity prices continue their gradually increasing trajectory, which has characterised the past seventeen years (2004–2020) [49].

UK domestic electricity prices are typically higher than the European average, whereas gas prices are lower [79,80]. This price differential when coupled with the poor building fabric tends to discourage electrification, and therefore the decarbonisation of heat, particularly in rural areas where electricity grids may be weak and there is no access to the gas grid for alternative fuels such as hydrogen. In the face of climate change perturbations and the complexity associated with fossil energy market forces, it is unclear whether such volatility will continue [81], since multiple factors such as supply chain issues, energy systems decarbonisation, and emissions trading scheme carbon prices affect costs [76].

Where energy costs are both high and volatile over a long period, on-site de-centralised energy generation becomes particularly important and AD at small scale using mainly wastes becomes increasingly viable, cushioning these important farm businesses against such volatility and helping the UK and other countries meet their COP26 Methane Pledge and wider climate goals.

However, unless there is a guarantee that prices will be sustained at such levels, AD projects are still unlikely to be considered worthy of investment, so some form of support for technology implementation would still be required, e.g., a minimum price guarantee or floor price. Energy based incentives which also value the non-energy contributions of AD to such farms could be an option [7,82]. These could, for example, combine a small-scale electricity/heat tariff to improve on-site/local energy utilisation. There is an added opportunity to consider the benefits of AD in the context of the post-Brexit agricultural policy support schemes.

3.6. Digestate Savings/Income

The value of digestate as a source of nutrients that can displace fossil fuel based synthetic fertilisers (which can be costed) and as a soil conditioner, which replenishes soil carbon (which is difficult to cost) was not taken into account in any of the above calculations. Digestate is often regarded as an expense to the business, particularly for large digesters where large volumes of digestate require longer transport distances in order not to cause soil nutrient overloading [83]. Whilst the ADAT tool can include the energetic cost of transport, this was not included within this economic case study, as digestate was not being transported any further than the slurry and wastes otherwise would have been.

Digestate can also be separated into liquid and solid fractions to facilitate differential nutrient application [16]. Prior to the introduction of end of waste criteria, digestate fibre mixes used to be sold to gardeners [21] as a peat-free option, but the costs now associated with meeting the requirements [84] are likely to preclude small digesters from this market, unless costs could be defrayed through, for example, an aggregation mechanism whereby several smaller operations could be considered as one for the purposes of regulation.

Against a backdrop of high fossil fuel energy prices, the value of digestate can offset the commensurately high price of fossil fertilisers [48]:

- Nitrogen—34.5% Ammonium Nitrate—£618 tonne^{-1}
- Phosphate (P_2O_5)—46% Phosphate TSP—£525 tonne^{-1}
- Potash (K_2O)—60% Muriate of Potash—£534 tonne^{-1}

Using the above fertiliser costs [48] and RB209 nutrient guide values [47], a tonne of digestate was calculated to be worth £9.39 or a total of £119,160 in this case study, potentially further improving the economics of the project under study. For a CAPM RHI digester using 70% heat with no subsidies and digestate at this price, the IRR is a respectable 12.76%.

The results thus confirm previous reports that recycling such organic materials back to land in the form of digestate can potentially add value [33,53] while improving the GHG credentials of the business [75,85].

The post-Brexit agricultural policy aims to reward public goods, particularly those related to the environment [86,87], including improved soil health, water quality improvement, and reduced GHG emissions. A well-run AD plant that recycles carbon back to soils as digestate and captures or avoids otherwise uncontrolled emissions occurring from biodegradation of organic matter or storage of materials (particularly methane from slurry stores) can contribute to these aims [1] and as such should be recognised within that policy [4].

4. Conclusions

As illustrated in this research, small farm-scale AD installations that utilise diffuse on-site or locally sourced wastes to produce both electricity and heat from a CHP can improve the overall efficiency and economics over those that do not valorise the heat. Additionally, they can provide heat at a scale that can often be beneficially utilised. AD can help cushion such food production businesses against high and/or volatile energy prices, as well as helping to decarbonise a sector that often uses carbon intensive fuels for heating. Where rural electricity grids are weak and/or electricity is expensive, on-site electricity production also provides a route to farm business electrification, increasing economic options for utilisation of electrical vehicles, agritech, robotics, and more.

The Leckford Estate is a large, diversified farming business owned by a major retailer with a commitment to sustainability and access to a wide range of local waste feedstocks, but under the current UK policy regime, which is characterised by a lack of support at this relatively small scale (145 kWe, 1500 m^3 digester), it is uneconomic for the estate to utilise these resources through AD in order to reduce its carbon footprint and improve their sustainability in the face of climate change.

As energy prices increase, the required level of support for such projects decreases, but projects also struggle to put an economic business case together where there is no long-term pricing clarity or alternative support mechanisms that value the many benefits of AD. The results of the current study can be used to provide an indication of recommended levels of support, and a basis for varying them in response to shifting energy prices.

Designing a policy that values these benefits is a challenge for all countries and regions [33], but with the introduction of a new agricultural policy, the UK has an opportunity to valorise these diffuse organic wastes through small AD, and support farm businesses in the face of the urgency of the climate crisis and high energy prices.

One potential support mechanism could be to include the 'public goods' benefits of on-farm AD (including greenhouse gas reduction, improved nutrient management, positive soil organic carbon impact, and strengthened rural development) in the UK post-Brexit agricultural policy support scheme.

Author Contributions: Conceptualization, A.B. and S.K.-B.; Methodology, A.B.; Software, A.B.; Validation, S.K.-B.; Formal analysis, A.B. and S.K.-B.; Investigation, A.B.; Data curation, A.B.; Writing—original draft preparation, A.B. and S.K.-B.; Writing—review and editing, A.B. and S.K.-B. All authors have read and agreed to the published version of the manuscript.

Funding: This research was funded by a Business Interaction Voucher (BIV2016020) from the Anaerobic Digestion Network (grant number BB/L013835/1), a Network in Industrial Biotechnology and Bioenergy (NIBB) funded by the Biotechnology and Biological Sciences Research Council.

Institutional Review Board Statement: Not applicable.

Informed Consent Statement: Not applicable.

Data Availability Statement: Data supporting this study are openly available from the University of Southampton repository at https://eprints.soton.ac.uk/ access on 10 December 2021.

Acknowledgments: The authors wish to thank the Leckford Estate and the John Lewis Partnership for giving us access to their estate for the purpose of this study, for making available data about potential AD feedstocks at their estate and for the assistance provided to collect the required data; Yue Zhang (University of Southampton) for scientific advice; James Murcott (Methanogen UK Ltd.),

Chris Morris (Fre-Energy Ltd.), Michael Chesshire (Lutra), Chris Cooper (CPN Projects Ltd.) for providing data about commercial costs in the AD sector; and the Environmental Biotechnology Network's (a BBSRC/EPSRC NIBB) Anaerobic Digestion Working Group for expert knowledge that contributed to ensure robustness of the assessments made. Special thanks to Sonia Heaven (University of Southampton) for scientific advice and discussions, which have enriched this work in a very valuable way.

Conflicts of Interest: The authors declare no conflict of interest. The funders had no role in the design of the study; in the collection, analyses, or interpretation of data; in the writing of the manuscript, or in the decision to publish the results.

References

1. ADBA. *ADBA Policy Report–October 2021*; Anaerobic Digestion and Bioresources Association: London, UK, 2021.
2. Lin, L.; Xu, F.; Ge, X.; Li, Y. Improving the sustainability of organic waste management practices in the food-energy-water nexus: A comparative review of anaerobic digestion and composting. *Renew. Sustain. Energy Rev.* **2018**, *89*, 151–167. [CrossRef]
3. Scott, A.; Blanchard, R. The role of anaerobic digestion in reducing dairy farm greenhouse gas emissions. *Sustainability* **2021**, *13*, 2612. [CrossRef]
4. ADBA. *Biomethane and Hydrogen: Two Green Gases, One Future*; Anaerobic Digestion and Bioresources Association: London, UK, 2021.
5. Agostini, A.; Battini, F.; Giuntoli, J.; Tabaglio, V.; Padella, M.; Baxter, D.; Marelli, L.; Amaducci, S. Environmentally sustainable biogas? The key role of manure co-digestion with energy crops. *Energies* **2015**, *8*, 5234–5265. [CrossRef]
6. Banks, C.; Salter, A.; Chesshire, M. Potential of anaerobic digestion for mitigation of greenhouse gas emissions and production of renewable energy from agriculture: Barriers and incentives to widespread adoption in Europe. *Water Sci. Technol.* **2007**, *55*, 165–173. [CrossRef] [PubMed]
7. Lovarelli, D.; Falcone, G.; Orsi, L.; Bacenetti, J. Agricultural small anaerobic digestion plants: Combining economic and environmental assessment. *Biomass Bioenergy* **2019**, *128*, 105302. [CrossRef]
8. Barłóg, P.; Hlisnikovský, L.; Kunzová, E. Effect of digestate on soil organic carbon and plant-available nutrient content compared to cattle slurry and mineral fertilization. *Agronomy* **2020**, *10*, 379. [CrossRef]
9. Finzi, A.; Mattachini, G.; Lovarelli, D.; Riva, E.; Provolo, G. Technical, economic, and environmental assessment of a collective integrated treatment system for energy recovery and nutrient removal from livestock manure. *Sustainability* **2020**, *12*, 2756. [CrossRef]
10. Harrison, J.; Ndegwa, P.; Pagliari, P.H.; He, Z. Anaerobic digestion of dairy and swine waste. In *Animal Manure: Production, Characteristics, Environmental Concerns and Management*; Waldrip, H.M., Pagliari, P.H., He, Z., Eds.; American Society of Agronomy: Madison, WI, USA, 2019; pp. 115–127.
11. Walsh, J.J.; Jones, D.L.; Edwards-Jones, G.; Williams, A.P. Replacing inorganic fertilizer with anaerobic digestate may maintain agricultural productivity at less environmental cost. *J. Plant Nutr. Soil Sci.* **2012**, *175*, 840–845. [CrossRef]
12. Adani, F.; D'Imporzano, G.; Tambone, F.; Riva, C.; Boccasile, G.; Orzi, V. Anaerobic digestion and renewable fertilizers: Case studies in Northern Italy. In *Biorefinery of Inorganics: Recovering Mineral Nutrients from Biomass and Organic Waste*; Meers, E., Velthof, G., Michels, E., Rietra, R., Eds.; John Wiley & Sons: Chichester, UK, 2020; pp. 215–229.
13. Walsh, J.J.; Jones, D.L.; Chadwick, D.; Williams, A. Repeated application of anaerobic digestate, undigested cattle slurry and inorganic fertilizer N: Impacts on pasture yield and quality. *Grass Forage Sci.* **2018**, *73*, 758–763. [CrossRef]
14. Burch, T.R.; Spencer, S.K.; Borchardt, S.S.; Larson, R.A.; Borchardt, M.A. Fate of manure-borne pathogens during anaerobic digestion and solids separation. *J. Environ. Qual.* **2018**, *47*, 336–344. [CrossRef]
15. Johansen, A.; Nielsen, H.B.; Hansen, C.M.; Andreasen, C.; Carlsgart, J.; Hauggard-Nielsen, H.; Roepstorff, A. Survival of weed seeds and animal parasites as affected by anaerobic digestion at meso-and thermophilic conditions. *Waste Manag.* **2013**, *33*, 807–812. [CrossRef]
16. Lukehurst, C.T.; Frost, P.; Al Seadi, T. *Utilisation of Digestate from Biogas Plants as Biofertilizer*; IEA Bioenergy: Kent, UK, 2010.
17. Banks, C.J.; Salter, A.M.; Heaven, S.; Riley, K. Energetic and environmental benefits of co-digestion of food waste and cattle slurry: A preliminary assessment. *Resour. Conserv. Recycl.* **2011**, *56*, 71–79. [CrossRef]
18. Wang, J. Decentralized biogas technology of anaerobic digestion and farm ecosystem: Opportunities and challenges. *Front. Energy Res.* **2014**, *2*, 10. [CrossRef]
19. Aui, A.; Li, W.; Wright, M.M. Techno-economic and life cycle analysis of a farm-scale anaerobic digestion plant in Iowa. *Waste Manag.* **2019**, *89*, 154–164. [CrossRef]
20. Zglobisz, N.; Castillo-Castillo, A.; Grimes, S.; Jones, P. Influence of UK energy policy on the deployment of anaerobic digestion. *Energy Policy* **2010**, *38*, 5988–5999. [CrossRef]
21. Letcher, J. *Farm Digesters*; UIT Cambridge: Cambridge, UK, 2016.
22. McCabe, B.; Kroebel, R.; Pezzaglia, M.; Lukehurst, C.; Lalonde, C.; Wellisch, M.; Murphy, J. *Integration of Anaerobic Digestion into Farming Systems in Australia, Canada, Italy, and the UK*; IEA Bioenergy Task 37: Brisbane, Australia, 2020.
23. Murcott, A.J.; Methanogen Ltd., Ludlow, Shropshire, UK. Personal Communication, 2021.

24. REF. Renewable Generators. Renewable Energy Foundation. Available online: https://www.ref.org.uk/generators/index.php (accessed on 30 December 2021).
25. BEIS. *The Renewable Heat Incentive: A Reformed Scheme*; Department for Business, Energy and Industrial Strategy: London, UK, 2016.
26. Defra. *The Future Farming and Environment Evidence Compendium-September 2019 Update*; Department for Environment Food & Rural Affairs: London, UK, 2019.
27. Mesa-Dominguez, E.; Styles, D.; Zennaro, K.; Thompson, P. *Evaluating Cost-Effective Greenhouse Gas Abatement By Small-Scale Anaerobic Digestion*; Bangor University: Bangor, UK; Renewable Energy Association: London, UK, 2015.
28. Styles, D.; Gibbons, J.; Williams, A.P.; Stichnothe, H.; Chadwick, D.R.; Healey, J.R. Cattle feed or bioenergy? Consequential life cycle assessment of biogas feedstock options on dairy farms. *Glob. Change Biol. Bioenergy* **2015**, *7*, 1034–1049. [CrossRef]
29. Beausang, C.; McDonnell, K.; Murphy, F. Assessing the environmental sustainability of grass silage and cattle slurry for biogas production. *J. Clean. Prod.* **2021**, *298*, 126838. [CrossRef]
30. Bywater, A.M. *A Review of Anaerobic Digestion Plants on UK Farms*; Royal Agricultural Society of England: Stoneleigh, UK, 2011.
31. European Commission (EC). *COM(2010)235 final: Communication from the Commission to the Council and the European Parliament on Future Steps in Bio-Waste Management in the European Union*; European Commission: Brussels, Belgium, 2010.
32. Gustafsson, M.; Ammenberg, J.; Murphy, J.D. (Eds.) *IEA Bioenergy Task 37–Country Reports Summaries 2019*; International Energy Agency: Cork, UK, 2020.
33. Winquist, E.; Rikkonen, P.; Pyysiäinen, J.; Varho, V. Is biogas an energy or a sustainability product?-Business opportunities in the Finnish biogas branch. *J. Clean. Prod.* **2019**, *233*, 1344–1354. [CrossRef]
34. Waitrose. Leckford Estate, Waitrose and Partners Farm. Available online: https://leckfordestate.co.uk/ (accessed on 30 December 2021).
35. AHDB. UK and EU Cow Numbers. Agriculture and Horticulture Development Board. Available online: https://ahdb.org.uk/dairy/uk-and-eu-cow-numbers (accessed on 30 December 2021).
36. Matthews, A. Measuring Changing Farm Structure in the EU. CAP Reform. Available online: http://capreform.eu/measuring-changing-farm-structure-in-the-eu/ (accessed on 30 December 2021).
37. Progressive Dairy. 2020 U.S. Dairy Statistics. Progressive Publishing. Available online: https://www.progressivedairy.com/site/stats/us-dairy-stats (accessed on 30 December 2021).
38. Zhang, Y.; Kusch-Brandt, S.; Salter, A.M.; Heaven, S. Estimating the methane potential of energy crops: An overview on types of data sources and their limitations. *Processes* **2021**, *9*, 1565. [CrossRef]
39. ADAT. *Anaerobic Digestion Assessment Tool: User Manual*; University of Southampton: Southampton, UK, 2017.
40. Zhang, W.; Heaven, S.; Banks, C.J. Thermophilic digestion of food waste by dilution: Ammonia limit values and energy considerations. *Energy Fuels* **2017**, *31*, 10890–10900. [CrossRef]
41. Suhartini, S.; Nurika, I.; Paul, R.; Melville, L. Estimation of biogas production and the emission savings from anaerobic digestion of fruit-based agro-industrial waste and agricultural crops residues. *Bioenerg. Res.* **2021**, *14*, 844–859. [CrossRef]
42. Hidayat, N.; Suhartini, S.; Utami, R.N.; Pangestuti, M.B. Anaerobic digestion of fungally pre-treated oil palm empty fruit bunches: Energy and carbon emission footprint. *IOP Conf. Ser.: Earth Environ. Sci.* **2020**, *524*, 012019. [CrossRef]
43. Zhang, W.; Venetsaneas, N.; Heaven, S.; Banks, C.J. Impact of low loading on digestion of the mechanically-separated organic fraction of municipal solid waste. *Waste Manag.* **2020**, *107*, 101–112. [CrossRef]
44. Ofgem. Feed-in Tariff (FIT): Tariff Table 1 April 2021. Available online: https://www.ofgem.gov.uk/publications/feed-tariff-fit-tariff-table-1-april-2021 (accessed on 15 November 2021).
45. Ofgem. Non-Domestic Renewable Heat Incentive (RHI). Available online: https://www.ofgem.gov.uk/environmental-and-social-schemes/non-domestic-renewable-heat-incentive-rhi (accessed on 15 November 2021).
46. Jones, P.; Salter, A. Modelling the economics of farm-based anaerobic digestion in a UK whole-farm context. *Energy Policy* **2013**, *62*, 215–225. [CrossRef]
47. AHDB. Nutrient Management Guide (RB209). updated January 2021. Available online: https://ahdb.org.uk/rb209 (accessed on 25 December 2021).
48. AHDB. GB Fertiliser Prices. Available online: https://ahdb.org.uk/GB-fertiliser-prices (accessed on 5 December 2021).
49. BEIS. Statistical Data Set–Gas and Electricity Prices in the Non-Domestic Sector. Updated 23 December 2021. Available online: https://www.gov.uk/government/statistical-data-sets/gas-and-electricity-prices-in-the-non-domestic-sector (accessed on 25 December 2021).
50. Boiler Juice. UK Average Home Heating Oil Prices. Available online: https://www.boilerjuice.com/heating-oil-prices/ (accessed on 15 November 2021).
51. Redman, G. *A Detailed Economic Assessment of Anaerobic Digestion Technology and Its Suitability to UK Farming and Waste Systems*; The Andersons Centre: York, UK, 2010.
52. Jain, S. *Cost of Abating Greenhouse Gas Emissions from UK Dairy Farms by Anaerobic Digestion of Slurry*; University of Southampton: Southampton, UK, 2013.
53. Lukehurst, C.; Bywater, A.M. *Exploring the Viability of Small Scale Anaerobic Digesters in Livestock Farming*; IEA Bioenergy: Brussels, Belgium, 2015.

54. LfL. Biogasausbeuten Verschiedener Substrate. Bayerische Landesanstalt für Landwirtschaft. Available online: https://www.lfl.bayern.de/iba/energie/049711/?sel_list=29%2Cb&anker0=substratanker#substratanker (accessed on 25 December 2021).
55. Dubrovskis, V.; Plume, I. Biogas from wastes of pumpkin, marrow and apple. *Agron. Res.* **2017**, *15*, 69–78.
56. KTBL. Wirtschaftlichkeitsrechner Biogas. Kuratorium für Technik und Bauwesen in der Landwirtschaft. Available online: https://daten.ktbl.de/biogas/showSubstrate.do?zustandReq=42#anwendung (accessed on 25 December 2021).
57. Li, W.; Khalid, H.; Amin, F.R.; Zhang, H.; Dai, Z.; Chen, C.; Liu, G. Biomethane production characteristics, kinetic analysis, and energy potential of different paper wastes in anaerobic digestion. *Renew. Energy* **2020**, *157*, 1081–1088. [CrossRef]
58. Takizawa, S.; Baba, Y.; Tada, C.; Fukuda, Y.; Nakai, Y. Pretreatment with rumen fluid improves methane production in the anaerobic digestion of paper sludge. *Waste Manag.* **2018**, *78*, 379–384. [CrossRef] [PubMed]
59. Pečar, D.; Goršek, A. Kinetics of methane production during anaerobic digestion of chicken manure with sawdust and miscanthus. *Biomass Bioenergy* **2020**, *143*, 105820. [CrossRef]
60. Gao, X.; Tang, X.; Zhao, K.; Balan, V.; Zhu, Q. Biogas production from anaerobic co-digestion of spent mushroom substrate with different livestock manure. *Energies* **2021**, *14*, 570. [CrossRef]
61. Luo, X.; Yuan, X.; Wang, S.; Sun, F.; Hou, Z.; Hu, Q.; Zhai, L.; Cui, Z.; Zou, Y. Methane production and characteristics of the microbial community in the co-digestion of spent mushroom substrate with dairy manure. *Bioresour. Technol.* **2018**, *250*, 611–620. [CrossRef] [PubMed]
62. Nwokolo, N.; Mukumba, P.; Obileke, K.; Enebe, M. Waste to energy: A focus on the impact of substrate type in biogas production. *Processes* **2020**, *8*, 1224. [CrossRef]
63. Morales-Polo, C.; Cledera-Castro, M.D.M.; Soria, B.Y.M. Reviewing the anaerobic digestion of food waste: From waste generation and anaerobic process to its perspectives. *Appl. Sci.* **2018**, *8*, 1804. [CrossRef]
64. Zhang, Y.; Kusch-Brandt, S.; Heaven, S.; Banks, C.J. Effect of pasteurisation on methane yield from food waste and other substrates in anaerobic digestion. *Processes* **2020**, *8*, 1351. [CrossRef]
65. Maranon, E.; Salter, A.M.; Castrillon, L.; Heaven, S.; Fernández-Nava, Y. Reducing the environmental impact of methane emissions from dairy farms by anaerobic digestion of cattle waste. *Waste Manag.* **2011**, *31*, 1745–1751. [CrossRef] [PubMed]
66. Shine, P.; Upton, J.; Sefeedpari, P.; Murphy, M.D. Energy consumption on dairy farms: A review of monitoring, prediction modelling, and analyses. *Energies* **2020**, *13*, 1288. [CrossRef]
67. WBA. World Biogas Association. Biogas Financial Calculator (Online Tool). Available online: https://www.worldbiogasassociation.org/financial-calculator/ (accessed on 20 November 2021).
68. Parsons Brinckerhoff. *Small-Scale Generation Cost Update. Report for the Department of Energy and Climate Change*; Department of Energy and Climate Change: London, UK, 2015.
69. Ofgem. Feed-in Tariffs (FIT)–Timeline of Changes to the Scheme. Available online: https://www.ofgem.gov.uk/environmental-and-social-schemes/feed-tariffs-fit/changes-fit-scheme (accessed on 27 December 2021).
70. ADBA. *ADBA Policy Report–April 2017*; Anaerobic Digestion and Bioresources Association: London, UK, 2017.
71. ADBA. *ADBA Policy Report–November 2017*; Anaerobic Digestion and Bioresources Association: London, UK, 2017.
72. Trimble, D. Managing the Cost of Water and Electricity in the New Dairy Unit at CAFRE. Available online: https://www.daera-ni.gov.uk/articles/managing-cost-water-and-electricity-new-dairy-unit-cafre (accessed on 25 December 2021).
73. Milam, K.; Coppock, C.; West, J.; Lanham, J.; Nave, D.; Labore, J.; Stermer, R.; Brasington, C. Effects of drinking water temperature on production responses in lactating Holstein cows in summer. *J. Dairy Sci.* **1986**, *69*, 1013–1019. [CrossRef]
74. Osborne, V.; Hacker, R.; McBride, B. Effects of heated drinking water on the production responses of lactating Holstein and Jersey cows. *Can. J. Anim. Sci.* **2002**, *82*, 267–273. [CrossRef]
75. Vida, E.; Tedesco, D.E.A. The carbon footprint of integrated milk production and renewable energy systems–A case study. *Sci. Total Environ.* **2017**, *609*, 1286–1294. [CrossRef] [PubMed]
76. Alvarez, C.F.; Molnar, G. What Is Behind Soaring Energy Prices and What Happens Next? Available online: https://www.iea.org/commentaries/what-is-behind-soaring-energy-prices-and-what-happens-next (accessed on 28 December 2021).
77. Energy-Stats UK. Wholesale Energy Costs. Available online: https://www.energy-stats.uk/wholesale-energy-pricing/ (accessed on 1 December 2021).
78. BEIS. *UK Energy in Brief 2021*; Department for Business, Energy and Industrial Strategy: London, UK, 2021.
79. Eurostat. Gas Prices by Type of User. Available online: https://ec.europa.eu/eurostat/web/products-datasets/-/ten00118 (accessed on 25 December 2021).
80. Ofgem. Bills, Prices and Profits (Online Publication). Available online: https://www.ofgem.gov.uk/publications/infographic-bills-prices-and-profits (accessed on 25 December 2021).
81. IEA. *Gas Market Report, Q3-2021*; International Energy Agency: Paris, France, 2021.
82. Arango-Osorio, S.; Vasco-Echeverri, O.; López-Jiménez, G.; González-Sanchez, J.; Isaac-Millán, I. Methodology for the design and economic assessment of anaerobic digestion plants to produce energy and biofertilizer from livestock waste. *Sci. Total Environ.* **2019**, *685*, 1169–1180. [CrossRef]
83. De la Fuente, C.; Alburquerque, J.; Clemente, R.; Bernal, M. Soil C and N mineralisation and agricultural value of the products of an anaerobic digestion system. *Biol. Fertil. Soils* **2013**, *49*, 313–322. [CrossRef]

84. WRAP. *Quality Protocol. Anaerobic Digestate. End of Waste Criteria for the Production and Use of Quality Outputs From Anaerobic Digestion of Source-Segregated Biodegradable Waste Quality Protocol Anaerobic Digestate*; Waste and Resources Action Programme: Banbury, UK, 2014.
85. Muskolus, A.; Salter, A.M.; Jones, P.J. Linking economic and energy modelling with environmental assessment when modelling the on-farm implementation of anaerobic digestion. In *Multi-Level Processes of Integration and Disintegration*; Schaft, F., Balmann, A., Eds.; Leibniz-Institut für Agrarentwicklung in Mittel-und Osteuropa: Halle, Germany, 2009; pp. 191–199.
86. Bateman, I.J.; Balmford, B. Public funding for public goods: A post-Brexit perspective on principles for agricultural policy. *Land Use Policy* **2018**, *79*, 293–300. [CrossRef]
87. Defra. *Health and Harmony: The Future for Food, Farming and the Environment in a Green Brexit*; Department for Environment, Food and Rural Affairs: London, UK, 2018.

processes

MDPI

Review

Recent Advances in Membrane-Based Biogas and Biohydrogen Upgrading

Cenit Soto [1,2], Laura Palacio [1,2], Raúl Muñoz [2,3], Pedro Prádanos [1,2,*] and Antonio Hernandez [1,2,*]

1 Surfaces and Porous Materials (SMAP), Associated Research Unit to CSIC, Facultad de Ciencias, University of Valladolid, Paseo Belén 7, E-47011 Valladolid, Spain
2 Institute of Sustainable Processes (ISP), Dr. Mergelina s/n, 47011 Valladolid, Spain
3 Department of Chemical Engineering and Environmental Technology, School of Industrial Engineering, University of Valladolid, Dr. Mergelina s/n, 47011 Valladolid, Spain
* Correspondence: ppradanos@uva.es (P.P.); antonio.hernandez@uva.es (A.H.)

Abstract: Biogas and biohydrogen, due to their renewable nature and zero carbon footprint, are considered two of the gaseous biofuels that will replace conventional fossil fuels. Biogas from anaerobic digestion must be purified and converted into high-quality biomethane prior to use as a vehicle fuel or injection into natural gas networks. Likewise, the enrichment of biohydrogen from dark fermentation requires the removal of CO_2, which is the main pollutant of this new gaseous biofuel. Currently, the removal of CO_2 from both biogas and biohydrogen is carried out by means of physical/chemical technologies, which exhibit high operating costs and corrosion problems. Biological technologies for CO_2 removal from biogas, such as photosynthetic enrichment and hydrogenotrophic enrichment, are still in an experimental development phase. In this context, membrane separation has emerged as the only physical/chemical technology with the potential to improve the performance of CO_2 separation from both biogas and biohydrogen, and to reduce investment and operating costs, as a result of the recent advances in the field of nanotechnology and materials science. This review will focus on the fundamentals, potential and limitations of CO_2 and H_2 membrane separation technologies. The latest advances on membrane materials for biogas and biohydrogen purification will be systematically reviewed.

Keywords: biogas; biomethane; biohydrogen; membrane separation; mixed matrix membranes; upgrading technologies; thermal rearrangement

Citation: Soto, C.; Palacio, L.; Muñoz, R.; Prádanos, P.; Hernandez, A. Recent Advances in Membrane-Based Biogas and Biohydrogen Upgrading. *Processes* **2022**, *10*, 1918. https://doi.org/10.3390/pr10101918

Academic Editors: Sonia Heaven, Sigrid Kusch-Brandt and Charles Banks

Received: 29 August 2022
Accepted: 18 September 2022
Published: 22 September 2022

1. Biogas and Biohydrogen as Green Energy Vectors

Biogas is produced via Anaerobic Digestion (AD) of residual biomass from diverse origins such as urban solid waste, livestock waste, agricultural waste, and wastewater. AD is a biological process (based on the action of micro-organisms) able to convert this residual biomass, by means of oxidations and reductions of organic carbon, to carbon dioxide and methane (CO_2 and CH_4, respectively) in the absence of oxygen [1,2]. This biological conversion is carried out through a sequence of hydrolysis, acidogenesis, acetogenesis and methanogenesis steps in an anaerobic digester [3]. Biogas is typically composed of CH_4 and CO_2 in a concentration range of 45–85% and 25–50%, respectively, and minor concentrations of other components such as H_2O (5–10%), N_2 (~0–1%), O_2 (~0–0.5%), H_2S (0–10,000 ppm), NH_3 (0–100 ppm) and hydrocarbons (0–200 mg Nm^{-3}) [4,5]. The biogas produced by AD represents an excellent alternative to fossil-based energy vectors [2], since biogas can be employed for the production of electricity, steam and heat, as a feedstock in fuel cells, as a green substitute of natural gas for domestic and industrial use or as a vehicle fuel [1]. The contribution of biogas in the European Union could account for 10% of the natural gas demand by 2030 and up to 30–40% by 2050.

Based on the latest report of the World Biogas Association [6], 50 million micro-scale digesters generating biogas for cooking or heating were in operation, mainly in China

(42 million) and India (4.9 million). On the other hand, 18,774 large-scale plants devoted to generating 11 GW (a biomethane plant produces an average of 36 GWh per year) of electricity were in operation in 2021 in Europe, Germany being the leader in the European market with 11,279 in 2020 plants (140 plants/1 Mio capita), followed by Italy (1666 in 2020) and France (833 new plants in 2020) [4,7]. China with 6972 large scale digesters and the USA with 2200 AD plants in 2015 represented the second and third largest biogas producer in the world, respectively. The global electricity generation from biogas increased by 90% in six years (from 46,108 GWh in 2010 to 87,500 GWh in 2016) and by 11.5 % from 2016 to 2020 (from 87,500 GWh in 2016 to 96,565 GWh in 2020) [6,8].

Biogas can be purified and converted into a high-quality biomethane via three sequential processes: desulfurization (elimination of the H_2S), CO_2 removal and biomethane polishing (removal of the minor biogas contaminants) [9]. The European EN-16723 Standard for biomethane introduction into natural gas networks (UNE-EN 16723-1-2016) and automotive/vehicle fuel (UNE-EN 16723-2-2017) requires an effective cleaning of biogas. This UNE-EN 16723-1-2016 standard has resulted in a specific Spanish standard for biomethane injection into the natural gas grid, requiring a minimum methane content of 90% and a maximum CO_2 content of 2% (v/v) [10]. In 2017, the number of biogas upgrading plants in the world accounted for 700 plants, Europe being the leading region with 540 upgrading plants in operation.

At the end of 2020 (the most recent data available), 880 biogas upgrading plants with a production capacity of 2.43 billion m^3 were in operation in Europe (161 additional plants relative to 2019) [4,7]. By 2021 the increase in the number of biomethane plants is expected to be even faster since 115 plants have started operation by August 2021 [7].

On the other hand, biologically produced hydrogen (commonly referred to as biohydrogen) generated via Dark Fermentation (DF) represents another alternative bioenergy source [11]. Biohydrogen (bioH_2) has the potential to become a relevant H_2 generation platform for the creation of a green economy [12]. In this context, hydrogen has multiple advantages as a clean energy vector such as: (i) the combustion of H_2 gas can be pollution-free in fuel cells, (ii) its energy efficiency in H_2 fuel cells is approx. 50% higher than that of gasoline, (iii) it has a specific energy content of 122 kJ/g (~2.75-fold larger than conventional fossil fuels), (iv) its conversion efficiency to electricity could be doubled using fuel cells instead of gas turbines, and finally (v) it can be stored as a metal hydride.

Dark fermentation is based on hydrogen and carbon dioxide (CO_2) production via anaerobic bacteria [13] and/or algae growing in the absence of light and with high carbohydrate content as substrate [14,15]. The biohydrogen produced is mainly composed of hydrogen (40–60%) and carbon dioxide (47–60%) with traces of methane and H_2S [16,17]. Currently, only 1% of hydrogen is produced from biomass [15]. This fact is probably due to the relatively late research on bioH_2 production by dark fermentation, where research is still conducted at a laboratory scale with a limited number of experiments at pilot scale [18]. Despite the fact that the H_2 yield from dark fermentation is higher than that of other processes, the main disadvantage of the gas generated during dark fermentation is its low hydrogen concentration (40–60%; v/v) [19], which hinders its direct use in fuel cells for electricity generation (where the purity of hydrogen is crucial to achieve high energy yields) [16]. Therefore, it is crucial to separate H_2 from the multiple gas by-products from DF, mainly CO_2, in order to obtain purified hydrogen. For instance, a hydrogen content of 73% can be obtained in a two-step gas membrane separation module [19].

The sustained use of non-polluting renewable energy vector such as biogas and bioH_2 is required to reduce the demand and dependence from fossil fuels [20]. Based on the International Energy Agency, the share of renewable and low-carbon transport fuels should increase up to 6.8% in 2030 in Europe, with advanced biofuels representing at least 3.6% of the total fuel consumption. The development of low footprint and cost technologies for the conversion of biogas to a purified biomethane and bioH_2 to pure H_2 is essential to guarantee the competitiveness of these green gas vectors as an energy source.

2. Biogas and Biohydrogen Purification with Membrane Technology

Nowadays, there are two main types of technologies for biogas purification, physicochemical and biological methods, while bioH_2 purification is only performed by physicochemical methods. Physicochemical technologies exhibit high energy and chemical demand, and therefore they present large operating costs and environmental impacts. As an example, this section will only focus on CO_2 removal technologies.

Pressure swing adsorption (PSA), cryogenic CO_2 separation, scrubbing with H_2O, chemical solutions or organic solvents, and membrane separation, dominate the biogas upgrading market nowadays [21], while cryogenic distillation, PSA and membrane separation are the most popular processes for H_2 purification at commercial scale [22–24].

Separation of gas mixtures through membranes has become a relevant unitary operation for the recovery of valuable gases and mitigation of atmospheric pollution, which offers several advantages over conventional gas separation methods [25]. Indeed, Membrane Separation (MS) is considered nowadays the most promising gas purification technology. Membrane separation relies on the interaction (physical or chemical) of certain gases with the membrane material [26]. The membranes used are selective physical barriers with certain components that permeate across them [27]. Gas separation by membrane technology is characterized by selectivity properties and flux, which supports a functional transport of the target gases across the barrier (permeability). This technology presents a low energy consumption, a simple operation, cost effectiveness, smaller footprint, a negligible chemical consumption and low environmental impacts [28,29]. The potential of MS to achieve high efficiencies of gas separation foster their use in different industrial applications including refineries and chemical industries, and recent advances in material science render MS a competitive technology [30]. The lifetime of commercial membranes account for 5–10 years [31]. Today, the use of membranes in industry includes the separation of N_2 or O_2 from air, separation of H_2 from gases such as CH_4, separation of CH_4 from biogas, separation of H_2S and CO_2 from natural gas, etc. The use of membranes in separation processes is rapidly growing, especially in Europe (Figure 1). Among the available technologies for the purification of biogas to biomethane, membrane separation is currently the most widely used technology (39%), followed by water scrubbing (22%) and chemical scrubbing (18%). Pressure swing adsorption (12%), cryogenic separation (1%) and physical washing (1%) complete the market share (with the exception of 7% of European biomethane plants, with no data available in the EBA database) [7]. For instance, Baker (2002), calculated that the market share of membrane gas separation processes in 2020 would be five times higher than that of year 2000 [32]. Indeed, the market share of MS for biogas upgrading application has increased from 10% in 2012 to 25% in 2017 [33]. Likewise, MS has grown exponentially since the initial commercial application of Prism membranes by Permea (Monsanto) for H_2 separation from the off-gas stream of NH_3 production plants [26].

A detailed economic study of the total costs of biogas purification is a difficult task nowadays due to the large number of parameters to be considered. However, Miltner and co-workers (2017) have published some general estimates and a comparison of the most common physicochemical technologies such as pressurized water scrubbing, amine scrubbing, pressure swing adsorption and gas permeation. This study included investment costs (15 years' depreciation), plant reliability of 98%, operational consumptions in terms of electricity and consumables (electricity price 15 €ct/kWh), as well as maintenance and overhaul (without engineering costs, taxes and revenues). Thus, the costs for an installation with a capacity of 250 $m^3{}_{STP}$/h are in the range of 25 €ct/$m^3{}_{STP}$, while these costs drop below 15 €ct/$m^3{}_{STP}$ for capacities above 2000 $m^3{}_{STP}$/h. This work concluded that gas permeation is slightly more advantageous for sizes below 1000 $m^3{}_{STP}$/h. Overall, small-scale biogas upgrading entails higher capital and operational costs [34].

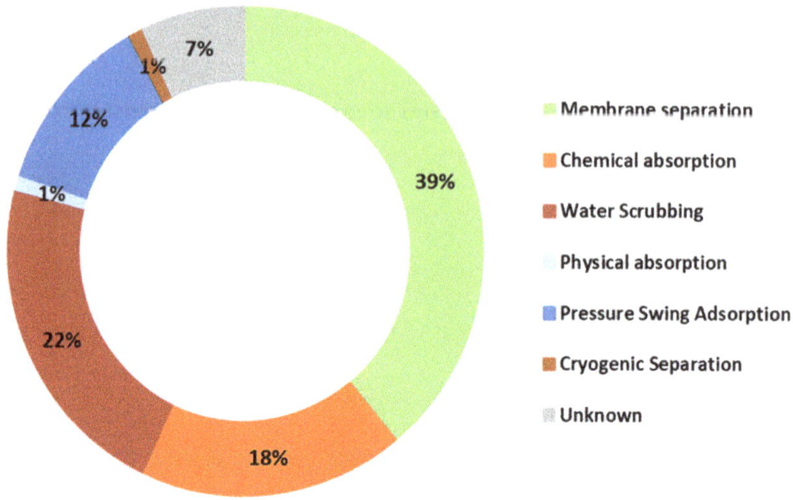

Figure 1. Market share of different upgrading technologies in Europe in 2020. Figure adapted from EBA, 2021.

Ideally, membrane materials for gas separation should exhibit a high selectivity and big fluxes, excellent chemical, mechanical and thermal stability, a defect-free production and be cost effective. Membranes are classified according to the type of material, configuration, structure, composition, support material and industrial reactions, among others (Figure 2) [35–37]. Four kinds of membranes are typically proposed for development and commercialization in hydrogen purification: (i) polymeric membranes (organic), (ii) porous membranes (ceramic, carbon, metal) (iii) dense metal membranes and (iv) ion conductive membranes, the last three also referred to as inorganic membranes [27]. In this context, dense-metal membranes and polymeric have experienced the largest advances in terms of scale-up [38]. The most commonly used polymeric membranes for gas separation are nonporous membranes, which are classified as glassy or rubbery. Of them, glassy polymers are most typically used for gas separation applications. These polymers include polysulfones (PSF), polycarbonates (PC) and polyimides (PI), which are often employed for the separation of H_2/CH_4, H_2/N_2 and O_2/N_2 [39]. On the other hand, membranes can be configured as hollow fibers, capillaries, flat sheets and tubular and can be installed in a suitable membrane module. The most commonly used modules are pleated cartridges, tubular and capillary, hollow-fiber and plate-and-frame and spiral-wound systems [40].

H_2 separation was one of the pioneered applications in gas separation membranes, DuPont (E. I. du Pont de Nemours and Co., Wilmington, DE, USA) being the pioneer in manufacturing small-diameter hollow-fiber membranes. Due to the limited productivity (or permeance) of these membranes and their high cost, Monsanto Co. (Monsanto Company, St. Louis, MO, USA) developed polysulfone hollow-fiber membranes, which considerably increased the transport through the fibers, and consequently were successfully implemented at industrial-scale for hydrogen recovery from ammonia purge gases [41]. Then, Separex Corp (Champigneulles, France) developed Separex® spiral-wound cellulose acetate membranes (including separations for natural gas and dehydration [41] providing better performance than hollow fiber membranes due to their high resistance of hydrogen impurities [42]. Polymeric membranes, especially polyimides, have been employed to separate hydrogen from gaseous mixtures (N_2, CO and hydrocarbons) based on their economic viability, easy processibility and satisfactory thermal stability (350–450 °C) [43]. Polyimide membranes with excellent heat resistances were introduced by Ube in Japan (Ube Industries, Ltd., Tokyo, Japan), and the refinery at Seibu Oils (Seibu Oil Company Limited, Onoba, Japan) was the first facility to apply them commercially [41]. Commercial membrane systems provide a H_2 purity of 90–95% during hydrogen purification with a moderate recovery of 85–90% [44].

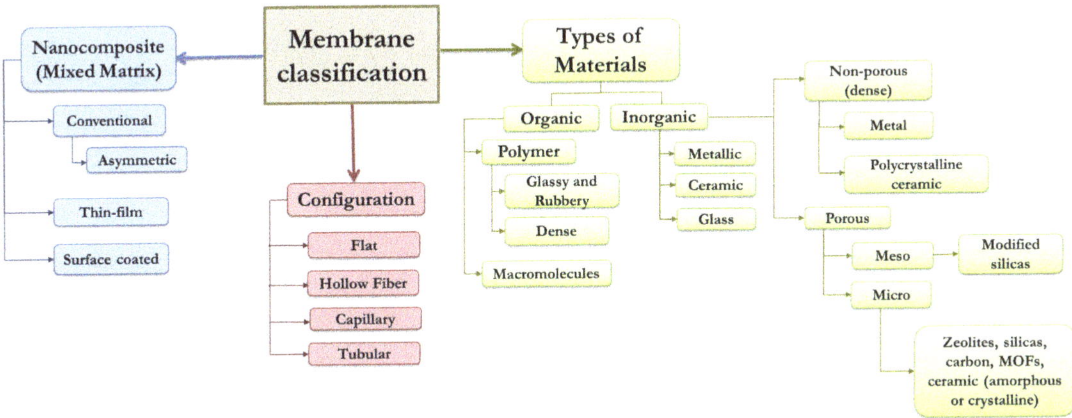

Figure 2. Classification of membranes for gas separation. Adapted from Vinoba and co-workers, 2017.

At the beginning of the 1990s, gas mixture separation membranes with a poor recovery of methane and low selectivity were installed for the upgrading of landfill biogas [45]. In 2007, Air Liquide Medal^{TM} further developed and tested new selective membranes combining high CH_4 recoveries with high CH_4 concentrations.

Today, membrane-based biogas upgrading can provide methane concentrations of 97–98% in the biomethane with a concomitant methane recovery above 98%, based on the high permeabilities of CO_2 in commercial membrane materials. The permeation rate mainly depends on the molecular size of the gas components and on the membrane construction material [46]. Membrane-based biogas upgrading at commercial scale is carried out at 6–20 bar, which entails energy consumption of 0.18–0.20 kWh/Nm^3 of raw biogas or 0.14–0.26 kWh/Nm^3 of biomethane [9].

In this regard, despite polymeric membranes having consistently demonstrated promising results and being commercially available at large-scale for hydrogen and biogas purification, their use is limited to 8–9 polymeric materials (e.g., cellulose acetate, polyimides, perfluoropolymer etc.) [47,48]. Therefore, further research in the field of material science needs to be conducted to achieve new membranes with superior gas separation properties: higher permeability, selectivity and stability (mainly restricted plasticization) [47].

3. Fundamentals of Membrane-Based Gas Separation

The membrane gas separation process is based on the separation of gases by selective permeation of one or more gaseous components through a thin membrane (porous or dense membrane) [49]. The separation potential of the membrane is governed by its transport properties of the components of a mixture. This transport rate is in turn dictated by the permeability and selectivity of the membrane material and its driving force [38].

Gas separation takes place according to the morphology of the membrane materials and can be based on three transport mechanisms depending on the porous size: solution-diffusion, molecular sieve and Knudsen diffusion (Figure 3). In this context, the transport of gases by Knudsen diffusion takes place in porous membranes (pore diameter in the range of 50–100 Å), with smaller pore size than the gas molecules. In this mechanism, gas molecules interact more frequently with the pore walls, colliding with each other, allowing diffusion of lighter molecules to occur through the pores. The molecular sieving mechanism, with pore size between 3.0–5.2 Å, is based on the size exclusion of gas molecules, leading to the separation of gas molecules of different kinetic sizes. Indeed, the pores only allow the passage of molecules smaller than that size, preventing the passage of larger ones [26,29].

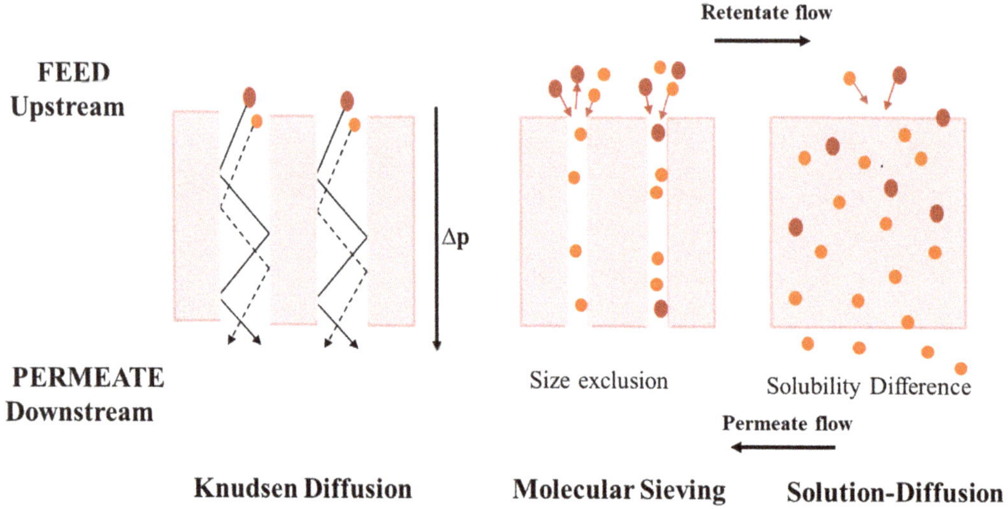

Figure 3. Schematic representation of the three mechanisms for gas mixture separation in membranes: diffusion—Knudsen diffusion, molecular sieving, and solution-diffusion. Orange circle (molecule A), Red circle (molecule B). The figure was adapted from Sridhar and co-workers, 2014, and Ismail and co-workers, 2015.

Gas transport in non-porous dense polymeric membranes is most commonly described by solution-diffusion mechanisms (used exclusively in current commercial devices), which allows gases to pass through the membrane free volume units and consists of three steps [50]: (i) sorption in upstream side; (ii) diffusion through the membrane and (iii) desorption at the downstream side. Figure 4 shows a schematic overview of mass transfer by solution-diffusion, where gas molecules sorb into the high-pressure face of the membrane, then diffuse through along the membrane and later desorb from the low-pressure face of the membrane [51]. This mechanism of solution-diffusion is determined by the occurrence of differences in the thermodynamic activities at the upstream and downstream faces of the membrane, and the interacting force working among the gas molecules, which depends on the membrane components and the permeate molecules [52].

A key parameter to evaluate membrane transport properties is the Permeability coefficient (P), which refers to the gas flux across a membrane considering the membrane thickness and pressure gradient ($p_{i,0}$-$p_{i,l}$) through the membrane (Equation (1)).

$$P = \frac{N_i\, l}{\Delta p} \tag{1}$$

where N_i is molar flux of a gas component i through the membrane, l is the membrane thickness and Δp is the pressure gradient, calculates as the difference between $p_{i,0}$ (the upstream pressure) and $p_{i,l}$ (the downstream pressure) [53].

The Permeability coefficient ranges from 10^{-4} to 10^4 Barrer as a function of the gas component and the polymer structure [52]. Permeability coefficients are expressed in mol ($m^2 \cdot s \cdot Pa$) in the international system of units. However, P is typically given in Barrer, where $1\ Barrer = 10^{-10} \frac{cm^3{}_{STP} \cdot cm}{cm^2 \cdot s \cdot cmHg}$.

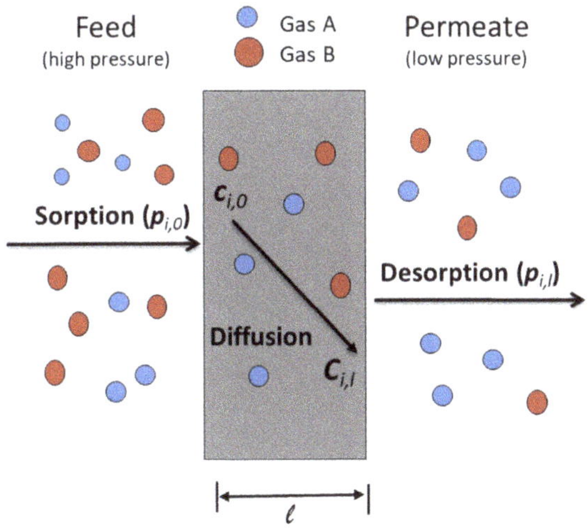

Figure 4. Detailed overview of mass transfer by solution-diffusion model.

The solution-diffusion model considers that the conditions of equilibrium between sorption and desorption are maintained. In this context, a solubility coefficient, S_i, is introduced, which is the ratio between the concentration of gas component dissolved in the membrane material, C_i, and the pressure of the gas, p_i, in contact with the polymer (Equation (2)). The solubility of a gas component i in the polymeric material depends mainly on the gas molecule condensability.

$$S_i = \frac{C_{i,0}}{p_{i,0}} = \frac{C_{i,l}}{p_{i,l}} \tag{2}$$

where $C_{i,0}$ and $C_{i,l}$ stand for concentration of the gas component i at the feed and permeate side, respectively.

On the other hand, the molar flux, N_i, can be expressed as a function of the diffusivity coefficient (D_i) described by the Fick's Law (Equation (3)):

$$N_i = D_i S_i \frac{p_{i,0} - p_{i,l}}{l} \tag{3}$$

According to the solution-diffusion model, the ability of a gas molecule to pass through the membrane depends on a kinetic factor, the diffusivity, (D_i), which characterizes the movement of the gas molecules diffusing through the polymer, and a thermodynamic factor, the solubility, (S_i), which characterizes the number of gas molecules passing through the membrane. Thus, P can be represented as the product of the diffusion coefficient, D_i, and gas solubility coefficient, S_i (Equation (4)) [53,54].

$$P = D_i\, S_i \tag{4}$$

On the other hand, a parameter characteristic of gas separation is the ability of a membrane to separate two gas components (A and B). Typically, selectivity is also treated as a material property of the polymer and is represented by Equation (5). The parameter α is defined as the permeability ratio of the faster permeable gas (P_A) between the slower permeable gas (P_B), so that $\alpha_{AB} > 1$ [52].

$$\alpha_{AB} = \frac{P_A}{P_B} \tag{5}$$

Usually, pure gas permeabilities are used in Equation (5) giving the so-called ideal selectivity (α). According with the solution-diffusion model, Equation (5) can be reworded using Equation (4), and the selectivity of diffusivity and solubility can be expressed using Equation (6):

$$\alpha_{AB} = \frac{D_A}{D_B}\frac{S_A}{S_B} = \alpha_{AB}^{D}\,\alpha_{AB}^{S} \tag{6}$$

where D_A/D_B stands for the diffusion coefficients ratio of gas components A and B, while S_A/S_B is the ratio of their solubility coefficients. Membrane selectivity determines the energy needed to support gas separation and directly impacts on the operating cost of a membrane system [55].

4. Challenges in Polymer Membranes for Gas Separation

The membrane transport properties are governed by factors such as the change in feed composition and the degree of swelling at the gas–membrane interface. In addition, other phenomena such as plasticization and ageing influence the transport properties of membranes. In this context, large contents of condensable gases such as CO_2 may plasticize the membrane material. Nowadays, research in membrane-based gas separation targets the development of new membranes with increasing permeabilities and selectivities, with increasing permeabilities without compromising the selectivity or improving the selectivity maintaining the permeability values. Indeed, the increase in permeability without compromising selectivity is typically considered one of the main target routes to expand the market share of membrane materials for gas separation [56].

4.1. Trade-Off Relationship

Membrane gas separation has been used for the purification of hydrogen (in H_2/CO_2, H_2/CH_4 and H_2/N_2 gas mixtures) in refineries and the petrochemical industry, for the separation of CO_2/CH_4 mixtures (in natural gas sweetening and biogas upgrading) and for the treatment of flue gas (CO_2/N_2) [24,30,57–59].

As stated above, permeability and selectivity represent key parameters for optimal gas separation. However, these parameters typically experience a trade-off relationship since highly permeable polymers tend to have less selectivity and vice versa. In this context, an experimental upper-bound relationship between selectivity and permeability was proposed by Robeson in order to benchmark membranes for gas separation [60,61]. This upper bound has been employed to relate gas permeability values in a different format. Later on, Robeson 1991 and Freeman provided a fundamental theory for this observation [62]. As more data on the gas separation characteristics of the polymers employed in the analysis published in 1991 were available, an updated compilation was published in 2008 [60], where the most significant changes were triggered by the information of perfluorinated polymers not reported in 1991. These data confirmed that when the permeability of a gas increases, the permeability of other gases also increases, since the diffusion coefficient of gases is related to the polymer free volume [53]. Figure 5, displays an example of a Robeson-type trade-off graph for CO_2/CH_4, where the CO_2/CH_4 selectivity is shown against the CO_2 permeable support material [60,61].

Swaidan reported in 2015 new permeability/selectivity "upper bounds" for commercial membrane modules for air and hydrogen separation (H_2/N_2, H_2/CH_4 and O_2/N_2) [63]. The Robeson upper bound behavior was redefined by Comesaña-Gandara in 2019 for CO_2/N_2 and CO_2/CH_4 separations using ultra-permeable Polymers of Intrinsic Microporosity (PIM) [64].

By transferring this trade-off relationship to the Robeson upper bound, the optimum balance involving a high selectivity in combination with a high permeability is determined. Nowadays, the research in this field is focused on developing new polymer materials capable of exceeding the upper bounds for the most relevant gas pairs. The key variables of the upper bound plots from the upper bound correlations $P_i = k\,\alpha_{ij}^{n}$ are tabulated in Table 1, for the present upper bound data against the previous upper bound data.

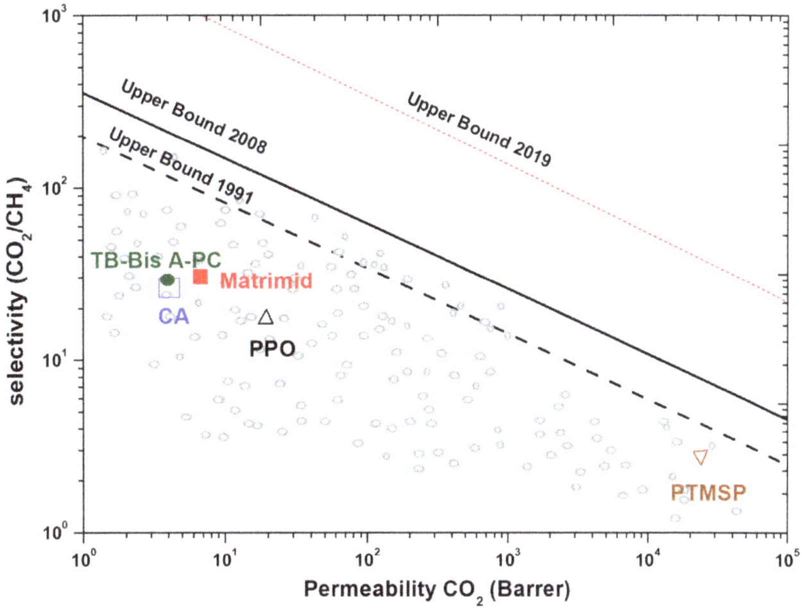

Figure 5. CO_2/CH_4 Robeson plot for conventional glassy polymers. CA: cellulose acetate; PPO: poly(phenylene oxide); PTMST: Poly(trimethylsilylpropyne); TB-Bis A-PC: tetrabromobisphenol A poly(carbonate), Matrimid®: commercial poliimide. Permeabilities for single gases were measured in the range 25–35 °C and pressures from 1 to 20 bar. The continuous line stands for the 2008 upper bound, while dashed line represents the 1991 upper bound (adapted from Galizia and co-workers, 2017).

Table 1. Tabulation of the values of upper bound slope *n* and the front factor *k*. Table adapted from Robeson, 2008.

Gas Pair	*k* (Barrer)	*n*
Before 2008 Robeson's bond [61,65]		
O_2/N_2	389,224	−5.800
CO_2/CH_4	1,073,700	−2.6264
H_2/N_2	52,918	−1.5275
H_2/CH_4	18,500	−1.2112
He/N_2	12,500	−1.0242
He/CH_4	5002	−0.7857
He/H_2	960	−4.9535
CO_2/N_2	NA	NA
N_2/CH_4	NA	NA
H_2/CO_2	1200	−1.9363
He/CO_2	705	−1.220
H_2/O_2	35,760	−2.277
He/O_2	4600	−1.295
2008 Robeson's bond [60]		
O_2/N_2	1,396,000	−5.666
CO_2/CH_4	5,369,140	−2.636
H_2/N_2	97,650	−1.4841
H_2/CH_4	27,200	−1.107
He/N_2	19,890	−1.017
He/CH_4	19,800	−0.809

Table 1. *Cont.*

Gas Pair	k (Barrer)	n
He/H_2	59,910	−4.864
CO_2/N_2	30,967,000	−2.888
N_2/CH_4	2570	−4.507
H_2/CO_2	4515	−2.302
He/CO_2	3760	−1.192
H_2/O_2	NA	NA
He/O_2	NA	NA

4.2. Physical Aging and Plasticization

Plasticization is a frequently observed problem affecting the performance of membranes for gas separation (mostly from glassy polymers) [66,67]. Plasticization occurs when the gas concentration inside a polymer increases, causing swelling. As a result, the free volume and chain movement in the polymer material increase and in turn, the coefficients of gas diffusion increase and diffusion selectivity decreases [53,68]. A typical phenomenon observed during plasticization of glassy polymers is the increase in the permeability of a pure (or mixed) gas as the partial pressure (upstream) of the gas increases [67] caused by the loss of the polymer selectivity. The permeability increase is driven by the increase in diffusion coefficient, which in turn is governed by the penetrant (upstream) pressure [69]. CO_2 is the gas most commonly investigated in plasticization studies [70–73]. Gas sorption is known to increase after exposing a glassy polymer to CO_2 at a given pressure for a certain timeframe, which can even affect the mechanical properties of the polymer [74]. For glassy polymers, plasticization typically occurs at pressures of 10–40 bars and CO_2 concentration of 38 ± 7 cm^3 (STP)/$cm^3_{polymer}$. Since pressure is related to CO_2 concentration in the polymer, it has been hypothesized that each polymer needs the same CO_2 concentration to induce plasticization but a different pressure to achieve it. As a rule of thumb, polymers that absorb more CO_2 are more likely to plasticize than those that absorb less CO_2 at a given pressure [53]. The thickness of a glassy polymer film (membrane) represents a key factor in the plasticization process because thinner films tend to be more sensitive to CO_2 pressure changes. Thus, a thin film tends to plasticize more quickly [75].

There is a wide variety of glassy polymers with outstanding performance in gas separations. These materials, by their nature, are not in equilibrium and have a high free volume due to their inefficient packing (caused by the movement of their chains), which avoids fully equilibrium properties to be reached [76]. This gradual approach to equilibrium influences various properties that change over time and consequently the material undergoes "physical aging". This frequent drawback affecting the membrane performance is a steady continuation of the glass transition that sets in around Tg. Thus, physical ageing influences all temperature-dependent properties that change significantly and sharply at Tg. Ageing can be explained by the free-volume theory (Figure 6). The free-volume concept assumes that the transport mobility of the particles depends mainly on the degree of packing of the system. If packing is efficient, the number and size of free volume elements are reduced, and thus the gas diffuses slower through the membrane over time [76]. The rate of physical ageing should then decrease over time because, when the free volume gradually decreases, the driving force governing physical ageing decreases, and also the pace of segmental movements that help reorganize the polymer chains decreases [53].

Physical ageing, apart from reducing gas permeability, also impacts other physical properties with an increase in internal energy concomitant with an increase in entropy [77]. Therefore, as the polymer ages, the free volume decreases along with permeability (although at slower rates as time goes on), which is accompanied by an increase in selectivity as a consequence of the reduction of membrane flux over time [53].

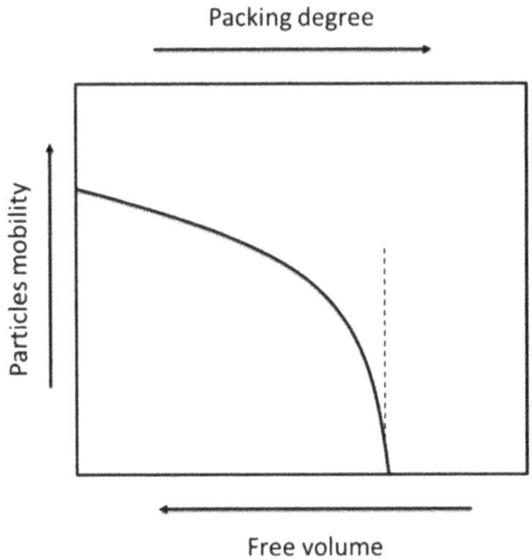

Figure 6. The qualitative free volume concept. adapted from Struik, 1978. An increasing degree of packing entails a decrease in the mobility. At a critical degree of packing, the mobility steeply falls to zero.

Membrane thickness represents another factor influencing physical ageing. According to Baker and Low (2014), the free volume elements migrate to the surface as bubbles, leaving a viscous liquid, with the migration distance being proportional to the square of the thickness of the membrane. Therefore, rearrangement and loss of permeability occurs in a short time in thin membranes [46]. In this context, Tiwari and co-workers investigated the impact of physical ageing on gas permeability in thin and thick membranes manufactured with "high free-volume" glassy polymers (e.g., PIM-1). The results of this study showed a dominant ageing effect in thin films, where even physical ageing overcame the CO_2 plasticization effects [71]. Figure 7 displays an example of the time course of the decrease in membrane permeability. This effect, using Matrimid® coated with polydimethylsiloxane (PDMS) membranes, was investigated by Rowe and co-workers (2009), who observed that ageing rapidly increases in thinner membranes [76]. Likewise, Xia and co-workers (2014) investigated both the effect of the membrane thickness on ageing and the influence of the ageing time on the plasticization using a commercial polyimide membrane, Matrimid®, for gas separation [78]. This study concluded that membranes become more vulnerable to CO_2 plasticization as their thickness decreases and the ageing time increases [78]. Finally, it is worth mentioning that the ageing process can be reversed by heating the membrane above Tg [79].

4.3. *Novel Polymeric Membrane Materials for Gas Separation*

Good mechanical strength, sorption capacity and chemical resistance rank among the most relevant criteria for selecting polymeric material for gas separation. However, the membrane permeability, the capacity of the polymer to withstand swelling mediated plasticization and the processability of the polymer into a useful asymmetric or thin film composite morphology have been identified as key properties of membrane materials. Moreover, the polymer material should exhibit a good interaction with at least one of the components of the mixture in order to induce an effective separation [29]. Today, research in the field of gas separation is devoted to the development of novel membranes materials with superior permeability and selectivity performance exceeding the latest published Robeson upper bound limit, and consequently overcome the trade-off effect of conventional membranes [60,61,63–65].

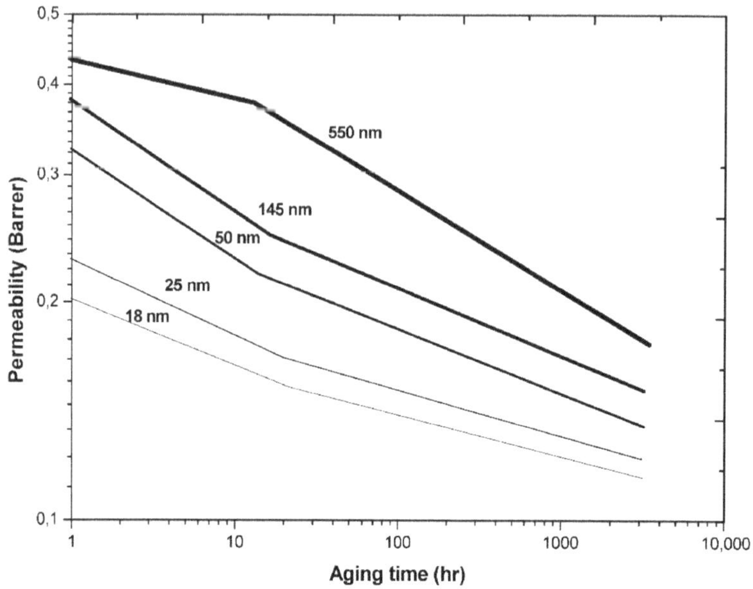

Figure 7. Effect of physical aging on CH_4 permeability in Matrimid® films as a function on time and thickness at 35 °C and 2 atm (Adapted from Rowe and co-workers, 2009).

According to Galizia and co-workers [55], most of the polymers developed for gas separation membranes in the last 30 years were evaluated without systematically proving their superior performance compared to the existing materials. Due to their high flexibility, one of the most synthesized families of materials for creating and understanding structure-property relationships are polyimides [55]. However, it has not been possible to significantly improve the structure-property balance of polyimides-based membranes. Therefore, despite polymeric membranes can be utilized in the separation of almost any gas mixtures such as O_2/N_2 separation, hydrogen purification (H_2/N_2, H_2/CH_4, and H_2/CO), CO_2/CH_4 biogas mixtures and vapor/gas separation, it is necessary to move beyond conventional polymers. In this context, new membrane materials for gas separation must provide higher permeabilities and permselectivities than conventional membranes. In addition, the production of new membranes for gas separation must consider good film-forming, good mechanical properties, absence of microdefects in the thin film, outstanding thermal and chemical stability, and absence of ageing [52].

Poly(benzimidazoles) (PBIs) often exhibit glass transition temperatures (Tg) greater than 400 °C, and a good thermal, mechanical and chemical stability, which is not typical among glassy polymers. Celazole® (PBI Performance Products, Inc., Charlotte, NC, USA) (sometimes named as m-PBI) is an example of membranes derived from PBIs that exhibit promising gas transport properties. However, Celazole® exhibits a low solubility in common solvents due to its structural features and intermolecular hydrogen bonding forces [80,81]. Borjigin and co-workers synthesized a novel PBI with sulfonyl moieties by performing a structural modification using 3,3′,4,4′-tetraamino-diphenylsulfone (TADPS) as monomer, which entailed a good solubility in common solvents such as N-methyl-2-Pyrrolidinone, NMP, N,N-dimethylacetamide, DMAc and dimethylsulphoxide, DMSO. Unfortunately, despite the good thermal stability and high permeabilities of PBIs, these materials are still susceptible to physical ageing [82].

Aromatic Polyamides (PA) were one of the first aromatic linear polymers considered thermally stable. PA typically exhibit a high cohesive energy density, a strong tendency for highly efficient polymer chain packing and a semicrystalline morphology [83]. Additionally, PA reported also a fair balance of properties: good chemical stability, high thermal resistance, good mechanical properties and an easier processability than aromatic polyimides [84,85]. However, PA support a low gas permeability of small molecules compared to polyimides. In recent years, there have been many attempts to improve PA gas separation performance by introducing bulky moieties, contoured structures or by introducing hexa-fluoropropane parts into the macromolecular chain, but with a limited success [86,87]. Likewise, Lozano and co-workers carried out in situ sialylation of diamines by adding trimethylchlorosilane (TMSCl) to the diamine solutions that, after the addition of a diacid chloride, resulted in high molecular weight aromatic polyamides, which guarantees high performance [88].

On the other hand, the so-called nanoporous polymers, as a result of their extremely fine nanoporous structure, have shown an outstanding performance in terms of gas separation. Examples of these materials are:

(i) Polymers of Intrinsic Microporosity (PIMs): PIMs were initially developed by McKeown and Budd [89] and have been demonstrated to be good candidates for gas separation due to their strong interactions with gas molecules and their nanometer pore size [89–91]. However, their physical ageing and the instability of their permeability properties over time are the major obstacles to their commercialization [46,70,71].

(ii) Thermally Rearranged polymers (TR): TRs were initially introduced by Park and co-workers in 2007 [92] and show a high selectivity and an extraordinarily high permeability. Additionally, TRs exhibit a good resistance to plasticization as well as a strong chemical and thermal resistance.

Recently, significant advances were achieved in the optimization of Mixed Matrix Membranes (MMMs) [35]. MMMs allow tuning the transport properties of conventional polymers for target applications by combining the high permeability of the polymer and the good selectivity of the filler materials.

5. Mixed Matrix Membranes for Gas Separation

Polymeric membranes have been successful in some gas separation processes such as natural gas sweetening but are still subject to the trade-off between permeability and selectivity, and the impact of physical ageing and plasticization, which makes them unstable for industrial applications. Recently, Barker (2014) reviewed the barriers limiting the development of membranes with high selectivity and permeance from the last 35 years and identified the need to develop new materials for new and future membrane applications [46]. Therefore, most research efforts are devoted nowadays to the development of new polymeric materials and membranes material such as zeolites, metal organic framework (MOF), carbon molecular sieves, carbon nanotubes and graphenes to improve the gas separation performance of membranes [93].

In this context, hybrid materials known as MMMs have been manufactured by adding inorganic materials as the disperse phase into polymers in order to take advantage of the processability of polymers and simultaneously overcome the trade-off between permeability and selectivity. The mixed matrix membranes concept has been described in multiple scientific publications. According to the most recent definitions, MMMs results from the combination of an inorganic or inorganic-organic hybrid material (micro or nanoparticles)—in the form of dispersed particles called additive or filler—and a polymeric matrix-continuous phase (Figure 8) [30,93]. PIMs and HPI are the most commonly used polymeric matrices, and zeolites the most common fillers. Moreover, MMMs have been recently thermally treated to obtain MMM-TR with outstanding gas transport properties for gas pairs such as CO_2/CH_4, O_2/N_2, H_2/CO_2, etc. [94–98].

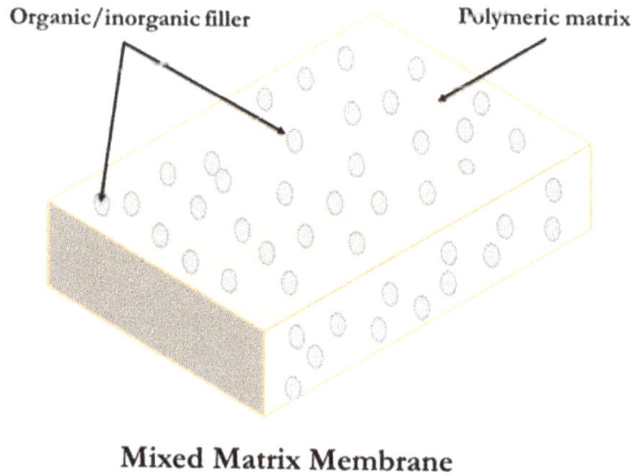

Figure 8. Illustration of a mixed matrix membrane (adapted from Lin and co-workers, 2018) [99].

MMMs have emerged as a promising material for gas separation in membrane technology. The main objective of the manufacture of MMMs is to provide solutions to the existing permeability versus selectivity trade-off relationship of gas separation polymeric membranes by taking advantage of the superior properties of inorganic particles [100,101]. In addition, MMMs compensate the unavoidable fragility limitation of inorganic membranes using a flexible polymer as the continuous matrix. These features provide MMMs with the potential to achieve a greater selectivity, permeability (caused by increasing the diffusion coefficients) or both, compared to existing polymeric membranes and to exceed the upper limit proposed by Robeson. These organic and inorganic materials employed as fillers could have a unique structure, surface chemistry and mechanical strength. Inorganic fillers contribute to enhanced diffusivity selectivity by acting as molecular sieves due to their precise pore size and shape and geometry, thus overcoming the properties of common polymeric membranes [55,102]. Overall, MMMs support unprecedented increases in permeability while maintaining selectivity by introducing fillers into the polymeric matrix, due to the increase in diffusion coefficients.

The first reports of the manufacture of MMMs were published in the 1970s. For instance, Paul and Kemp (1973) added a commercial zeolite (Molecular Sieve Type 5A) as a filler to a PDMS rubber used as polymer matrix [103]. A good interaction between the polymer and the zeolite was observed due to the flexibility of the rubber polymer and a large increase of a delayed diffusion time lag effect. However, high fluxes of gas in the polymer matrix can result in a low improvement in the selectivity [30]. In the last decade, manufacture of MMMs, researcher on of their mechanical and transport properties, as well as the investigation of their nanostructure have increased a significant attention in the membrane research field [52].

5.1. *Factors Influencing Mixed-Matrix Membrane Manufacture*

Multiple factors during the preparation of MMMs can cause: interfacial defects caused by particle sedimentation (due to the differences in physical properties and density with the polymer), migration of filler particles or agglomeration in the surface, especially when the fillers load is high due to the fact that this scenario increases the diffusion distance within the solid phase agglomerate [100].

According to Noble (2011), the compatibility between the disperse and continuous phases in terms of permeability is an important factor to consider due to the fact that the resistance to mass transfer is typically much higher in phases with much lower gas permeability [104]. In addition, there is a relationship between the filler particle size and membrane thickness, as smaller particles provide a higher surface area/volume ratio, which supports a greater mass transfer between phases. Finally, an effective contact between the two phases is necessary to prevent any gaps between them that could block the access to the pores [104].

Today, the achievement of the desired morphology, mechanical/chemical stability, and gas separation properties in MMMs requires overcoming multiple manufacturing challenges such as: obtaining a flawless interface to guarantee a good separation performance of the membranes, obtaining a homogeneous dispersion between the two phases, avoiding agglomerations responsible for low selectivity and finally selecting materials with excellent separation properties and good compatibility between the phases [102,105].

5.1.1. Morphologies of the Mixed-Matrix Membrane

The desired morphology of MMMs would include the absence of defects in the polymer–particle interface and must ensure gas transport across the dispersed phase instead through the continuous phase (polymeric matrix) (Figure 9) [101]. The advantages of morphology can be understood in terms of the ideal Maxwell model that represents the simplest case for mixed matrix transport properties [106]. This model, described by Robeson as a dilute suspension of particles in a polymeric matrix, was mainly developed for estimating dielectric properties of composites and describes the effective permeability of MMMs, P_{eff} as follows [107,108]:

$$P_{eff} = P_c \left[\frac{P_d + 2P_c - 2\,\Phi_d\,(P_c - P_d)}{P_d + 2\,P_c + \Phi_d\,(P_c - P_d)} \right] \tag{7}$$

where P_c is the continuous phase permeability (i.e., polymer matrix), represents the dispersed phase permeability (i.e., filler) and Φ_d is the dispersed phase volume fraction. Note that Equation (7) goes to the appropriate value of P in the limits as Φ_d = 0 or 1. Maxwell's model can be complicated by assuming that the dispersed phase, being uniformly distributed, is encapsulated by an "interface" (region between polymer matrix and inorganic fillers) with characteristics different from both the dispersed and continuous phases [104,106]. The formation of the interface is attributed to the inhibition of the mobility of the polymer chains in compressive stress near the polymer–particle interface. Figure 10 shows a representation of the polymer matrix, the dispersed phase and the rigidified interface (three-phase MMM system) [109].

One of the disadvantages of this model for MMMs is the need to determine the transport properties (e.g., through kinetic sorption in monodispersed crystals) in order to obtain a good characterization of the dispersed phase [106]. Moreover, it is also only applicable to low filler loadings with free volume fractions lower than 0.2. In this context, high values of Φ_d render the ideal Maxwell model useless. In addition, the Maxwell model does not consider the morphological properties of the filler such as particle shape, particle size distribution or the aggregation of filler particles [100].

Thus, the preparation of ideal MMMs entails a difficult procedure as a result of the formation of defects at the polymer–particle interface, which are typically caused by a weak particle-polymer adhesion, induced by the difference in properties between both phases [102]. These interface defects between the continuous and dispersed phases can impact membrane properties such as the membrane separation performance.

The most common factors responsible for interfacial defects can be divided into three main categories: (i) Interfacial voids or sieves-in-a-cage, (ii) Rigidified polymer layer around the inorganic fillers, and (iii) Particle pore blockage [100,105,106].

A low linkage between the continuous phase and the dispersed phase could lead to the formation of non-selective voids in the interfacial region (Figure 11, case i). Other factors responsible for interfacial voids formation are the modification of the polymer packing in the vicinity of the dispersed phase, the repulsive force between the two phases, the different thermal expansion coefficients and the elongation stress during fiber spinning [100]. In addition, interfacial voids or sieves-in-a-cage are attributed to the de-wetting of the polymeric chains on the external surface of the particles [101]. Moore and Koros (2005) observed that solvent evaporation, thermal effects and the resulting stresses at the polymer-disperse phase interface cause defects such as interface void formation, due to the partial or apparent clogging of the dispersed phase [106]. The formation of these defects allows the gases to pass and, hence, deteriorates the apparent selectivity and increases the permeability of MMMs.

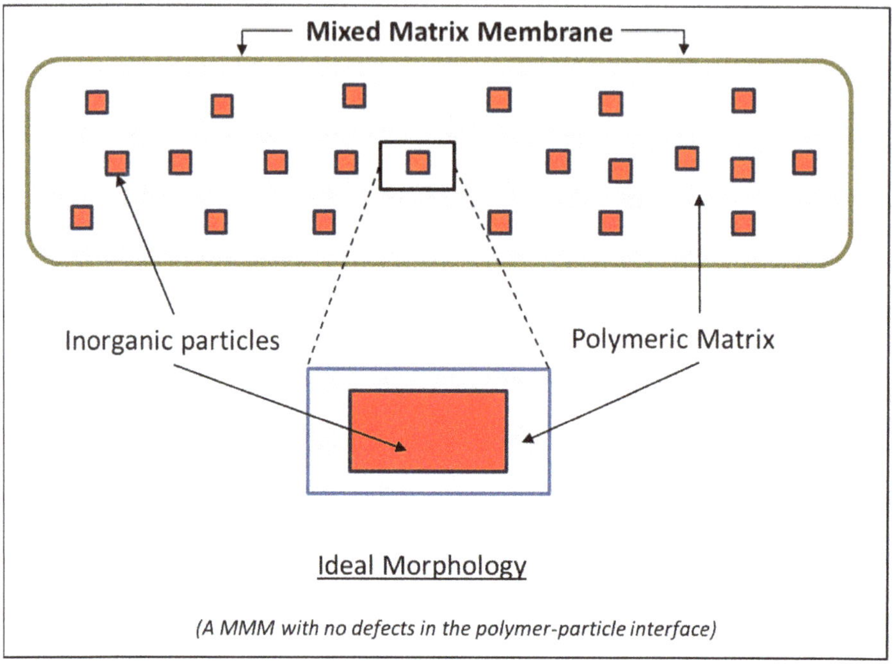

Figure 9. Schematic diagram of an ideal MMM. This figure was adapted from Aroon and co-workers, 2010.

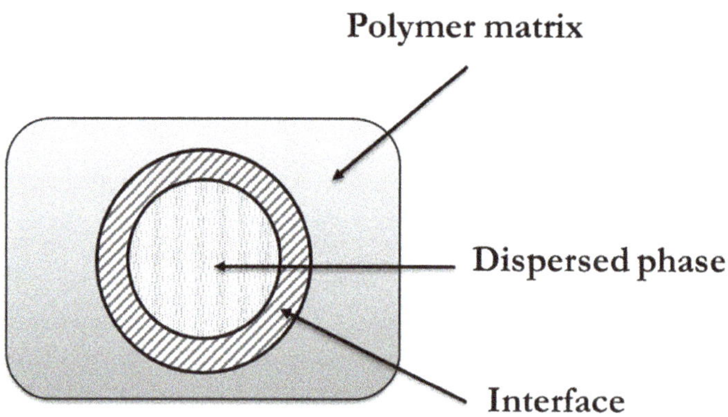

Figure 10. Schematic representation of polymer matrix, dispersed phase, and their interphase.

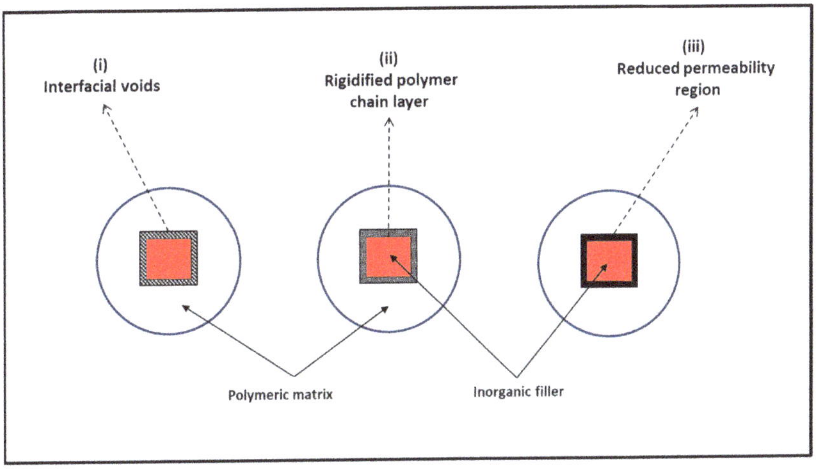

Figure 11. Schematic diagram of an interface void (i), rigidified polymer (ii) and partial blockage (iii) in MMMs. (Adapted from Aroon and co-workers, 2010).

Rigidified polymer layer around the inorganic fillers occurs when the polymer matrix chains, in direct contact with the filler surface, are rigidified as compared with the bulk polymer chains, which reduces the free volume and is related to a uniform tension around the particles [102,105]. Moore and Koros (2005) hypothesized that polymer rigidification (Figure 11, case ii) enhanced the diffusive selectivity and decreased membrane permeability [106].

Particle pore blockage occurs when the surface pores of the filler are partially blocked by the rigidified polymer chains (Figure 11, case iii). This clogging is usually generated by the presence of sorbent, solvent traces, a contaminant or a minor component in the feed gas, before, during and after the manufacture of the MMMs [105,106,108,110]. However, there is no accurate methodology to differentiate the influence of these factors. Based on previous research, if the pores are completely blocked, the gas cannot pass through the particle fillers, and no enhancement in selectivity over the neat polymer is reached as in the case of MMMs filled with nonporous particles.

The formation of a rigidified polymer layer around the inorganic fillers and particle pore blockage are caused by sorption of a strongly retained molecule. In the first case, the strongly retained molecule completely prevents the penetrants of interest from entering the dispersed phase, while in the second case, the penetrants of interest enter or pass through the dispersed phase more slowly than usual [105,106].

In summary, poor adhesion, mobility of polymer matrix chains and pore clogging by the matrix are just some of the phenomena observed when incorporating a dispersed phase into a continuous phase during the fabrication of MMMs.

Methods for Manufacturing Defect-Free Membranes

Poor adhesion and repulsive forces between the continuous and disperse phases, incompatibility between polymer and filler, solvent evaporation during membrane formation, polymer packing disruption in the vicinity of the inorganic phase, and different thermal expansion coefficients for polymer and filler can induce multiple interfacial defects and non-ideal morphologies in MMMs [102]. In order to avoid these interfacial defects and manufacture defect-free MMMs, the following methodological strategies have been applied:

An important factor during the manufacture of an ideal MMM with optimal performance is the homogeneous distribution (or dispersion) of the filler within the continuous phase in order to guarantee an effective filler/polymer contact [101]. In fact, a poor filler distribution can affect membrane performance by agglomeration, which leads to the formation of non-selective interfacial voids [99]. Unfortunately, high filler loadings can sometimes result in particle aggregation, which can form voids within the particle aggregates that cannot be reached by polymer chain segments and act as channels facilitating gas molecules transport, thus reducing the selectivity of the MMMs. Similarly, high filler loadings can cause sedimentation, which also contributes to the poor dispersion of the filler into the continuous phase [101]. This filler agglomeration entails the creation of pinholes that cannot be reached by polymer segments, resulting in non-selective defects in MMMs [105].

In this context, the so called "priming" method created by Mahajan and Koros (2002) is the most common strategy to avoid filler agglomeration [111]. This technique can reduce the stress at the filler/polymer interface, thus resulting in an improved interaction between the polymer primed filler and the bulk polymer, concomitantly with a reduced agglomeration of the filler [101,102]. This prime method consists in dispersing the particles in a suitable solvent, subjecting them to sonication followed by coating the surface of the filler in suspension. This coating is carried out by adding a small percentage of homogeneous polymer solution prior to the dispersion in the bulk polymer solution [110]. On the other hand, the preparation of polymer diluted solutions to increase the viscosity and decrease membrane thickness have been proposed to avoid agglomeration since this methodology can reduce particle sedimentation. Alternatively, the membrane can be cast "quickly" so that the fillers do not have time to precipitate.

Finally, another approach to achieve flexibility during membrane formation is to mimic the use of a low Tg polymer by forming the membrane close to the Tg of the polymer matrix used as a precursor of the MMMs. An obvious limitation of this strategy is the common tendency to use suitable casting solvents that boil at temperatures below the Tg of a typical rigid polymer such as Matrimid® [112].

5.1.2. Polymer Materials

The optimum selection of materials for both the continuous phase and the dispersed phase is a key factor during the development of MMMs since the properties of the precursor materials can affect the morphology and separation performance of membranes [105]. Despite the selection of optimum fillers being the major concern in the early manufacture of MMMs, the selection of the polymer used as the matrix greatly impacts the gas separation performance of MMMs [105].

In the field of gas separation using membranes, rubbery and glassy polymers have been traditionally used. Rubbery polymers contain flexible polymer chain structures and have the ability to stretch the chains apart, the chains returning to their original position when tension is released. Rubbery polymers also exhibit a high permeability and a low selectivity for the separation of common gas pairs, as a result of the different condensability of the gas components [30]. On the contrary, glassy polymers possess rigid chain structures with restricted segmental motion. This rigid chain structure offers desirable separation properties such as high selectivity combined with medium/low permeability [26]. The high selectivity of glassy polymers can be attributed to their lower free volume, the narrower distribution of the free volume and the lower flexibility of the polymer chains compared to their rubbery counterparts.

Due to the high degree of mobility, rubbery polymers ensure good adhesion between the polymeric matrix and the fillers, which can avoid interfacial voids and facilitate the manufacture of defect-free MMMs. However, a high mobility also entails a high permeability, which suggests that gas transport is dominated mainly by the polymer matrix and only a small portion is attributed to the filler. On the other hand, although glassy polymers exhibit superior properties to rubbery polymers, their rigid chain structure typically results in a poor adhesion of the pair polymer-filler, thus generating voids at the interface [101]. Therefore, the gas transport properties of the materials and adhesion between the phases should be carefully considered when selecting the polymer matrix [102]. In this context, novel polymers capable of separating gas mixtures by solubility selectivity are needed.

Material selection to manufacture MMMs is a difficult task, especially for glassy polymers. However, a considerable number of glassy polymers are being employed as continuous phase in MMMs such as cellulose acetate (CA), polyimide (PI), polysulfone (PSU), polyamide (PA), polypropylene (PP), polyethersulfone (PES), poly-vinylidene fluoride (PVDF) and perfluorinated materials, etc. [95,97,113–116]. Polymers such as PMP (4-methyl-2-pentyne), PTBA (*tert*-butylacetylene) and PTMSP (1-trimethylsilyl-1-propyne), namely "reverse-selective polymers", have also been used as continuous phase due to their high fractional free volume. In the latter membranes, the gas transport mechanism shifts from being controlled by diffusivity to being controlled by solubility (contrarily to the observations in traditional low-free volume glassy polymers), and therefore, transport properties are favored for more condensable species (e.g., CO_2) than for smaller molecules [117].

In recent years, the most common materials developed for the manufacture of MMMs are divided into three groups: (i) Advanced high permeability polymers (PIM, Polyimides and TR polymers), (ii) Polymers with moderate permeability and high selectivity and (iii) Ionic liquid/poly ionic liquids with high permeability and high selectivity [118]. For instance, a limited number of researchers have studied the transport properties using PIM-1 as a continuous phase for the separation of CO_2/CH_4 with a notable increase in permeabilities compared to the matrix. These studies also demonstrated that the introduction of a filler (ZIF-8) to this polymeric matrix mediates an increase in free volume, as a result of the combination of the contributing cavities and looser polymeric chains at the boundary between the filler and the PIM-1 matrix [98,119,120]. These membranes represent good candidates for CO_2 removal from biogas, although they suffer from severe physical ageing. On the other hand, the introduction of TR materials (e.g., hydroxypolyimide, HPI, and hydroxypolyamide, HPA) as a continuous phase has been proposed as a promising alternative since TR polymers show superior gas separation properties and can help to reduce non-selective voids during the manufacture of MMMs [97,121–123].

The permeation properties of MMMs are mainly determined by the shape and size of the filler, its pore size, pore size distribution, sedimentation and agglomeration properties and the gas separation operational conditions (gas composition, pressure, and temperature). In addition, the permeability of both the continuous and disperse phase should be comparable since a continuous phase with a high permeability reduces the contribution of the filler to gas separation [100].

5.1.3. Advanced Functional Fillers

The major challenges encountered during the manufacture of MMMs are the selection of adequate fillers that provide a good interaction with the polymer for the enhancement of gas separation properties. Indeed, the addition of suitable fillers in the polymer matrix results in a significant increase in the overall separation efficiency and therefore in a superior gas selectivity performance by MMMs [35].

There is a great variety of fillers that have been used in the development of MMMs as disperse phase. In recent years, the synthesis of novel organic/inorganic membrane materials has yielded in emerging materials used as high-performance fillers in MMMs for gas separation. Here, the most studied fillers (with a good compatibility with polymers) to date are briefly reviewed, particularly Zeolites, Metal Organic frameworks (MOFs) [124], Covalent Organic Frameworks (COFs) [125], Porous Aromatic Framework (PAFs) [126] and Porous Polymer Networks (PPNs) [127], recently named Porous Organic Polymers (POPs) [128] (Figure 12). Due to the fact that only few of the fillers used in the field of mixed matrix membranes have been mentioned, this article will only focus on representative work for the separation of CO_2 and H_2 from biogas and biohydrogen.

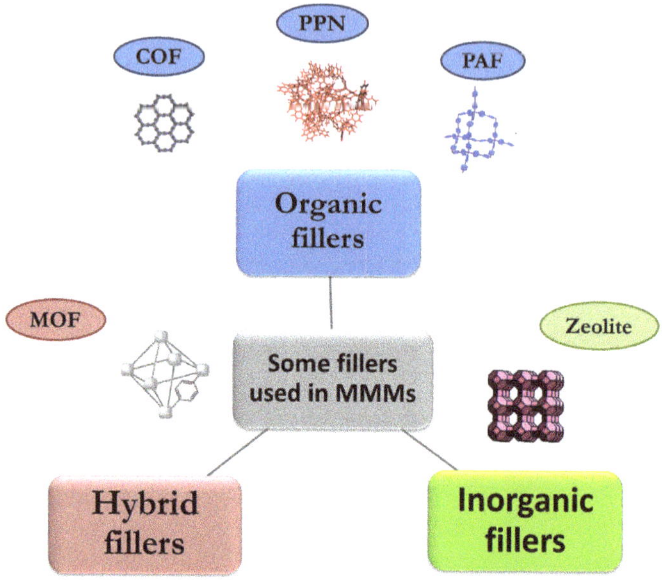

Figure 12. Some types of fillers used in MMM preparation (Adapted from Chakrabarty and co-workers, 2022) [129].

- Zeolites

Based on structural features, zeolites are an inorganic material frequently used as disperse phase for the manufacture of MMMs. Zeolites are hydrated aluminosilicate materials with opened three-dimensional framework structures that possess regular intracrystalline cavities and channels of molecular dimension (3–12 Å). Its structure is formed by SiO_4 and AlO_4 tetrahedral, by sharing an oxygen ion. Due to the presence of the tetrahedron AlO_4, the chemical characteristics of the frameworks are determined, which tend to have negative charge compensated by alkali or earth alkali cations, located in the micropores [130]. Moreover, zeolites are materials with shape/size-selective nanopores [131,132]. The pore sizes of zeolites range between 0.3 and 1 nm, with pore volumes of about 0.10–0.35 cc/g [133]. There are 176 types of zeolite structures approved by the IZA Structures Commission (IZA-SC) in February 2007 and assigned with a 3-letter code [134], of which, according to Bastani and co-workers, the most common are: 4A (3.8 Å), 5A (4.3 Å), 13X (7.4 Å), NaY (7.4 Å), ZSM-5 (5.1 × 5.5 Å and 5.3 × 5.6 Å), SSZ-13 (3.8Å), etc. [135].

Zeolites have received increasing attention due to the fact that they have a wide range of structures with different chemical compositions and physicochemical properties. Zeolites are widely used as catalysts, adsorbents and ion exchange media. The transport of gas molecules starts by molecular adsorption into the pores, diffusion onto the zeolite surface and desorption into the permeate. The gas molecules that have the strongest attractive force with the zeolite pores are those with the highest dipole moment, which is why CO_2 is adsorbed most strongly on zeolites, followed by H_2, CH_4 and N_2 [135]. The success of zeolite-based MMMs is attributed to the type of zeolites used and their adsorption capacity [133]. For example, zeolite 5A has a CO_2 adsorption capacity of 222 mg CO_2/g adsorbent at 0.1 MPa. Moreover, due to their good selectivity and adsorbent selection parameters, zeolite 5A turns out to be a better adsorbent for removing CO_2 and N_2O from air and for separating CO_2/CH_4 gas mixture compared with MOFs [136]. Likewise, NaX zeolites have an adsorption capacity of 263 mg CO_2/g adsorbent, which renders it an excellent candidate for separating CO_2 from CH_4 [137]. Similarly, zeolite 13X is another kind of zeolite with a great CO_2 adsorption capacity of 324 mg CO_2/g adsorbent, making it an excellent candidate for the purification of methane from natural gas [138].

One of the most relevant properties of this material is sorption and diffusion due to the different sizes of its channels and cavities, which determines the free space or void volume of the MMMs [132]. Zeolites possess interesting thermal and chemical stability, a well-defined microstructure and high mechanical strength, which makes them suitable candidates to be used as a dispersed phase in the manufacture of MMMs [135,139]. Interestingly, the low packing density of zeolites makes them an unsuitable material for gas separation, however their use as a dispersed phase in the fabrication of MMMs provides an opportunity to overcome this problem [101]. In addition, zeolites exhibit a permeability and selectivity superior to polymeric materials due to their unique molecular sieving characteristics.

Zeolites have traditionally received attention as potential fillers due to their thermal stability and promising separation and transport properties. Thus, the specific sorption properties and shape selectivity of zeolites, when applied to polymers with easy processability, provide superior gas separation properties to MMMs [132]. Several investigations have shown that the transport properties of MMMs are affected by the type of zeolite used. For instance, MMMs prepared with zeolite 4A support an effective O_2/N_2 separation due to their adequate pore size (3.8–4.0 Å), with selectivities up to 37. Membranes with zeolite 5A as filler exhibit much higher H_2/N_2 and O_2/N_2 selectivity than membranes with zeolite 4A as filler. In this context, zeolites are still of interest for membrane investigations despite providing low permeabilities for O_2 (0.8 Barrer) [111,140]. Ahmad and co-workers (2021) investigated the CO_2/CH_4 separation properties behavior of MMMs fabricated with SSZ-16 zeolite at different loading ratios as filler and 6FDA-DAM:DABA as polymeric matrix. As a result, MMMs loaded at 5 wt.% SSZ-16 supported up to two-fold higher CO_2 permeability with respect to pristine membranes, while maintaining the CO_2/CH_4 selectivity. In addition, these authors found that a 5 wt.% loading provides an excellent filler dispersion [141]. Zhang and co-workers (2008) prepared MMMs based on Matrimid and ZSM-5 zeolite, increasing the H_2/N_2 separation from 79 for Matrimid and 143 at 10% load, while the ideal H_2/CH_4 separation factor increased from 83 to 169 at 20% load, further confirming the excellent interactions between the particles and the polymer [142]. Ebadi Amooghin and co-workers (2016) modified zeolite-Y by introducing silver cations (via ion-exchange method) to form Ag+ zeolite and use it as filler on Matrimid® to form novel Matrimid®/AgY MMMs. In this particular study, CO_2 permeability increased from 8.34 Barrer for the pure membrane up to 18.62 Barrer for Matrimid/AgY (15 wt.%) without affecting CO_2/CH_4 selectivity, which increased from 36 to 60 for pure membrane and MMMs, respectively [143]. Finally, Montes Luna and co-workers modified the natural zeolite Clinoptilolite (CLINO) with $CaCl_2$ in an aqueous solution to replace Na+ ions with Ca^{2+} ions, thus enhancing gas separation properties for $CH_4/N_2/CO_2$ gas mixtures [144].

Despite the promising results obtained in the laboratory, MMMs with zeolites as the dispersed phase have not been commercially exploited due to the poor adhesion at the zeolite–polymer interface (especially glassy polymers), resulting in a "sieve-in-a-cage" morphology. This defect is responsible for the non-selective penetration of gas molecules, the reduction in selectivity and poor mechanical properties, especially in the formation of thin films. In addition, high zeolite loadings often result in non-uniform dispersions in MMMs [55].

- Metal Organic Frameworks

Metal Organic Frameworks, MOFs, are hybrid materials prepared by combining organic ligands with metal ions or metal-oxide clusters. Ligands play a key role in defining the final framework of MOFs, while metal ions also influence the structure of MOFs due to their tunable geometries [145]. MOFs are highly porous chemically mutable materials, with unique properties, different pore sizes and shapes, and multiple functional sites and high specific surface areas that allow creating a wide variety of crystals [93,118]. Compared to zeolites, the tunable structure of MOFs results in well-dispersed fillers, which allows high affinity organic linkers in MOFs and polymer chains, thus reducing non-selective defects at the polymer–filler interface. The partially organic nature of MOFs supports a better polymer-filler interaction, which represents a structural advantage compared to other porous materials [55].

In order to optimize gas diffusion and selectivity, new strategies for the formation of high-performance MMMs using MOFs as dispersed phase have been assessed. A wide variety of MOFs subfamilies with ultrasmall aperture sizes have been chosen as potential fillers. The most typically studied MOFs are Zeolitic Imidazolate Frameworks (ZIFs), copper-based MOFs (Cu-MOFs), Materials Institute Lavoisier (MIL) series, MOF-74 series, and University of Oslo-66 (UiO-66) series [146]. ZIFs possess a similar topology to zeolites with tunable pore structures and with high thermal and chemical stabilities [35]. In this context, ZIF-8, HKUST-1, MIL-53, MIL-101, MOF-74, and UiO-66 have been specifically tested. For instance, ZIFs-8 are a new class of porous crystals (3.4 Å pore aperture and 11.6 Å cages) [147] composed of tetrahedral metal ions (typically zinc or cobalt) forming extended three-dimensional structures bridged by imidazolate (Im) [148].

Khdhayyer and co-workers studied the gas transport properties of MMMs based on PIM-1 as polymeric matrix and three isoreticular MOFs (UiO-66, UiO-66-NH_2 and UiO-66-$(COOH)_2$) as fillers, confirming the good prospects of these MMMs for CO_2 removal from biogas [98]. Ahmad and co-workers investigated the gas separation properties of MMMs using three types of Zr-based MOFs (UiO-66 and its functionalized derivatives, UiO-66-NH_2 and UiO-66-NH-$COCH_3$) as fillers on 6FDA-DAM as a polymeric matrix. The addition of these particles improved both CO_2 permeability and CO_2/CH_4 selectivity. For instance, permeabilities of the polymer 6FDA-DAM and its 14–16 wt.% Zr MOFs MMMs, tested with binary (30:70 vol.%; CO_2:CH_4) feed mixture, were 231, 541, 359 y 291 Barrer, and for tertiary (30:5:65 vol.%; CO_2:H_2S:CH_4) feed mixture, permeabilities of 167, 385, 243 and 193 Barrer were recorded for neat membranes, UiO-66-based MMMs, UiO-66-NH_2-based MMMs and UiO-66- NH-$COCH_3$-based MMMs, respectively [149].

Recently, Kertik and co-workers (2017) created in-situ molecular sieves with controlled heat treatment up to 350 °C for 24 h for Matrimid®/ZIF-8, obtaining excellent selectivity for CO_2/CH_4 gas mixtures due to the excellent interfacial filler-polymer adhesion [150]. Matrimid®/ZIF-8 (40 wt.%) thermally treated MMMs exhibited a CO_2 permeability of ~1.9 Barrer and a CO_2/CH_4 selectivity of ~134 at 40 bar, 35 °C with gas mixtures containing 50 vol.% CO_2/50 vol.% CH_4 [55,150]. ZIF-8 as inorganic filler was added into 6FDA-durene diamine, obtaining a notable increase in CO_2 permeability from 1468 Barrer to 2185 Barrer for pure membrane and 30 wt.% loaded ZIF-8 MMM, respectively, and 17.1 of selectivity for CO_2/CH_4 gas pair [151].

Finally, it should be stressed that the preparation of membranes with well-dispersed fillers, good filler-polymer interfacial adhesion and a defect-free membrane surface represent nowadays the major challenges of MOF-based MMMs [152].

- Covalent Organic Frameworks

Covalent Organic Frameworks, COFs, developed by Côté and co-workers in 2005 [125], have been recently proposed as a type of porous organic material used as a filler for the fabrication of MMMs. COFs are crystalline porous materials synthesized by the covalent combination of rigid and stable organic monomers (phenyl diboronic acid and hexahydroxytriphenylene), which offer superior chemical and thermal stability compared with MOFs [153,154]. COF materials have well-defined and predictable 2D or 3D crystalline structures as a result of the formation of strong covalent bonds [155]. COFs are classified into three groups, based on their uptake capacities, pore size and structural dimensions: (i) 2D structures featuring small 1D pores (9 Å for COF-1 and -6); (ii) 2D structures with large 1D pores (27, 16 and 32 Å for COF-5, -8 and -10, respectively) and (iii) 3D structures containing medium-sized 3D pores (12 Å for COF-102 and -103) [154]. Three-dimensional COFs, COF-1 and COF-5 presented a high hydrothermal stability, and regular and stable porosity, with surface areas ranging from 700 and 1600 m^2 g^{-1} [125], while two-dimensional COFs, COF-6, -8, and -10 showed structures with pore sizes ranging from 6.4 to 34.1 Å and exhibited high thermal stability, low densities and high porosity with specific surface areas of 980, 1400, and 2100 m^2 g^{-1} for COF-6, -8, and -10, respectively [156]. The highest reported surface area for a COF is 4210 m^2 g (BET) in COF-103 [157]. Due to their properties such as low crystal density, ultrahigh accessibility and rich electronic lattice, COFs can be efficiently used for gas storage and selective adsorption. According to theoretical studies performed through grand canonical Monte-Carlo simulated calculations, the H_2 storage capacity with COF has been predicted, showing about 10% excess H_2 storage with COF-105 and 108 at 77 K, being the best-known organic materials for H_2 storage [154]. Han and co-workers (2008) conducted a study focused on the H_2 uptake capacities with experimental H_2 loading data for COF-5, achieving a total evacuation of the pores at 3.4 wt.% at 50 bar and 77 K. In the same study, a H_2 storage capacity of 8.9 wt.% at 77 K for COF-108 was observed, while the highest volumetric yield was shown for COF-102 (40.4 g L^{-1} of H_2 at 77 K). [158].

Due to their variable structures, easily modifiable scaffold and high affinity to the polymeric matrix, good thermal stability, appropriate solvent compatibility, COFs have demonstrated to be excellent candidates in the field of gas separation [118,159–161]. Despite the advantages offered by COFs, a limited research has been conducted with COF-based MMMs. For instance, Wu and co-workers (2017) incorporated COFs as particles into PIM-1 as a polymeric matrix, obtaining a remarkable improvement in CO_2 permeability and CO_2/CH_4 and CO_2/N_2 selectivity compared to pure PIM-1 [153]. Likewise, Biswal and co-workers (2016) manufactured MMMs incorporating TpBD into polybenzimidazole (PBI), resulting in permeabilities above 18 Barrer for CO_2 and selectivities of ~48 and 23 for CO_2/CH_4 and CO_2/N_2, respectively [162]. COF (imine-based COF with a two-dimensional network) was also incorporated into poly(vinylamine) (PVAm) to enhance membrane performance for CO_2/H_2 separation. As a result, a MMM at 10 wt.% COF load showed a CO_2/H_2 selectivity of 15 and a CO_2 permeance of 396 GPU at 0.15 MPa, which further suggested that COFs possess good compatibility with polymers, thus enabling the fabrication of MMMs with a superior performance [163].

- Porous Aromatic Framework

Porous Aromatic Framework, PAFs, are a subfamily of Covalent Organic Frameworks (COFs) that, unlike traditional COFs and MOFs, are stronger and more stable and exhibit a good physical-chemical stability [118]. PAFs are synthesized via irreversible cross-coupling reactions by aromatic rigid linkers [154] and constructed from carbon−carbon-bond-linked aromatic-based building units [164]. Moreover, compared to conventional porous materials (such as zeolites and MOFs), PAFs exhibit specificity in their chemistry and functionalities due to their strong carbon–carbon bonding, which makes them stable under severe chemical treatment [164]. Due to their covalent backbone, PAFs are chemically robust materials, although with a high irregular internal structure that reduces their porosity and associated crystallinity [93,154]. Indeed, these fillers exhibit a high porosity, narrow pore-size distributions for amorphous solids and Brunauer−Emmett−Teller (BET) surface areas as high as 5200 m^2 g^{-1}, which typically results in high affinities for adsorption of CO_2 and other gases [93,165]. PAF surface area and CO_2 capture may vary depending on the batch, tetrahedral core, phenyl chain length, functionalization and also the arrangement of the nanoparticles in the fillers [166]. PAFs, which are porous materials, have voids that serve to accommodate gas molecules, making them excellent absorbents. These PAFs are prepared with ultrahigh surface areas to enhance their H_2 storage capacity. For instance, the first reported PAF, PAF-1 with ultrahigh specific surface area (BET: 5600 m^2 g^{-1}, Langmuir: 7100 m^2 g^{-1}) [167], exhibited a hydrogen adsorption capacity of 7.0 wt.% at 48 bar and 77 K [164]. On the other hand, due to their high surface area and stability, the capacity of PAFs as CH_4 sorbents has also been investigated. For instance, the CH_4 uptake capacity of PAF-1 is 18 cm^3 g^{-1} at 14 KJ mol^{-1} heat adsorption, while PAF-3 (BET surface area of 2932 m^2 g^{-1}) showed the highest uptake capacity at 27 cm^3 g^{-1} and 15 KJ mol^{-1} heat adsorption and PAF-4 (BET surface area of 2246 m^2 g^{-1}) presented a similar capacity to PAF-1 but at 23.2 KJ mol^{-1} heat adsorption. With their well-defined networks, PAFs also offer a high potential for CO_2 capture at low pressure. For example, CO_2 sorption capacities of 46 $cm^3\ g^{-1}$ (9.1 wt.%) for PAF-1, 78 $cm^3\ g^{-1}$ (15.3 wt.%) for PAF-3 and 54 $cm^3\ g^{-1}$ (10.7 wt.%) for PAF-4 were recorded at 273 K and 1 atm. [168].

However, despite their exceptional surface areas and good thermal and hydrothermal stability, PAFs exhibit weak interactions with gases, which limits their gas storage capacity and operating temperature [169]. However, Hou and co-workers (2022) added PAF-1 into PIM-1, which improved gas separation performance following the conventional method to manufacture MMMs and combining the filler with a post UV irradiation treatment. As a result, MMMs permeability showed a high permeability (i.e., $P(H_2)$ = 4800 Barrer) as well as a remarkable improvement in the separation factor (i.e., improvement for H_2/CH_4 selectivity, from 5.4 to 90), surpassing the 2008 upper bounds for H_2/CO_2 and CO_2/CH_4 and 2015 upper bounds for H_2/N_2 and H_2/CH_4 [170]. Ben and co-workers (2009) synthesized a porous aromatic framework PAF-1 via phenyl-phenyl coupling with a Langmuir surface area of 7100 $m^2\ g^{-1}$ [171]. Likewise, Lau and co-workers (2014) demonstrated that the addition of PAF-1 as disperse phase into PTSMP and poly(methylpentyne) (PMP) can mitigate the permeability loss associated with physical ageing of these super glazed polymers [165].

- Porous Polymer Networks

Recent investigations have attributed new merits for gas separation to this family of adsorbents as a result of their high thermal and chemical stability, easy processing and low cost [172,173]. PPNs are synthesized by the homocoupling of tetrahedral monomers via the oxidative Eglinton coupling or Yamamoto-type Ullmann coupling reaction, exhibit high thermal and chemical stability and are insoluble in conventional solvents. PPNs possess Langmuir surface areas as high as 5323 $m^2\ g^{-1}$. Between the first reported PPNs (ie. PPN-1, 2 and 3), PPN-1 showed the highest gas affinity and exhibited more micropores of less than 1 nm diameter than PPN-2 and PPN-3. Despite the lowest surface area (827 $m^2\ g^{-1}$), PPN-1 showed the best CO_2/CH_4 selectivity. On the other hand, PPN-3 exhibited the highest H_2 uptake capacity (4.28 wt.%, 77 K) among these three (3.30 and 3.76 wt. % H_2 uptake for PPN-1 and PPN2, respectively) [172].

A new generation of PPNs, namely Porous Organic Polymers, POPs, was recently developed by reacting rigid trifunctional aromatic monomers with ketones exhibiting electron-withdrawing groups, in superacidic media via acid-catalyzed condensation (Lewis or Brönsted) at low temperatures. PPNs and POPs are microporous materials with Brunauer–Emmett–Teller (BET) surface areas ranging from 580 to 790 $m^2\ g^{-1}$ and from 760 to 935 $m^2\ g^{-1}$, respectively, and have attractive properties such as: excellent CO_2 uptake capacity as solid adsorbents (up to 207 mg g^{-1} (105 cm^3 (STP) g^{-1}) at 0 °C and 1 bar), ability to regenerate by vacuum without heating and an exceptional chemical and thermal stability [114,127]. Their ease of synthesis and high conversion render PPNs as materials easy to scale-up. In addition, these materials present a selective adsorption of CO_2 (32.7) superior to N_2 (22.5) under postcombustion conditions, which are higher when compared to other high-performance microporous materials [114]. In this context, Aguilar-Lugo and co-workers (2019) added PPNs (at different loads) as filler into Matrimid®, resulting in an improvement in the permeability of up to 700% for the gases tested without significantly affecting selectivity (CO_2/N_2 and CO_2/CH_4 selectivities decreased by 4% and 12%, respectively). These authors also observed a good filler-polymer adhesion, which was supported by the increase in the Tg of the MMMs compared to the pure polymer matrix [114]. Likewise, Rico-Martínez and coworkers incorporated bipyridine moieties based on POPs into aromatic polyimides at different loads, which supported four- and seven-fold increases in CO_2 and CH_4 permeability, respectively [115].

6. Thermally Rearranged Polymers

As previously mentioned, new materials with superior gas separation performance, increased chemical/thermal resistance to aggressive feed conditions and high selectivity are needed. Significant advances have been generated in the chemistry of polymeric membranes for gas separation, mainly aimed at increasing the molecular stiffness and improving the free volume fraction (FVF) of membranes, leading to a high permeability without a significant decrease in selectivity [174,175]. In this context, glassy polymers such as polybenzoxazoles (PBO), polybenzothiazoles (PBT), polypyrrolones (PPL) or benzimidazoles (PBI), represent a class of rigid-rod ordered polymers with outstanding mechanical and thermal properties, and extreme rigidity [176]. However, these materials are unattractive in gas separation because their efficient packing entails few free volume elements accessible to gas penetration, which hinders their manufacture in the form of flexible and tough films. Moreover, the above-mentioned glassy polymers are soluble only in strong acids, and consequently not suitable candidates for membrane fabrication [177]. Therefore, the new strategies for the synthesis of rigid-rod polymers are mainly focused on enhancing solubility and processability.

In this regard, Park and co-workers (2007), based on the thermal conversion of imides containing hydroxides to benzoxazoles performed by Tullos and co-workers (1999) [92,177], demonstrated the occurrence of free-volume structures in dense glassy polymers that can be systematically modified by thermal rearrangement. This process enables an extraordinary gas separation performance and constitutes a novel method to prepare high-performance polymers for molecular-scale separations [92]. This successful research based on poly(1,3-benzoazole)s membranes was carried out by subjecting membranes to a thermal treatment in solid state of poly(o-hydroxyimide)s, containing *ortho* positioned functional groups (with respect to the amino group) [55,178]. This thermal rearrangement process involves a thermal cyclization step subjected to temperatures of 350–450 °C for a certain duration of time and under inert atmosphere or vacuum. The need for thermal processing to manufacture these materials is responsible of their name as 'thermally rearranged', or TR polymers. Depending on the functional group in the *ortho* position (-OH, -SH, or -NH_2) of the precursor, the structures resulting from the cyclization process are PBO, PBT, PPL or PBI [92,179,180]. Since polybenzoxazoles may be a source of possible cross-linking as a consequence of the high temperature used during their conversion, which would also explain their insolubility, this material cannot be processed. In this sense, TR-precursors during the manufacture of these membranes can be *ortho*-hydroxyl polyimides (HPI) and *ortho*-hydroxyl polyamides (HPA) (also called α-TR and β-TR polymers, respectively, HPIs being the most studied [178]. Figure 13 shows the solid-state mechanism of a poly(hydroxyimide) (PI) and a poly(hydroxyamide) (PA) to form a TR-polymer with the proposed PBOs structure.

In both cases a final polyheterocycle of the polybenzoxazole type is reached by a cyclization process, where the heat treatment is carried out at different temperatures, depending on the TR-precursor. Additionally, the solid-state conversion process involves decarboxylation when the precursor is a polyimide, while the thermal reorganization phenomenon takes place through the loss of water molecules, or cyclodehydration, when using a polyimide as a precursor. The final PBO materials possess a chemically stable structure, resistance to CO_2 plasticization (likely due to their cross-linked structure) and excellent permeability and selectivity values due to the formation of a desirable free volume element distribution during thermal conversion [175,181–183].

Although in both cases the final structure of the PBO is similar, membranes exhibit different characteristics, especially in terms of gas transport properties. The TR precursor HPI can efficiently separate condensable gases, while TR precursor HPA has an outstanding ability to separate light gases. These aromatic polyamides exhibit an appropriate balance of properties such as good mechanical and chemical stability, high thermal resistance and easier processability when their precursor monomers are adequately selected [85].

Park and co-workers (2007) demonstrated that polymers with a medium cavity size, with a narrow cavity size distribution and a shape reminiscent of bottlenecks connecting adjacent chambers, possess high selectivity and high permeability [92]. Thus, TR polymers provide an increase in FFV as a consequence of the generation of microcavities with controlled size bimodal distribution in the range of 0.3–0.4 nm (which is beneficial for selective transport of gas molecules such as CO_2) and 0.7–0.9 nm (which entails an enhanced gas diffusion) [178]. The above bimodal cavity size distribution is governed by the structure of the precursor and the protocol of thermal treatment used to produce the TR-PBO [184].

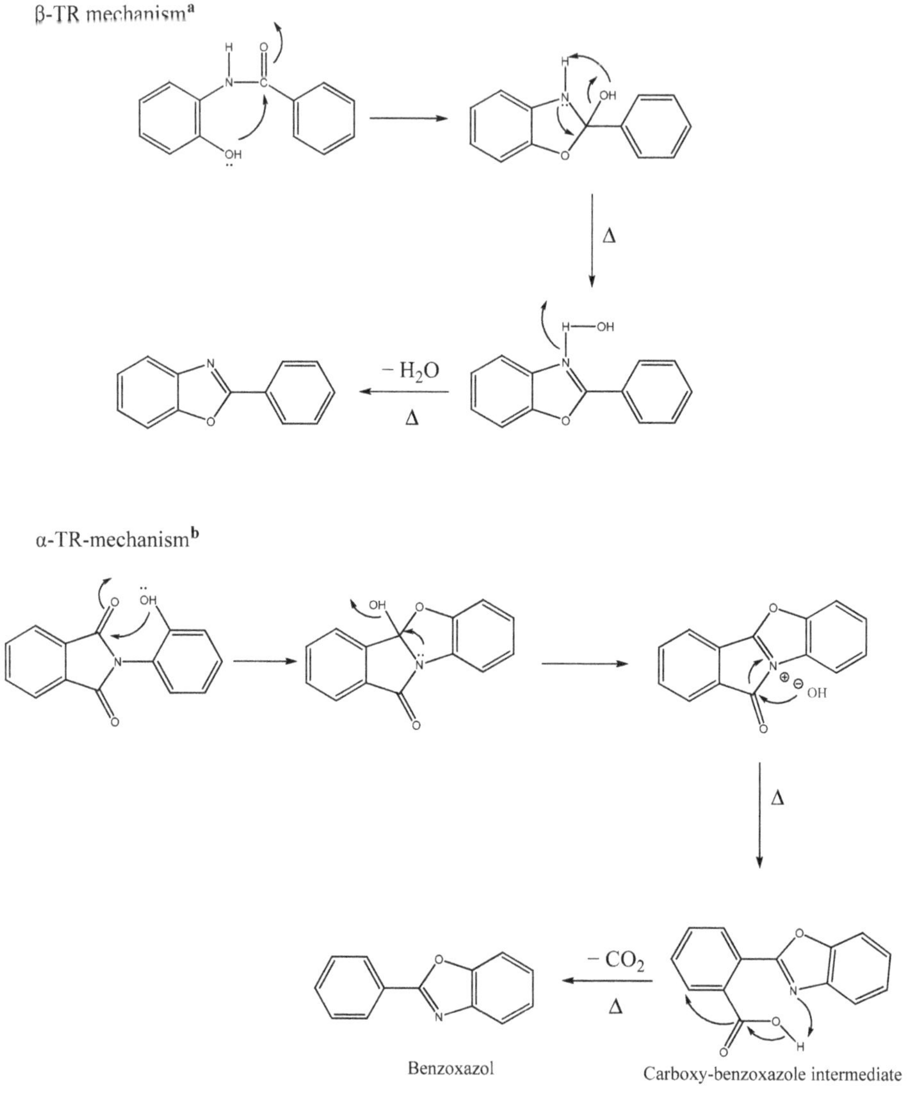

Figure 13. Mechanism of thermal conversion of a cyclodehydration of a hydroxypoliamide to polybenzoxazole (β–TR–PBO)[a] and thermal conversion of a hydroxypolyimide to polybenzoxazole (α–TR–PBO)[b].

Gas transport in TR-polymer membranes depends on the degree of thermal conversion, the nature of the free volume elements and their size distribution [185]. It is assumed that the newly created micropores mediating the transport of gases in TR polymers are responsible for the usual molecular screening in the separation of gases by glassy polymeric membranes. The narrowest part of these micropores plays a role of a molecular size caliber. Today, there are consistent empirical proofs confirming the exceptional selective molecular transport performance and high permselectivity in small molecules because of the free volume structure of these polymers.

7. Thermally Rearranged Mixed Matrix Membranes

Recent investigations in membrane gas separations are focused on taking advantage of the MMMs and the Thermally Rearranged polymers' properties, in order to improve the performance of gas transport properties, mainly for CO_2/CH_4, H_2/CO_2, CO_2/N_2 separation. MMMs manufactured with TR-able polymers are known as Thermally Rearranged Mixed Matrix Membranes (TR-MMMs). Membranes manufactured with thermally rearranged (TR) polymers result in unusually high selectivities and permeabilities, attributed to their unique hourglass configuration, while the addition of particles can add selective pathways for gas transport [97,185,186]. Despite this field of research being very recent, promising results in gas separations mixtures have been obtained. For instance, in 2017, Brunetti and co-workers manufactured the first TR-MMMs loaded with 0.5 wt.% of oxidized multi-wall carbon nanotubes (MWCNT) for CO_2 separation with an enhanced permselectivity and conducted an aging study. The addition of the nanotube entailed the increase of H_2 permeability followed by CO_2, N_2 and CH_4 compared to the neat TR (increasing from 171 a 201 Barrer for H_2, 105 to 126 for CO_2, 9.2 to 9.3 for N_2 and 4.4 to 4.9 for CH_4). Additionally, the influence of addition of nanotubes on aging resulted in a decrease in CO_2 permeability after 150 days of 13% [187]. Kim and co-workers (2019) fabricated TR-MMM for hydrogen separation using a TR-able Polyimide (HPI: HBA-DAM-6FDA) as polymeric matrix and a zeolitic imidazolate framework-8 (ZIF-8) as filler. As a result, MMMs loaded with 20% of ZIF-8 and thermally rearranged for 90 min Dwell time, exhibited excellent H_2 separation properties, with an increase from 365 to 1206 Barrers for H_2 permeability, before and after thermal treatment, respectively, and selectivity of 22.3 and 25.7 for H_2/N_2 and H_2/CH_4 gas pairs, respectively [121]. Smith and co-workers also carried out a pioneer study on TR-MMMs prepared by adding PAF-1 into 6FDA-HAB_5DAM_5 (DAM) TR-able polyimide in order to improve permeation properties. As a result, TR-MMMs showed an increase of 37-fold H_2 permeability and 55-fold for CO_2 gas permeability with similar gas selectivities [97]. Soto and co-workers (2020) developed a new family of TR-MMMs to enhance CO_2/CH_4 permselectivity using recent porous polymer networks (PPNs) as fillers on a polyamide, 6FCl-APAF, capable of producing benzoaxazoles, as a polymeric matrix. In this study, TR-MMMs showed a notable increase in gas permeability. For example, CO_2 permeability increased 34-fold for TR-MMM at 30% of filler compared to MMM at 30%, with a slight decrease in CO_2/CH_4 selectivity (from 27.75 pristine membrane to 24.02 for TR-MMM). Similarly, TR-MMMs with PPNs as a filler and polyimides (*ortho*-hydroxypolyimide, PIOH, or an *ortho*-acetylpolyimide, PIOAc) as polymer matrix have been recently carried out by Aguilar-Lugo and co-workers (2021), where membranes loaded at 30% of filler showed 1036 Barrer for CO_2 permeability with a CO_2/CH_4 selectivity of 28 for PIOAc-based TR-MMMs [122].

In general, TR-MMMs offer improvements in gas transport properties favored by the use of microporous materials with a high thermal stability. In addition, thermal treatment at high temperature contributes to eliminate the interfacial voids between the filler and the polymer matrix, leading to an increase in the gas selectivity of the membranes [121]. However, the excellent results in gas permeability can be accompanied by the loss of other desirable properties such as anti-aging permeability and pressure resistance, which requires further research [97].

8. Membrane Modules and System Design

8.1. Membrane Modules

The separation units in which the membrane surface is fitted are called membrane modules, which refer to the central part of a membrane device. The module should allow a separate conduction of the feed and permeate gas streams on both sides of the membrane. Figure 14 shows the schematic of the simplest design in which a single module is used.

A feed stream with a given composition and flow rate is introduced into the separation module (Feed), divided into two streams, one of which enters through the membrane (permeate) and the other (retentate) leaves the module in a smaller proportion [26,188]. The feed composition and flow rate within the module will change as a function of distance, as the membrane has the ability to transport one component more readily than another [188].

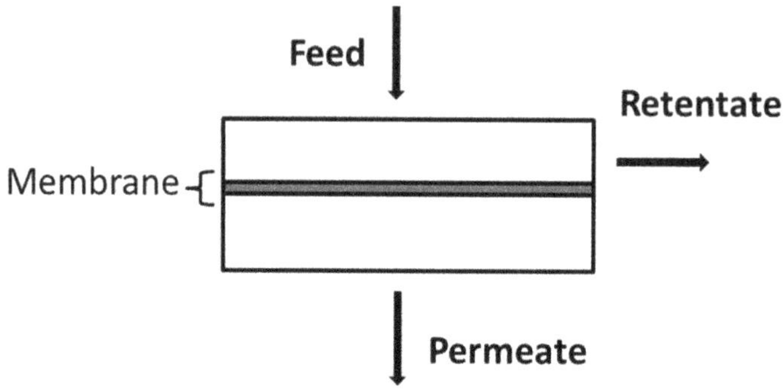

Figure 14. Module illustration of a membrane process.

The modules engineered to date are based on the membrane configuration, classified in two geometries: (1) flat sheet membranes, which include plate-and-frame and spirally wound modules and (2) tubular membranes, including tubular hollow fiber and modules based on fine capillaries or tubes housed such as a shell and tube heat exchanger [29]. The main difference between these types of membranes is based on their dimensions: tubular membranes exhibit diameters larger than 10 mm, diameters below 0.5 mm for hollow fibers and diameters between 10 mm and 0.5 mm for capillary membranes [189].

8.2. Plate-and-Frame Module

Plate-and-frame modules represent the pioneer types of membrane unit, whose design is based on the conventional filter-press [190]. Plate and frame modules are the most common setups, which are similar to the flat membranes used in the laboratory. They can be mounted in plate, bag or spiral-wound form [26]. These module membranes are separated by a feed spacer, with the separate layers stacked towards each other, like a sandwich. These spacers serve to seal the module and allow the flow of material through the drilled holes and alternate channels [29]. The membrane surfaces per module volume (packing density) range from 100–400 m^2/m^3. A stop disc is used in order to favor the flow over the surface membrane and reduce the formation of preferential channels [189]. Plate-and-frame modules present advantages such as: exchange ability of single membranes, low sensitivity to particulate blocking of the feed channels and usage of flat membranes without the usage of glue. In addition, they exhibit disadvantages such as: need of several sealings, high pressure drop and low packing density [191]. Currently, this kind of module is still used in ultrafiltration and pervaporation processes and represents the only plate-and-frame configuration used in solution-diffusion membranes [192]. Since plate-and-frame modules present smaller membrane surface area per unit volume, they are effective in pervaporation applications [192]. However, compared to hollow fiber and spiral-wound modules, plate-and-frame modules are less applied in gas separation [190]. For instance, oxygen enrichment from air, organic vapor recovery and even medical applications are among the commercial applications of these modules in gas separation [193].

8.3. Spirally Wound Modules

The spiral-wound format was the first to be commercialized [26] and was initially developed for reverse osmosis applications. Spiral-wound modules are typically applied in CO_2 removal from natural gas and vapor/gas separations [32]. Currently, this configuration is also used in ultrafiltration and gas permeation applications, which render it an important module in membrane applications [194]. This kind of module is used when countercurrent flow is not required to increase separation efficiency and when pressure drop must be taken into account [50].

This configuration consists of a plate-and-frame system that wraps around a central collection tube, similar to a sandwich roll. A spacer material is placed between the membranes to prevent contact of the feed and permeate, as well as to allow free space for the interaction of gas molecules with the membrane [29]. The interleaved sheets are spirally wound around a central permeate collection channel [26,188]. The feed stream flows along the center tube in axial direction, while the permeate flows in radial direction towards the center tube and is collected on the inside of the envelope. The packing density of this spirally wound module (300–1000 m^2/m^3) is greater than that of the plate-and-frame module. However, this parameter depends on the channel height, which in turn is determined by the permeate and feed-side spacer material. According to Caro and co-workers (2007), spiral-wound modules exhibit a good mass transfer due to feed spacers, are simple, and present a cost-effective fabrication and relatively high packing density/membrane area-to-volume ratio (up to 1000 m^2/m^3). However, spirally wound modules exhibit disadvantages such as difficultly to be cleaned and long permeate pathway [195]. Less than 20% of gas separation membranes nowadays are manufactured as spiral-wound modules. Currently, spiral-wound modules are industrially used in natural gas processing.

8.4. Tubular Modules

Tubular membrane modules are based on cylindrical membranes, which consist of thin layers of selective membrane deposited on the two membrane faces of a porous stainless steel, ceramic, or plastic tubular support with a diameter superior to 10 mm. Tubular membranes can be manufactured with inner diameters ranging from 5–25 mm, with 12.5 mm being the most common diameter. Although the number of tubes placed in the module is not limited, it can vary from 4 to 18 tubes [26,188]. The feed flows through the center of the membrane tubes and the permeate moves across the membrane from the inside to the outside, subsequently flowing into the larger tube [189]. Ceramic membranes are mainly assembled in such tubular module configurations. The packing density is rather low, typically <300 m^2/m^3 [29,188]. The main advantages of this module are: membrane fouling can be easily controlled, which reduces operating costs, as well as concentration polarization effects [26]. Thus, given their resistance to fouling (due to the effect of good fluid hydrodynamics), the use of tubular modules is often restricted to ultrafiltration applications [196].

8.5. Capillary Module

The capillary module consists of a large number of capillaries assembled together in a module with an inner diameter of 0.2–3 mm arranged in parallel as a bundle in a shell tube [26]. They are self-supporting and the free ends of the capillaries are encapsulated with agents such as epoxy resins, silicone rubber or polyurethanes. There are two types of module arrangements: (1) membranes where the feed passes through the bore of the capillary and the permeate exits through the side of the membrane and (2) membranes where the feed enters the module on the shell side and the permeate exits through the bores of the membrane [188,189]. The selection of the module arrangement will depend on the application, and parameters such as operating pressure, pressure drop, type of membrane material available, etc. Packing densities range from 600 to 1200 m^2/m^3 [188].

8.6. Hollow Fiber

The hollow fiber module is similarly to the capillary module. Spiral-wound and hollow fiber modules are commercially available for gas separation. Hollow fibers are based on a porous, non-selective support layer (~200 μm) and an active layer (actual membrane) (<40 nm). As a result of the small thickness of the active layer, this must be supported by a thicker layer in order to obtain mechanical strength, to withstand the pressure difference between the feed and permeate side [197].

The hollow fiber membrane module consists of a large number of hollow fibers assembled together into a bundle, which is encapsulated at the ends to prevent leakage between the feed and permeate chambers [29]. The fibers, arranged in parallel to pass through the tubular sheets or one or both ends of the device, range between 1.0 and 1.5 mm outside diameter and the bore of the fibers has a typical diameter of 0.5–1 mm. Two types of module arrangement can be distinguished: (1) membranes where the feed enters through the bore of the fiber ("inside-out") and the permeate is collected outside the membrane in the housing or (2) membranes where the feed enters on the outside ("outside-in") and the permeate passes into the membrane bore. Hollow fiber modules exhibit the highest packing density among all module configurations, reaching values of up to 30,000 m^2/m^3 [188]. The high membrane area-to-volume ratio, together with their high packing density and cheap fabrication cost, are the main advantage of the hollow fiber module. The low-pressure resistance and mostly laminar flows, which increases mass transfer limitations, rank among the main disadvantages of this membrane module [190].

8.7. Module Selection Criteria

Gas separation systems are commercially available as hollow fiber or spirally wound modules and, in some applications, also in plate-and-frame modules. The selection of the appropriate membrane module is typically determined by cost considerations. Hollow fiber modules are more economical per square meter, however the fabrication of very thin selective layers in the form of hollow fibers is a difficult process. As a result, the permeance in this type of membrane tends to be lower than in flat sheet membranes based on the same polymer. Hollow fiber modules require more membrane surface area to achieve the same separation factor. They also require more feed pretreatment than spirally wound modules for the removal of particles, oil residues and other fouling components [32,59]. According to Ismail and co-workers (2015), the manufacturing cost ($\$/m^2$) for hollow fiber ranges from 2 to 10 \$ per m^2, from 5 to 50 \$ per m^2 for capillary fibers, from 5 to 50 \$ per m^2 for spirally wound, and from 50 to 200 \$ per m^2 for plate-and-frame and tubular membranes [26,32]. However, capital costs are not the only factor to consider when selecting membranes modules. Therefore, it is necessary to consider that the choice of membrane module will also depend on the application (Table 2) [32,197].

In gas separation plants, especially refinery and petrochemical operations, the cost of the modules corresponds to only 10–25% of the total cost of gas separation. Indeed, even if the cost of the membrane modules was reduced, the total cost of the plant would decrease significantly [32].

The economics of the process of membrane-based separation is determined by process design. Single-stage configurations entail low capital costs and are only suitable when the required purity and product recovery are moderate. More demanding applications require multiple stages of separation and recycling. The design of a membrane system involves the configuration of the permeator network and the operating conditions of the individual permeator systems [198]. A key part of the membrane gas separation design is the selection of the separation configuration. Single-stage configurations without gas recycling are the most common and simplest design. However, the demand for higher product purity (for instance methane contents of 98–99.5% in biomethane) and the need for recovery target products makes the use of recycle streams as well as multi-stage configurations a must [199]. These multi-stage systems are typically designed using two, three or four stages [200]. Figure 15 displays the main process configurations.

Table 2. Characteristics and typical applications of the different modules for gas separation.

Module Configuration	Features	Typical Applications	Used by
• High Pressure • Shell-Side Feed • Hollow Fiber	- Cross-flow - Feed gas require pretreatment - Good feed flow distribution	- H_2 recovery in refineries - CO_2 removal from natural gas	Medal Cynara Other
• Low Pressure • Bore-Side Feed • Hollow Fibers	- Counter-flow - No Fouling issues	- N_2 from air - Dehydration of air	Medal Air products Parker
• Spirally Wound Modules	- Cross-flow - No fouling issues - Wide range of membrane can be used	- CO_2 removal from natural gas - Vapor/gas separations	MTR

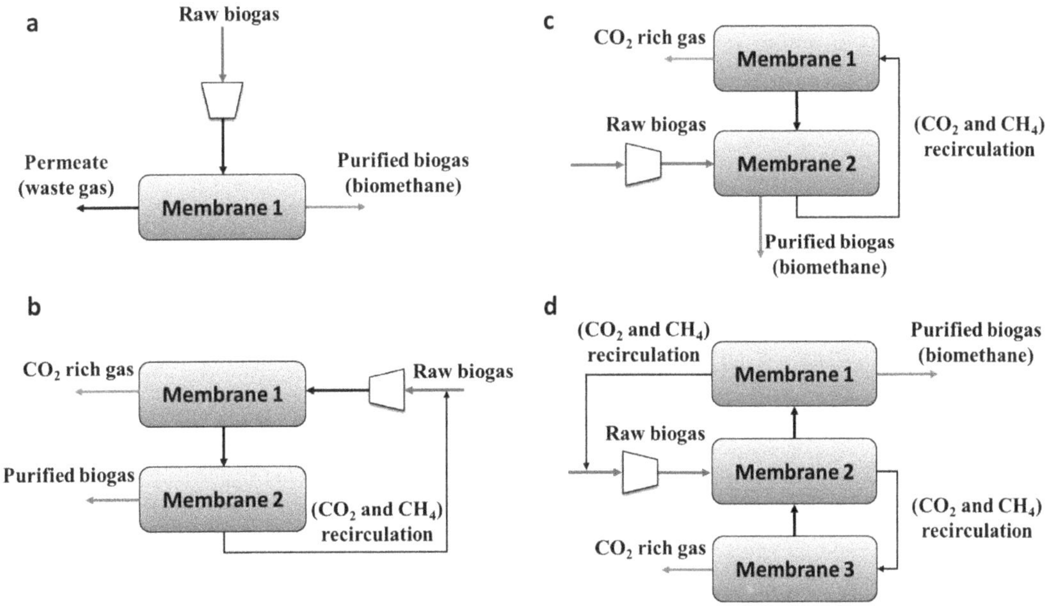

Figure 15. Different design configurations for biogas upgrading (**a**) single-stage configuration, (**b**) two-stage configuration with a recirculation loop, (**c**) two-stage configuration with sweep and (**d**) three-stage configuration with sweep. Adapted from Angelidaki and co-workers, 2018, and Bauer and co-workers, 2013.

The one-step system uses a single membrane and compressor (Figure 15a), which entails a low energy consumption, with no internal recirculation of the rejected gas [9,31,201]. This configuration needs less maintenance and reduces the operational cost compared to multistage membrane units [9]. The second-step system (Figure 15b) involves a gas recirculation loop for the gas retained to a second membrane installed to increase the purity of biomethane and the recovery of methane [9]. The third-step (Figure 15c) system is also based on two membranes, where the rejected gas from the first membrane is purified in a second membrane and recirculated to the first membrane [9]. The most complex configuration (Figure 15d) involves the purification of the permeate from the first membrane in a sequential membrane in order to increase the efficiency of the process, and the recovery of the CH_4 from the gas rejected by the first membrane (using a third membrane) and from the rejected gas of the polishing membrane via recirculation [31].

9. Conclusions

The development of compact and low-cost biogas-to-biomethane and biohydrogen-to-high purity H_2 conversion technologies is crucial to ensure the competitiveness of these green energy vectors, and to promote the implementation of anaerobic digestion and dark fermentation for organic waste treatment. Nowadays, the removal of CO_2 from biogas at the industrial scale is carried out by physical/chemical technologies, which exhibit high operating costs and corrosion problems. In fact, CO_2 removal at the commercial scale is performed using very energy-intensive technologies that require a prior removal of H_2S, such as pressurized water scrubbers, chemical and organic solvent scrubbers, PSA adsorption systems or cryogenic CO_2 separators. On the other hand, biological technologies for CO_2 removal from biogas are still in an experimental development phase and require large areas of land or the availability of renewable hydrogen. In this context, the energy demand and effectiveness of membrane-based CO_2 separation from biogas and biohydrogen is gradually decreasing as a result of the rapid advance in material science. In the last decades, a wide variety of polymeric materials have been developed to increase the gas transport performance of membranes. However, several challenges remain in the field, such as the trade-off between permeability and selectivity (which often prevents overcoming the Robeson limits), the physical aging of membranes and material plasticization (which visibly affects membrane performance). In this context, novel inorganic materials, with outstanding chemical and thermal properties (superior to polymeric materials) and excellent performance in gas separation, have been recently synthesized. However, despite these materials being difficult to process, their combination with polymeric materials in order to develop MMMs has resulted in unprecedented gas separation performance. In addition, polymeric materials capable of producing benzoaxazoles have been recently used to develop thermally rearranged MMMs, leading to excellent gas separation properties, exceeding the Robeson limit, as well as delaying physical aging. Thus, the development of new materials with enhanced physical and chemical properties compared with conventional organic and inorganic membranes, providing a superior performance in terms of permeability and selectivity, represents the cornerstone in biogas and biohydrogen upgrading.

Author Contributions: Conceptualization, methodology, investigation, writing-original draft preparation, C.S.; validation and visualization, C.S., L.P., R.M., P.P. and A.H.; writing—review and editing and supervision, L.P., R.M., P.P. and A.H.; supervision, R.M., P.P. and A.H.; funding acquisition, L.P. All authors have read and agreed to the published version of the manuscript.

Funding: This work was supported by the Spanish Government (AEI) through projects PID2019-109403RB-C21/AEI/10.13039/501100011033; and by the Regional Government of Castilla y León and the EU-FEDER programme (CLU2017-09, UIC082, CL-EI-2021-07 and UIC 315) and CDTI (ECLOSION PROJECT).

Acknowledgments: C.S. acknowledges the Regional Government of Castilla y León for her Ph. D. contract.

Conflicts of Interest: The authors declare no conflict of interest.

References

1. Andriani:, D.; Wresta, A.; Atmaja, T.D.; Saepudin, A. A review on optimization production and upgrading biogas through CO_2 removal using various techniques. *Appl. Biochem. Biotechnol.* **2014**, *172*, 1909–1928. [CrossRef]
2. Kougias, P.G.; Angelidaki, I. Biogas and its opportunities—A review. *Front. Environ. Sci. Eng.* **2018**, *12*, 14. [CrossRef]
3. Zhang, Q.; Hu, J.; Lee, D.J. Biogas from anaerobic digestion processes: Research updates. *Renew. Energy* **2016**, *98*, 108–119. [CrossRef]
4. EBA. *2020 Statical Report of the European Biogas Association 2020*; EBA: Brussels, Belgium, 2021.
5. Toledo-Cervantes, A.; Estrada, J.M.; Lebrero, R.; Muñoz, R. A comparative analysis of biogas upgrading technologies: Photosynthetic vs. physical/chemical processes. *Algal Res.* **2017**, *25*, 237–243. [CrossRef]

6. WBA Global Potential of Biogas. 2019. Available online: https://www.worldbiogasassociation.org/wp-content/uploads/2019/07/WBA-globalreport-56ppa4_digital.pdf (accessed on 28 August 2022).
7. EBA. *2021 Statistical Report of the European Biogas Association 2021*; EBA: Brussels, Belgium, 2021.
8. IRENA. 2022. Available online: https://www.irena.org/bioenergy (accessed on 14 September 2022).
9. Angelidaki, I.; Treu, L.; Tsapekos, P.; Luo, G.; Campanaro, S.; Wenzel, H.; Kougias, P.G. Biogas upgrading and utilization: Current status and perspectives. *Biotechnol. Adv.* **2018**, *36*, 452–466. [CrossRef]
10. BOE. *Resolución de 8 de Octubre de 2018, de La Dirección General de Política Energética y Minas, Por La Que Se Modifican Las Normas de Gestión Técnica Del Sistema NGTS-06, NGTS-07 y Los Protocolos de Detalle PD-01 y PD-02*; Boletín Oficial Del Estado: Madrid, Spain, 2018; pp. 102917–102948.
11. Bakonyi, P.; Nemestóthy, N.; Bélafi-Bakó, K. Biohydrogen purification by membranes: An overview on the operational conditions affecting the performance of non-porous, polymeric and ionic liquid based gas separation membranes. *Int. J. Hydrog. Energy* **2013**, *38*, 9673–9687. [CrossRef]
12. Ramírez-Morales, J.E.; Tapia-Venegas, E.; Toledo-Alarcón, J.; Ruiz-Filippi, G. Simultaneous production and separation of biohydrogen in mixed culture systems by continuous dark fermentation. *Water Sci. Technol.* **2015**, *71*, 1271–1285. [CrossRef] [PubMed]
13. Rittmann, S.; Herwig, C. A comprehensive and quantitative review of dark fermentative biohydrogen production. *Microb. Cell Fact.* **2012**, *11*, 20–25. [CrossRef]
14. Das, D.; Veziroglu, T.N. Advances in biological hydrogen production processes. *Int. J. Hydrogen Energy* **2008**, *33*, 6046–6057. [CrossRef]
15. Bharathiraja, B.; Sudharsanaa, T.; Bharghavi, A.; Jayamuthunagai, J.; Praveenkumar, R. Biohydrogen and biogas—An overview on feedstocks and enhancement process. *Fuel* **2016**, *185*, 810–828. [CrossRef]
16. Ramírez-Morales, J.E.; Tapia-Venegas, E.; Nemestóthy, N.; Bakonyi, P.; Bélafi-Bakó, K.; Ruiz-Filippi, G. Evaluation of two gas membrane modules for fermentative hydrogen separation. *Int. J. Hydrog. Energy* **2013**, *38*, 14042–14052. [CrossRef]
17. IEA. *CO2 Emmisions from Fuel Combustion Highlights*; International Energy Agency: Paris, France, 2019.
18. Tapia-Venegas, E.; Ramirez-Morales, J.E.; Silva-Illanes, F.; Toledo-Alarcón, J.; Paillet, F.; Escudie, R.; Lay, C.H.; Chu, C.Y.; Leu, H.J.; Marone, A.; et al. Biohydrogen production by dark fermentation: Scaling-up and technologies integration for a sustainable system. *Rev. Environ. Sci. Biotechnol.* **2015**, *14*, 761–785. [CrossRef]
19. Mona, S.; Kumar, S.S.; Kumar, V.; Parveen, K.; Saini, N.; Deepak, B.; Pugazhendhi, A. Green technology for sustainable biohydrogen production (waste to energy): A review. *Sci. Total Environ.* **2020**, *728*, 138481. [CrossRef]
20. Elbeshbishy, E.; Dhar, B.R.; Nakhla, G.; Lee, H.S. A critical review on inhibition of dark biohydrogen fermentation. *Renew. Sustain. Energy Rev.* **2017**, *79*, 656–668. [CrossRef]
21. Muñoz, R.; Meier, L.; Diaz, I.; Jeison, D. A review on the state-of-the-art of physical/chemical and biological technologies for biogas upgrading. *Rev. Environ. Sci. Biotechnol.* **2015**, *14*, 727–759. [CrossRef]
22. Liemberger, W.; Groß, M.; Miltner, M.; Harasek, M. Experimental analysis of membrane and Pressure Swing Adsorption (PSA) for the hydrogen separation from natural gas. *J. Clean. Prod.* **2017**, *167*, 896–907. [CrossRef]
23. Hinchliffe, A.B.; Porter, K.E. A comparison of membrane separation and distillation. *Chem. Eng. Res. Des.* **2000**, *78*, 255–268. [CrossRef]
24. Ockwig, N.W.; Nenoff, T.M. Membranes for hydrogen separation. *Chem. Rev.* **2007**, *107*, 4078–4110. [CrossRef] [PubMed]
25. Sridhar, S.; Smitha, B.; Aminabhavi, T.M. Separation of carbon dioxide from natural gas mixtures through polymeric membranes—A review. *Sep. Purif. Rev.* **2007**, *36*, 113–174. [CrossRef]
26. Ismail, A.F.; Khulbe, K.C.; Matsuura, T. *Gas Separation Membranes: Polymeric and Inorganic*; Springer: Ottawa, ON, Canada, 2015; ISBN 9783319010953.
27. Adhikari, S.; Fernando, S. Hydrogen membrane separation techniques. *Ind. Eng. Chem. Res.* **2006**, *45*, 875–881. [CrossRef]
28. Chen, H.Z.; Chung, T.S. CO_2-selective membranes for hydrogen purification and the effect of carbon monoxide (CO) on its gas separation performance. *Int. J. Hydrog. Energy* **2012**, *37*, 6001–6011. [CrossRef]
29. Sridhar, S.; Bee, S.; Bhargava, S. Membrane-based gas separation: Principle, applications and future potential. *Chem. Eng. Dig.* **2014**, *1*, 1–25.
30. Bernardo, P.; Drioli, E.; Golemme, G. Membrane gas separation: A review/state of the art. *Ind. Eng. Chem. Res.* **2009**, *48*, 4638–4663. [CrossRef]
31. Bauer, F.; Hulteberg, C.; Persson, T.; Tamm, D. *Biogas Upgrading—Review of Commercial Technologies*; SGC Rapport; Svenskt Gastekniskt Center AB: Malmö, Sweden, 2013; Volume 270.
32. Baker, R.W. Future directions of membrane gas separation technology. *Ind. Eng. Chem. Res.* **2002**, *41*, 1393–1411. [CrossRef]
33. EBA. *Statistical Report of the European Biogas Association 2018*; EBA: Brussels, Belgium, 2018.
34. Miltner, M.; Makaruk, A.; Harasek, M. Review on available biogas upgrading technologies and innovations towards advanced solutions. *J. Clean. Prod.* **2017**, *161*, 1329–1337. [CrossRef]
35. Vinoba, M.; Bhagiyalakshmi, M.; Alqaheem, Y.; Alomair, A.A.; Pérez, A.; Rana, M.S. Recent progress of fillers in mixed matrix membranes for CO_2 separation: A review. *Sep. Purif. Technol.* **2017**, *188*, 431–450. [CrossRef]
36. Al-Mufachi, N.A.; Rees, N.V.; Steinberger-Wilkens, R. Hydrogen selective membranes: A review of palladium-based dense metal membranes. *Renew. Sustain. Energy Rev.* **2015**, *47*, 540–551. [CrossRef]

37. Sazali, N.; Salleh, W.N.W.; Ismail, A.F. Synthetic polymer-based membranes for hydrogen separation. In *Synthetic Polymeric Membranes for Advanced Water Treatment, Gas Separation, and Energy Sustainability*; Elsevier: Amsterdam, The Netherlands, 2020; pp. 273–292.
38. Edlund, D. Hydrogen membrane technologies and application in fuel processing. In *Hydrogen and Syngas Production and Purification Technologies*; Liu, K., Song, C., Subramani, V., Eds.; Wiley-Blackwell: Hoboken, NJ, USA, 2009; pp. 357–384, ISBN 978-0-471-71975-5.
39. Jeon, Y.W.; Lee, D.H. Gas membranes for CO_2/CH_4 (Biogas) separation: A review. *Environ. Eng. Sci.* **2015**, *32*, 71–85. [CrossRef]
40. Strathmann, H. Membrane separation processes: Current relevance and future opportunities. *AIChE J.* **2001**, *47*, 1077–1087. [CrossRef]
41. Perry, J.D.; Nagai, K.; Koros, W.J. Polymer membranes for hydrogen separations. *MRS Bull.* **2006**, *31*, 745–749. [CrossRef]
42. Kaboorani, A.; Riedl, B.; Blanchet, P.; Fellin, M.; Hosseinaei, O.; Wang, S. Nanocrystalline Cellulose (NCC): A renewable nano-material for Polyvinyl Acetate (PVA) adhesive. *Eur. Polym. J.* **2012**, *48*, 1829–1837. [CrossRef]
43. Freeman, B.D.; Pinnau, I. Gas and Liquid Separations Using Membranes: An Overview. In *Advanced Materials for Membrane Separations*; ACS Symp., Ser.; Freeman, B.D., Pinnau, I., Eds.; American Chemical Society: Washington, DC, USA, 2004; Volume 876, pp. 1–23.
44. Peramanu, S.; Cox, B.G.; Pruden, B.B. Economics of hydrogen recovery processes for the purification of hydroprocessor purge and off-gases. *Int. J. Hydrog. Energy* **1999**, *24*, 405–424. [CrossRef]
45. Petersson, A.; Wellinger, A. Biogas upgrading technologies—Developments and innovations task 37—Energy from biogas and landfill gas IeA bioenergy aims to accelerate the use of environmental sound and cost-competitive bioenergy on a sustainable basis, and thereby achieve a substant. *IEA Bioenergy* **2009**, *13*, 1–19.
46. Baker, R.W.; Low, B.T. Gas separation membrane materials: A perspective. *Macromolecules* **2014**, *47*, 6999–7013. [CrossRef]
47. Basu, S.; Khan, A.L.; Cano-Odena, A.; Liu, C.; Vankelecom, I.F.J. Membrane-based technologies for biogas separations. *Chem. Soc. Rev.* **2010**, *39*, 750–768. [CrossRef]
48. Ozturk, B.; Demirciyeva, F. Comparison of biogas upgrading performances of different mixed matrix membranes. *Chem. Eng. J.* **2013**, *222*, 209–217. [CrossRef]
49. Kohl, A.L.; Nielsen, R.B. *Gas Purification*, 5th ed.; Gulf Publishing Company: Houston, TX, USA, 1997. [CrossRef]
50. Koros, W.J.; Fleming, G.K. Membrane-based gas separation. *J. Membr. Sci.* **1993**, *83*, 1–80. [CrossRef]
51. Wu, A.X.; Drayton, J.A.; Smith, Z.P. The perfluoropolymer upper bound. *AIChE J.* **2019**, *65*, e16700. [CrossRef]
52. Yampolskii, Y. Polymeric gas separation membranes. *Macromolecules* **2012**, *45*, 3298–3311. [CrossRef]
53. Sanders, D.F.; Smith, Z.P.; Guo, R.; Robeson, L.M.; McGrath, J.E.; Paul, D.R.; Freeman, B.D. Energy-efficient polymeric gas separation membranes for a sustainable future: A review. *Polymer* **2013**, *54*, 4729–4761. [CrossRef]
54. Matteucci, S.; Yampolskii, Y.; Freeman, B.D.; Pinnau, I. Transport of Gases and Vapors in Glassy and Rubbery Polymers. In *Materials Science of Membranes for Gas and Vapor Separation*; Yampolskii, Y., Pinnau, I., Freeman, B.D., Eds.; John Wiley & Sons: Chichester, UK, 2006; pp. 1–47, ISBN 0-470-85345-X.
55. Galizia, M.; Chi, W.S.; Smith, Z.P.; Merkel, T.C.; Baker, R.W.; Freeman, B.D. 50th anniversary perspective: Polymers and mixed matrix membranes for gas and vapor separation: A review and prospective opportunities. *Macromolecules* **2017**, *50*, 7809–7843. [CrossRef]
56. Adewole, J.K.; Ahmad, A.L.; Ismail, S.; Leo, C.P. Current challenges in membrane separation of CO_2 from natural gas: A review. *Int. J. Greenh. Gas Control* **2013**, *17*, 46–65. [CrossRef]
57. Baker, R.W.; Lokhandwala, K. Natural gas processing with membranes: An overview. *Ind. Eng. Chem. Res.* **2008**, *47*, 2109–2121. [CrossRef]
58. Brunetti, A.; Scura, F.; Barbieri, G.; Drioli, E. Membrane technologies for CO_2 separation. *J. Membr. Sci.* **2010**, *359*, 115–125. [CrossRef]
59. Bernardo, P.; Drioli, E. Membrane gas separation progresses for process intensification strategy in the petrochemical industry. *Pet. Chem.* **2010**, *50*, 271–282. [CrossRef]
60. Robeson, L.M. The upper bound revisited. *J. Membr. Sci.* **2008**, *320*, 165–185. [CrossRef]
61. Robeson, L.M. Correlation of separation factor versus permeability for polymeric membranes. *J. Membr. Sci.* **1991**, *62*, 165–185. [CrossRef]
62. Freeman, B.D. Basis of permeability/selectivity tradeoff relations in polymeric gas separation membranes. *Macromolecules* **1999**, *32*, 375–380. [CrossRef]
63. Swaidan, R.; Ghanem, B.; Pinnau, I. Fine-tuned intrinsically ultramicroporous polymers redefine the permeability/selectivity upper bounds of membrane-based air and hydrogen separations. *ACS Macro Lett.* **2015**, *4*, 947–951. [CrossRef]
64. Comesaña-Gándara, B.; Chen, J.; Bezzu, C.G.; Carta, M.; Rose, I.; Ferrari, M.C.; Esposito, E.; Fuoco, A.; Jansen, J.C.; McKeown, N.B. Redefining the robeson upper bounds for CO_2/CH_4 and CO_2/N_2 separations using a series of ultrapermeable benzotriptycene-based polymers of intrinsic microporosity. *Energy Environ. Sci.* **2019**, *12*, 2733–2740. [CrossRef]
65. Robeson, L.M.; Burgoyne, W.F.; Langsam, M.; Savoca, A.C.; Tien, C.F. High performance polymers for membrane separation. *Polymer* **1994**, *35*, 4970–4978. [CrossRef]

66. Comesaña-Gandara, B.; Ansaloni, L.; Lee, Y.M.; Lozano, A.E.; De Angelis, M.G. Sorption, diffusion, and permeability of humid gases and aging of Thermally Rearranged (TR) polymer membranes from a novel Ortho-Hydroxypolyimide. *J. Membr. Sci.* **2017**, *542*, 439–455. [CrossRef]
67. Wind, J.D.; Staudt-Bickel, C.; Paul, D.R.; Koros, W.J. The effects of crosslinking chemistry on CO_2 plasticization of polyimide gas separation membranes. *Ind. Eng. Chem. Res* **2002**, *41*, 45. [CrossRef]
68. Paul, D.R.; Pixton, M.R.; Paul, D.R. Relationships between structure and transport properties for polymers with aromatic backbones. In *Polymeric Gas Separation Membranes*; CRC Press: Boca Raton, FL, USA, 1994; pp. 83–154.
69. Singh, A.; Freeman, B.D.; Pinnau, I. *Pure and Mixed Gas Acetone/Nitrogen Permeation Properties of Polydimethylsiloxane [PDMS]*; John Wiley & Sons, Inc.: Hoboken, NJ, USA, 1998; Volume 36.
70. Swaidan, R.; Ghanem, B.; Litwiller, E.; Pinnau, I. Physical aging, plasticization and their effects on gas permeation in "rigid" polymers of intrinsic microporosity. *Macromolecules* **2015**, *48*, 6553–6561. [CrossRef]
71. Tiwari, R.R.; Jin, J.; Freeman, B.D.; Paul, D.R. Physical aging, CO_2 sorption and plasticization in thin films of Polymer with Intrinsic Microporosity (PIM-1). *J. Membr. Sci.* **2017**, *537*, 362–371. [CrossRef]
72. Ying, Y.; Cheng, Y.; Peh, S.B.; Liu, G.; Shah, B.B.; Zhai, L.; Zhao, D. Plasticization resistance-enhanced CO_2 separation at elevated pressures by mixed matrix membranes containing flexible metal-organic framework fillers. *J. Membr. Sci.* **2019**, *582*, 103–110. [CrossRef]
73. Chen, C.C.; Qiu, W.; Miller, S.J.; Koros, W.J. Plasticization-resistant hollow fiber membranes for CO_2/CH_4 separation based on a thermally crosslinkable polyimide. *J. Membr. Sci.* **2011**, *382*, 212–221. [CrossRef]
74. Chiou, J.S.; Paul, D.R. Effects of CO_2 exposure on gas transport properties of glassy polymers. *J. Membr. Sci.* **1987**, *32*, 195–205. [CrossRef]
75. Horn, N.R.; Paul, D.R. Carbon dioxide plasticization and conditioning effects in thick vs. thin glassy polymer films. *Polymer* **2011**, *52*, 1619–1627. [CrossRef]
76. Rowe, B.W.; Freeman, B.D.; Paul, D.R. Physical aging of ultrathin glassy polymer films tracked by gas permeability. *Polymer* **2009**, *50*, 5565–5575. [CrossRef]
77. Struik, L.C.E. Physical aging in amorphous polymers and other materials. *Polym. Eng. Sci.* **1978**, *17*, 165–173.
78. Xia, J.; Chung, T.S.; Paul, D.R. Physical aging and carbon dioxide plasticization of thin polyimide films in mixed gas permeation. *J. Membr. Sci.* **2014**, *450*, 457–468. [CrossRef]
79. McCaig, M.S.; Paul, D.R. Effect of film thickness on the changes in gas permeability of a glassy polyarylate due to physical aging: Part I. experimental observations. *Polymer* **2000**, *41*, 629–637. [CrossRef]
80. Vogel, H.; Marvel, C.S. Polybenzimidazoles, new thermally stable polymers. *J. Polym. Sci.* **1961**, *50*, 511–539. [CrossRef]
81. Borjigin, H.; Liu, Q.; Zhang, W.; Gaines, K.; Riffle, J.S.; Paul, D.R.; Freeman, B.D.; McGrath, J.E. Synthesis and characterization of Thermally Rearranged (TR) polybenzoxazoles: Influence of isomeric structure on gas transport properties. *Polymer* **2015**, *75*, 199–210. [CrossRef]
82. Borjigin, H.; Stevens, K.A.; Liu, R.; Moon, J.D.; Shaver, A.T.; Swinnea, S.; Freeman, B.D.; Riffle, J.S.; McGrath, J.E. Synthesis and characterization of polybenzimidazoles derived from tetraaminodiphenylsulfone for high temperature gas separation membranes. *Polymer* **2015**, *71*, 135–142. [CrossRef]
83. Ghosal, K.; Freeman, B.D.; Chern, R.T.; Alvarez, J.C.; de la Campa, J.G.; Lozano, A.E.; de Abajo, J. Gas separation properties of aromatic polyamides with sulfone groups. *Polymer* **1995**, *36*, 793–800. [CrossRef]
84. De Abajo, J.; De la Campa, J.G.; Lozano, A.E.; Espeso, J.; García, C. Designing aromatic polyamides and polyimides for gas separation membranes. *Macromol. Symp.* **2003**, *199*, 293–306. [CrossRef]
85. De Abajo, J.; de la Campa, J.G.; Lozano, A.E.; Alvarez, J.C. Thermally stable polymers: Novel aromatic polyamides. *Adv. Mater.* **1995**, *148*, 151. [CrossRef]
86. Espeso, J.; Lozano, A.E.; de la Campa, J.G.; de Abajo, J. Effect of substituents on the permeation properties of polyamide membranes. *J. Membr. Sci.* **2006**, *280*, 659–665. [CrossRef]
87. Maya, E.M.; García-Yoldi, I.; Lozano, A.E.; De La Campa, J.G.; De Abajo, J. Synthesis, characterization, and gas separation properties of novel copolyimides containing adamantyl ester pendant groups. *Macromolecules* **2011**, *44*, 2780–2790. [CrossRef]
88. Lozano, A.E.; de Abajo, J.; de la Campa, J.G. Synthesis of aromatic polyisophthalamides by in situ silylation of aromatic diamines. *Macromolecules* **1997**, *30*, 2507–2508. [CrossRef]
89. Mc Keown, N.B.; Budd, P.M. Polymers of intrinsic microporosity (PIMs): Organic materials for membrane separations, heterogeneous catalysis and hydrogen storage. *Chem. Soc. Rev.* **2006**, *35*, 675–683. [CrossRef]
90. Budd, P.M.; Msayib, K.J.; Tattershall, C.E.; Ghanem, B.S.; Reynolds, K.J.; McKeown, N.B.; Fritsch, D. Gas separation membranes from polymers of intrinsic microporosity. *J. Membr. Sci.* **2005**, *251*, 263–269. [CrossRef]
91. Lanč, M.; Pilnáček, K.; Mason, C.R.; Budd, P.M.; Rogan, Y.; Malpass-Evans, R.; Carta, M.; Gándara, B.C.; McKeown, N.B.; Jansen, J.C.; et al. Gas sorption in polymers of intrinsic microporosity: The difference between solubility coefficients determined via time-lag and direct sorption experiments. *J. Membr. Sci.* **2019**, *570*, 522–536. [CrossRef]
92. Park, H.B.; Jung, C.H.; Lee, Y.M.; Hill, A.J.; Pas, S.J.; Mudie, S.T.; Van Wagner, E.; Freeman, B.D.; Cookson, D.J. Polymers with cavities tuned for fast selective transport of small molecules and ions. *Science* **2007**, *318*, 254–258. [CrossRef]
93. Dechnik, J.; Gascon, J.; Doonan, C.J.; Janiak, C.; Sumby, C.J. Mixed-matrix membranes. *Angew. Chem. Int. Ed.* **2017**, *56*, 9292–9310. [CrossRef]

94. Kim, S.; Hou, J.; Wang, Y.; Ou, R.; Simon, G.P.; Seong, J.G.; Lee, Y.M.; Wang, H. Highly permeable thermally rearranged polymer composite membranes with a graphene oxide scaffold for gas separation. *J. Mater. Chem. A* **2018**, *6*, 7668–7674. [CrossRef]
95. Kim, S.; Shamsaei, E.; Lin, X.; Hu, Y.; Simon, G.P.; Seong, J.G.; Kim, J.S.; Lee, W.H.; Lee, Y.M.; Wang, H. The enhanced hydrogen separation performance of mixed matrix membranes by incorporation of two-dimensional zif-l into polyimide containing hydroxyl group. *J. Membr. Sci.* **2018**, *549*, 260–266. [CrossRef]
96. Wang, Y.; Low, Z.X.; Kim, S.; Zhang, H.; Chen, X.; Hou, J.; Seong, J.G.; Lee, Y.M.; Simon, G.P.; Davies, C.H.J.; et al. Functionalized boron nitride nanosheets: A thermally rearranged polymer nanocomposite membrane for hydrogen separation. *Angew. Chem. Int. Ed.* **2018**, *57*, 16288–16293. [CrossRef]
97. Smith, S.J.D.; Hou, R.; Lau, C.H.; Konstas, K.; Kitchin, M.; Dong, G.; Lee, J.; Lee, W.H.; Seong, J.G.; Lee, Y.M.; et al. Highly permeable Thermally Rearranged Mixed Matrix Membranes (TR-MMM). *J. Membr. Sci.* **2019**, *585*, 260–270. [CrossRef]
98. Khdhayyer, M.R.; Esposito, E.; Fuoco, A.; Monteleone, M.; Giorno, L.; Jansen, J.C.; Attfield, M.P.; Budd, P.M. Mixed matrix membranes based on UiO-66 MOFs in the polymer of intrinsic microporosity PIM-1. *Sep. Purif. Technol.* **2017**, *173*, 304–313. [CrossRef]
99. Lin, R.; Villacorta Hernandez, B.; Ge, L.; Zhu, Z. Metal organic framework based mixed matrix membranes: An overview on filler/polymer interfaces. *J. Mater. Chem. A* **2018**, *6*, 293–312. [CrossRef]
100. Vinh-Thang, H.; Kaliaguine, S. Predictive models for mixed-matrix membrane performance: A review. *Chem. Rev.* **2013**, *113*, 4980–5028. [CrossRef]
101. Dong, G.; Li, H.; Chen, V. Challenges and opportunities for mixed-matrix membranes for gas separation. *J. Mater. Chem. A* **2013**, *1*, 4610–4630. [CrossRef]
102. Aroon, M.A.; Ismail, A.F.; Matsuura, T.; Montazer-Rahmati, M.M. Performance studies of mixed matrix membranes for gas separation: A review. *Sep. Purif. Technol.* **2010**, *75*, 229–242. [CrossRef]
103. Paul, D.R.; Kemp, D.R. Containing adsorptive fillers. *J. Polym. Sci. Polym. Symp.* **1973**, *93*, 79–93.
104. Noble, R.D. Perspectives on mixed matrix membranes. *J. Membr. Sci.* **2011**, *378*, 393–397. [CrossRef]
105. Chung, T.S.; Jiang, L.Y.; Li, Y.; Kulprathipanja, S. Mixed Matrix Membranes (MMMs) comprising organic polymers with dispersed inorganic fillers for gas separation. *Prog. Polym. Sci.* **2007**, *32*, 483–507. [CrossRef]
106. Moore, T.T.; Koros, W.J. Non-ideal effects in organic-inorganic materials for gas separation membranes. *J. Mol. Struct.* **2005**, *739*, 87–98. [CrossRef]
107. Tena, A.; de la Viuda, M.; Palacio, L.; Prádanos, P.; Marcos-Fernández, Á.; Lozano, Á.E.; Hernández, A. Prediction of gas permeability of block-segregated polymeric membranes by an effective medium model. *J. Membr. Sci.* **2014**, *453*, 27–35. [CrossRef]
108. Zimmerman, C.M.; Singh, A.; Koros, W.J. Tailoring mixed matrix composite membranes for gas separations. *J. Membr. Sci.* **1997**, *137*, 145–154. [CrossRef]
109. Shimekit, B.; Mukhtar, H.; Murugesan, T. Prediction of the relative permeability of gases in mixed matrix membranes. *J. Membr. Sci.* **2011**, *373*, 152–159. [CrossRef]
110. Mahajan, R.; Koros, W.J. Factors controlling successful formation of mixed-matrix gas separation materials. *Ind. Eng. Chem. Res.* **2000**, *39*, 2692–2696. [CrossRef]
111. Mahajan, R.; Koros, W.J. Mixed matrix membrane materials with glassy polymers. Part 1. *Polym. Eng. Sci.* **2002**, *42*, 1420–1431. [CrossRef]
112. Mahajan, R.; Burns, R.; Schaeffer, M.; Koros, W.J. Challenges in forming successful mixed matrix membranes with rigid polymeric materials. *J. Appl. Polym. Sci.* **2002**, *86*, 881–890. [CrossRef]
113. Nasir, R.; Mukhtar, H.; Man, Z.; Mohshim, D.F. Material advancements in fabrication of mixed-matrix membranes. *Chem. Eng. Technol.* **2013**, *36*, 717–727. [CrossRef]
114. Aguilar-Lugo, C.; Suárez-García, F.; Hernández, A.; Miguel, J.A.; Lozano, Á.E.; De La Campa, J.G.; Álvarez, C. New materials for gas separation applications: Mixed matrix membranes made from linear polyimides and porous polymer networks having lactam groups. *Ind. Eng. Chem. Res.* **2019**, *58*, 9585–9595. [CrossRef]
115. Rico-Martínez, S.; Álvarez, C.; Hernández, A.; Miguel, J.A.; Lozano, Á.E. Mixed matrix membranes loaded with a porous organic polymer having bipyridine moieties. *Membranes* **2022**, *12*, 547. [CrossRef]
116. Etxeberria-Benavides, M.; David, O.; Johnson, T.; Łozińska, M.M.; Orsi, A.; Wright, P.A.; Mastel, S.; Hillenbrand, R.; Kapteijn, F.; Gascon, J. High performance Mixed Matrix Membranes (MMMs) composed of ZIF-94 filler and 6FDA-DAM polymer. *J. Membr. Sci.* **2018**, *550*, 198–207. [CrossRef]
117. Merkel, T.C.; He, Z.; Pinnau, I.; Freeman, B.D.; Meakin, P.; Hill, A.J. Sorption and transport in Poly(2,2-Bis(Trifluoromethyl)-4,5-Difluoro-1,3-Dioxole-Co-Tetrafluoroethylene) containing nanoscale fumed silica. *Macromolecules* **2003**, *36*, 8406–8414. [CrossRef]
118. Ebadi Amooghin, A.; Mashhadikhan, S.; Sanaeepur, H.; Moghadassi, A.; Matsuura, T.; Ramakrishna, S. Substantial breakthroughs on function-led design of advanced materials used in Mixed Matrix Membranes (MMMs): A new horizon for efficient CO_2 separation. *Prog. Mater. Sci.* **2019**, *102*, 222–295. [CrossRef]
119. Bushell, A.F.; Attfield, M.P.; Mason, C.R.; Budd, P.M.; Yampolskii, Y.; Starannikova, L.; Rebrov, A.; Bazzarelli, F.; Bernardo, P.; Carolus Jansen, J.; et al. Gas permeation parameters of mixed matrix membranes based on the polymer of intrinsic microporosity PIM-1 and the zeolitic imidazolate framework ZIF-8. *J. Membr. Sci.* **2013**, *427*, 48–62. [CrossRef]
120. Tien-Binh, N.; Vinh-Thang, H.; Chen, X.Y.; Rodrigue, D.; Kaliaguine, S. Crosslinked MOF-polymer to enhance gas separation of mixed matrix membranes. *J. Membr. Sci.* **2016**, *520*, 941–950. [CrossRef]

121. Kim, J.S.; Moon, S.J.; Wang, H.H.; Kim, S.; Lee, Y.M. Mixed matrix membranes with a thermally rearranged polymer and ZIF-8 for hydrogen separation. *J. Membr. Sci.* **2019**, *582*, 381–390. [CrossRef]
122. Aguilar-Lugo, C.; Lee, W.H.; Miguel, J.A.; De La Campa, J.G.; Prádanos, P.; Bae, J.Y.; Lee, Y.M.; Álvarez, C.; Lozano, Á.E. Highly permeable mixed matrix membranes of thermally rearranged polymers and porous polymer networks for gas separations. *ACS Appl. Polym. Mater.* **2021**, *3*, 5224–5235. [CrossRef]
123. Soto, C.; Aguilar Lugo, C.; Rodríguez, S.; Palacio, L.; Lozano, E.; Prádanos, P.; Hernandez, A. Enhancement of CO_2/CH_4 permselectivity via thermal rearrangement of mixed matrix membranes made from an o-hydroxy polyamide with an optimal load of a porous polymer network. *Sep. Purif. Technol.* **2020**, *247*, 116895. [CrossRef]
124. Seoane, B.; Coronas, J.; Gascon, I.; Benavides, M.E.; Karvan, O.; Caro, J.; Kapteijn, F.; Gascon, J. Metal-organic framework based mixed matrix membranes: A solution for highly efficient CO_2 capture? *Chem. Soc. Rev.* **2015**, *44*, 2421–2454. [CrossRef]
125. Coté, A.P.; Benin, A.I.; Ockwig, N.W.; O'keeffe, M.; Mtzger, A.J.; Yaghi, O.M. Porous, crystalline, covalent organic frameworks. *Science* **2005**, *310*, 1166–1170. [CrossRef]
126. Lau, C.H.; Konstas, K.; Thornton, A.W.; Liu, A.C.Y.; Mudie, S.; Kennedy, D.F.; Howard, S.C.; Hill, A.J.; Hill, M.R. Gas-separation membranes loaded with porous aromatic frameworks that improve with age. *Angew. Chem.* **2015**, *127*, 2707–2711. [CrossRef]
127. Lopez-Iglesias, B.; Suárez-García, F.; Aguilar-Lugo, C.; González Ortega, A.; Bartolomé, C.; Martínez-Ilarduya, J.M.; De La Campa, J.G.; Lozano, Á.E.; Álvarez, C. Microporous polymer networks for carbon capture applications. *ACS Appl. Mater. Interfaces* **2018**, *10*, 26195–26205. [CrossRef] [PubMed]
128. Esteban, N.; Ferrer, M.L.; Ania, C.O.; De La Campa, J.G.; Lozano, Á.E.; Álvarez, C.; Miguel, J.A. Porous organic polymers containing active metal centers for Suzuki-Miyaura heterocoupling reactions. *ACS Appl. Mater. Interfaces* **2020**, *12*, 56974–56986. [CrossRef] [PubMed]
129. Chakrabarty, T.; Giri, A.K.; Sarkar, S. Mixed-matrix gas separation membranes for sustainable future: A mini review. *Polym. Adv. Technol.* **2022**, *33*, 1747–1761. [CrossRef]
130. Millini, R.; Bellussi, G. Zeolite Science and Perspectives. In *Zeolites in Catalysis*; Royal Society of Chemistry: London, UK, 2017; pp. 1–36, ISBN 9781788010610.
131. Kim, W.G.; Zhang, X.; Lee, J.S.; Tsapatsis, M.; Nair, S. Epitaxially grown layered MFI-Bulk MFI hybrid zeolitic materials. *ACS Nano* **2012**, *6*, 9978–9988. [CrossRef] [PubMed]
132. Goh, P.S.; Ismail, A.F.; Sanip, S.M.; Ng, B.C.; Aziz, M. Recent advances of inorganic fillers in mixed matrix membrane for gas separation. *Sep. Purif. Technol.* **2011**, *81*, 243–264. [CrossRef]
133. Castro-Muñoz, R.; Fíla, V. Progress on incorporating zeolites in Matrimid® 5218 mixed matrix membranes towards gas separation. *Membranes* **2018**, *8*, 30. [CrossRef]
134. McCusker, L.B.; Olson, D.H.; Baerlocher, C. *Atlas of Zeolite Framework Types*; 6th Revised Edition; Elsevier: Amsterdam, The Netherlands, 2007; 308p, ISBN 9780444530646.
135. Bastani, D.; Esmaeili, N.; Asadollahi, M. Polymeric mixed matrix membranes containing zeolites as a filler for gas separation applications: A review. *J. Ind. Eng. Chem.* **2013**, *19*, 375–393. [CrossRef]
136. Saha, D.; Bao, Z. Adsorption of CO_2, CH_4, N_2O and N_2 on MOF-5, MOF-177, and Zeolite 5A. *Environ. Sci. Technol.* **2010**, *44*, 1820–1826. [CrossRef]
137. Li, Y.; Yi, H.; Tang, X.; Li, F.; Yuan, Q. Adsorption separation of CO_2/CH_4 gas mixture on the commercial zeolites at atmospheric pressure. *Chem. Eng. J.* **2013**, *229*, 50–56. [CrossRef]
138. Cavenati, S.; Grande, C.A.; Rodrigues, A.E. Adsorption equilibrium of methane, carbon dioxide, and nitrogen on zeolite 13X at high pressures. *J. Chem. Eng. Data* **2004**, *49*, 1095–1101. [CrossRef]
139. Sanaeepur, H.; Kargari, A. Modeling and analysis cellulose acetate/nano-porous zeolite mixed matrix membrane for CO_2 separation. *Greenh. Gas Sci. Technol.* **2015**, *5*, 291–304. [CrossRef]
140. Muhammad Hussain, A.K. Mixed-matrix membrane for gas separation: Polydimethylsiloxane filled with zeolite. *Chem. Eng. Technol.* **2012**, *35*, 561–569. [CrossRef]
141. Ahmad, M.Z.; Martin-Gil, V.; Supinkova, T.; Lambert, P.; Castro-Muñoz, R.; Hrabanek, P.; Kocirik, M.; Fila, V. Novel MMM using CO_2 selective SSZ-16 and high-performance 6FDA-Polyimide for CO_2/CH_4 separation. *Sep. Purif. Technol.* **2021**, *254*, 117582. [CrossRef]
142. Zhang, Y.; Balkus, K.J.; Musselman, I.H.; Ferraris, J.P. Mixed-matrix membranes composed of Matrimid® and mesoporous ZSM-5 nanoparticles. *J. Membr. Sci.* **2008**, *325*, 28–39. [CrossRef]
143. Ebadi Amooghin, A.; Omidkhah, M.; Sanaeepur, H.; Kargari, A. Preparation and characterization of ag+ ion-exchanged Zeolite-Matrimid®5218 mixed matrix membrane for CO_2/CH_4 separation. *J. Energy Chem.* **2016**, *25*, 450–462. [CrossRef]
144. Montes Luna, A.D.J.M.; de León, G.; Rodríguez, S.P.; López, N.C.; López, N.C.; Camacho, O.P.; Mercado, Y.A.P. $Na+/Ca^{2+}$ aqueous ion exchange in natural clinoptilolite zeolite for polymer-zeolite composite membranes production and their $CH_4/CO_2/N_2$ separation performance. *J. Nat. Gas Sci. Eng.* **2018**, *54*, 47–53. [CrossRef]
145. Kuppler, R.J.; Timmons, D.J.; Fang, Q.R.; Li, J.R.; Makal, T.A.; Young, M.D.; Yuan, D.; Zhao, D.; Zhuang, W.; Zhou, H.C. Potential applications of metal-organic frameworks. *Coord. Chem. Rev.* **2009**, *253*, 3042–3066. [CrossRef]
146. Cheng, Y.; Ying, Y.; Japip, S.; Jiang, S.D.; Chung, T.S.; Zhang, S.; Zhao, D. Advanced porous materials in mixed matrix membranes. *Adv. Mater.* **2018**, *30*, 1802401. [CrossRef]

147. Park, K.S.; Ni, Z.; Cô, A.P.; Choi, J.Y.; Huang, R.; Uribe-Romo, F.J.; Chae, H.K.; O'keeffe, M.; Yaghi, O.M. Exceptional Chemical and Thermal Stability of Zeolitic Imidazolate Frameworks. *Proc. Natl. Acad. Sci. USA* **2006**, *103*, 10186–10191. [CrossRef]
148. Phan, A.; Doonan, C.J.; Uribe-Romo, F.J.; Knobler, C.B.; Okeeffe, M.; Yaghi, O.M. Synthesis, structure, and carbon dioxide capture properties of zeolitic imidazolate frameworks. *Acc. Chem. Res.* **2009**, *43*, 58–67. [CrossRef]
149. Ahmad, M.Z.; Peters, T.A., Konnertz, N.M.; Visser, T.; Téllez, C.; Coronas, J.; Fila, V.; de Vos, W.M.; Benes, N.E. High-pressure CO_2/CH_4 separation of Zr-MOFs based mixed matrix membranes. *Sep. Purif. Technol.* **2020**, *230*, 115858. [CrossRef]
150. Kertik, A.; Wee, L.H.; Pfannmöller, M.; Bals, S.; Martens, J.A.; Vankelecom, I.F.J. Highly selective gas separation membrane using in situ Amorphised metal-organic frameworks. *Energy Environ. Sci.* **2017**, *10*, 2342–2351. [CrossRef]
151. Nafisi, V.; Hägg, M.B. Gas separation properties of ZIF-8/6FDA-durene diamine mixed matrix membrane. *Sep. Purif. Technol.* **2014**, *128*, 31–38. [CrossRef]
152. Zhang, Y.; Feng, X.; Yuan, S.; Zhou, J.; Wang, B. Challenges and recent advances in MOF-Polymer composite membranes for gas separation. *Inorg. Chem. Front.* **2016**, *3*, 896–909. [CrossRef]
153. Wu, X.; Tian, Z.; Wang, S.; Peng, D.; Yang, L.; Wu, Y.; Xin, Q.; Wu, H.; Jiang, Z. Mixed matrix membranes comprising polymers of intrinsic microporosity and covalent organic framework for gas separation. *J. Membr. Sci.* **2017**, *528*, 273–283. [CrossRef]
154. Díaz, U.; Corma, A. Ordered covalent organic frameworks, COFs and PAFs: From preparation to application. *Coord. Chem. Rev.* **2016**, *311*, 85–124. [CrossRef]
155. Ding, S.-Y.; Wang, W. Covalent Organic Frameworks (COFs): From design to applications. *Chem. Soc. Rev.* **2013**, *42*, 548. [CrossRef]
156. Côté, A.P.; El-Kaderi, H.M.; Furukawa, H.; Hunt, J.R.; Yaghi, O.M. Reticular synthesis of microporous and mesoporous 2D covalent organic frameworks. *J. Am. Chem. Soc.* **2007**, *129*, 12914–12915. [CrossRef]
157. El-Kaderi, H.M.; Hunt, J.R.; Mendoza-Cortés, J.L.; Côté, A.P.; Taylor, R.E.; O'Keeffe, M.; Yaghi, O.M. Designed synthesis of 3D covalent organic frameworks. *Science* **2007**, *316*, 268–273.
158. Sang, S.H.; Furukawa, H.; Yaghi, O.M.; Goddard, W.A. Covalent organic frameworks as exceptional hydrogen storage materials. *J. Am. Chem. Soc.* **2008**, *130*, 11580–11581. [CrossRef]
159. Gao, X.; Zou, X.; Ma, H.; Meng, S.; Zhu, G. Highly selective and permeable porous organic framework membrane for CO_2 capture. *Adv. Mater.* **2014**, *26*, 3644–3648. [CrossRef] [PubMed]
160. Huang, N.; Chen, X.; Krishna, R.; Jiang, D. Two-dimensional covalent organic frameworks for carbon dioxide capture through channel-wall functionalization. *Angew. Chem.—Int. Ed.* **2015**, *54*, 2986–2990. [CrossRef]
161. Hao, D.; Zhang, J.; Lu, H.; Leng, W.; Ge, R.; Dai, X.; Gao, Y. Fabrication of a COF-5 membrane on a functionalized α-Al_2O_3 ceramic support using a microwave irradiation method. *Chem. Commun.* **2014**, *50*, 1462–1464. [CrossRef]
162. Biswal, B.P.; Chaudhari, H.D.; Banerjee, R.; Kharul, U.K. Chemically stable Covalent Organic Framework (COF)-Polybenzimidazole hybrid membranes: Enhanced gas separation through pore modulation. *Chem. Eur. J.* **2016**, *22*, 4695–4699. [CrossRef]
163. Cao, X.; Qiao, Z.; Wang, Z.; Zhao, S.; Li, P.; Wang, J.; Wang, S. Enhanced performance of mixed matrix membrane by incorporating a highly compatible covalent organic framework into Poly(Vinylamine) for hydrogen purification. *Int. J. Hydrog. Energy* **2016**, *41*, 9167–9174. [CrossRef]
164. Tian, Y.; Zhu, G. Porous Aromatic Frameworks (PAFs). *Chem. Rev.* **2020**, *120*, 8934–8986. [CrossRef]
165. Lau, C.H.; Nguyen, P.T.; Hill, M.R.; Thornton, A.W.; Konstas, K.; Doherty, C.M.; Mulder, R.J.; Bourgeois, L.; Liu, A.C.Y.; Sprouster, D.J.; et al. Ending aging in super glassy polymer membranes. *Angew. Chem. Int. Ed.* **2014**, *53*, 5322–5326. [CrossRef]
166. Lau, C.H.; Konstas, K.; Doherty, C.M.; Kanehashi, S.; Ozcelik, B.; Kentish, S.E.; Hill, A.J.; Hill, M.R. Tailoring physical aging in super glassy polymers with functionalized porous aromatic frameworks for CO_2 capture. *Chem. Mater.* **2015**, *27*, 4756–4762. [CrossRef]
167. Yuan, Y.; Zhu, G. Porous aromatic frameworks as a platform for multifunctional applications. *ACS Cent. Sci.* **2019**, *5*, 409–418. [CrossRef]
168. Ben, T.; Pei, C.; Zhang, D.; Xu, J.; Deng, F.; Jing, X.; Qiu, S. Gas storage in Porous Aromatic Frameworks (PAFs). *Energy Environ. Sci.* **2011**, *4*, 3991–3999. [CrossRef]
169. Konstas, K.; Taylor, J.W.; Thornton, A.W.; Doherty, C.M.; Lim, W.X.; Bastow, T.J.; Kennedy, D.F.; Wood, C.D.; Cox, B.J.; Hill, J.M.; et al. Lithiated porous aromatic frameworks with exceptional gas storage capacity. *Angew. Chem. Int. Ed.* **2012**, *51*, 6639–6642. [CrossRef] [PubMed]
170. Hou, R.; Smith, S.J.D.; Konstas, K.; Doherty, C.M.; Easton, C.D.; Park, J.; Yoon, H.; Wang, H.; Freeman, B.D.; Hill, M.R. Synergistically improved PIM-1 membrane gas separation performance by PAF-1 incorporation and UV irradiation. *J. Mater. Chem. A* **2022**, *10*, 10107–10119. [CrossRef]
171. Ben, T.; Ren, H.; Shengqian, M.; Cao, D.; Lan, J.; Jing, X.; Wang, W.; Xu, J.; Deng, F.; Simmons, J.M.; et al. Targeted synthesis of a porous aromatic framework with high stability and exceptionally high surface area. *Angew. Chem. Int. Ed.* **2009**, *48*, 9457–9460. [CrossRef]
172. Lu, W.; Yuan, D.; Zhao, D.; Schilling, C.I.; Plietzsch, O.; Muller, T.; Brase, S.; Guenther, J.; Blumel, J.; Krishna, R.; et al. Porous polymer networks: Synthesis, porosity, and applications in gas storage/separation. *Chem. Mater.* **2010**, *22*, 5964–5972. [CrossRef]
173. Davankov, V.; Tsyurupa, M. Hypercrosslinked polymers—A novel class of polymeric materials. *Compr. Anal. Chem.* **2011**, *56*, 315–358. [CrossRef]

174. Comesaña-Gándara, B.; Calle, M.; Jo, H.J.; Hernández, A.; de la Campa, J.G.; de Abajo, J.; Lozano, A.E.; Lee, Y.M. Thermally rearranged polybenzoxazoles membranes with biphenyl moieties: Monomer isomeric effect. *J. Membr. Sci.* **2014**, *450*, 369–379. [CrossRef]
175. Smith, Z.P.; Hernández, G.; Gleason, K.L.; Anand, A.; Doherty, C.M.; Konstas, K.; Alvarez, C.; Hill, A.J.; Lozano, A.E.; Paul, D.R.; et al. Effect of polymer structure on gas transport properties of selected aromatic polyimides, polyamides and TR polymers. *J. Membr. Sci.* **2015**, *493*, 766–781. [CrossRef]
176. Hu, X.D.; Jenkins, S.E.; Min, B.G.; Polk, M.B.; Kumar, S. Rigid-rod polymers: Synthesis, processing, simulation, structure, and properties. *Macromol. Mater. Eng.* **2003**, *288*, 823–843. [CrossRef]
177. Tullos, G.L.; Powers, J.M.; Jeskey, S.J.; Mathias, L.J. Thermal conversion of hydroxy-containing imides to benzoxazoles: Polymer and model compound study. *Macromolecules* **1999**, *32*, 3598–3612. [CrossRef]
178. Han, S.H.; Kwon, H.J.; Kim, K.Y.; Seong, J.G.; Park, C.H.; Kim, S.; Doherty, C.M.; Thornton, A.W.; Hill, A.J.; Lozano, Á.E.; et al. Tuning microcavities in thermally rearranged polymer membranes for CO_2 capture. *Phys. Chem. Chem. Phys.* **2012**, *14*, 4365–4373. [CrossRef]
179. Han, S.H.; Lee, J.E.; Lee, K.J.; Park, H.B.; Lee, Y.M. Highly gas permeable and microporous polybenzimidazole membrane by thermal rearrangement. *J. Membr. Sci.* **2010**, *357*, 143–151. [CrossRef]
180. Choi, J.I.; Jung, C.H.; Han, S.H.; Park, H.B.; Lee, Y.M. Thermally Rearranged (TR) poly(benzoxazole-co-pyrrolone) membranes tuned for high gas permeability and selectivity. *J. Membr. Sci.* **2010**, *349*, 358–368. [CrossRef]
181. Smith, Z.P.; Czenkusch, K.; Wi, S.; Gleason, K.L.; Hernández, G.; Doherty, C.M.; Konstas, K.; Bastow, T.J.; Álvarez, C.; Hill, A.J.; et al. Investigation of the chemical and morphological structure of thermally rearranged polymers. *Polymer* **2014**, *55*, 6649–6657. [CrossRef]
182. Díez, B.; Cuadrado, P.; Marcos-Fernández, Á.; de la Campa, J.G.; Tena, A.; Prádanos, P.; Palacio, L.; Lee, Y.M.; Alvarez, C.; Lozano, Á.E.; et al. Thermally rearranged polybenzoxazoles made from Poly(Ortho-Hydroxyamide)s. characterization and evaluation as gas separation membranes. *React. Funct. Polym.* **2018**, *127*, 38–47. [CrossRef]
183. Calle, M.; Chan, Y.; Jo, H.J.; Lee, Y.M. The relationship between the chemical structure and thermal conversion temperatures of Thermally Rearranged (TR) polymers. *Polymer* **2012**, *53*, 2783–2791. [CrossRef]
184. Ye, L.; Wang, L.; Jie, X.; Yu, C.; Kang, G.; Cao, Y. Effect of hexafluoroisopropylidene group contents and treatment temperature on the performance of thermally rearranged Poly(Hydroxyamide)s membranes. *J. Membr. Sci.* **2020**, *595*, 117540. [CrossRef]
185. Park, H.B.; Han, S.H.; Jung, C.H.; Lee, Y.M.; Hill, A.J. Thermally Rearranged (TR) polymer membranes for CO_2 separation. *J. Membr. Sci.* **2010**, *359*, 11–24. [CrossRef]
186. Thornton, A.W.; Doherty, C.M.; Falcaro, P.; Buso, D.; Amenitsch, H.; Han, S.H.; Lee, Y.M.; Hill, A.J. Architecturing nanospace via thermal rearrangement for highly efficient gas separations. *J. Phys. Chem. C* **2013**, *117*, 24654–24661. [CrossRef]
187. Brunetti, A.; Cersosimo, M.; Kim, J.S.; Dong, G.; Fontananova, E.; Lee, Y.M.; Drioli, E.; Barbieri, G. Thermally rearranged mixed matrix membranes for CO_2 separation: An aging study. *Int. J. Greenh. Gas Control* **2017**, *61*, 16–26. [CrossRef]
188. Mulder, M. Module and process design. In *Basic Principles of Membrane Technology*; Springer: Dordrecht, The Netherland, 1997; pp. 465–520, ISBN 978-94-009-1766-8.
189. Kluiters, S.C.A. Status review on membrane systems for hydrogen separation. In *Intermediate Report EU Project MIGREYD NNE5-2001-670*; Energy Center of the Netherlands: Petten, The Netherlands, 2004.
190. Baker, R.W.; Cussler, E.L.; Eykamp, W.; Koros, W.J.; Riley, R.L.; Baker, E.L.R.W.; Cussler, W.; Eykamp, W.J.; Koros, R.L.; Riley, H.S. *Membrane Separation System: Recent Developments and Future Directions*; Noyes Data Corp.: Park Ridge, NJ, USA, 1991; Volume 451, pp. 1–464.
191. Blackmer, R.H.; Hedman, J.W. Membrane Oxygen Enricher Apparatus. U.S. Patent 4174955-1979, 20 November 1979.
192. Balster, J. Plate and frame membrane module. In *Encyclopedia of Membranes*; Drioli, E., Giorno, L., Eds.; Springer: Berlin/Heidelberg, Germany, 2013; pp. 1–3.
193. Lemanski, J.; Lipscomb, G.G. Effect of shell-side flows on the performance of hollow-fiber gas separation modules. *J. Membr. Sci.* **2002**, *195*, 215–228. [CrossRef]
194. Balster, J. Spiral wound membrane module. In *Encyclopedia of Membranes*; Drioli, E., Giorno, L., Eds.; Springer: Berlin/Heidelberg, Germany, 2016; ISBN 9783662443248.
195. Caro, J.; Caspary, K.J.; Hamel, C.; Hoting, B.; Kölsch, P.; Langanke, B.; Nassauer, K.; Schiestel, T.; Schmidt, A.; Schomäcker, R.; et al. Catalytic membrane reactors for partial oxidation using perovskite hollow fiber membranes and for partial hydrogenation using a catalytic membrane contactor. *Ind. Eng. Chem. Res.* **2007**, *46*, 2286–2294. [CrossRef]
196. Balster, J. Tubular membrane module. In *Encyclopedia of Membranes*; Drioli, E., Giorno, L., Eds.; Springer: Berlin/Heidelberg, Germany, 2016.
197. Scholz, M.; Wessling, M.B.J. Design of membrane modules for gas separations. In *Membrane Engineering for the Treatment of Gases: Gas-Separation Problems with Membranes*; Drioli, E., Barbiere, G., Eds.; RSC Publishing: Cambridge, UK, 2011; Volume 5.
198. Qi, R.; Henson, M.A. Optimal design of spiral-wound membrane networks for gas separations. *J. Membr. Sci.* **1998**, *148*, 71–89. [CrossRef]
199. Rojo, E.; Carmona, A.; Soto, C.; Díaz, I.; Fernández-Polanco, M.; Palacio, L.; Muñoz, R.; Bolado, S. Environment and material science technology for anaerobic digestion-based circular bioeconomy. In *Biomass, Biofuels, Biochemicals: Circular Bioeconomy-Current Developments and Future Outlook*; Elsevier: Amsterdam, The Netherlands, 2021; pp. 25–55, ISBN 9780128218785.

200. Lababidi, H.; Al-Enezi, G.A.; Ettouney, H.M. Optimization of module configuration in membrane gas separation. *J. Membr. Sci.* **1996**, *112*, 185–197. [CrossRef]
201. Makaruk, A.; Miltner, M.; Harasek, M. Membrane biogas upgrading processes for the production of natural gas substitute. *Sep. Purif. Technol.* **2010**, *74*, 83–92. [CrossRef]

processes

MDPI

Article

Potential for Biomethanisation of CO_2 from Anaerobic Digestion of Organic Wastes in the United Kingdom

Angela Bywater *, Sonia Heaven *, Yue Zhang and Charles J. Banks

Water and Environmental Engineering Group, University of Southampton, Southampton SO16 7QF, UK; y.zhang@soton.ac.uk (Y.Z.); c.j.banks@soton.ac.uk (C.J.B.)
* Correspondence: a.m.bywater@soton.ac.uk (A.B.); s.heaven@soton.ac.uk (S.H.)

Abstract: The United Kingdom (UK) has a decarbonisation strategy that includes energy from both hydrogen and biomethane. The latter comes from the growing anaerobic digestion (AD) market, which in 2020 produced 23.3 TWh of energy in the form of biogas. According to the strategy, this must be upgraded to biomethane by removal of carbon dioxide (CO_2): a goal that could also be fulfilled through CO_2 biomethanisation, alleviating the need for carbon capture and storage. Results are presented from a survey of publicly available datasets coupled with modelling to identify potential scale and knowledge gaps. Literature data were used to estimate maximum biomethane concentrations by feedstock type: these ranged from 79% for food wastes to 93% for livestock manures. Data from various government sources were used to estimate the overall potential for CO_2 biomethanisation with current AD infrastructure. Values for the uplift in biomethane production ranged from 57% to 61%, but the need for more consistent data collection methodologies was highlighted. On average, however, if CO_2 biomethanisation was applied in all currently operating UK AD plants an energy production uplift of 12,954 GWh could be achieved based on 2020 figures. This is sufficient to justify the inclusion of CO_2 biomethanisation in decarbonisation strategies, in the UK and worldwide.

Keywords: in-situ biomethanisation; power-to-gas; anaerobic digestion; biomethane production; United Kingdom policy; energy security

Citation: Bywater, A.; Heaven, S.; Zhang, Y.; Banks, C.J. Potential for Biomethanisation of CO_2 from Anaerobic Digestion of Organic Wastes in the United Kingdom. *Processes* **2022**, *10*, 1202. https://doi.org/10.3390/pr10061202

Academic Editor: Pietro Bartocci

Received: 31 May 2022
Accepted: 14 June 2022
Published: 16 June 2022

Publisher's Note: MDPI stays neutral with regard to jurisdictional claims in published maps and institutional affiliations.

1. Introduction

CO_2 biomethanisation is the microbially mediated transformation of carbon dioxide (CO_2) to methane (CH_4) via the addition of exogenous hydrogen (H_2), according to the overall reaction shown in Equation (1):

$$4\,H_2 + CO_2 \rightarrow CH_4 + 2\,H_2O \quad (1)$$

The process has clear potential applications in the anaerobic digestion (AD) industry, which utilises microbial communities to transform a wide variety of organic materials into biogas, a mixture of biomethane and CO_2 [1]. These mixed communities already contain the hydrogenotrophic methanogens which catalyse the direct CO_2 biomethanisation route shown in Equation (1), as well as syntrophic organisms able to mediate indirect routes, e.g., via homoacetogenesis and acetoclastic methanogenesis [2,3]. The combination of AD with H_2 addition to promote CO_2 biomethanisation can thus improve the methane productivity of digesters fed on organic feedstocks [4], increasing the energy output (and therefore, viability) [5,6] and enhancing the carbon utilisation efficiency [7].

The United Kingdom (UK) has a well-established AD market, with initial incentives for distributed small-scale (<5 MWe) electricity production introduced in the 2002 Renewables Obligation (RO) and the 2010 Feed-In Tariff (FIT). It is, however, becoming increasingly common practice in many countries to upgrade biogas to biomethane by removal of CO_2 and other impurities [4,8]. This is because biomethane is valuable as a low carbon fuel that can be utilised locally or upgraded to share existing natural gas storage and distribution networks for use in heating, transport and the centralised generation of electricity [4].

Biomethane production in the UK has been promoted by the Renewable Heat Incentive (RHI), which accepted applicants between 2011 and 2021. This was followed by the Green Gas Support Scheme (GGSS), introduced in late 2021 and designed to encourage the production of biomethane for gas grid injection, with a minimum of 50% of the gas coming from wastes. The GGSS reflects the UK Government's ongoing commitment to increasing the growth of biomethane production through AD, based on the advice and projections of the Climate Change Committee (CCC) in its Sixth Carbon Budget [9]. The CCC calculates that biomethane/biogas could displace up to 10% of UK demand for natural gas and predicts that by 2030, biomethane production could more than treble from 2020 levels. Furthermore, by 2035, it could abate the equivalent of 1.5 million tonnes year^{-1} of CO_2 through fossil gas displacement [9].

Hydrogen also plays an important part in the UK's decarbonisation scenario, with plans to kick-start a mass market for hydrogen by reformation of fossil gas accompanied by carbon capture and storage (CCS). In addition, electrolytic hydrogen production from excess renewables will also be developed in a 'Balanced Pathway' scenario, in which reformed hydrogen will provide 60% of the requirement by 2035, after which, this proportion will decrease in favour of electrolytic hydrogen production, which will make up almost 50% of the rising target for production by 2050 [9].

The current UK strategy proposes that biomethane targets are met through refining biogas to remove the biogenic CO_2, which will subsequently require permanent storage, i.e., through carbon capture and storage (CCS) solutions. As biomethane and electrolytic hydrogen production are already part of the UK's energy strategy, however, it also makes sense to consider the potential role of CO_2 biomethanisation of organic wastes. This could address the upgrading requirement, utilising existing AD assets at a relatively low capital cost [5,10] with the added bonus of increasing overall methane yields, and the prospect that the process may be more energetically efficient and cheaper than CCS with its as-yet-unknown costs. CO_2 biomethanisation thus offers both a means to support the transition from carbon-based fuels to hydrogen, and a rational long-term solution to maximise the energetic value of biomass carbon-based renewable fuels in their own right.

Nevertheless, future policy which considers this option cannot be formulated and then prescribed without the necessary data for modelling and assessment. CO_2 biomethanisation is still in its early stages as a commercial process, with only a few examples of plants operating at scale [7], almost all of which are based on the use of separate dedicated bioreactors fed on gaseous inputs in ex situ processes [3,5]. There is, however, a growing body of research on in situ and hybrid processes in which H_2 alone or with additional CO_2 is injected directly into a digester, with a variety of organic feedstocks, equipment configurations and operating conditions. There is thus a need to assess the available data with regard to anticipated improvements in process performance, and to the scale of existing AD resources in the UK. The outcomes can then be considered in the context of dynamic economic, technology and infrastructure developments in renewable power and hydrogen production, and use.

As a first step in this complex journey, the current work has carried out a high-level review of the potential scale of application in the UK, supported by an assessment of reported data on in-situ CO_2 biomethanisation performance according to feedstock type. This provides an essential starting point for identifying research and policy needs for the integration of CO_2 biomethanisation as a contributor to the UK future energy mix and for assessment of its potential role in the transition from a fossil-based energy system.

Alternative routes to CO_2 methanisation via thermal catalytic conversion are also currently under development [4,7] but are not considered in this paper as the principle focus is on conversion of biogas from AD of organic wastes.

2. Materials and Methods

2.1. Performance of CO_2 Biomethanisation of Organic Feedstocks

The objective of this part of the work was to establish values for the increases in methane production and biogas methane content that could be achieved through CO_2 biomethanisation of a range of organic feedstocks. To achieve this, experimental results were collated from relevant studies, focusing primarily on addition of exogenous H_2: a small number of studies examining approaches such as syngas addition were also considered, but bioelectrochemical systems were not included as, although promising, the technology is some way from large-scale application [4,11].

Parameters taken from the literature included substrate type, operating temperature (°C), digester configuration and characteristics, organic loading rate (OLR) expressed as g volatile solids (VS) L^{-1} day^{-1} or g Chemical Oxygen Demand (COD) L^{-1} day^{-1}, hydraulic retention time (HRT, days) and operating pH. Reported and/or calculated values for specific methane potential (SMP: L CH_4 g^{-1} VS or L CH_4 g^{-1} COD), volumetric methane production (VMP: L CH_4 L^{-1} digester day^{-1}), H_2 input (L H_2 L^{-1} day^{-1}) and output gas composition (% volume) were used to calculate the change in SMP and VMP due to CO_2 biomethanisation. Note that the term SMP is used here to refer to the total volume of methane produced, from both anaerobic degradation of organic material and biomethanisation of CO_2 where applicable, per unit of organic feed added. Similarly, the VMP is based on the total volume of methane produced from both sources, per unit working volume of reactor per unit time. The difference between VMP with and without addition of H_2 (or H_2 and exogenous CO_2) is assumed to be methane produced from CO_2 biomethanisation and is referred to as the Methane Evolution Rate (MER: L CH_4 L^{-1} day^{-1}). The CO_2 removal rate (CRR) is similarly defined as the difference between volumetric CO_2 output (L CO_2 L^{-1} day^{-1}) with and without H_2 addition, adjusted for any additional exogenous CO_2 input. The terms volumetric biogas production and volumetric gas production (VBP and VGP, L L^{-1} day^{-1}) were used to distinguish between the sum of methane and carbon dioxide outputs, and the total gas output including any residual H_2, respectively. More detailed nomenclature and definitions are provided in Supplementary Materials S1.

A number of other parameters based on the above data were also evaluated. The H_2 transfer efficiency E (%) refers to the proportion of H_2 successfully transferred into the system, i.e., not leaving in gaseous form in the output gas; and is calculated from (H_2 input-H_2 output)/H_2 input, with inputs and outputs in L H_2 L^{-1} day^{-1}. The expected MER is the amount of H_2 transferred divided by 4, based on reaction stoichiometry, and is equal to the expected CRR; these values can then be compared with the actual observed MER and CRR. The ratios between actual MER and actual CRR; actual MER and original volumetric CO_2 output (without CO_2 biomethanisation); VBP with and without CO_2 biomethanisation; H_2 transferred to actual MER and actual CRR; and H_2 input to original volumetric CO_2 output (without CO_2 biomethanisation) were also determined.

Where control reactors without H_2 addition were operated, the change in SMP or VMP due to CO_2 biomethanisation was taken as the difference between control and experimental values during the same period; in trials without controls, experimental values were compared with those of a baseline period without H_2 addition. Where multiple sets of conditions were tested, the best performance in terms of SMP and/or biogas methane content under apparently stable conditions is shown. As far as possible, calculations were carried out in a standard manner. In some cases, this means values differ slightly from those reported in the original paper. Wherever possible, the consistency of results was checked by comparing reported and calculated values: e.g., under steady state conditions SMP $\times$ OLR = VMP. If some aspect appeared unclear or inconsistent, the authors were contacted to request additional information. More details of calculation methods are provided in the Supplementary Materials.

Digestion systems operating in the temperature range 35–40 °C are referred to as mesophilic, and between 50–60 °C as thermophilic. Unless specified, standard temperature and pressure (STP) values of 0 °C and 101.325 kPa were assumed and applied throughout this work.

To provide indicative values for modelling, outputs from individual studies were grouped into the following feedstock types: livestock manures, crop materials and agro-wastes, food wastes, the organic fraction of municipal solid waste (OFMSW) and sewage sludges. In some cases, there was no single study that directly demonstrated the maximum achievable SMP and biogas methane content for a given feedstock type. Where possible, this was estimated from baseline pH and biogas CO_2 concentrations (i.e., without CO_2 biomethanisation) and observed or assumed maximum pH values for stable operation with CO_2 biomethanisation, using an equation derived and tested for this purpose [12].

2.2. AD Feedstock/Energy Production Data and Calculations

Several potential sources of data on AD feedstock quantities and energy production are available and were assessed for use in the current work. Those selected for use were derived from freely available public sources. The most detailed is the dataset which accompanies the Renewable Obligation's Annual Report (2019–20) [13], produced by the Office of Gas and Electricity Markets (Ofgem), the regulator for the electricity and natural gas markets in Great Britain. All RO biomass electricity generators over 50 kWe are required to report on feedstock sustainability criteria in two areas: greenhouse gas (GHG) criteria for national emissions data collection; and land criteria, to assess land use and potential change [14]. The resulting data are expressed as volume of biogas produced for each of 10 feedstock categories aggregated from the RO feedstock consignment sustainability data (RO-SUS) [15]. In order to derive fresh feedstock tonnages, biomethane content and the energetic value of the biomethane from the biogas data, these feedstock categories were characterised using data from the University of Southampton's Anaerobic Digestion Assessment Tool (ADAT) [16], and from the literature review.

A similar feedstock characterisation and energy calculation exercise was carried out on AD feedstock volume data provided by the UK Government's Department for Environment, Food and Rural Affairs (Defra) [17]. Results were compared with the RO data where possible.

Values used in the RO/Defra feedstock calculations were applied to the energy production data in the 2021 UK Digest of UK Energy Statistics (DUKES) [18] for anaerobic digestion and sewage gas. The DUKES data, which are produced by the UK Government's Department for Business, Energy and Industrial Strategy (BEIS) from a wide array of other sources, were also used to provide a wider overview.

Maximum CH_4 values for CO_2 biomethanisation of organic feedstocks derived from the literature review were then utilised to calculate a potential uplift in energy for the RO, Defra and DUKES data [13,17,18].

3. Results and Discussion

3.1. Performance of CO_2 Biomethanisation by Feedstock Type

Several excellent reviews on CO_2 biomethanisation have recently been published, from the wider perspective of biogas upgrading [4,19] to those more specifically focused on the biological process [3,11] and ranging from technical assessment to fundamental biochemical aspects [2,5,6]. This part of the current work considered the same data from the viewpoint of assessing the achievable performance according to feedstock type. An ideal study for this purpose accurately simulates large-scale operation, but in practice, this can be challenging. Common issues include the fact that laboratory-scale reactors may be fed only once per day and/or 5 days per week, while feeding of commercial digesters is usually more frequent. Arrangements for the supply and recirculation of external and headspace gases are also different at large and small scale, with laboratory experiments relying on a variety of approaches from continuous or intermittent input with or without recycling to daily injection under pressure. These affect the availability of exogenous gases over time, and further interact with different modes organic waste feeding which in turn influence the in-situ gas composition. Perhaps most importantly, reactor mixing and mass transfer parameters can vary considerably with scale. All of these factors may affect the biochemical environment, microbial community and metabolic pathways which determine the achievable methane yields and concentrations. In addition, experiments are designed for a range of different purposes and do not necessarily provide data on the maximum stable methane yield or concentration; nonetheless, these studies provide an indication of the potential performance and what is known to date.

In some cases, issues were encountered when attempting to put data from different sources into a comparable format. Some studies include residual H_2 in the reported biogas production, while others do not: not all specify this. Most studies indicate that reported gas volumes have been normalised to STP, but not all state which STP conditions are used. When gas compositions are reported, the main components (CH_4, CO_2, H_2) do not always sum to 100%: authors do not always state whether they are reporting dry and/or normalised gas volumes, or whether any discrepancies are due, e.g., to variability in experimental measurement or to the inclusion of water vapour and other trace gases. A range of different terms are used to describe the proportion of input H_2 that is consumed, and the resulting increase or decrease in CH_4 and CO_2 produced. There is a clear need for more consistent reporting, and a comprehensive set of standards for this purpose has been proposed [20].

3.1.1. Livestock Manures

Starting with the earliest work on in situ CO_2 biomethanisation of waste feedstocks [1], several studies have been conducted using cattle manure (CM) as the sole or main substrate. Luo et al. [1] ran duplicate CSTR digesters for 3 HRT at an OLR of around 6 g VS L^{-1} day^{-1} using sieved CM diluted by a factor of 2 for ease of small-scale operation. H_2 was added continuously to one digester via two ceramic diffusers at around four times CO_2 production (H_2/CO_2 ratio 4:1 v/v), while the other digester continued operating as a control. The experiment aimed to demonstrate the principle rather than to optimise the system: specific methane production per unit of organic feed (SMP) rose from 60 to 73 L CH_4 kg^{-1} VS but the biogas methane concentration only increased from 62% to 65% due to unconverted H_2 in the output gas.

Two-stage systems in which the first digester was fed twice daily on CM were tested by Bassani et al. [21]. The second digester received biogas and digestate from the first, with exogenous H_2 added at four times the initial CO_2 volume. One system was operated under mesophilic and one under thermophilic conditions. As expected, the thermophilic system gave higher specific and volumetric methane yields than the mesophilic. Both systems had higher specific methane yields than those obtained by Luo et al. [1], probably at least in part due to the longer overall HRT; but VMP in the mesophilic system was lower than in Luo et al. [1] because of the requirement for a second reactor. The methane concentration was 88.9% in the mesophilic system and 85.1% for the thermophilic, reflecting a higher residual H_2 content in the thermophilic system. The same system configuration and operating conditions were used to investigate changes in performance and microbial community over an extended period [22]. Results for the early stages of stable operation were similar to those of Bassani et al. [21]; but after two years of operation, biogas methane content was increased to around 99%, with an associated rise in both pH and volatile fatty acid (VFA) concentration. The VFA was primarily acetate, and was attributed to enhanced homoacetogenesis accompanied by changes in the microbial population.

Wahid and Horn [23] used a 2-stage system with two 6-L CSTRs operating in series at 55 °C. The first reactor was fed on CM at an HRT of 15 days, while the second received biogas and digestate from the first at an HRT of 20 days. After 120 days of stable operation, H_2 was added initially at a 2/1 H_2/CO_2 ratio, and subsequently at 4/1. Mixing speeds and gas recirculation rates were varied. There are some inconsistencies between reported values for OLR, SMP and VMP but the highest observed increase in SMP was 90%, similar to the value found by Treu et al. [22], though H_2 transfer efficiency was lower, resulting in a higher residual H_2 content.

Lebranchu et al. [24] operated a 142-L CSTR with a working volume of 100 L at 40 °C. The digester was fed continuously on CM with a VS content of 11% at an HRT of 28 days; the source of inoculum is not stated. H_2 was added continuously via a silicone tube diffuser without gas recirculation, and digester contents were mixed by helical and Archimedes screw mixers. The trial ran for 2.5 HRT with H_2 addition rates increasing from zero to 0.17, 0.29 and 0.45 L H_2 L^{-1} day^{-1}. This was followed by trials at different mixing speeds, then additional exogenous CO_2 was injected at 0.05 L CO_2 L^{-1} day^{-1} with a matching stoichiometric increase in H_2. SMP rose from 0.186 to 0.221 L CH_4 g^{-1} VS with H_2 addition, and to 0.236 L CH_4 g^{-1} VS with exogenous CO_2 and H_2. The relatively large scale, simple configuration and operating mode of this study make these results particularly useful in estimating the possible performance of conventional commercial systems.

Diluted pig manure was digested at mesophilic and thermophilic temperatures under different mixing conditions in CSTRs operating at an OLR of 2 g VS L^{-1} day^{-1} and an HRT of 25 days [25–27]. Under the same operating conditions, thermophilic digestion gave higher specific and volumetric methane yields than mesophilic. The highest SMP was observed in thermophilic conditions with continuous mixing and addition of H_2 and sodium formate [25], although the biogas methane concentration was lower than without the H_2 addition due to the presence of residual H_2 in the output gas. The greatest increase in SMP was under mesophilic conditions with H_2 addition and intermittent mixing [26]; in this case, methane production decreased with continuous stirring. The different response of the two systems to a change from intermittent to continuous stirring was attributed to different effects on alkalinity, TAN, VFA concentration and pH associated with CO_2 dissolution in the liquid phase.

No studies appear to have used other livestock manures as a mono-substrate, although digestate from a 2-stage system fed on chicken manure was used to inoculate reactors receiving exogenous H_2 and CO_2 and to replenish buffering capacity during the enrichment stage [28].

Systems with CM as a major feedstock component have been studied by several researchers. Whey has been used as an acidogenic co-substrate to counteract the pH increase associated with in situ CO_2 conversion [29]. An increase of 80% in SMP was achieved in a single-stage thermophilic CSTR with ceramic diffuser and magnetic stirring [30]; the highest stirring speed led to a small reduction in methane production. The best results were achieved using a hollow fibre membrane (HFM) for H_2 transfer, with an increase in SMP from 0.288 to 0.516 L CH_4 g^{-1} VS, a biogas methane content of 96.1%, and no residual H_2 detected. A two-stage thermophilic system previously operated with CM as a mono-substrate [23] was also used to test CM + whey at different feedstock ratios, feeding frequencies and stirring speeds [31]: SMP increased by up to 23% with H_2 addition but residual CO_2 and H_2 contents were high, indicating poor mass transfer and leading to a rather low biogas methane concentration.

CM and potato starch were used as co-substrates in a 2-stage system consisting of a CSTR followed by an upflow anaerobic sludge blanket (UASB) reactor, with H_2 added to the first stage and gas recirculated between the stages [32]. H_2 transfer efficiencies were around 98.5% but VFA accumulation occurred, and stable operation with H_2 addition was not achieved: the results are therefore not shown in Table S1.

Mesophilic digestion of CM and vegetable wastes was carried out in a 2-stage reactor at a reported OLR of 3.5 g VS L^{-1} day^{-1} and HRT of 10 days, under a range of gas recirculation and H_2 addition rates [33]. H_2 addition gave increases in SMP of around 67% and 157% with and without optimised gas recirculation, respectively. The 2-stage system appeared capable of eliminating residual H_2 and was able to achieve biogas methane concentrations of 92% without and 99% with recirculation, respectively.

For studies using other co-substrates with CM as the minor component, see the sections on crops and agro-wastes and on food wastes below.

Several other studies not shown in Table S1 have used manure or manure-based digestates and feedstocks and/or inoculum. Batch experiments with CM as substrate were carried out to investigate the effect of H_2 addition at different ammonia concentrations [34,35]. Garcia-Robledo et al. [36] used micro-sensors to study the dynamics of H_2 and CO_2 conversion in fresh CM and CM digestate, but the experiment was not designed to establish likely yields in a typical commercial process. Lab-scale breeder reactors inoculated from CM digesters and fed on digestate from these were used to generate material for batch testing with H_2 addition [37]. The work was part of a study investigating the effect of inoculum source on metabolic pathway, and also tested wastewater biosolids digestate; but again, the aim was not to simulate performance in a full-scale process.

The results in Table S1 confirm that useful increases in SMP can be achieved by biomethanisation of CO_2 in CM digestion: this is of particular interest for this substrate, which is known for its low energy potential, making small-scale on-farm digestion economically challenging [38]. High rates of H_2 transfer and conversion were demonstrated in several of the studies. Membrane systems (ceramic and HFM) generally achieved good transfer without gas recirculation, while digester mixing also had an effect, although high mixing rates did not always improve methane yields [26,29,31]. Mixing and mass transfer performance are strongly affected by scale: effective mixing is more difficult to achieve in a full-scale plant, while depth and pressure effects may contribute to improved gas transfer [39–41]. Understanding of these effects and how they interact with system biology is still in its early stages, and more research at pilot and full scale is clearly needed if industry is to have confidence in adopting these technologies [3,4].

As can be seen from Table S1, the change in biogas methane content brought about by CO_2 biomethanisation is highly dependent on the applied H_2 loading and the effectiveness of the transfer method, as well as requiring the presence of an appropriate microbial population. Residual unconverted H_2 reduces the methane concentration in the output gas. Depending on the intended end-use, this may be a significant issue, while low transfer and conversion rates arguably represent a waste of H_2. Two-stage systems were able to achieve a high methane content, but provision of a second digester is expensive, especially if heating and mixing are applied. In the systems described here, the primary digester generally received both organic feed and H_2, with biogas and digestate then passed to the secondary digester to improve the H_2 conversion with or without further gas recirculation. This approach is already a step towards ex-situ biomethanisation, in which case, a reactor type with more efficient gas transfer may be preferred. Many manure digesters have unheated static tanks for digestate storage, however, so there may also be scope for research on whether and how these could be adapted to fulfil a similar role in increasing H_2 conversion.

Values for VMP and MER varied considerably between the studies in Table S1, reflecting differences in substrate properties and concentration/dilution, as well as the presence of residual unconverted H_2 and CO_2. These parameters are commercially significant, however, and care is therefore needed when interpreting or reporting them. Co-digestion of Whey + CM may be attractive as a means of increasing the volumetric and specific methane productivity as well as reducing the pH, while both substrates may be locally available in dairy areas.

Conclusions for pig manure were broadly similar to those for CM, despite some differences in the typical characteristics for these substrates [16].

Most of the studies reported in Table S1 were carried out under thermophilic conditions, which generally give a higher SMP at a shorter HRT than mesophilic manure digestion [21]. Thermophilic digestion of CM is not widely practiced in the UK, however; and in combination with the different conditions applied in different studies, this makes it difficult to choose appropriate values for use in the current modelling. It was therefore decided to take typical baseline values for mesophilic CM digestion of SMP 0.190 L CH_4 g^{-1} VS, biogas composition 60:40% CH_4: CO_2 and pH 7.5 [16,24], and apply the pH/pCO_2 relationship developed in Tao et al. [12]. Assuming stable mesophilic operation with H_2 addition is possible at pH 8.2 (Table S1, [21]), this equates to a biogas methane content of 95% and SMP around 0.300 L CH_4 g^{-1} VS with CO_2 biomethanisation. These values are summarised in Table 1 for application in the modelling stage of this work.

Table 1. Estimated maximum biogas methane content and SMP for use in modelling.

Substrate Type	Without CO_2 Biometh				With CO_2 Biometh		
	SMP [a]	CH_4 [a]	pH	a [b]	Assumed Max pH [a]	Predicted Max CH_4	Predicted Max SMP
	L CH_4 g^{-1} VS	%				%	L CH_4 g^{-1} VS
Livestock manures	0.190	60	7.5	7.60×10^{-8}	8.2	93	0.295
Crop & agro-wastes	0.350	55	7.5	6.75×10^{-8}	8.2	92	0.587
Post-consumer food wastes	0.450	55	7.9	2.54×10^{-8}	8.2	79	0.649
OFMSW	0.300	55	7.5	6.75×10^{-8}	8.2	92	0.503
Sewage sludge	0.260	65	7.5	8.68×10^{-8}	8.0	90	0.359

[a] Based on literature values (see Tables S1–S4), cross-checked with ADAT [16] where possible. [b] Coefficient for equation relating pH and pCO_2 under mesophilic conditions [12].

3.1.2. Crops and Agro-Wastes

Relatively few CO_2 biomethanisation studies to date have focused on these substrates. A comparison of different possible scenarios for in-situ and ex-situ CO_2 biomethanisation was carried out based on experimental data for grass silage [42]. Only in-situ results are included in Table S1, although ex-situ systems may correspond to the second stage of a two-stage process in some operating configurations. In this part of the study, the grass silage was digested thermophilically in a single-stage CSTR at an OLR of 4 g VS L^{-1} day^{-1} and HRT of 46 days using two types of diffuser. The findings demonstrated once again the importance of effective mass transfer, with the ceramic diffuser achieving a 68% increase in SMP compared to 19% for the low-capacity diffuser, and quite creditable VMP and MER values, although still with a significant residual H_2 content.

130-L anaerobic filters (AF) fed on maize silage hydrolysate were trialled at two H_2/CO_2 ratios, using a venturi nozzle for gas injection and with both liquid and gas recirculation [43]. The AF acted as the methanogenic reactors in a 2-stage system, in which the first stage was a continuously stirred acidification reactor fed on maize silage with a small component of sugar beet silage effluent. The fixed bed in the AF consisted of high-density polyethylene bio-media with a specific surface area of 312 m^2 m^{-3}. The AF configuration offers potential advantages in biomass retention and gas transfer, but previous studies have focused on ex-situ biomethanisation. It should be noted that some values presented in Table S1 differ from those reported in the paper due to differences in definition and calculation methods. Maize silage was also the feedstock in a 2-phase thermophilic system consisting of hydrolysis and fixed-bed methanogenic reactors [44]. The authors trialled a range of hydrolysis conditions, with H_2 added to the methanogenic reactor in some runs. Specific methane production is only reported for the whole system, and the control and experimental methanogenic reactors were, respectively, 145 and 180 L, making it difficult to compare performance on a volumetric basis; but methane concentrations were successfully increased to over 90%.

Digestate from an AD plant processing mixed agro-wastes was used as inoculum and feed in a single-stage mesophilic CSTR trial at different H_2 loadings [45]. The feedstock of the main AD plant is not described in detail, but is believed to include pig manure, deep litter, slaughterhouse residues and some high-lipid wastes [46]. H_2 was added to the headspace for up to 5 consecutive days, followed by a pause of 10–21 days, in a process described as pulsed injection. While the use of digestate in this way may be viewed as replenishing an inoculum rather than adding a substrate, the slow anaerobic biodegradation rates of some of the original feedstock components mean there is likely to be residual methane potential even after conventional digestion: in this case, the control reactor without H_2 addition had an SMP of 0.293 L CH_4 g^{-1} VS, higher than typical values for CM. This increased to 0.571 L CH_4 g^{-1} VS at the best-performing H_2/CO_2 ratio tested. Although the experimental method did not simulate a conventional AD process, it confirmed the potential for CO_2 biomethanisation to raise methane yields for this material, and provided some insights on process and microbial community parameters. The same set-up was used to explore the effect of H_2 additions at higher and lower headspace CO_2 concentrations across a range of OLR [47], with the results also showing acclimatisation over successive pulsed injection cycles (Table S2).

Two important studies using a similar mixed-agrowaste feedstock containing manure, straw, deep litter, grass and maize silage were carried out in a 1110 m^3 CSTR operating at 52 °C with a HRT of 13 days [39,40]. Some details of the substrate components, which varied over the experimental period, are provided in the papers. H_2 was injected using a venturi system over relatively short periods in order to allow evaluation of the mass transfer potential at different flow rates and with/without gas recirculation. These studies are not included in Table S2 as they represent short-term tests without equilibration and acclimatisation of the system; but the very large size supported by detailed analyses of performance provides valuable reassurance on the potential for full-scale application.

A full-scale thermophilic digester fed on a mixture of pig and cattle manure, maize silage and deep litter was used to provide inoculum for batch reactors. These were fed once only on maize leaf then periodically supplemented with H_2, in an experiment designed to investigate metabolic pathways and microbial community structure [48]. Fed batch operation was also employed in a serum-bottle trial using inoculum from a mesophilic AD plant fed on maize and sweet sorghum silage and pig manure, with cellulose as the trial feedstock [49]. Sarker et al. [50] operated a 5-L CSTR at 39 °C, using inoculum from a food waste digester with a 0.55% *w*/*w* admixture of CM. From day 39–71 the digester was fed intermittently on CM without digestate removal. Increasing volumes of H_2 were injected sequentially between days 40 and 71 into the headspace without gas recirculation. VFA concentrations increased and effective CO_2 biomethanisation was not achieved, which the authors attributed to mass transfer limitations. Although these studies provide insights on the respective feedstocks and inoculums, they did not attempt to simulate conventional digestion and thus do not appear in Table S2.

Details for the two types of crop material investigated are summarised in Table S2. Digestion of grass silage in a conventional CSTR with H_2 addition via a ceramic diffuser was an effective means of increasing SMP [42]: values for VMP and MER were also relatively high, thanks to the good baseline SMP and applied OLR for this substrate. The output gas contained a significant percentage of residual unconverted H_2; but ratios between H_2 consumed and methane produced were close to stoichiometric, giving confidence in the experimental results. In situ CO_2 biomethanisation in fixed bed reactors processing maize hydrolysates was also effective at increasing SMP [43,44], though here too, there was relatively little improvement in biogas methane content under the operating modes tested. CH_4 production and CO_2 consumption ratios in Illi et al. [43] were lower than would be expected based on H_2 transferred, possibly indicating dissolution in the liquid phase; no detailed analysis was possible for Schönberg and Busch [44] as results for the two stages could not be disaggregated.

Studies on CO_2 biomethanisation of digestates from mixed agro-wastes clearly demonstrated the potential for increasing methane yields towards those typical of commercially-interesting substrates (Table S2, [45,47]). The high biogas methane concentrations achieved here reflected the pulsed addition mode of H_2 operation, however, and cannot be transferred directly to conventionally operated full-scale digestion.

Based on the results in Table S2 and the ADAT database of feedstock properties [16] a baseline SMP of 0.35 L CH_4 g^{-1} VS with a biogas methane content of 55% was selected for both crops and mixed agro-wastes, giving an assumed maximum methane content of 95% (Table 1).

3.1.3. Food Wastes

Mixed food waste from a commercial AD plant, including wastes from food processing, catering, dairy production and selected abattoir waste fractions, was the substrate for a mesophilic trial with biomethanisation of both endogenous and exogenous CO_2 [12]. Conversion of the internally produced CO_2 increased the SMP from 0.561 to 0.776 L CH_4 g^{-1} VS with a methane concentration of around 90% and no residual H_2. Attempts to further increase the methane content led to VFA accumulation, and it was concluded that stable operation was possible up to pH 8.5. Addition of exogenous CO_2 with H_2 enabled a further increase in SMP to 1.215 L CH_4 g^{-1} VS with the same biogas composition. The system achieved an impressive VMP of 5 L CH_4 L^{-1} day^{-1}, and the authors commented that the maximum exogenous CO_2 addition was determined by the capacity of the experimental equipment rather than by any biological limitation in this case.

Source-separated domestic food waste has been widely studied because of its rising popularity as an AD feedstock. A mesophilic trial carried out using FW from a source characterised in previous work [51,52] demonstrated an increase in SMP from 0.446 to 0.719 L CH_4 g^{-1} VS and stable operation at pH 8.1. Around 45% of CO_2 was converted, but the methane content of the output gas was only around 74% due to the presence of residual H_2 [53]. A combination of H_2 addition and auto-generated pressure was tested as a means of increasing biogas methane concentration using a synthetic FW composed of tinned pork and beans with white bread [54]. In the example in Table S3, the quantity of H_2 added was only sufficient to convert around 9% of the endogenous CO_2 on a stoichiometric basis; but when combined with additional dissolution of CO_2 caused by the 5-bar operating pressure, the biogas methane concentration increased to 90.6%.

Food waste from a University cafeteria was the organic substrate in a trial involving injection of a synthetic syngas consisting of H_2 and carbon monoxide (CO) at a 5:4 v/v ratio [55]. Both mesophilic and thermophilic conditions were tested and the digestates were also pyrolysed to determine the properties of the resulting syngas, with a view to creating an integrated process. The authors noted that methane productivity was higher than expected from stochiometric conversion of syngas, and attributed this to synergistic effects; although the difference decreased as the trial progressed. There is some inconsistency, however, between the reported feedstock properties, OLR and HRT: a feedstock VS of 25.1 g L^{-1} would give an HRT of 7 days at the reported OLR. A value of 251 g VS L^{-1} is more typical for FW, and without dilution, would correspond to an HRT of around 71 days: the reported HRT is 20 days, however, so feedstock dilution may have been carried out to facilitate thermophilic operation without ammonia inhibition [56]. Digesters typically require around 3 HRT to approach steady-state conditions with regard to organic loading, and this washout process might offer an additional explanation for observed changes in VS and COD content if the HRT was set at 20 days. In any case, stable operation was achieved with high syngas utilisation, a methane concentration of around 64%, and an increase in SMP of around 31% and 33% in mesophilic and thermophilic conditions, respectively.

CO_2 biomethanisation of thermally-treated FW digestate from a laboratory-scale digester was carried out in a trickle-bed reactor operating at 37 °C with a 10-day HRT, at different H_2 loadings and recirculation rates [57]. The SMP increased from 0.248 to 0.450 L CH_4 g^{-1} VS under the best conditions trialled, but biogas methane content was limited by the presence of residual H_2, and dosing with HCl was introduced after the pH reached 8.45.

Batch experiments using food waste from a university dining hall were conducted to investigate the effect of acclimatisation and other parameters [58–60]. The trials used headspace injection over consecutive periods and were not designed to simulate conventional AD. There were some errors in the mass balance calculations and the H_2 addition proposed is far from stochiometric needs or typical literature values (Tables S1–S4) [59]; but inoculum acclimatisation was clearly demonstrated.

Other CO_2 biomethanisation studies have used specific waste streams from food processing industries. Cheese whey and related dairy wastes were used as the sole or main substrate by several researchers. Treu et al. [61] attempted to digest whey at mesophilic and thermophilic temperatures with and without CO_2 biomethanisation, but were unable to achieve stable operation in thermophilic conditions. $NaHCO_3$ was added to provide buffering during mesophilic digestion without H_2 addition; and consumption of accumulated VFA may have affected methane production values in some periods. A feedstock of cheese whey permeate and cheese powder was successfully digested at 54 °C using NaOH for pH control [62]. In both studies, SMP increased with biomethanisation, but residual H_2 reduced the biogas methane concentration. Co-digestion of cheese whey with CM was successful as a means of improving process stability in thermophilic conditions [61,63], though increases in methane content and SMP were relatively small under the conditions applied. The co-digestion trial begun in Treu et al. [61] was continued in a study focusing on the effects of bioaugmentation [64]: no control digester without H_2 addition was run and the results are therefore not presented in Table S3.

Other food industry substrates trialled include potato starch wastewaters [65] and bioethanol distillery effluents [66]. The former was tested in a thermophilic UASB reactor with different diffuser types and gas and liquid recirculation rates. As can be seen from the results in Table S3, H_2 transfer efficiency was higher with a ceramic diffuser and gas recirculation but some residual H_2 was still present in the output gas. The latter study used a 148-L mesophilic anaerobic membrane bioreactor, and achieved a biogas methane content around 98% with minimal residual CO_2 or H_2. Both studies contained minor inconsistencies between reported values for some parameters, but clearly demonstrated that CO_2 biomethanisation could successfully enhance SMP from organic wastes in these systems.

A synthetic feed containing yeast extract, sucrose and a range of nutrients was used in trials to assess the influence of OLR [67] and of total ammonia nitrogen (TAN) concentration [12] on CO_2 biomethanisation performance. Mesophilic digesters at OLR of 2 and 3 g COD L^{-1} day^{-1} also received additional exogenous CO_2 and H_2, increasing the original SMP by more than 2-fold and giving biogas methane contents of around 90%. The limiting pH for this substrate and inoculum appeared to be around 7.9 at lower TAN concentrations, and 8.2 at higher TAN. The authors noted that further studies are needed to identify the factors determining maximum operating pH in different systems [12].

Glucose was used as a model feedstock in mesophilic reactors in a trial investigating the effect of different H_2 loadings [68]. H_2 was injected once per day into the headspace, which was sealed and allowed to pressurise until venting before the next injection. The inoculum used came from a digester fed primarily on cattle manure, and nutrients were provided by occasional dosing with diluted inoculum. The applied OLR was very low and pH control was required, but the mode of operation enabled a reported increase in biogas methane content from 66% to 94%. A two-stage mesophilic UASB system with glucose as the sole carbon source was tested at different OLR, H_2 loadings and gas recirculation rates [69], and achieved H_2 transfer efficiencies of up to 98.8% with biogas methane contents between 92–94%. Glucose was also used as the organic substrate in the trial of CO injection into a mesophilic UASB reactor [70]. The SMP increased from 0.312 to 0.536 L CH_4 g^{-1} COD added but biogas methane content fell due to additional CO_2 production, as indicated by the stoichiometric relationship shown in Equation (2):

$$4CO + 2\,H_2O \rightarrow CH_4 + 3CO_2 \quad (2)$$

Synthetic substrates are generally adopted to provide controlled conditions for laboratory studies. Where the substrate is chemically defined, the theoretical SMP and expected biogas methane content are known or can be estimated; while parameters such as operating pH may be determined by the substrate composition or be controlled for experimental purposes. In any case these materials are rarely encountered as real-world feedstocks, and thus, values for modelling purposes were not required.

The characteristics of individual waste streams from food processing industries vary widely and it is clearly not possible to choose representative values; but information on baseline digestion conditions is often available from other studies. The minimum CO_2 and corresponding maximum achievable methane content for stable operation is more difficult to determine, and may depend in part on reactor type and operating conditions. In CO_2 biomethanisation trials with potato starch wastewater in UASB reactors (Bassani et al., 2016), stable operation was reported at a pH 8.38 with a CO_2 content 10% in the output gas. For ethanol distillery wastewater treated in an anaerobic membrane bioreactor (AnMBR) [66], the pH had reached 7.9 at a biogas methane content of 97.9% CH_4 and 1.4% CO_2, with a corresponding increase in SMP from 0.297 to 0.389 L CH_4 g^{-1} VS: values close to these can therefore be taken as the maximum for this substrate for modelling purposes.

For post-consumer domestic food wastes in Europe, the SMP value is typically around 0.450 L CH_4 g^{-1} VS [71] with a methane content of around 54%. There are no published studies to confirm the minimum biogas CO_2 content or maximum pH for stable operation, so for the purposes of modelling, a maximum pH of 8.2 was conservatively assumed [53], corresponding to a biogas methane content of 79% (Table 1).

3.1.4. OFMSW

Very little work has been done on CO_2 biomethanisation of OFMSW. When H_2 and CO_2 were batch-fed to samples from real and simulated landfill wastes, methane production was observed but homoacetogenesis was found to be the dominant pathway under the conditions used [72]. Mixtures of CM and synthetic OFMSW (composed of paper, bread and fruit and vegetable wastes) were batch digested at 55 °C at five CM:OFMSW ratios from 100:0 to 50:50 on a mass basis [73]. A nutrient medium containing glucose was added but the only source of inoculum was the fresh CM. After 24 h of fermentation, H_2 and CO_2 were injected continuously at a $4/1$ v/v ratio for 20 days. The results confirmed that increases in VMP could be achieved using this approach, with biogas methane contents up to 97%; the values are not included in Table S2 as the study was not designed to simulate a conventional AD process. In the absence of suitable experimental findings, a baseline SMP of 0.35 L CH_4 g^{-1} VS with a biogas methane content of 60% was adopted [16] for the purposes of modelling, with assumed baseline and maximum pH values of 7.5 and 8.2, respectively, giving a maximum methane content of 92% [53].

3.1.5. Sewage Sludges

AD is a well-established treatment technology for sewage sludges, with existing infrastructure across the water industry, making CO_2 biomethanisation of these substrates an interesting prospect. Alfaro et al. [74] carried out an extended trial using two 20-L mesophilic CSTRs fed on thickened primary and secondary wastewater biosolids from a full-scale wastewater treatment plant (WWTP). H_2 was injected via a HFM and the system was tested at different gas recirculation rates. The baseline SMP varied slightly with different batches of feedstock, but SMP was increased by more than 40% with a biogas methane content of 70–73% and good H_2 utilisation at higher recirculation rates. A study in a mesophilic fermenter with H_2 addition via bubbling and gas recirculation produced a similar improvement in SMP, and demonstrated stable operation at a biogas methane content of 90% with a maximum pH in the range 7.9–8.0 [12]. A trial using the same equipment and a similar sludge from a different wastewater treatment plant achieved a biogas methane content of 85% at pH 7.9. Addition of exogenous CO_2 and H_2 enabled a four-fold increase in the original SMP, but the biogas methane content fell due to the presence of residual CO_2 and H_2 [53].

An enrichment trial was carried out in three mesophilic CSTR digesters fed on mixed primary and secondary sludge [75]. The low feedstock solids content led to a relatively low OLR of around 1 g VS L^{-1} day^{-1} at an HRT of 15 days. Batchwise addition of H_2 to the headspace produced a maximum biogas methane content of 80% with a near-stoichiometric ratio between H_2 consumption and CH_4 production. Another trial using sewage sludge from the same source and continuous H_2 addition achieved a biogas methane content of 90% at a H_2/CO_2 ratio of 7/1 without any adverse effect on VS degradation or significant VFA accumulation [76].

A study combining H_2 addition and pressurisation was carried out in a 35-L mesophilic digester treating mixed wastewater biosolids under a range of operating pressures and at two H_2 loadings [77]. The experimental design did not include a control without H_2 addition, but the best performance was achieved at the highest pressure and H_2 loading tested, with a SMP of 0.418 L CH_4 g^{-1} VS and a biogas methane content of 92.9%. Based on VS removal, it was estimated that around 0.37 L CH_4 g^{-1} VS of the SMP was due to organic load and 0.13 L CH_4 g^{-1} VS to CO_2 biomethanisation. Digester pH remained below 7.2 presumably due to CO_2 dissolution, and there were no signs of process instability or VFA accumulation. Pressurisation is likely to require more extensive modifications to reactor design and operation than H_2 addition under ambient pressures, but the authors noted that process efficiency in terms of VS removal was unaffected by the changing conditions and that a pressure increase could be an effective way to enhance H_2 mass transfer without incurring high energy costs.

Addition of CO was trialled in a thermophilic digester fed on mixed primary and secondary wastewater biosolids [78]. The highest SMP achieved was more than twice the value without CO addition, but the biogas methane content was low (<20%) due to the presence of both residual CO and additional CO_2 generated in accordance with Equation (2). The highest SMP without residual CO was around 1.7 times the baseline value with a biogas methane content of around 30%. The CO was added via a HFM module, ensuring good dissolution despite its rather low solubility, and no signs of process inhibition were observed. A HFM module was also used to inject simulated coke-oven gas (SCOG) consisting of 92% H_2 and 8% CO into a mesophilic CSTR digester fed on mixed primary and secondary sludge [79]. The maximum SMP achieved was 0.604 L CH_4 g^{-1} VS, compared to a baseline value of 0.256 L CH_4 g^{-1} VS without SCOG addition. Maximum biogas methane content was 98.8% with 0.3% CO_2, and pH controlled to 8.0.

In other work with wastewater biosolids, Hu et al. [80] carried out batch tests using nano-scale zero-valent iron and waste iron scraps as a means of generating H_2, with waste activated sludge (WAS) as an organic substrate in some set-ups. Vechi et al. [37] used lab-scale reactors inoculated with sewage sludge digestate and fed on primary and secondary sewage sludge to produce material for batch testing of H_2 additions, for comparison with inoculum from CM digestion. Inoculum from a range of sources including digestion of sewage sludge, paper mill sludge, cattle and poultry manures and FW, plus aerobic wastewater sludge, was tested to investigate the immediate response of different microbial communities to exposure to high H_2 partial pressures [81]. These studies did not attempt to simulate typical operating regimes but provide a variety of insights on microbial populations and metabolic pathways.

Apart from one study of with CO addition [78], all of the work presented in Table S4 was carried out under mesophilic conditions, reflecting the current widespread use of this temperature range by the water industry. CO_2 biomethanisation gave useful increases of 40–50% in SMP. VMP remained low, reflecting the dilute nature of this substrate, but high values were achieved when additional CO_2 and H_2 were added [53], reinforcing the observation that at sites with multiple digesters a single digester retrofitted for H_2 injection should be capable of processing biogas from several others [12]. Good H_2 transfer was achieved at the scales tested even without membrane diffusers, and ratios between H_2 consumed and methane produced were close to stoichiometric values. The results indicate that CO_2 biomethanisation of this type of substrate can operate stably in a pH range of 7.9–8 [12,74] corresponding to a max biogas methane content around 90% [12]. Baseline SMP values without CO_2 biomethanisation vary depending on factors such as HRT, OLR and the proportion of primary and secondary wastewater biosolids: for the conventional CSTR trials reported here, they ranged from 0.21 to 0.3 L CH_4 g^{-1} VS with a biogas methane content of 60–65%. These correspond well to typical values of 0.260 L CH_4 g^{-1} VS and 60% CH_4 given by ADAT [16], and were therefore taken forward into Table 1.

3.1.6. Conclusions from Performance Analysis for CO_2 Biomethanisation of Organic Feedstocks

Consideration of the literature shows a wide range of study types, though there are also obviously many gaps. Table 1 contains a group of parameters for each feedstock type, selected specifically for the purposes of the current modelling work. In Tables S1–S4, a wider selection of parameters and indices to facilitate comparison between results is presented: the significance of these is briefly discussed in the Supplementary Materials. Tables S1–S4 focus on the most successful outcomes in which stable operation was apparently achieved: the studies considered cover many other sets of operating conditions, including some where failure occurred as evidenced, e.g., by irreversible VFA accumulation. Many studies also now include analysis of microbial populations, which when combined with operational data contributes to a growing understanding of the links between community structure, functionality and performance. The majority of studies reported are still at laboratory scale, however, and more work is needed on the effects of scale-up and the interaction of system biology and engineering parameters such as mixing and mass transfer behaviour. For industry to be able to adopt this technology with confidence, a better understanding of the mechanisms affecting performance is needed, together with the development of simple and robust control strategies to avoid instability or suboptimal conditions.

While there are issues in attempting to compare data from different studies, Tables S1–S4 provide some useful insights and parameter values. VMP, MER and H_2 transfer efficiency are important for techno-economic assessment, although values are likely to vary considerably with operating scale and system configuration. SMP is an indicator of the efficiency of conversion for a particular feedstock, while CH_4 and H_2 content of the output gas affect potential end uses.

In many studies, though not all, the ratio of H_2 transferred to CH_4 produced and to CO_2 removed settled at close to the stoichiometric value of 4. Values diverging from this can be explained by utilisation of transferred H_2 for other purposes such as VFA production or biomass growth. Most studies that calculate ratios of this type also choose to ignore the very small amount of dissolved H_2 leaving in the digestate [1,77,82]. Intriguingly, in several cases, the introduction of CO_2 biomethanisation is associated with an increase in VBP [21,22,24,27,29,33,43,45,47,57,61,66,68,74], meaning a larger amount of CH_4 is being produced and/or there is more CO_2 to biomethanise, per unit of organic feed.

It is difficult to assess the significance of this as other factors can also vary: since these studies use real organic feedstocks, there may be minor changes in feedstock properties both day-to-day and between batches in longer studies. HRT are relatively long and not all reported values are from periods where stable steady state operation (often defined as 3 HRT under the same conditions) has been achieved. Accurate determination of H_2 concentrations can also be problematic and as noted, there are different ways of dealing with variability in experimental measurements, with some studies normalising gas composition to 100% and others not. In some cases, however, the observed MER is greater than the volumetric CO_2 production without CO_2 biomethanisation [21,30,33,45,66,68], strongly supporting an increase in VBP. A similar phenomenon of increases in VBP or VMP has been observed in other work on gas recirculation [83], though no clear mechanisms have been confirmed. Many studies are not designed to separate the impacts of gas recirculation and H_2 addition: but Khan et al. [33] found that the increase from H_2 addition and recirculation was greater than from recirculation only. The simple method adopted in the current work, of calculating the MER and the increase in SMP from the difference between experimental and control or baseline values, conflates these effects as well as other factors; but several authors have suggested that addition of H_2 or syngas may have synergistic effects beyond stoichiometric CO_2 conversion [25,33], and more research is clearly needed in this area.

3.2. *UK Feedstock Data Extraction and Analysis of CO_2 Biomethanisation Potential*

Several data sources, in addition to those used, were considered as a basis for assessing the potential for CO_2 biomethanisation from organic wastes in the UK. This included three

UK government sources based on specific incentive schemes and expressed in terms of energy generation capacity: the FIT, RHI and Renewable Transport Fuel Obligation (RTFO). In addition, several other organisations collate data on AD plant operations, including the Anaerobic Digestion and Bioresources Association (ADBA), the Renewable Energy Foundation (REF) and the National Non-Food Crops Centre (NNFCC). The different datasets are collected for different purposes, do not cover all materials and use different reporting formats and assumptions, making comparison and generalisation difficult. Brief details on these sources and their advantages and disadvantages are presented in Section S2 of Supplementary Materials.

3.2.1. Estimation of CO_2 Biomethanisation Based on Ofgem RO Data

Although only a proportion of UK AD plants generating electricity are covered by the RO, the RO dataset on feedstock and biogas production is arguably the most detailed freely available. The Ofgem 2019–2020 RO Annual Report [13] aggregates the feedstock consignment data provided by 157 AD plants operating under the RO scheme into 10 broad feedstock categories: Silage; Food, garden and plant waste; Manures and slurries; Distillery waste; DAF sludge/wastewater; Crops; Glycerol; Dairy waste; Municipal waste; and Other. For this part of the study, it was decided to use the aggregated categories and not the original RO-SUS data [15] because it was not clear how the consignment data had been mapped to the categories and because some of the individual RO-SUS consignment data were coded, so the organic source could not be ascertained. The Ofgem feedstock categories also did not map precisely onto those used in the literature survey. The following assumptions were therefore made and confirmed where possible by cross-checking with uncoded RO-SUS data.

Tables S1–S4 demonstrate that, under suitable conditions, biogas methane contents of 90% and above can be achieved for a range of feedstocks commonly used in commercial AD plants. Where possible, expected values for methane content after CO_2 biomethanisation of the Ofgem feedstock categories were therefore taken from Table 1; where no values were available (i.e., Distillery waste, Glycerol, Dairy waste and Other, detailed below), a methane content of 90% was assumed.

The majority of biogas in the 'Silage' category derives from maize, although RO-SUS data also shows the presence of grass silage, and crop silages such as those from rye and wheat. This category was therefore assumed to map onto 'Crop and agro-wastes' in Table 1 and to have the TS and VS content of maize silage in the ADAT database [16].

'Food, garden and plant waste' likely included post-consumer food waste (e.g., from households and restaurants), as well as food production and processing wastes such as reject potatoes, fruit and vegetable waste. It could also include garden and plant materials such as 'Food/garden waste', bulbs, grass and sugar beet pulp. The majority of named plant materials were high energy (e.g., sweetcorn and beet), however, and consignments mentioning 'Food/garden waste' only appeared to make up around 5% of the total for this category. 'Post-consumer food waste' values from Table 1 and source separated food waste from ADAT [16] were therefore adopted for this category.

The 'Manures and slurries' category was conservatively assumed to have the characteristics of cattle slurry, since no indication was given of the proportion of manure to slurry, or of different slurry types. Data for 'Livestock manures' from Table 1 were used, with feedstock characteristics for 'Cattle Slurry' from ADAT [16].

'Distillery waste' may include feedstocks such as draff, pot ale syrup and malt effluent. Again, a conservative approach was taken and, as aggregates, these were assumed to be low TS wastes. There was no directly applicable category in Table 1, so a default methane content of 90% was used. Feedstock characteristics were not available from ADAT [16], so data from the literature were used [84].

The 'DAF sludge/wastewater' category was assumed to equate to the feedstock characterisation for 'Sewage sludge' in Table 1 and in ADAT [16].

The definition used by Ofgem [13] for the 'Crop' category is unclear, but might include feedstocks such as whole crop and maize meal. For these, the maximum methane content was taken from 'Crop and agro-wastes' in Table 1 with TS and VS values based on an average for maize silage and maize corn in the ADAT database [16].

'Glycerol' could be clearly identified in the RO-SUS data [15] and was characterised using standard ADAT values [16], with a default maximum methane content of 90%, as Table 1 had no direct equivalent.

The 'Dairy waste' category could include RO-SUS data for milk whey, dairy waste, liquid food/dairy waste, AD whey permeate and dairy sludge, and was assigned the characteristics of 'Whey' in the ADAT database [16]. With no direct equivalent in Table 1, a default maximum methane content of 90% was assumed.

'Municipal waste' was characterised as 'OFMSW' as this category appears in both Table 1 and ADAT [16].

The narrative in the Ofgem annual report identified 'Other' consignments as including blood, viscera, tallow, fishery waste and plant oils [13] which are all high-strength wastes, and were assumed to have the same properties as the Ofgem 'Food, garden and plant waste' category.

The dataset accompanying the Ofgem report [13] lists 'Quantity [of biogas] burnt (million m^3)'. Biogas volumes for individual feedstocks in the dataset are reported in million m^3, to a numerical total of 525,191,780. This does not match the report narrative, which notes production of 525 million m^3; and is also three orders of magnitude larger than the 364.06 million m^3 biogas in the 2018–2019 Annual Report and dataset [85]. The units in the database were therefore corrected from 'million m^3' to 'm^3'.

To calculate the gross energetic value of the biogas, biogas volume was multiplied by the estimated methane content and the calorific value (lower heating value) of methane, taken as 35.82 MJ m^{-3} CH_4 at STP, and converted to GWh as shown in Table 2. The gross energetic value with CO_2 biomethanisation was then calculated by multiplying the original gross energetic value by the ratio of biogas methane contents with and without CO_2 biomethanisation for the given feedstock, i.e., assuming no change in biogas volume. Table 2 shows the resulting improvement in gross energetic values through the increased methane production due to the addition of H_2. This equates to a 57% uplift in the overall energy value of the biogas. While this assumes CO_2 biomethanisation could be applied in every participating AD plant, which is clearly an over-optimistic scenario, it nevertheless represents a very significant potential uplift that warrants further investigation.

In order to estimate the tonnages of fresh feedstock giving rise to the produced biogas, typical values for the SMP, total solids (TS) and volatile solids (as a percentage of TS) for each feedstock were taken from Table 1 and from the ADAT database [16]. Tonnages for each feedstock category were then calculated from these values as shown in Table 2. The assumptions used to derive the feedstock parameters are provided in Table 2, and result in a total estimated tonnage of 6.8 million tonnes of FM, of which 1.3 million tonnes consists of silage and crops. Together, silage (18.1%), food/garden/plant waste (31.5%) and manures/slurries (12.3%) produce 85.6% of the biogas shown in Table 2.

Table 2. RO AD plant feedstock characteristics with and without CO_2 biomethanisation.

Feedstock Category [1]	Biogas Produced [1] m^3	Without CO_2 Biometh			With CO_2 Biometh			TS [2] %FM	VS [2] %TS	Feedstock [7] Tonnes FM
		CH_4 [2] % vol	SMP [2] L CH_4 kg^{-1} VS	Gross Energetic Value [3] GWh	CH_4 [4] % vol	SMP [5] L CH_4 kg^{-1} VS	Gross Energetic Value [6] GWh			
(a)	(b)	(c)	(d)	(e)	(f)	(g)	(h)	(i)	(j)	(k)
Silage	219,462,510	55	350	1201	92	585	2009	30	94	1,222,942
Food, garden and plant waste	165,291,843	55	450	905	79	646	1299	24	92	914,961
Manures and slurries	64,654,159	60	190	386	93	295	598	9	83	2,733,213
Distillery waste	16,055,060	60	300	96	90	450	144	7.1	73	619,528
DAF sludge/wastewater	14,960,483	65	260	97	90	360	134	6	65	959,005
Crops	14,141,474	55	350	77	92	585	129	60	93	39,825
Glycerol	9,919,812	60	425	59	90	638	89	99.5	99.5	14,146
Dairy waste	7,921,015	51	450	40	90	794	71	6.1	90	163,518
Municipal waste	6,271,886	55	300	34	92	502	57	24	92	52,076
Other	6,513,539	50	350	32	90	630	58	24	92	42,142
TOTAL	525,191,781			2928			4589			6,761,357

[1] Based on Ofgem data [13]. [2] From ADAT [16] and values in Table 1 and literature review data, using the feedstock mapping assumptions described above. [3] (e) = (b) ∗ (c) ∗ 35.82 MJ m^{-3}/(100% ∗ 3.6 ∗ 10^6 MJ GWh^{-1}). [4] From Table 1 and associated literature data; categories distillery waste, glycerol, dairy waste and other assumed as 90%. [5] (g) = (d) ∗ (f)/(c). [6] (h) = (e) ∗ (f)/(c). [7] (k) = (b) ∗ (c) ∗ 100/((d) ∗ (i) ∗ (j)).

3.2.2. Estimation of CO_2 Biomethanisation Uplift Based on Defra's UK AD Feedstock Data

The 2019–2020 Defra feedstock data for anaerobic digestion [17] are derived from the (paywalled) NNFCC report on Anaerobic Digestion Deployment in the UK [86] which records annual feedstock usage in AD plants operating under a range of AD incentives. It therefore covers a wider range than the Ofgem data [13], which only include plants covered by the Renewables Obligation, in which the biogas produced is burned to generate electricity for on-site use and/or feeding into the national electricity grid. The values exclude feedstock data from the wastewater sector (i.e., sludges from WWTP).

The Defra data [17] are presented as percentages of a total tonnage across five feedstock categories: Crops, Food waste, Manures and slurries, Crop wastes and Others. These were mapped using the same assumptions on feedstock characteristics as in Table 2. Specifically, it was assumed that 'Crops' were likely to be silage so this category was treated as maize silage, even though there will be a proportion of other silages and crops. As with the Ofgem data [13], 'Manures and slurries' were conservatively characterised as slurry, as the proportions were unknown. TS, VS and methane production values for 'Crop waste' were aligned with those for 'Crops' in Table 2. Feedstock tonnages shown in bold in Table 3 were taken directly from the Defra report [17].

Table 3. Potential uplift in gross energetic value through CO_2 biomethanisation of reported UK feedstocks (Defra data).

Feedstock Category [1]	Feedstock [2] Tonnes FM	TS [3] %FM	VS [3] %TS	Without CO_2 Biometh			With CO_2 Biometh		
				CH_4 [4] % vol	SMP [4] L CH_4 kg^{-1} VS	Gross Energetic Value [5] GWh	CH_4 [4] % vol	SMP [4] L CH_4 kg^{-1} VS	Gross Energetic Value [6] GWh
(a)	(b)	(c)	(d)	(e)	(f)	(g)	(h)	(i)	(j)
Crops (treated as silage)	4,163,000	30	94	55	350	4088	92	585	6839
Food waste	4,084,000	24	92	55	450	4038	79	646	5799
Manures and slurries	2,359,000	9	83	60	190	333	93	295	516
Crop waste	525,000	60	93	55	350	1020	92	585	1707
Other	2,727,000	24	92	50	350	2097	90	630	3774
TOTAL	13,858,000					11,576			18,635

[1] Based on Defra data [17]. [2] Tonnage FM basis. [3] From ADAT [16] using the assumptions described above. [4] From Table 2. [5] (g) = (b) ∗ (c) ∗ (d) ∗ (f) ∗ 35.82 MJ m^{-3}/(100% ∗ 100% ∗ 3.6 ∗ 10^6 MJ GWh^{-1}). [6] (j) = (g) ∗ (h)/(e)

Table 3 shows the estimated gross energy value with and without CO_2 biomethanisation. The results indicate an overall energy uplift of 61% for this set of feedstocks-further indicating that CO_2 biomethanisation has the potential to significantly increase the energy value of existing feedstocks, as well as the viability of the digesters that produce the gas.

The overall gross energetic value is significantly larger than that derived from the Ofgem RO data as the Defra data covers more than twice the total tonnage of feedstock (Table 3). Additionally, the derived uplift value of 61% differs slightly from the Ofgem RO value of 57% due to differences in the relative proportion of feedstocks.

To estimate the potential biomethane increase from CO_2 biomethanisation, analysis of the available feedstock data in conjunction with data from the scientific literature is a logical approach. It is also possible to apply the overall feedstock-derived energy uplift to generation data when the feedstock underlying it is unknown.

The 2021 DUKES data [18] show energy production from UK AD plants for the year 2020 of 953, 68 and 490 thousand tonnes of oil equivalent (ktoe) for electricity generation, heat generation and grid injection of methane, respectively; no figures are specifically reported for use of biomethane as a vehicle fuel. Applying the DUKES data conversion factor of 11.63 GWh $ktoe^{-1}$ to this total of 1511 ktoe gives a total gross energy production from anaerobic digestion of 17,572 GWh. This is considerably more than the 11,576 GWh shown in Table 3, reflecting the degree of uncertainty associated even with official data. Timing differences could account for some part of this, e.g., the reporting year for incentives tends to run from 1 April to 31 March whereas DUKES [18] reports by calendar year. Further differences could be due to the conservative feedstock assumptions made above; to DUKES load factor (LF) assumptions when deriving gross energy values from net electricity/heat generation; and/or to variations in data sources and rounding errors. Differences on this scale, however, strongly support the need for more unified and consistent data collection.

Nevertheless, applying the overall energy uplift of 61% derived from Table 3 would result in a gross energy value for AD of 28,317 GWh based on DUKES data [18]. It should be noted this value does not include biomethane used in vehicle fuel.

3.2.3. Potential for CO_2 Biomethanisation Using Generation Derived from Wastewater Treatment

Neither the Ofgem/RO [13] nor the Defra [17] feedstock data shown above include the energy contribution from the AD of sludges from municipal wastewater treatment.

This approach could be used to ascertain improved energy production from DUKES data on sewage gas [18] which show electricity, heat and gas grid injection energy figures of 350, 90 and 54 ktoe, respectively, equating to a total of 5744 GWh. If the maximum methane content from Table 1 is applied, equating to a 38% uplift in gross energy output, generation of 7953 GWh could potentially be achieved if CO_2 biomethanisation were introduced across the wastewater sector.

In the UK, 93% of sewage sludge is treated by AD or advanced AD [87] at approximately 170 AD plants [88]. Unlike commercial and industrial plants where feedstocks are often mixed, and thus, characteristics can be variable, sewage sludge is a relatively consistent material. Additionally, the water industry has ambitious decarbonisation plans [89], as its energy use is considerable (e.g., in 2009 this sector accounted for up to 3% of total energy use in the UK [90]). A relatively small number of AD plants treating a large quantity of a minimally variable feedstock in a sector that has high energy use, relevant technical expertise and ambitious decarbonisation plans makes it relatively feasible to introduce this technology and achieve a significant proportion of the theoretical uplift.

3.3. *Advantages of CO_2 Biomethanisation at Scale*

Despite some variability between sources, the existing feedstock and energy data, supported by the scientific literature, indicate that CO_2 biomethanisation has significant potential to increase the energy contribution of AD to the UK's energy mix, while also increasing the efficiency of utilisation of carbon from organic feedstocks. Applying biomethanisation technology to current DUKES [18] figures of 23,316 GWh (sewage gas and AD) could produce 36,270 GWh, an uplift of 55%. AD and sewage gas currently account for 15% of the UK's bioenergy contribution and this could increase to ~22%. This potential uplift of 12,954 GWh is equivalent to the annual energy use of 858,000 households [91], e.g., the domestic properties in a city the size of Leeds.

To effect a similar increase in current gas grid injection alone would require the growth of another 251,000 ha of maize (~4.4% of UK arable land), more than double the 121,000 ha used for all UK bioenergy and more than three times the 75,000 ha used for AD maize in 2020 [92]. Maximizing the carbon utilisation efficiency from existing feedstocks thus makes sense in terms of land use and the food/energy/climate nexus.

Although not explicitly included in the 2021 DUKES [18] figures, the role of biomethane for transport also is growing, particularly in the heavy goods vehicle (HGV) sector where it is currently the only commercial decarbonisation option, until suitable fuel current battery and hydrogen technologies can be developed. Transport is difficult to decarbonise: as energy systems have become less carbon-intensive, transport became the UK's largest emitting sector of GHG emissions in 2016 [93]. If the uplift were used for transport, it could replace nearly 1.1 million litres of diesel, with a fossil CO_2 equivalent of 3.4 million tonnes [94].

This approach could eliminate the energy and capital expenditure required to retrofit CO_2 carbon capture and storage (CCS), allowing more effective utilisation of short-cycle carbon rather than, for example, pumping it into permanent underground storage. This would have the added advantage of displacing its equivalent in fossil gas and associated imports: the Sixth Carbon Budget accounts for abatement from the additional use of biomethane to displace fossil gas [9] and CO_2 biomethanisation could offer an effective way to achieve this.

The longer-term value of incorporating CO_2 biomethanisation into AD systems is further strengthened by the CCC's projected growth in the hydrogen economy, in particular, green hydrogen from water electrolysis powered by renewables, since equipment and installation costs tend to fall as deployment increases. As noted earlier, achieving the benefits of CO_2 biomethanisation requires the use of renewables-based H_2, and this may be seen as a barrier in the short term since competition for H_2 from grid-based resources is likely to intensify [95]. AD is used at a wide range of scales, however, from on-site slurry treatment at a single dairy farm to processing the municipal wastes of a city. It is therefore flexible in its needs and could be coupled with electrolytic H_2 production across a similar range. In their simplest form, small-scale AD applications can provide an opportunity for hybridisation of on-site renewable power with waste processing, to provide short carbon-cycle biomethane for local or on-site use in locations where export of electricity to grid is technically challenging or economically unattractive. Larger AD installations could be used, with relatively low capital expenditure [2,5,10], to transform excess intermittent power production from large-scale renewable facilities into biomethane for gas grid injection, hence providing buffer storage and allowing a more rapid expansion in the renewables sector. By taking electricity from multiple sources and across different time spans, the technology could thus make a major contribution in supporting the energy transition from carbon-based to non-carbon-based gaseous fuels [7,20].

Given the potential contribution of CO_2 biomethanisation, it is surprising that it does not appear to be mentioned in policy documents and remains under the policy radar, even where other technologies-some with technology readiness levels that are similar or lower-are actively discussed. For example, the BEIS November 2021 Biomass Policy Statement [96] which 'provides a strategic view on the role of biomass across the economy in the medium- to long-term' mentions the 'clear opportunity' for 'material processing of biomass into high value products'; but does not consider CO_2 biomethanisation as either an interim or long-term strategy. Indeed, although it identifies AD as 'the only commercially scalable technology currently available for greening the gas grid', the sole approach for AD's biogenic carbon is Bioenergy Carbon Capture and Storage (BECCs), which will require the development of an effective market for greenhouse gas removals. The UK market for biomethane-to-grid is already well-developed and growing, so maximizing the carbon utilisation in such plants instead of permanently storing it underground could potentially be a better option in the decarbonisation pathway. This, of course, would need to happen alongside the increase in hydrogen production outlined by the CCC.

4. Conclusions

Biomethane has an important role to play in the UK's energy decarbonisation plans, due to its flexible use in transport, heat and electricity production. A considerable amount of work has been conducted at laboratory scale on CO_2 biomethanisation of organic wastes. The results show clearly that significant improvements can be achieved in methane yield per unit of organic feed and in biogas methane content. Interestingly, the overall survey also appears to indicate that underlying biogas productivity can be increased in some systems. Many scientific and technical questions remain to be addressed, particularly those associated with the effects of scale-up and of differences in operating practice, which need to be resolved if industry is to adopt this technology with confidence. While approaches such as ex-situ CO_2 biomethanisation may be more suitable in some applications, however, it appears there is clear potential for in situ or hybrid conversion of CO_2 generated by anaerobic digestion of organic materials.

Data on existing UK AD feedstocks are highly fragmented, and it is not always clear how values in different sources have been arrived at or how they relate to one another. In the absence of coherent information, it is difficult for government bodies to reach conclusions on overall policy, especially when individual sectors and technologies are lobbying for their own interests. The availability of better and more consistent data sources as a basis for evaluation and policy making is thus a clear priority. Based on the available UK data, however, significant increases in biomethane productivity could potentially be achieved, ranging from 38–68% for different feedstock types and equivalent to an overall uplift in the contribution of AD to UK bioenergy from 15 to 22%. Again, there are many issues to consider: the current survey only looked at data on the highest level, and for realistic assessments of the potential scale and impact of technology application it will be necessary to take into account both the end-uses of biomethane and techno-economic viability on individual sites. The potential contribution from CO_2 biomethanisation of organic wastes is large enough, however, to warrant consideration in both short and long-term planning. There is thus a clear need for more work looking at both the research issues and the policy needs to optimise the contribution of this approach and integrate it with national energy and sustainability strategies.

Supplementary Materials: The following supporting information can be downloaded at: https://www.mdpi.com/article/10.3390/pr10061202/s1. Text S1: Assessment of performance data for CO_2 biomethanisation of organic feedstocks (including nomenclature); Table S1: Performance data for CO_2 biomethanisation of livestock manures; Table S2: Performance data for CO_2 biomethanisation of crops and agro-wastes; Table S3: Performance data for CO_2 biomethanisation of food wastes; Table S4: Performance data for CO_2 biomethanisation of sewage sludges; Text S2: Sources of Anaerobic digestion and Feedstock data.

Author Contributions: Conceptualization, A.B. and S.H.; methodology, A.B. and S.H.; validation, A.B., S.H., Y.Z. and C.J.B.; formal analysis, A.B., S.H. and Y.Z.; investigation, A.B., S.H. and Y.Z.; data curation, A.B. and S.H.; writing—original draft preparation, A.B.; writing—review and editing, A.B., S.H., C.J.B. and Y.Z.; funding acquisition: C.J.B., S.H. and Y.Z. All authors have read and agreed to the published version of the manuscript.

Funding: This research was supported by the IBCat H2AD project (EP/M028208/1) funded by the Engineering and Physical Sciences Research Council (EPSRC), by a grant from the Carbon Recycling Network (POC-02-ZHANG-CCnet) funded by the Biotechnology and Biological Sciences Research Council (BBSRC), and by in-kind support from the Environmental Biotechnology Network (BB/S009795/1) funded by BBSRC and EPSRC.

Institutional Review Board Statement: Not applicable.

Informed Consent Statement: Not applicable.

Data Availability Statement: Data supporting this study are openly available from the University of Southampton repository at http://doi.org/10.5258/SOTON/D2241 (accessed on 30 May 2022).

Acknowledgments: The authors wish to acknowledge the help of Alam Khan of the Sustainable Bioenergy and Biorefinery Laboratory, Department of Microbiology, Quaid-i-Azam University, Pakistan for provision of additional information on his work, and Bing Tao, formerly of the University of Southampton, for additional data. Thanks are also due to the AD Working Group of the Environmental Biotechnology Network (BB/S009795/1) for comments on the text.

Conflicts of Interest: The authors declare no conflict of interest. The funders had no role in the design of the study; in the collection, analyses, or interpretation of data; in the writing of the manuscript, or in the decision to publish the results.

References

1. Luo, G.; Angelidaki, I. Integrated biogas upgrading and hydrogen utilization in an anaerobic reactor containing enriched hydrogenotrophic methanogenic culture. *Biotechnol. Bioeng.* **2012**, *109*, 2729–2736. [CrossRef] [PubMed]
2. Fu, S.; Angelidaki, I.; Zhang, Y. In situ biogas upgrading by CO_2-to-CH_4 bioconversion. *Trends Biotechnol.* **2020**, *39*, 336–347. [CrossRef] [PubMed]
3. Rafrafi, Y.; Laguillaumie, L.; Dumas, C. Biological Methanation of H_2 and CO_2 with Mixed Cultures: Current Advances, Hurdles and Challenges. *Waste Biomass Valorization* **2020**, *12*, 5259–5282. [CrossRef]

4. Angelidaki, I.; Treu, L.; Tsapekos, P.; Luo, G.; Campanaro, S.; Wenzel, H.; Kougias, P.G. Biogas upgrading and utilization: Current status and perspectives. *Biotechnol. Adv.* **2018**, *36*, 452–466. [CrossRef] [PubMed]
5. Aryal, N.; Kvist, T.; Ammam, F.; Pant, D.; Ottosen, L.D. An overview of microbial biogas enrichment. *Bioresour. Technol.* **2018**, *264*, 359–369. [CrossRef]
6. Lecker, B.; Illi, L.; Lemmer, A.; Oechsner, H. Biological hydrogen methanation—A review. *Bioresour. Technol.* **2017**, *245*, 1220–1228. [CrossRef]
7. Zavarkó, M.; Imre, A.R.; Pörzse, G.; Csedő, Z. Past, Present and Near Future: An Overview of Closed, Running and Planned Biomethanation Facilities in Europe. *Energies* **2021**, *14*, 5591. [CrossRef]
8. Sarker, S.; Lamb, J.J.; Hjelme, D.R.; Lien, K.M. Overview of recent progress towards in-situ biogas upgradation techniques. *Fuel* **2018**, *226*, 686–697. [CrossRef]
9. CCC. *The Sixth Carbon Budget*; Committee on Climate Change: London, UK, 2020.
10. D'Silva, T.C.; Isha, A.; Chandra, R.; Vijay, V.K.; Subbarao, P.M.V.; Kumar, R.; Chaudhary, V.P.; Singh, H.; Khan, A.A.; Tyagi, V.K. Enhancing methane production in anaerobic digestion through hydrogen assisted pathways–A state-of-the-art review. *Renew. Sustain. Energy Rev.* **2021**, *151*, 111536. [CrossRef]
11. Zhao, J.; Li, Y.; Dong, R. Recent progress towards in-situ biogas upgrading technologies. *Sci. Total Environ.* **2021**, *800*, 149667. [CrossRef]
12. Tao, B.; Zhang, Y.; Heaven, S.; Banks, C.J. Predicting pH rise as a control measure for integration of CO_2 biomethanisation with anaerobic digestion. *Appl. Energy* **2020**, *277*, 115535. [CrossRef]
13. Ofgem. *Renewables Obligation (RO) Annual Report 2019-20*; Office of Gas and Electricity Markets: London, UK, 2021.
14. BEIS. *Sustainability Standards for Electricity Generation from Biomass*; Department for Business, Energy & Industrial Strategy: London, UK, 2013; Volume 2022.
15. Ofgem. *Biomass Sustainability*; Office of Gas and Electricity Markets: London, UK, 2021.
16. ADAT. *Anaerobic Digestion Assessment Tool: User Manual*; University of Southampton: Southampton, UK, 2017.
17. Defra. *Official Statistics Section 3: Anaerobic Digestion*; Department for Environment Food & Rural Affairs: London, UK, 2021.
18. BEIS. *Digest of UK Energy Statistics (DUKES)*; Department for Business, Energy & Industrial Strategy: London, UK, 2021.
19. Khan, M.U.; Lee, J.T.E.; Bashir, M.A.; Dissanayake, P.D.; Ok, Y.S.; Tong, Y.W.; Shariati, M.A.; Wu, S.; Ahring, B.K. Current status of biogas upgrading for direct biomethane use: A review. *Renew. Sustain. Energy Rev.* **2021**, *149*, 111343. [CrossRef]
20. Thema, M.; Weidlich, T.; Hörl, M.; Bellack, A.; Mörs, F.; Hackl, F.; Kohlmayer, M.; Gleich, J.; Stabenau, C.; Trabold, T. Biological CO_2-methanation: An approach to standardization. *Energies* **2019**, *12*, 1670. [CrossRef]
21. Bassani, I.; Kougias, P.G.; Treu, L.; Angelidaki, I. Biogas upgrading via hydrogenotrophic methanogenesis in two-stage continuous stirred tank reactors at mesophilic and thermophilic conditions. *Environ. Sci. Technol.* **2015**, *49*, 12585–12593. [CrossRef]
22. Treu, L.; Kougias, P.; de Diego-Díaz, B.; Campanaro, S.; Bassani, I.; Fernández-Rodríguez, J.; Angelidaki, I. Two-year microbial adaptation during hydrogen-mediated biogas upgrading process in a serial reactor configuration. *Bioresour. Technol.* **2018**, *264*, 140–147. [CrossRef]
23. Wahid, R.; Horn, S.J. The effect of mixing rate and gas recirculation on biological CO_2 methanation in two-stage CSTR systems. *Biomass Bioenergy* **2021**, *144*, 105918. [CrossRef]
24. Lebranchu, A.; Blanchard, F.; Fick, M.; Pacaud, S.; Olmos, E.; Delaunay, S. Pilot-scale biomethanation of cattle manure using dense membranes. *Bioresour. Technol.* **2019**, *284*, 430–436. [CrossRef]
25. Zhu, X.; Cao, Q.; Chen, Y.; Sun, X.; Liu, X.; Li, D. Effects of mixing and sodium formate on thermophilic in-situ biogas upgrading by H_2 addition. *J. Clean. Prod.* **2019**, *216*, 373–381. [CrossRef]
26. Zhu, X.; Chen, L.; Chen, Y.; Cao, Q.; Liu, X.; Li, D. Differences of methanogenesis between mesophilic and thermophilic in situ biogas-upgrading systems by hydrogen addition. *J. Ind. Microbiol. Biotechnol.* **2019**, *46*, 1569–1581. [CrossRef]
27. Zhu, X.; Chen, L.; Chen, Y.; Cao, Q.; Liu, X.; Li, D. Effect of H_2 addition on the microbial community structure of a mesophilic anaerobic digestion system. *Energy* **2020**, *198*, 117368. [CrossRef]
28. Jiang, H.; Wu, F.; Wang, Y.; Feng, L.; Zhou, H.; Li, Y. Characteristics of in-situ hydrogen biomethanation at mesophilic and thermophilic temperatures. *Bioresour. Technol.* **2021**, *337*, 125455. [CrossRef] [PubMed]
29. Luo, G.; Angelidaki, I. Co-digestion of manure and whey for in situ biogas upgrading by the addition of H_2: Process performance and microbial insights. *Appl. Microbiol. Biotechnol.* **2013**, *97*, 1373–1381. [CrossRef] [PubMed]
30. Luo, G.; Angelidaki, I. Hollow fiber membrane based H_2 diffusion for efficient in situ biogas upgrading in an anaerobic reactor. *Appl. Microbiol. Biotechnol.* **2013**, *97*, 3739–3744. [CrossRef] [PubMed]
31. Wahid, R.; Horn, S.J. Impact of operational conditions on methane yield and microbial community composition during biological methanation in in situ and hybrid reactor systems. *Biotechnol. Biofuels* **2021**, *14*, 170. [CrossRef] [PubMed]
32. Corbellini, V.; Kougias, P.G.; Treu, L.; Bassani, I.; Malpei, F.; Angelidaki, I. Hybrid biogas upgrading in a two-stage thermophilic reactor. *Energy Convers. Manag.* **2018**, *168*, 1–10. [CrossRef]
33. Khan, A.; Akbar, S.; Okonkwo, V.; Smith, C.; Khan, S.; Shah, A.A.; Adnan, F.; Ijaz, U.Z.; Ahmed, S.; Badshah, M. Enrichment of the hydrogenotrophic methanogens for, in-situ biogas up-gradation by recirculation of gases and supply of hydrogen in methanogenic reactor. *Bioresour. Technol.* **2022**, *345*, 126219. [CrossRef]
34. Wang, H.; Zhang, Y.; Angelidaki, I. Ammonia inhibition on hydrogen enriched anaerobic digestion of manure under mesophilic and thermophilic conditions. *Water Res.* **2016**, *105*, 314–319. [CrossRef]

35. Wang, H.; Zhu, X.; Yan, Q.; Zhang, Y.; Angelidaki, I. Microbial community response to ammonia levels in hydrogen assisted biogas production and upgrading process. *Bioresour. Technol.* **2020**, *296*, 122276. [CrossRef]
36. Garcia-Robledo, E.; Ottosen, L.D.; Voigt, N.V.; Kofoed, M.; Revsbech, N.P. Micro-scale H_2–CO_2 dynamics in a hydrogenotrophic methanogenic membrane reactor. *Front. Microbiol.* **2016**, *7*, 1276. [CrossRef]
37. Vechi, N.T.; Agneessens, L.M.; Feilberg, A.; Ottosen, L.D.M.; Kofoed, M.V.W. In situ biomethanation: Inoculum origin influences acetate consumption rate during hydrogen addition. *Bioresour. Technol. Rep.* **2021**, *14*, 100656. [CrossRef]
38. Lukehurst, C.; Bywater, A. *Exploring the Viability of Small Scale Anaerobic Digesters in Livestock Farming*; IEA Bioenergy: Paris, France, 2015.
39. Jensen, M.B.; Jensen, B.; Ottosen, L.D.M.; Kofoed, M.V.W. Integrating H_2 injection and reactor mixing for low-cost H_2 gas-liquid mass transfer in full-scale in situ biomethanation. *Biochem. Eng. J.* **2021**, *166*, 107869. [CrossRef]
40. Jensen, M.B.; Kofoed, M.V.W.; Fischer, K.; Voigt, N.V.; Agneessens, L.M.; Batstone, D.J.; Ottosen, L.D.M. Venturi-type injection system as a potential H_2 mass transfer technology for full-scale in situ biomethanation. *Appl. Energy* **2018**, *222*, 840–846. [CrossRef]
41. Nock, W.J.; Serna-Maza, A.; Heaven, S.; Banks, C.J. Evaluation of microporous hollow fibre membranes for mass transfer of H_2 into anaerobic digesters for biomethanization. *J. Chem. Technol. Biotechnol.* **2019**, *94*, 2693–2701. [CrossRef]
42. Voelklein, M.; Rusmanis, D.; Murphy, J. Biological methanation: Strategies for in-situ and ex-situ upgrading in anaerobic digestion. *Appl. Energy* **2019**, *235*, 1061–1071. [CrossRef]
43. Illi, L.; Lecker, B.; Lemmer, A.; Müller, J.; Oechsner, H. Biological methanation of injected hydrogen in a two-stage anaerobic digestion process. *Bioresour. Technol.* **2021**, *333*, 125126. [CrossRef]
44. Schönberg, V.; Busch, G. Steigerung des Methangehaltes durch biologische Wasserstoffumsetzung. *Bornim Agrar. Ber* **2012**, *79*, 66–75.
45. Agneessens, L.M.; Ottosen, L.D.M.; Voigt, N.V.; Nielsen, J.L.; de Jonge, N.; Fischer, C.H.; Kofoed, M.V.W. In-situ biogas upgrading with pulse H_2 additions: The relevance of methanogen adaption and inorganic carbon level. *Bioresour. Technol.* **2017**, *233*, 256–263. [CrossRef]
46. Frydendal-Nielsen, S.; Jørgensen, U.; Hjorth, M.; Felby, C.; Gislum, R. Comparing methods for measuring the digestibility of miscanthus in bioethanol or biogas processing. *GCB Bioenergy* **2017**, *9*, 168–175. [CrossRef]
47. Agneessens, L.M.; Ottosen, L.D.M.; Andersen, M.; Olesen, C.B.; Feilberg, A.; Kofoed, M.V.W. Parameters affecting acetate concentrations during in-situ biological hydrogen methanation. *Bioresour. Technol.* **2018**, *258*, 33–40. [CrossRef]
48. Mulat, D.G.; Mosbæk, F.; Ward, A.J.; Polag, D.; Greule, M.; Keppler, F.; Nielsen, J.L.; Feilberg, A. Exogenous addition of H_2 for an in situ biogas upgrading through biological reduction of carbon dioxide into methane. *Waste Manag.* **2017**, *68*, 146–156. [CrossRef]
49. Szuhaj, M.; Ács, N.; Tengölics, R.; Bodor, A.; Rákhely, G.; Kovács, K.L.; Bagi, Z. Conversion of H_2 and CO_2 to CH_4 and acetate in fed-batch biogas reactors by mixed biogas community: A novel route for the power-to-gas concept. *Biotechnol. Biofuels* **2016**, *9*, 102. [CrossRef] [PubMed]
50. Sarker, S.; Wijnsma, S.N.; Lien, K.M. Lessons Learned from an Experimental Campaign on Promoting Energy Content of Renewable Biogas by Injecting H_2 during Anaerobic Digestion. *Energies* **2020**, *13*, 3542. [CrossRef]
51. Song, H.; Zhang, Y.; Kusch-Brandt, S.; Banks, C.J. Comparison of variable and constant loading for mesophilic food waste digestion in a long-term experiment. *Energies* **2020**, *13*, 1279. [CrossRef]
52. Zhang, W.; Alessi, A.M.; Heaven, S.; Chong, J.P.; Banks, C.J. Dynamic changes in anaerobic digester metabolic pathways and microbial populations during acclimatisation to increasing ammonium concentrations. *Waste Manag.* **2021**, *135*, 409–419. [CrossRef]
53. Zhang, Y. *Unpublished Work on CO_2 Biomethanisation of Organic Feedstocks*; Univerity of Southampton: Southampton, UK, 2022.
54. Kim, S.; Mostafa, A.; Im, S.; Lee, M.-K.; Kang, S.; Na, J.-G.; Kim, D.-H. Production of high-calorific biogas from food waste by integrating two approaches: Autogenerative high-pressure and hydrogen injection. *Water Res.* **2021**, *194*, 116920. [CrossRef]
55. Yang, Z.; Liu, Y.; Zhang, J.; Mao, K.; Kurbonova, M.; Liu, G.; Zhang, R.; Wang, W. Improvement of biofuel recovery from food waste by integration of anaerobic digestion, digestate pyrolysis and syngas biomethanation under mesophilic and thermophilic conditions. *J. Clean. Prod.* **2020**, *256*, 120594. [CrossRef]
56. Zhang, W.; Heaven, S.; Banks, C.J. Thermophilic digestion of food waste by dilution: Ammonia limit values and energy considerations. *Energy Fuels* **2017**, *31*, 10890–10900. [CrossRef]
57. Thapa, A.; Park, J.-G.; Yang, H.-M.; Jun, H.-B. In-situ biogas upgrading in an anaerobic trickling filter bed reactor treating a thermal post-treated digestate. *J. Environ. Chem. Eng.* **2021**, *9*, 106780. [CrossRef]
58. Okoro-Shekwaga, C.K.; Ross, A.B.; Camargo-Valero, M.A. Improving the biomethane yield from food waste by boosting hydrogenotrophic methanogenesis. *Appl. Energy* **2019**, *254*, 113629. [CrossRef]
59. Okoro-Shekwaga, C.K.; Ross, A.B.; Camargo-Valero, M.A. Enhanced in-situ biomethanation of food waste by sequential inoculum acclimation: Energy efficiency and carbon savings analysis. *Waste Manag.* **2021**, *130*, 12–22. [CrossRef]
60. Okoro-Shekwaga, C.K.; Ross, A.; Camargo-Valero, M.A. Enhancing bioenergy production from food waste by in situ biomethanation: Effect of the hydrogen injection point. *Food Energy Secur.* **2021**, *10*, e288. [CrossRef]
61. Treu, L.; Tsapekos, P.; Peprah, M.; Campanaro, S.; Giacomini, A.; Corich, V.; Kougias, P.G.; Angelidaki, I. Microbial profiling during anaerobic digestion of cheese whey in reactors operated at different conditions. *Bioresour. Technol.* **2019**, *275*, 375–385. [CrossRef] [PubMed]

62. Fontana, A.; Kougias, P.G.; Treu, L.; Kovalovszki, A.; Valle, G.; Cappa, F.; Morelli, L.; Angelidaki, I.; Campanaro, S. Microbial activity response to hydrogen injection in thermophilic anaerobic digesters revealed by genome-centric metatranscriptomics. *Microbiome* **2018**, *6*, 194. [CrossRef] [PubMed]
63. Lovato, G.; Alvarado-Morales, M.; Kovalovszki, A.; Peprah, M.; Kougias, P.G.; Rodrigues, J.A.D.; Angelidaki, I. In-situ biogas upgrading process: Modelling and simulations aspects. *Bioresour. Technol.* **2017**, *245*, 332–341. [CrossRef]
64. Palù, M.; Peprah, M.; Tsapekos, P.; Kougias, P.; Campanaro, S.; Angelidaki, I.; Treu, L. In-situ biogas upgrading assisted by bioaugmentation with hydrogenotrophic methanogens during mesophilic and thermophilic co-digestion. *Bioresour. Technol.* **2022**, *348*, 126754. [CrossRef]
65. Bassani, I.; Kougias, P.G.; Angelidaki, I. In-situ biogas upgrading in thermophilic granular UASB reactor: Key factors affecting the hydrogen mass transfer rate. *Bioresour. Technol.* **2016**, *221*, 485–491. [CrossRef]
66. Deschamps, L.; Imatoukene, N.; Lemaire, J.; Mounkaila, M.; Filali, R.; Lopez, M.; Theoleyre, M.-A. In-situ biogas upgrading by bio-methanation with an innovative membrane bioreactor combining sludge filtration and H_2 injection. *Bioresour. Technol.* **2021**, *337*, 125444. [CrossRef]
67. Tao, B.; Alessi, A.M.; Zhang, Y.; Chong, J.P.; Heaven, S.; Banks, C.J. Simultaneous biomethanisation of endogenous and imported CO_2 in organically loaded anaerobic digesters. *Appl. Energy* **2019**, *247*, 670–681. [CrossRef]
68. Wahid, R.; Mulat, D.G.; Gaby, J.C.; Horn, S.J. Effects of H_2: CO_2 ratio and H_2 supply fluctuation on methane content and microbial community composition during in-situ biological biogas upgrading. *Biotechnol. Biofuels* **2019**, *12*, 104. [CrossRef]
69. Xu, H.; Wang, K.; Zhang, X.; Gong, H.; Xia, Y.; Holmes, D.E. Application of in-situ H_2-assisted biogas upgrading in high-rate anaerobic wastewater treatment. *Bioresour. Technol.* **2020**, *299*, 122598. [CrossRef]
70. Jing, Y.; Campanaro, S.; Kougias, P.; Treu, L.; Angelidaki, I.; Zhang, S.; Luo, G. Anaerobic granular sludge for simultaneous biomethanation of synthetic wastewater and CO with focus on the identification of CO-converting microorganisms. *Water Res.* **2017**, *126*, 19–28. [CrossRef] [PubMed]
71. Banks, C.; Heaven, S.; Zhang, Y.; Baier, U. *Food Waste Digestion: Anaerobic Digestion of Food Waste for a Circular Economy*; IEA Bioenergy: Paris, France, 2018.
72. Chen, A.C.; Ohashi, A.; Harada, H. Acetate synthesis from H_2/CO_2 in simulated and actual landfill samples. *Environ. Technol.* **2003**, *24*, 435–443. [CrossRef] [PubMed]
73. Sekoai, P.T.; Engelbrecht, N.; du Preez, S.P.; Bessarabov, D. Thermophilic Biogas Upgrading via ex Situ Addition of H_2 and CO_2 Using Codigested Feedstocks of Cow Manure and the Organic Fraction of Solid Municipal Waste. *ACS Omega* **2020**, *5*, 17367–17376. [CrossRef] [PubMed]
74. Alfaro, N.; Fdz-Polanco, M.; Fdz-Polanco, F.; Díaz, I. H_2 addition through a submerged membrane for in-situ biogas upgrading in the anaerobic digestion of sewage sludge. *Bioresour. Technol.* **2019**, *280*, 1–8. [CrossRef] [PubMed]
75. Corbellini, V.; Catenacci, A.; Malpei, F. Hydrogenotrophic biogas upgrading integrated into WWTPs: Enrichment strategy. *Water Sci. Technol.* **2019**, *79*, 759–770. [CrossRef] [PubMed]
76. Corbellini, V.; Feng, C.; Bellucci, M.; Catenacci, A.; Stella, T.; Espinoza-Tofalos, A.; Malpei, F. Performance Analysis and Microbial Community Evolution of In Situ Biological Biogas Upgrading with Increasing H_2/CO_2 Ratio. *Archaea* **2021**, 8894455. [CrossRef]
77. Díaz, I.; Fdz-Polanco, F.; Mutsvene, B.; Fdz-Polanco, M. Effect of operating pressure on direct biomethane production from carbon dioxide and exogenous hydrogen in the anaerobic digestion of sewage sludge. *Appl. Energy* **2020**, *280*, 115915. [CrossRef]
78. Luo, G.; Wang, W.; Angelidaki, I. Anaerobic digestion for simultaneous sewage sludge treatment and CO biomethanation: Process performance and microbial ecology. *Environ. Sci. Technol.* **2013**, *47*, 10685–10693. [CrossRef]
79. Wang, W.; Xie, L.; Luo, G.; Zhou, Q.; Angelidaki, I. Performance and microbial community analysis of the anaerobic reactor with coke oven gas biomethanation and in situ biogas upgrading. *Bioresour. Technol.* **2013**, *146*, 234–239. [CrossRef]
80. Hu, Y.; Hao, X.; Zhao, D.; Fu, K. Enhancing the CH_4 yield of anaerobic digestion via endogenous CO_2 fixation by exogenous H_2. *Chemosphere* **2015**, *140*, 34–39. [CrossRef]
81. Braga Nan, L.; Trably, E.; Santa-Catalina, G.; Bernet, N.; Delgenès, J.-P.; Escudié, R. Biomethanation processes: New insights on the effect of a high H_2 partial pressure on microbial communities. *Biotechnol. Biofuels* **2020**, *13*, 141. [CrossRef] [PubMed]
82. Alfaro, N.; Fdz-Polanco, M.; Fdz-Polanco, F.; Díaz, I. Evaluation of process performance, energy consumption and microbiota characterization in a ceramic membrane bioreactor for ex-situ biomethanation of H_2 and CO_2. *Bioresour. Technol.* **2018**, *258*, 142–150. [CrossRef] [PubMed]
83. Fernández, Y.B.; Soares, A.; Villa, R.; Vale, P.; Cartmell, E. Carbon capture and biogas enhancement by carbon dioxide enrichment of anaerobic digesters treating sewage sludge or food waste. *Bioresour. Technol.* **2014**, *159*, 1–7. [CrossRef]
84. Mikucka, W.; Zielińska, M. Distillery stillage: Characteristics, treatment, and valorization. *Appl. Biochem. Biotechnol.* **2020**, *192*, 770–793. [CrossRef] [PubMed]
85. Ofgem. *Renewables Obligation (RO) Annual Report 2018–19*; Office of Gas and Electricity Markets: London, UK, 2020.
86. NNFCC. *Anaerobic Digestion Deployment in the UK*; National Non-Food Crops Centre: York, UK, 2021.
87. Ofwat. *Bioresources Market Review: Jacobs Report*; Office of Water Services: Birmingham, UK, 2021.
88. ADBA. *ADBA Policy Report October 2021*; Anaerobic Digestion and Bioresources Association: London, UK, 2021.
89. UKWIR. *BQ10: How Do We Remove More Carbon than We Emit by 2050?* UK Water Industry Research: London, UK, 2022.
90. EA. *Renewable Energy Potential for the Water Industry*; Environment Agency-Evidence Directorate: Bristol, UK, 2009.

91. Ofgem. *Decision on Revised Typical Domestic Consumption Values for Gas and Electricity and Economy 7 Consumption Split*; Office of Gas and Electricity Markets: London, UK, 2020.
92. Defra. *Key Messages for 2021*; Department for Environment & Rural Affairs: London, UK, 2021.
93. DfT. *Transport and Environment Statistics: Autumn 2021*; Department for Transport: London, UK, 2021.
94. BEIS. *Greenhouse Gas Reporting: Conversion Factors 2018*; Department for Business, Energy & Industrial Strategy: London, UK, 2018.
95. Zhang, L.; Kuroki, A.; Tong, Y.W. A mini-review on in situ biogas upgrading technologies via enhanced hydrogenotrophic methanogenesis to improve the quality of biogas from anaerobic digesters. *Front. Energy Res.* **2020**, *8*, 69. [CrossRef]
96. BEIS. *Biomass Policy Statement*; Department for Business, Energy & Industrial Strategy: London, UK, 2021.

processes MDPI

Article

Experimental Evaluation of Continuous In-Situ Biomethanation of CO_2 in Anaerobic Digesters Fed on Sewage Sludge and Food Waste and the Influence of Hydrogen Gas–Liquid Mass Transfer

Davide Poggio [1,*], Arman Sastraatmaja [1], Mark Walker [2], Stavros Michailos [2], William Nimmo [1] and Mohamed Pourkashanian [1,*]

1 Energy2050, Department of Mechanical Engineering, Faculty of Engineering, University of Sheffield, Sheffield S10 2TN, UK
2 School of Engineering, University of Hull, Hull HU6 7RX, UK
* Correspondence: d.poggio@sheffield.ac.uk (D.P.); m.pourkashanian@sheffield.ac.uk (M.P.)

Abstract: In-situ biomethanation combines conventional biogas production from the anaerobic digestion (AD) of organic matter with the addition of hydrogen to produce a higher quality biomethane gas. However, challenges surrounding its performance and control could hinder its uptake. To investigate this, an automated rig was designed and operated to study in-situ biomethanation with sewage sludge (SS) and food waste (FW) feedstocks. The effects that were experimentally investigated included the biogas recirculation rate, stirring intensity, and organic loading rate (OLR). All the results highlighted the rate-limiting effect of H_2 gas–liquid mass transfer (measured k_La in the range of 43–82 day^{-1}), which was implied by a lack of evidence of hydrogen-induced biological inhibition and a high average equilibrium hydrogen content in the biogas (a volume of 7–37%). At an OLR of 2 g VS L^{-1}day^{-1}, increasing biogas recirculation and mechanical stirring rates improved the methane evolution rate up to 0.17 and 0.23 L L^{-1}day^{-1} and the H_2 conversion up to 80 and 66% for sewage sludge and food waste, respectively. A lower OLR of 1 g VS L^{-1}day^{-1} allowed for increased hydrogen conversion but at a lower level of methane productivity. A process model, validated on experimental data, predicted that improving the k_La to at least 240 day^{-1} would be required for in-situ biomethanation at OLRs common in AD systems in order to achieve a drop-in quality in terms of the biogas, with further downstream treatment required for certain applications.

Keywords: biomethanation; in-situ; biogas upgrading; hydrogen; gas–liquid mass transfer

Citation: Poggio, D.; Sastraatmaja, A.; Walker, M.; Michailos, S.; Nimmo, W.; Pourkashanian, M. Experimental Evaluation of Continuous In-Situ Biomethanation of CO_2 in Anaerobic Digesters Fed on Sewage Sludge and Food Waste and the Influence of Hydrogen Gas–Liquid Mass Transfer. *Processes* **2023**, *11*, 604. https://doi.org/10.3390/pr11020604

Academic Editors: Sonia Heaven, Sigrid Kusch-Brandt and Charles Banks

Received: 18 December 2022
Revised: 10 February 2023
Accepted: 13 February 2023
Published: 16 February 2023

1. Introduction

Interest in the methanation of hydrogen in a power-to-gas concept has increased, since methane currently has a broader range of drop-in applications where natural gas is the existing fuel [1] (e.g., in industrial and domestic heating, power generation, and vehicle fuel) and is therefore often preferred to hydrogen as an energy vector. The biological methanation of hydrogen and carbon dioxide can be applied in such a concept where electricity (e.g., produced by intermittent renewables) is used to produce hydrogen through electrolysis, which is then converted through the action of hydrogenotrophic methanogens [2] to produce biomethane in the following reaction:

$$CO_2 + 4H_2 \rightarrow CH_4 + 2H_2O \quad (1)$$

The biomethanation reaction can take place in either a dedicated reactor (ex-situ) or, as studied in this work, within anaerobic digesters (in-situ) [3,4] fed on both biomass and hydrogen, where the carbon dioxide requirement is satisfied by the biogas from the anaerobic digestion (AD) process itself. By this method, the methane content in the biogas can be increased from 60–70% to >90% [5].

operational pH for food waste AD that was reported during biomethanation by [9]. The scheduling equations, setpoints/constraints, and gain parameters used are shown in Table 1 (chosen from earlier testing, documented in [29]).

$$G_{H2\_MFC} = MIN(G_{H2\_est}, G_{H2\_CH4}, G_{H2\_CO2}, G_{H2\_H2}, G_{H2\_pH}) \quad (3)$$

Table 1. Gain scheduling parameters and equations.

Scheduling Parameter (Measured)	Unit	Setpoint Type	Setpoint Name and Value	Gain Parameter and Value	Calculated Hydrogen Injection Value ($mL\ min^{-1}$) and Scheduling Equation
CH_4 conc. (χ_{CH4})	[% vol].	Setpoint	S_{CH4_sp}= 90	k_{CH4}= 0.3	$G_{H2\_CH4} = k_{CH4}(S_{CH4\_sp} - \chi_{CH4})$
CO_2 conc. (χ_{CO2})	[% vol.]	Min	S_{CO2_min}= 5	k_{CO2}= 0.3	$G_{H2\_CO2} = k_{CO2}(\chi_{CO2} - S_{CO2\_min})$
H_2 conc. (χ_{H2})	[% vol.]	Max	S_{H2_max}= 40	k_{H2}= 0.3	$G_{H2\_H2} = k_{H2}(S_{H2\_max} - \chi_{H2})$
pH	[-]	Max	S_{ph_max}= 8.2	k_{pH}= 5.0	$G_{H2\_pH} = k_{pH}(S_{pH\_max} - pH)$

If any condition resulted in a negative hydrogen injection value (or below the minimum range of the MFC), then the value passed to the MFC was simply set to zero for that period. The actual hydrogen injection setpoint was updated for each cycle of the online GC.

2.7. In Situ Biomethanation—Experimental Design

For each feedstock (SS, FW), operation with biomethanation (i.e., hydrogen injection) was performed in duplicate (SS1, SS2 and FW1, FW2), alongside a single control reactor with no hydrogen injection (Control SS and FW), as summarised in Table 2.

Table 2. Reactor conditions during biomethanation experiments.

Reactor number	**1**	**2**	**3**	**4**	**5**	**6**
Reactor name	Control SS	SS1	SS2	Control FW	FW1	FW2
Feedstock	Sewage Sludge (SS)			Food Waste (FW)		
Biomethanation	No	Yes		No	Yes	

The experiment was split into two main stages (R1–R3 and O1–O3), which are summarised in Table 3. In the first stage (R1–R3), the study was focused on the effect of the recirculation rate on in-situ biomethanation. In this stage, the gas recirculation was varied into three different recirculation rates at 20, 120, and 280 rpm or the equivalent to 12, 67, and 155 $L\ L^{-1}\ day^{-1}$. In the second stage (O1–O3), the process optimisation was carried out by modifying an operational condition, including an additional sparger on the recirculation line, a higher liquid mixing rate, and operating in a lower OLR, all the while maintaining the biogas recycling rate at the median value tested in the previous stage.

Liquid samples were taken twice a week to analyse the biological process indicators, such as TS, VS, alkalinity, ammonia, and volatile fatty acids. Prior to each period, all reactors were operated without hydrogen injection to establish the baseline biogas production and composition for at least 1 week or until stable biogas production was obtained.

2.8. In Situ Biomethanation—Calculated Parameters

The following are the calculated parameters that were used to characterise performance during experiments.

The gas retention time (RT_G) is an important process parameter which directly influences hydrogen conversion. The longer the hydrogen is in contact with the liquid phase, the higher the amount that will be dissolved and finally converted to methane. It is the ratio between the overall system gas headspace (V_H) and total biogas outflow (Q_{biogas}), expressed by:

$$RT_G = \frac{V_H}{Q_{biogas}} \tag{4}$$

Table 3. Summary of experimental periods.

Experimental Period	R1	R2	R3	O1	O2	O3
Variable of interest and short description	Recirculation flow rate			As R2 with sparger on recirculation line	As O1 with increased mixing rate	As O2 with reduced OLR
SS trial length (days)	17	12	21	19	20	18
FW trial length (days)	14	9	21	32	20	18
Biogas recirc. rate (RPM)	20	120	240	120	120	120
Biogas recirc. rate (L $L^{-1}d^{-1}$)	12	67	115	67	67	67
With/without 10μm sparger on recirculation line		No		Yes	Yes	Yes
Mixing stirring rate (RPM)		60		60	110	110
OLR (g VS L^{-1} d^{-1})		2		2	2	1
SS HRT (days)		14		14	14	28
FW HRT (days)		68		68	68	136

Hydrogen conversion describes how much of the hydrogen injected is actually converted in the biomethanation process and is defined as:

$$X_{H2} = \frac{\sum Q_{H2,inj} - \sum Q_{H2,out} + \Delta H_2}{\sum Q_{H2,inj}} \tag{5}$$

taken over the various experimental periods, where $\sum Q_{H2,inj}$ is the sum of the hydrogen injected into the reactor and $\sum Q_{H2,out}$ is the sum of the hydrogen leaving the reactor, while ΔH_2 is the volumetric variation in the H_2 amount contained in the headspace in a given time interval.

The methane evolution rate (MER) expresses the increase in the volumetric methane production rate, over the methane production from the background AD process, resulting from the in-situ biomethanation process and can be calculated using:

$$\text{MER} = \frac{\sum Q_{H2,inj} - \sum Q_{H2,out} + \Delta H_2}{4\, V_R} \tag{6}$$

The biomethanation extent expresses the extent to which biomethanation has increased the ratio of methane to carbon dioxide in the biogas and ignores any hydrogen content. It can be calculated using Equation (7), where ∅ represents the volume or molar fraction:

$$\text{Biomethanation extent} = \frac{\chi_{CH4}}{\chi_{CH4} + \chi_{CO2}} \tag{7}$$

The volumetric gas–liquid mass transfer coefficient ($k_L a$) is expressed in Equation (8) for hydrogen mass transfer, in a molar form, derived from the two-film theory [5,30]:

$$\dot{n}_{G/L} = V_R\, k_L a \left(C^*_{H2,l} - C_{H2,l} \right) \tag{8}$$

where $\dot{n}_{G/L}$ is the molar transfer rate (mol d^{-1}), V_R is the reactor working volume, $C^*_{H2,l}$ and $C_{H2,l}$ are the (molar) equilibrium and actual liquid concentrations, respectively, and $k_L a$ is the gas–liquid transfer coefficient (d^{-1}).

Due to difficulties surrounding the measurement of $C_{H2,l}$, and the fact that in a healthy biogas system, the concentration of dissolved hydrogen is maintained at a very low value by the action of hydrogenotrophic methanogens, it is common to assume that the C_l value is negligible compared to C_l^*, e.g., [5]. This is equivalent to the assumption that the gas–liquid transfer rate of H_2 is the limiting step of the biomethanation reaction.

$C_{H2,l}^*$ can be evaluated knowing the concentration of the gas in the bulk phase $C_{H2,g}$ using Henry's Law:

$$C_{H2,l}^* = \frac{C_{H2,g}}{H} \tag{9}$$

where H is the dimensionless Henry's constant; for hydrogen and water at 35 °C, the value is 50 [5].

Since the concentration of hydrogen, $C_{H2,g}$, changes spatially in the system (e.g., injected as pure hydrogen and mixed with other gasses in the headspace), a mean logarithmic concentration, shown in Equation (10), was used as the $C_{H2,g}$, as in [30]:

$$C_{H2,g} = \frac{C_{H2,\ g,injection} - C_{H2,\ g,headspace}}{\ln(C_{H2,\ g,injection}) - \ln\left(C_{g,headspace}\right)} \tag{10}$$

Combining the previous equations, it is possible to calculate the value of $k_L a$ for a given set of experimental measurements:

$$k_L a = \frac{H\,\dot{n}_{G/L}}{V_R\,C_{H2,g}} = \frac{H(\sum Q_{H2,inj} - \sum Q_{H2,out} + \Delta H_2)}{V_R\,C_{H2,g}}\left(\frac{p}{RT}\right) \tag{11}$$

where the molar flow, $\dot{n}_{G/L}$, is transformed into the measured volumetric flow (L d^{-1}) at STP conditions (temperature (T) 273.15 K, pressure (p) 1 bar, and gas constant R 0.08314 L bar K^{-1} mol^{-1}).

2.9. Mathematical Modelling

A mathematical model was produced to better understand the interaction between various operational conditions and observed experimental results. An overview of the material fluxes considered is shown in Figure 3. The model was not designed to fully describe the biomethanation process but instead to support hypotheses relating to underlying mechanisms. The model is based on a single AD reaction that produces only methane and carbon dioxide:

$$\text{Feedstock} \rightarrow CH_{4,\ AD} + CO_{2,\ AD} + \text{Digestate} \tag{12}$$

The biomethanation follows Equation (1) and relies on a number of simplifying assumptions:

1. The background AD process (the degradation of biomass into biogas) occurs at a constant rate and fixed stoichiometry.

2. The CO_2 produced by the background AD process is dissolved and available for the biomethanation of injected hydrogen.

3. The reactor has fixed liquid and headspace volumes (i.e., the fedfeedstock volume is equal to the removed digestate volume) and the headspace acts as a fully mixed reactor compartment containing the gasses produced by the AD and biomethanation processed

4. As per Section 2.8, the liquid concentration of dissolved hydrogen is negligible and therefore the process is mass transfer limited.

For ease of comparison with experimental data, a volumetric balance was made around the digester headspace, assuming all gasses behave ideally, at a fixed temperature and pressure. The dissolved gases fluxes were calculated as their equivalent in volume for convenience, although they themselves occupied no volume.

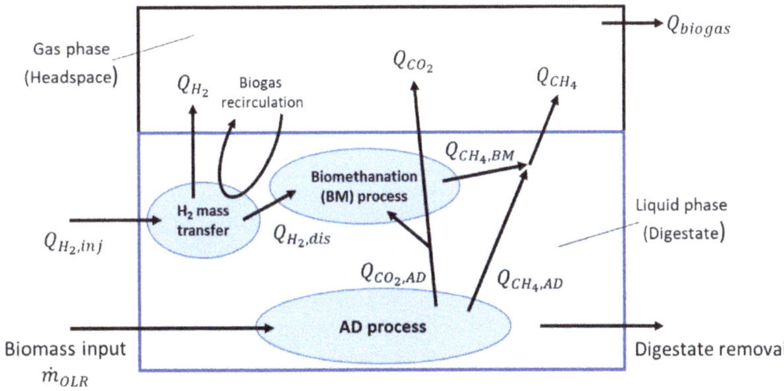

Figure 3. Overview of the mathematical model of the in-situ biomethanation process.

The background volumetric production of methane $\left(Q_{CH_4,AD}\right)$ and carbon dioxide $(Q_{CO_2,AD})$ from the AD process are proportional to the OLR ($\dot{m}_{OLR}$) using:

$$Q_{CH_4,AD} = Y_{CH4}\dot{m}_{OLR}V_R \tag{13}$$

$$Q_{CO_2,AD} = Y_{CO2}\dot{m}_{OLR}V_R \tag{14}$$

where k_{CH4} and k_{CO2} are the specific methane and carbon dioxide production constants taken from the experimental data measured on the control reactors.

Based on the control system design as described in Section 2.6, the volumetric hydrogen injection, when it is not constrained by pH or gas composition, can be expressed by its stoichiometric requirement:

$$Q_{H_2,\ inj} = 0.9\left(4\ Q_{CO_2,AD}\right) \tag{15}$$

Based on Equation (8), the dissolved hydrogen flux, expressed volumetrically $(Q_{H_2,\ dis})$, at STP conditions, can be calculated using:

$$Q_{H_2,dis} = k_L a\left(\frac{RT}{p}\right)V_R C_L^* \tag{16}$$

where C_L^* is calculated as in Equations (9) and (10).

The hydrogen (Q_{H_2}) and carbon dioxide $\left(Q_{CO_2}\right)$ entering the headspace can be calculated by molar balance assuming that all the dissolved hydrogen reacts with carbon dioxide and that the unreacted carbon dioxide enters the headspace:

$$Q_{H_2} = Q_{H_2,inj} - Q_{H_2,dis} \tag{17}$$

$$Q_{CO_2} = Q_{CO_2,AD} - \frac{1}{4}Q_{H_2,dis} \tag{18}$$

Methane production from biomethanation $(Q_{CH_4,BM})$ is calculated based on reaction stoichiometry (19), and the total methane produced is the sum of that from AD and biomethanation (20), with the subsequent total biogas outflow (Q_{biogas}) being calculated using Equation (21):

$$Q_{CH_4,BM} = \frac{1}{4}Q_{H_2,\ dis} \tag{19}$$

$$Q_{CH_4} = Q_{CH_4,BM} + Q_{CH_4,AD} \tag{20}$$

$$Q_{biogas} = Q_{CH_4} + Q_{CO_2} + Q_{H_2} \tag{21}$$

The resulting equilibrium biogas composition can be calculated, based on a component volumetric balance by finding the steady state solution ($d\chi_{gas}/dt = 0$), where χ_{gas} is the volumetric composition in the headspace, to Equations (22)–(24) for all three component gasses (CH_4, CO_2, H_2), noting that the biogas recirculation is ignored during this balance since all mass transfer is accounted for in Equation (16) and that otherwise the recirculation does not result in a net transfer to or from the gas phase.

$$\frac{d\chi_{CH_4}}{t} = \frac{1}{V_H}\left(Q_{CH_4} - \chi_{CH_4} Q_{biogas}\right) \quad (22)$$

$$\frac{d\chi_{CO_2}}{t} = \frac{1}{V_H}\left(Q_{CO_2} - \chi_{CO_2} Q_{biogas}\right) \quad (23)$$

$$\frac{d\chi_{H_2}}{t} = \frac{1}{V_H}\left(Q_{H_2} - \chi_{H_2} Q_{biogas}\right) \quad (24)$$

where V_H is the overall gas phase volume.

3. Results and Discussion

3.1. BMP and Baseline AD Process

The results of the characterisation of the inoculum and feedstock samples are presented in the Appendix A.1 (Tables A1 and A2). The BMP results are shown in Figure 4. The average methane production of the SS and FW during the BMP test was 0.402 ± 0.005 and 0.471 ± 0.020 L g^{-1} VS, respectively. These are within the ranges given in the literature of 0.220–0.460 [31–33] and 0.460–0.530 [34–36] for SS and FW, respectively. Using the values of the elemental analysis for both feedstocks (Table A2) and the calculated theoretical methane potential [37], the BMP of SS and FW resulted in 76.04% and 87.95% of their respective theoretical potentials. The methane potential of the cellulose as a positive control using the two inoculums gave values of 0.370 and 0.374 L g^{-1} VS, equivalent to 89.5% and 90.4% of the theoretical values.

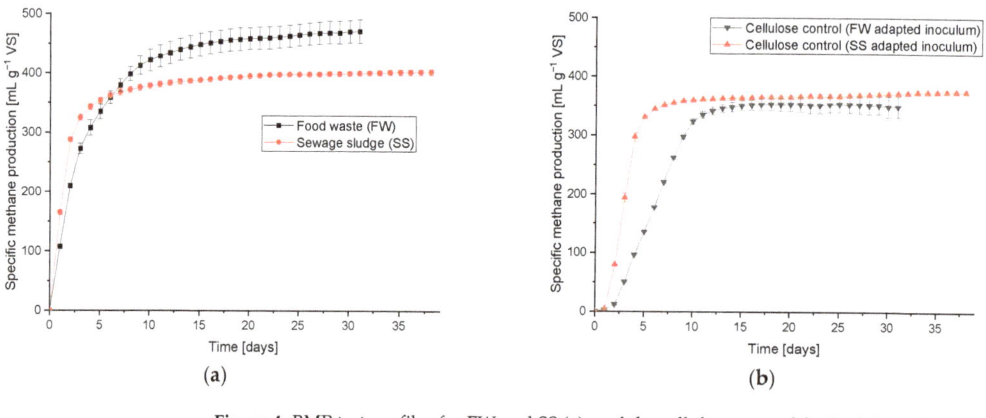

Figure 4. BMP test profiles for FW and SS (**a**), and the cellulose control for both inoculums (**b**).

Figure 5 shows the biogas composition from the initial baseline period prior to the biomethanation periods, which was similar in all replicates and validates the repeatability of the experimental approach; further baseline results are given in Appendix A.1 Table A3. The average methane and carbon dioxide specific yields were 0.24 and 0.12 L g^{-1} VS for SS and 0.42 and 0.28 L g^{-1} VS for FW, respectively. The baseline AD achieved 58 and 86% of the BMP specific methane yields, which is expected when comparing batch to continuous AD processes [38]. The lower result for SS can be attributed to its relatively short hydraulic retention time (HRT) (14 days, at OLR 2 g^{-1} VS L^{-1} day^{-1}) compared with the FW reactors (~68 days), since, in general, a longer HRT can be associated with a greater degree of feedstock degradation, as shown for FW in [36]. During the baseline AD testing, the alkalinity ratio (IA/PA) for all reactors was stable, with it being approximately 0.40 (SS) and 0.38 (FW).

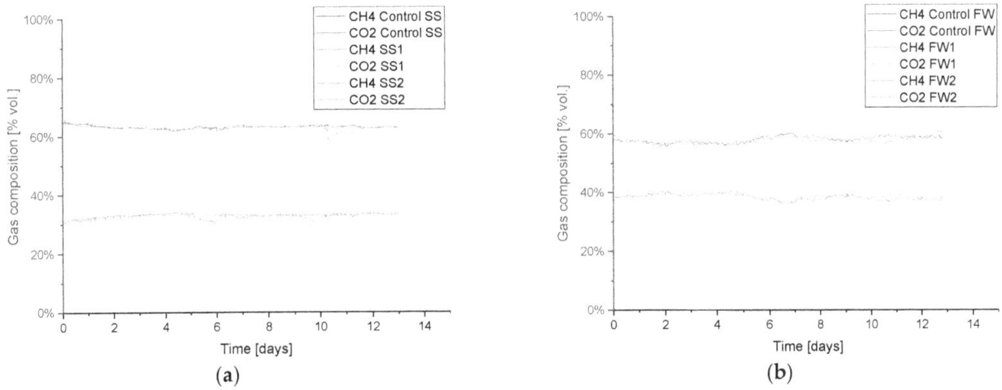

Figure 5. Biogas composition profile during a baseline AD testing prior to hydrogen injection, for sewage sludge (SS) (**a**) and food waste (FW) (**b**).

3.2. In-Situ Biomethanation Testing

For each experimental period (R1–R3, O1–O3), prior to hydrogen injection, it was ensured that all replicates had a similar baseline in terms of gas composition, gas content, pH, and alkalinity ratio. The hydrogen was then injected into the four biomethanation reactors (FW11, FW2, SS1, and SS2) in accordance with the specific experimental design of each period. The estimated hydrogen injection stoichiometric requirement (G_{H2_est}) was updated before each experimental period based on the carbon dioxide production of the control prior to the start of hydrogen injection. The following sections will present the results for each of these periods, focusing on the results during hydrogen injection.

The volume of data collected, the amount of possible detailed discussion in terms of describing results, and the operational challenges would make a comprehensive discussion of all the experimental data too lengthy. Instead, a single experimental period (R1) and feedstock (SS) will be described in detail to exemplify the data gathered as well as the challenges and complexities faced. Subsequently, average data taken over the whole experimental periods will be used to assess the overall trends and subsequent implications. The average data also include the transient period at the initial hydrogen injection, which generally lasts up to two days; therefore, the average data can be considered a good approximation of the steady state performance of the various experimental conditions. The full dataset, containing the detailed experimental outputs for the six experimental periods for the all the replicates and controls, is available on a separate data repository [39], and for a more detailed treatment of individual experimental conditions the reader is directed to [29].

3.2.1. In Situ Biomethanation Dynamics (Example Period R1 with SS)

Period R1 (gas recirculation 12 L $L^{-1}day^{-1}$) was monitored for 17 days, and the average measured OLR during the experiment for control SS, SS1, and SS2 were 1.91, 1.88 and 1.92 g VS. L^{-1} day^{-1}, respectively. Hydrogen injection remained stable at its setpoint value (1.02 mL min^{-1}, equivalent to approximately 0.86 L L^{-1} day^{-1}), excluding short periods of activation regarding the hydrogen gain loop (i.e., G_{H2_H2}) of the feedback control on days 0 and 9 and a technical problem with the hydrogen injection system on day 6. The biogas output and OLR data for this period are shown in Figure 6.

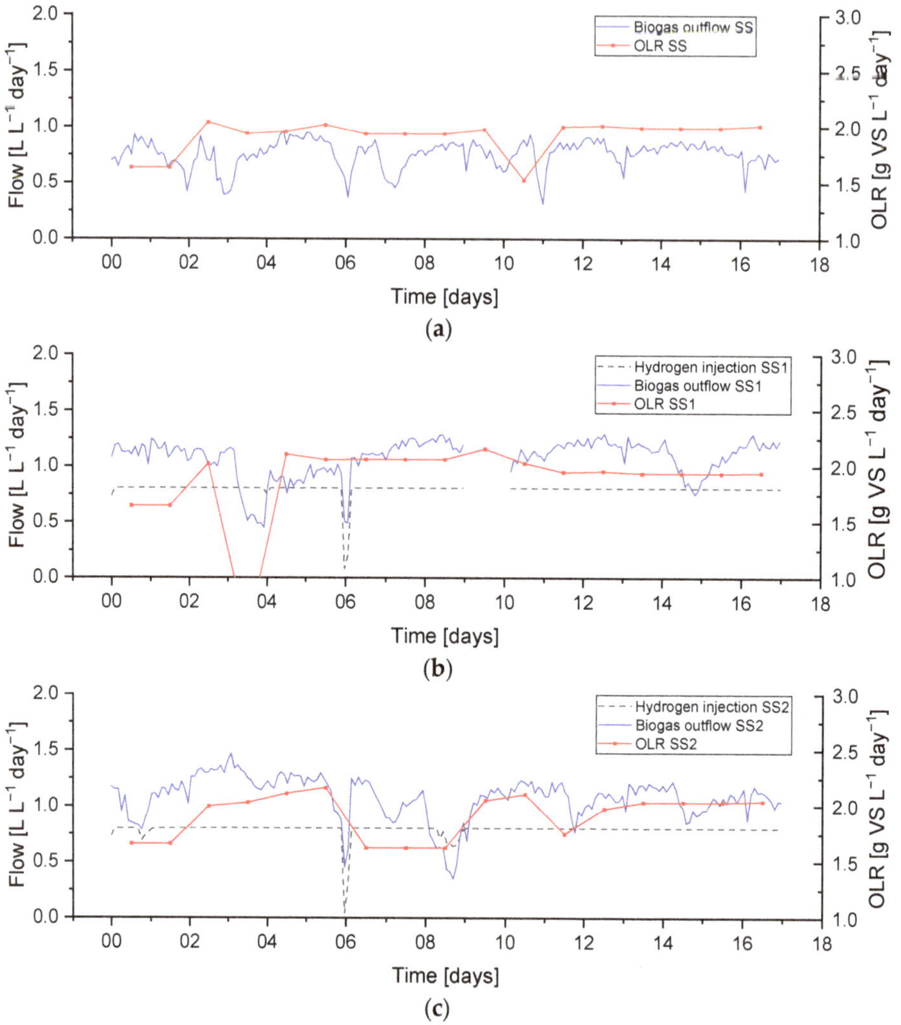

Figure 6. Biogas outflow, OLR, and hydrogen injection rate during period R1 for sewage sludge (SS)-fed reactors: (**a**) Control SS, (**b**) biomethanation reactor SS1, (**c**) biomethanation reactor SS2.

The initial pH, in all reactors, at the beginning of the experiment was around 7.1 (Figure 7). Indicative of the initiation of hydrogen injection in all experimental periods, the pH in the reactor SS1 and SS2 increased immediately after the addition of hydrogen, which was expected and previously reported due to bicarbonate consumption by the biomethanation reaction and the resulting predominance of the ammonia buffer on the carbonate buffer system [3]. Both duplicates showed similar pH profiles, with the average values for SS1 and SS2 being 7.41 and 7.44, respectively, compared with 7.20 in the case of the control reactor.

The duplicate reactors showed a similar biomethanation extent, which reached a value of approximately 80%, with the control value being approximately 70% (Figure 8), while the H_2 conversions were between 50–70%, with an average of 64 and 59% for SS1 and SS2, respectively. The CO_2 content decreased to approximately 15% in both biomethanation reactors and a 30% level in the control reactor. The methane also decreased in the biomethanation reactors, reaching an average of 56 and 54% in SS1 and SS2, respectively, due to the dilution by hydrogen in the headspace, which reached a maximum content of 33 and 36%, respectively.

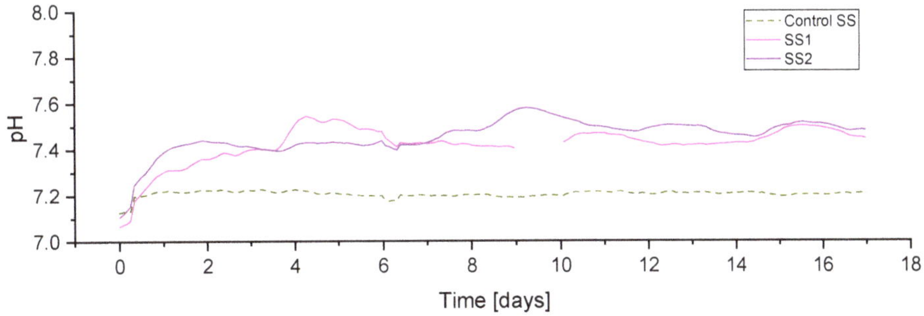

Figure 7. pH profile during period R1 for sewage sludge (SS) fed reactors.

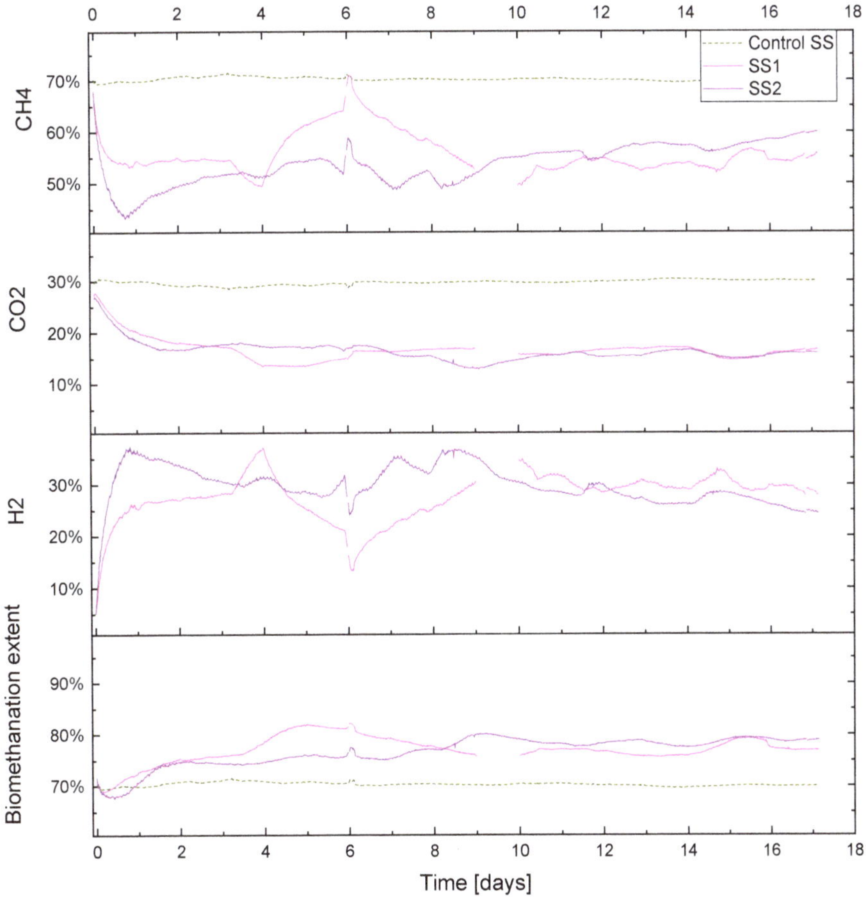

Figure 8. Gas composition (CH_4, CO_2, H_2 % vol.) and biomethanation extent (%) during period R1 for sewage sludge (SS) fed reactors.

A feedstock feeding pump failure on day 3 on reactor SS1 (the effect on the OLR can be seen in Figure 6) caused the feeding not to be delivered properly. This feeding failure caused an evident drop in biogas production, while the flow of hydrogen injection remained at its constant setpoint value. The lower biogas production and constant hydrogen injection caused an increase in the hydrogen concentration of up to 37%, while the methane and carbon dioxide concentrations dropped to 50 and 14%, respectively. In addition, the reduction in biogas production resulted in a higher retention time, leading to higher H_2 conversion, evident between days 3–4 in Figure 9.

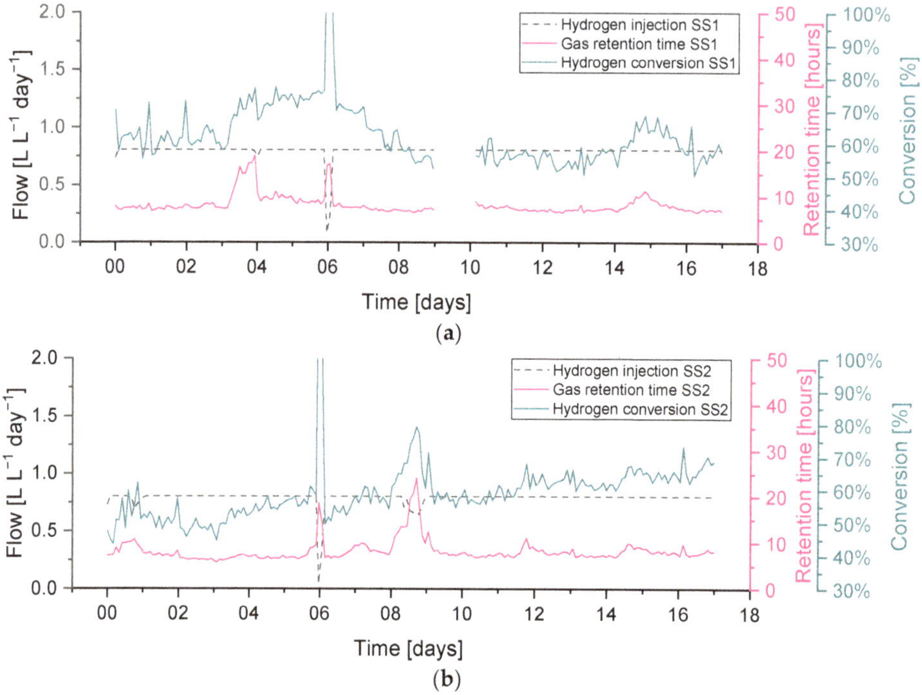

Figure 9. Hydrogen conversion in relation to gas retention time and hydrogen injection rate during period R1 for sewage sludge (SS)-fed reactors: (**a**) biomethanation reactor SS1, (**b**) biomethanation reactor SS2.

Abrupt changes in hydrogen injection also influence the process performance; this can be seen on day 6 in both reactors SS1 and SS2 when hydrogen had to be stopped for approximately five hours due to a technical issue. Biogas outflow immediately follows the change in hydrogen injection, and a similar reduction can be seen in Figure 6. This leads to a reduction in the gas flow and an increase in the gas residence time RT, which leads to a higher hydrogen conversion. The resulting changes in gas composition can be seen in Figure 8, where the hydrogen concentration diminishes and methane increase towards the value of the baseline; the carbon dioxide, on the other hand, has a slower response, and its content in the headspace remains more stable due to its increased solubility and residual buffering capacity in the liquid phase.

The hydrogen gain loop (i.e., G_{H2_H2}) of the feedback controller was activated on days 0 and 9 on the reactor SS2, as shown as a small drop in the hydrogen injection flow (see Figure 6). The gas composition of methane and carbon dioxide remained far from the constraints, while the hydrogen was near the upper constraint (40%).

The hydrogen conversion tended to increase from 50% at the beginning to 70% at the end of the experiment in both reactors (Figure 9). This trend is also confirmed by the gas composition data (Figure 8), where on the final six days of the experimental period, the methane content was observed to slightly increase in reactor SS2 along with a decrease in the hydrogen content. The H_2 conversion trend can be explained by microbial acclimation and the growth of the hydrogenotrophic population during the experiment. Microbial activity can, in fact, increase the gas–liquid mass transfer rate compared to a purely physical process in abiotic liquid by converting the absorbed gas into the stagnant liquid layer surrounding the gas bubble, thereby increasing the diffusional gradient. This phenomenon is reported in the literature as the microbial enhancement of the gas–liquid mass transfer [30].

3.2.2. Variation in Biogas Recirculation Rate (Periods R1–R3)

A summary of the average values of the main process parameters across periods R1–R3 (i.e., the variation in the biogas recirculation rate) for both SS and FW can be found in Figure 10. These data are supplemented by more detailed (per reactor) results presented in the Appendix A.2 (Tables A4 and A6). During periods R1–R3, the general observations and trends are similar for the SS and FW fed reactors, despite the generally higher all gas flow values in the FW reactors due to the high BMP of FW compared with SS. Consequently, the discussion will focus on SS whilst highlighting deviations from these for the FW reactors.

The general trend is that the increasing recirculation rate improved the hydrogen consumption rate. On average, the hydrogen conversion increased from 0.84 to 1.09 L day^{-1} between periods R1–R3 and increased the specific methane production rate by 26% compared with the control reactor. The hydrogen conversion rates constituted with the hydrogen injection rate in all periods were between 60 and 75%, where the highest value was achieved at R3–SS. This conversion rate was lower than that obtained previously [12] using a UASB reactor and a ceramic sponge diffuser (86.8%), but much larger than that obtained from a large-scale reactor (10–26%) [15].

Increasing the biogas recirculation rate also improves the methane evolution rate (MER) from 0.12 L L^{-1} day^{-1} to 0.15 L L^{-1} day^{-1}. In general, the MER of in-situ biomethanation is in the range of 0.08 to 0.39 L L^{-1} day^{-1} [40].

An anomaly appears when analysing the consumption of hydrogen and the volume of methane that is produced additional to the control, which should theoretically be 0.25 L CH_4 per L of H_2 consumed. For example, in R1, the ratio of the additional methane enrichment with hydrogen converted was 0.15 L CH_4 L^{-1} H_2. In R3, the ratio increased to 0.23 L CH_4 L^{-1} H_2 (0.27 for FW). Beyond any undetected experimental error, an explanation for this could be that hydrogen is being consumed for microbial growth [5,17]; a H_2:CH_4 ratio above 4 has been suggested to account for microbial biomass growth [41]. Related to this, the sum of the theoretical consumption of CO_2 by biomethanation (calculated from converted H_2) and the volumetric CO_2 from the output gas in all periods was 35%–39% higher compared to the CO_2 produced in the reactor control. In this case, the extra CO_2 might have come from bicarbonate consumption and reduced final dissolved CO_2, as has been observed previously [12], or could be the increased biochemical production of carbon dioxide due to differing process conditions between the control reactor and biomethanation reactors (cf. average pH in Appendix A), meaning that the assumed parity between the background AD process is invalid. The experimental overestimation of hydrogen consumption could explain both of these observations but a thorough examination of all measurements and calculations did not yield an opportunity for such an error.

The average pH values on the SS1 and SS2 increased compared to the pH at the reactor control. The average pH tends to increase along with the increase in the recirculation rate, owing to the increase in hydrogen consumption and therefore reduced carbon dioxide concentration that buffers the digestate pH. The average alkalinity during the experiment in SS1 and SS2 was improved along with the increasing recirculation rates. The average alkalinity ratio (IA/PA) in R1, 2, and 3 were 0.41, 0.35, and 0.28, respectively, with the recommended threshold of ratio being 0.3 [26]. The average total VFA equally decreased from an average of 2 to 1.2 g L^{-1}.

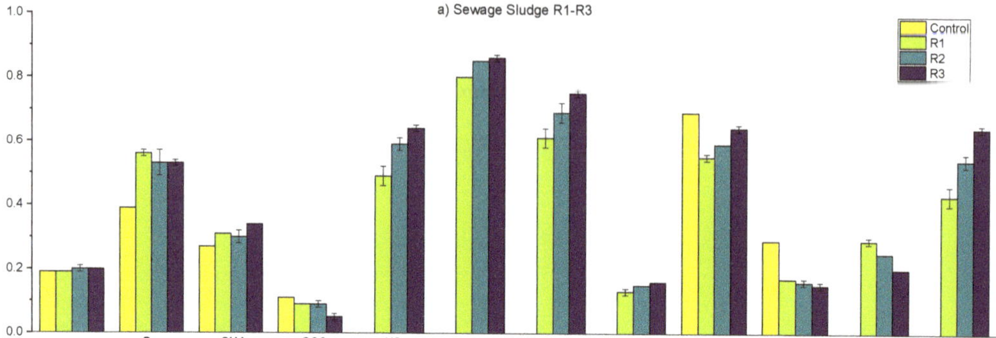

Indicator	OLR	Gas outflow	CH4 production	CO2 production	H2 consumed	H2 injected	H2 conversion	MER	CH4 content	CO2 content	H2 content	kLa
Unit	g VS L−1 day−1 E−1	L g−1 VS	L g−1 VS	L g−1 VS	L L−1 day	L L−1 day	% E−2	L L−1 day−1	% E−2	% E−2	% E−2	day−1 E−2
Control	0.19	0.39	0.27	0.11	0.00	0.00	0.00	0.00	0.69	0.29	0.00	0.00
R1	0.19	0.56	0.31	0.09	0.49	0.80	0.61	0.13	0.55	0.17	0.29	0.43
R2	0.2	0.53	0.3	0.09	0.59	0.85	0.69	0.15	0.59	0.16	0.25	0.54
R3	0.2	0.53	0.34	0.05	0.64	0.86	0.75	0.16	0.64	0.15	0.2	0.64

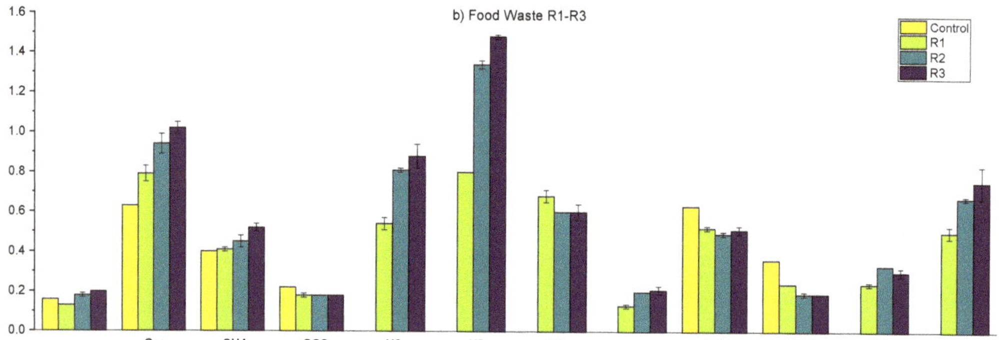

Indicator	OLR	Gas outflow	CH4 production	CO2 production	H2 consumed	H2 injected	H2 conversion	MER	CH4 content	CO2 content	H2 content	kLa
Unit	g VS L−1 day−1 E−1	L g−1 VS	L g−1 VS	L g−1 VS	L L−1 day	L L−1 day	% E−2	L L−1 day−1	% E−2	% E−2	% E−2	day−1 E−2
Control	0.16	0.63	0.40	0.22	0.00	0.00	0.00	0.00	0.63	0.36	0.00	0.00
R1	0.13	0.79	0.41	0.18	0.54	0.80	0.68	0.13	0.52	0.24	0.24	0.50
R2	0.18	0.94	0.45	0.18	0.81	1.34	0.6	0.2	0.49	0.19	0.33	0.67
R3	0.2	1.02	0.52	0.18	0.88	1.48	0.6	0.21	0.51	0.19	0.3	0.75

Figure 10. Summarised biomethanation experimental results; periods R1–R3 using SS (**a**) and FW (**b**). Error bars show duplicate reactor averages, and E notation is used to scale results to the same vertical axis.

As expected, the calculated k_La increased through the experimental periods due to the increased flow rate in the recirculation stream (greater gas holdup in the liquid phase and therefore higher specific area in terms of the bubbles). While this increase in mass transfer allowed for the significant consumption of hydrogen by the reactors, the presence of high concentrations of remaining hydrogen in the biogas outflow indicated that the process was mass transfer limited in all cases.

In the case of FW, despite similar trends, the hydrogen content was observed to be stable at a higher level compared to the equivalent SS reactors. The hydrogen gain loop was activated for the majority of the time during R1–R3, mainly caused by the larger hydrogen injection requirement of the feedstock, but with similar mass transfer characteristics. The methane concentration in the biogas outflow was lower for FW due to dilution with a higher amount of hydrogen but also the lower methane content from the background FW digestion.

To explore the variations in the gathered data over each operational period, in recognition of the broad variation in the measured data and performance parameters and the difficulty in terms of comparing the different operating conditions, the gas retention time (RT_G) was plotted against the hydrogen conversion for each biomethanation reactor and experimental period, as seen in Figure 11. Both variables were calculated as two hours average, covering the whole experimental period. Trendlines have been added of the form:

$$y = kx/(1 + kx) \tag{25}$$

as suggested for a gas–liquid mass transfer limited process [42] since they offer a good representation of the observed trends, i.e., that the hydrogen conversion eventually appears to saturate with respect to an increased gas retention time (RT_G). The curves were fitted using OriginPro® to elucidate trends from the highly scattered data.

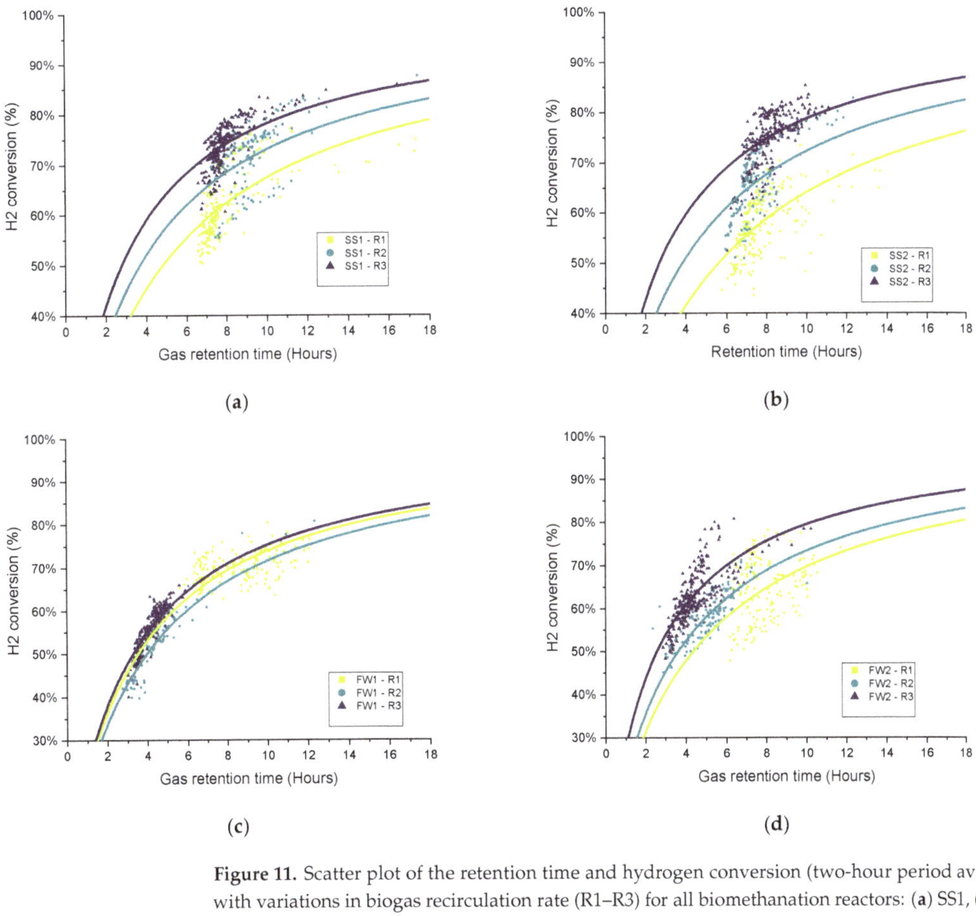

Figure 11. Scatter plot of the retention time and hydrogen conversion (two-hour period averages) with variations in biogas recirculation rate (R1–R3) for all biomethanation reactors: (**a**) SS1, (**b**) SS2, (**c**) FW1, and (**d**) FW2.

In general, the data distribution shows that hydrogen conversion increases along with the increase in the recirculation rate since the trendlines are also "ranked" in the graph following the same, from lowest to highest recirculation rates (R1–R3), with this result being previously reported [17]. There is a certain amount of deviation between the results of duplicates FW1 and FW2, which can also be seen in the hydrogen consumption and k_La results for FW (Figure 10).

3.2.3. Further Optimisation of In-Situ Biomethanation (O1–O3)

As per the previous results, a summary of the average process parameters for periods O1–O3 for both SS and FW reactors are shown in Figure 12, with a more detailed breakdown in the Appendix A.2 (Tables A5 and A7). The expectation for the further optimisation periods was that all three interventions considered, additional sparger on the biogas recirculation line (O1), an increase in the mechanical mixing rate (O2), and a reduction in the OLR (O3), should improve the overall performance of the biomethanation process. The mechanism for this would be through increased mass transfer for the recirculation stream in O1 and both the injection and recirculation stream in O2, while for O3, the improved performance would result from the increased gas retention time due to generally lower biogas production and hydrogen injection.

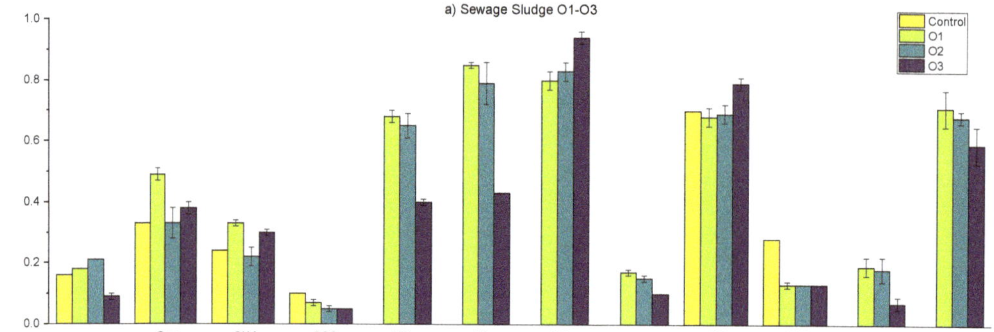

Indicator	OLR	Gas outflow	CH4 production	CO2 production	H2 consumed	H2 injected	H2 conversion	MER	CH4 content	CO2 content	H2 content	kLa
Unit	g VS L−1 day−1 E−1	L g−1 VS	L g−1 VS	L g−1 VS	L L−1 day	L L−1 day	% E−2	L L−1 day−1	% E−2	% E−2	% E−2	day−1 E−2
Control	0.16	0.33	0.24	0.10	0.00	0.00	0.00	0.00	0.70	0.28	0.00	0.00
O1	0.18	0.49	0.33	0.07	0.68	0.85	0.80	0.17	0.68	0.13	0.19	0.71
O2	0.21	0.33	0.22	0.05	0.65	0.79	0.83	0.15	0.69	0.13	0.18	0.68
O3	0.09	0.38	0.3	0.05	0.4	0.43	0.94	0.1	0.79	0.13	0.07	0.59

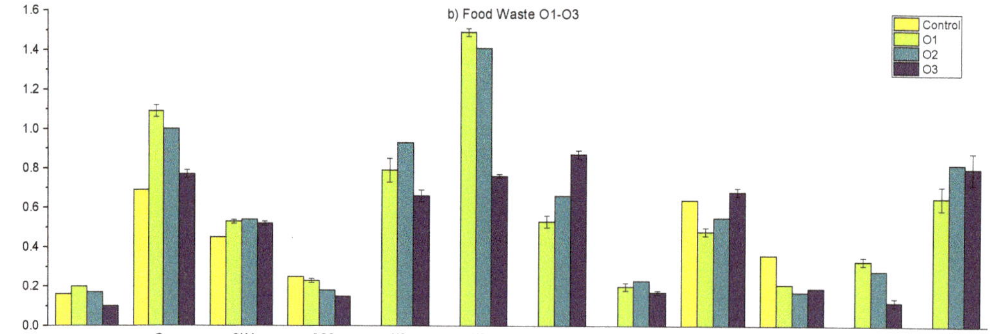

Indicator	OLR	Gas outflow	CH4 production	CO2 production	H2 consumed	H2 injected	H2 conversion	MER	CH4 content	CO2 content	H2 content	kLa
Unit	g VS L−1 day−1 E−1	L g−1 VS	L g−1 VS	L g−1 VS	L L−1 day	L L−1 day	% E−2	L L−1 day−1	% E−2	% E−2	% E−2	day−1 E−2
Control	0.16	0.69	0.45	0.25	0.00	0.00	0.00	0.00	0.64	0.36	0.00	0.00
O1	0.20	1.09	0.53	0.23	0.79	1.49	0.53	0.20	0.48	0.21	0.33	0.65
O2	0.17	1	0.54	0.18	0.93	1.41	0.66	0.23	0.55	0.17	0.28	0.82
O3	0.1	0.77	0.52	0.15	0.66	0.76	0.87	0.17	0.68	0.19	0.12	0.8

Figure 12. Summarised biomethanation experimental results; periods O1-O3 using SS (**a**) and FW (**b**). Error bars show duplicate reactor averages, and E notation is used to scale results to the same vertical axis.

For the SS reactors in period O1, this prediction was confirmed, by a comparison with the results from R2 (equivalent biogas recirculation rate), through an increase in k_La to 71 day^{-1} (c.f. 54) and hydrogen conversion to 80% (c.f. 60%). In period O2, however, the results show no improvement from the increase in the mechanical mixing rate, and even a slight, unexplained reduction in performance for SS2, except for the fact that the hydrogen conversion increased in O2 to 83% (c.f. 80% in O1). The periods are difficult to compare since there is a variation in the OLR from 1.8 to 2.0 gVS L^{-1} day^{-1} in O1 and O2, respectively. A higher OLR would lead towards shorted gas retention times and a reduction in hydrogen consumption, with other conditions remaining the same.

The best biomethanation performance in terms of desirable output biogas composition (high methane (79%), low carbon dioxide (13%), and hydrogen (7%)) and highest hydrogen conversion (94%) was achieved during O3 with a reduced in the OLR of 0.9 g VS L^{-1} day^{-1}. This is despite a reduction in the observed k_La, which was expected due to the reduced hydrogen injection rate (in proportion with the reduced OLR) and therefore reduced gas hold-up. This increased performance is traded off against a lower production rate in terms of methane, expressed by a MER of 0.1 L L^{-1} day^{-1} (c.f. 0.15 in O2), and, due to the very low OLR used, this is unlikely to be a practical solution for improved biomethanation performance in a real-world scenario.

With regard to the FW experiments, O1–O3, a comparison between O1-FW (with the additional sparger) and R2-FW (with equivalent biogas recirculation rate) shows a slight reduction in hydrogen consumption, k_La, and the outflowing biogas composition. It is possible that the increased OLR in O1 (2.0 c.f. 1.8 g VS L^{-1} day^{-1} in R2) is masking any improvements in performance. It was also observed during O1 that there was an increase in the ammonia concentration (from 3.62 to 3.95 gTAN kg^{-1}) compared with previous experimental runs and, also, for the first time, foaming was detected. There are numerous causes for foaming in AD systems, such as improper mixing, fluctuations in the OLR, and substrate types [43]. The levels of VFA and alkalinity ratio in the biomethanation reactors were, on average, comparable to the levels in the control reactors, so no indication of biological instability was noted.

The addition of mechanical mixing in O2 did lead to improvements in biomethanation performance compared with R2 in terms of a higher hydrogen conversion (66 c.f. 60%), MER (0.23 cf. 0.20 L L^{-1} day^{-}), and k_La (82 c.f. 67 day^{-1}), with the trend being similar to that of O2 to R2 for the SS reactors. As per the SS reactors, the best biomethanation performance for FW in terms of hydrogen conversion and desirable output biogas composition was at the reduced OLR in O3.

The distribution of the hydrogen conversion as a function of the gas retention time for experimental periods O1–O3 for both SS and FW is shown in Figure 13. Similar to Figure 11, the fitted curves were added only for illustration rather than to imply quality of fit. On the whole, the trends observed and discussed above can be confirmed in the scatter plots, in that the biomethanation performance ranked in terms of hydrogen conversion for SS was R2 $<$ O1 $\approx$ O2 $<$ O3. For FW, O1 (additional sparger) did not see the process gains in one of the duplicates (FW1), and the ranking can be ordered as R2 $\approx$ O1 $<$ O2 $<$ O3 (as discussed above).

3.3. *In-Situ Biomethanation Modelling*

The model developed in Section 2.9 was used to simulate the conditions for the in-situ biomethanation experiments for two periods, O1 for SS and O2 for FW, as these configurations showed the overall best results in terms of biomethanation performance considering a combination of product quality (i.e., a high hydrogen consumption rate and a high/low concentration of methane/carbon dioxide in biogas) and quantity (the methane evolution rate).

Only minimal experimental data were required for the simulation owing to the low model complexity, i.e., no reaction kinetics, inhibition, and biochemical considerations. OLR and hydrogen injection were the model input. Only five parameters were required by the model structure: the specific yields of methane and carbon dioxide produced by the background AD process (Y_{CH4}, Y_{CO2}) from the control reactor, the volumetric gas–liquid mass transfer coefficient ($k_L a$) from the biomethanation experimental data, and the reactor and headspace volumes (V_R, V_H).

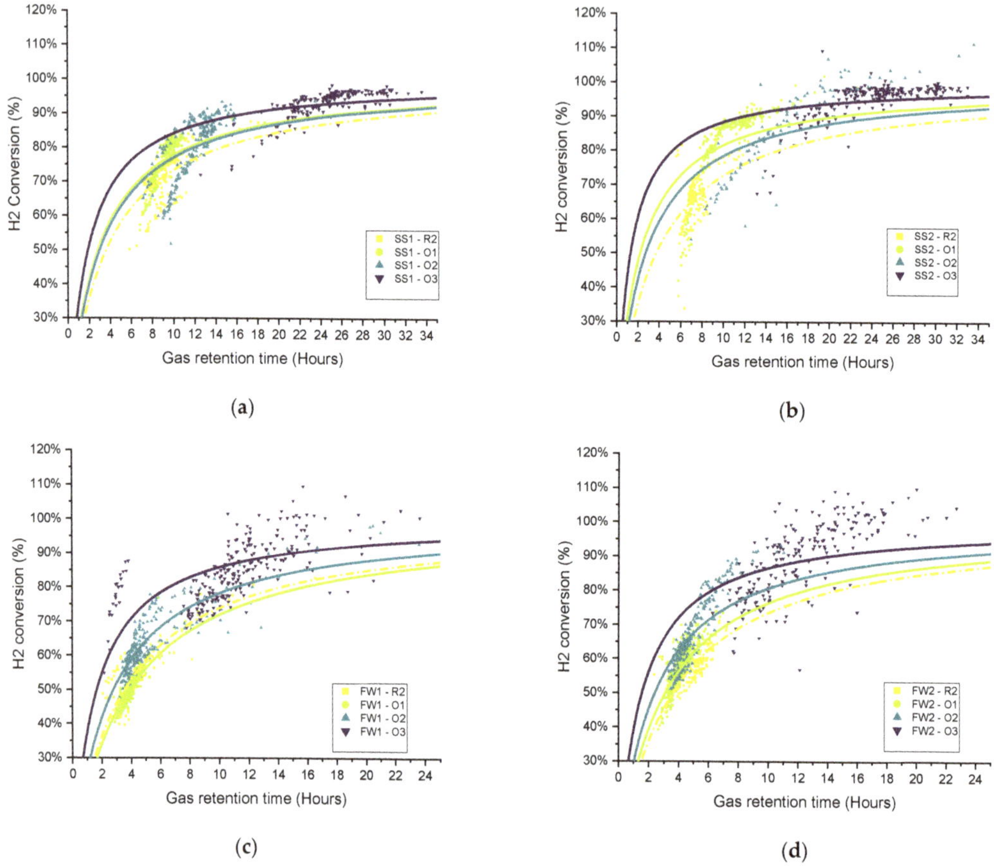

Figure 13. Scatter plot of the retention time and hydrogen conversion (two-hour periods average) with variations in operational conditions (O1–O3) for all biomethanation reactors:(**a**) SS1, (**b**) SS2, (**c**) FW1, and (**d**) FW2).

Despite a simplified process model, comparison between the simulated and experimental data show a good fit in terms of reproducing the main experimental average outputs, as shown in Figure 14. Apart from carbon dioxide specific yield and concentration in biogas, all the main results were fitted with a relative error below 10 and 11% for SS and FW, respectively. Across both feedstocks, carbon dioxide production and content were poorly represented, which can be related back to the discussion in 3.2.2 surrounding the mass balance between the observed hydrogen consumption and carbon dioxide production. Any mechanism that leads to increased observed carbon dioxide production (e.g., release through alkalinity/pH change or changes to the background AD process) was not included in the model, and therefore this experimental anomaly was not replicated.

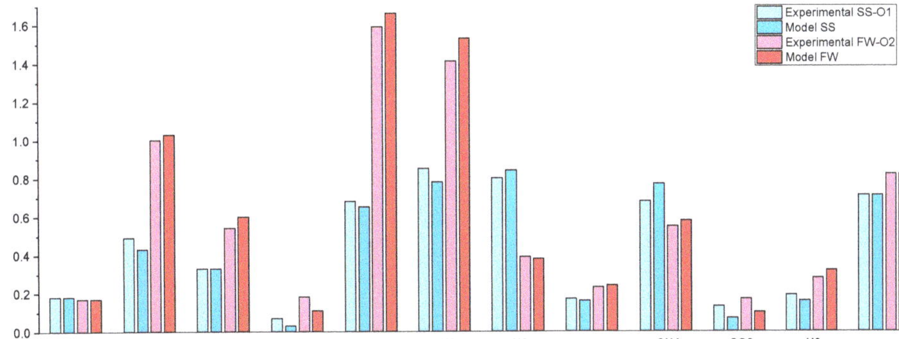

Indicator	OLR	Gas outflow	CH4 production	CO2 production	H2 consumed	H2 injected	H2 conversion	MER	CH4 content	CO2 content	H2 content	kLa
Unit	g VS L-1 day-1 E-1	L g-1 VS	L g-1 VS	L g-1 VS	L L-1 day	L L-1 day	% E-2	L L-1 day-1	% E-2	% E-2	% E-2	day-1 E-2
Experimental SS-O1	0.18	0.49	0.33	0.07	0.68	0.85	0.80	0.17	0.68	0.13	0.19	0.71
Model SS	0.18	0.43	0.33	0.03	0.65	0.78	0.84	0.16	0.77	0.07	0.16	0.71
Experimental FW-O2	0.17	1	0.54	0.18	1.59	1.41	0.39	0.23	0.55	0.17	0.28	0.82
Model FW	0.17	1.03	0.6	0.11	1.66	1.53	0.38	0.24	0.58	0.1	0.32	0.82

Figure 14. Comparison of the experimental data and modelled data for SS period O1 and FW period O2.

The qualitative comparison between the modelling and experimental data also validates that the assumptions made during model formulation are also likely to be valid for the experiment. The background AD process appears to be relatively unaffected by the addition of additional hydrogen injection, and the process appears to be mass transfer limited.

The model can be used more broadly to explore the performance of an in-situ biomethanation system, as long as these founding assumptions hold. For demonstration, process contour maps were generated (Figure 15) showing the effect of variations in the OLR and hydrogen injection ratio, S (defined as $S = Q_{H_2,inj}/(4Q_{CO_2,AD})$, on a range of process characteristics and performance indicators, for three levels of mass-transfer: 60 day^{-1}, corresponding approximately to the conditions explored in this paper, and then 120 and 240 day^{-1}. The delivery of these improvements in mass transfer characteristics would require the redesign of the equipment, e.g., the use of membranes, increasing the aspect ratio (to increase the bubble path length), or a reduction in bubble size, but, given other works in this area, these values are not considered unrealistic.

Generally speaking, the model predicts that, within the explored conditions, in the best case, the process cannot deliver high performance single reactor in-situ biomethanation in a practical and sensible range in terms of the OLR commonly found in AD systems. For example, even at the highest k_La explored (240 day^{-1}), at an OLR of 5 g VS L^{-1} day^{-1} (Point A, Figure 15), the process is predicted to produce biogas upgraded to contain ~10% both carbon dioxide and hydrogen with ~80% methane. This gas would require further treatment for most current applications. For use as a biomethane (as a natural gas drop-in replacement), it is likely that some kind of CO_2 removal would be required, either through a further methanation step (e.g., ex situ biomethanation) or through a physical separation. For use as a vehicle fuel (as a CNG drop-in replacement), this gas would likely require hydrogen removal through, e.g., membrane separation.

Depending on the mass transfer capabilities of the system, the available quantity of hydrogen, the targeted product gas application, and/or to match the composition of the product gas with the available downstream purification options, it may be beneficial to tune the in-situ process, which can be facilitated using the contour maps in Figure 15. The produced process contours can be helpful to explore this. For example, a system with a mass transfer capability of 120 day^{-1} may be optimised for 'near complete' in-situ biomethanation via operation at an OLR of 1 g VS L^{-1} day^{-1} (point B), with these conditions being able to produce a high-quality biomethane (~94% CH_4 ~5% CO_2, ~1% H_2) but only at a low productivity (MER) of 0.1 day^{-1} and while significantly underutilising the biomass treatment capacity of the system. Another option for the same system may be to accept a low biomethanation extent, but to operate at a reduced stoichiometric ratio (S = 0.25) and an increased OLR (Point C) to partially upgrade the biogas (~75% CH_4 ~24% CO_2, ~1% H_2) whilst minimising hydrogen contamination at a modest value of MER (0.75 L L^{-1} day^{-1}). In another application, it may be better to maximise the consumption of carbon dioxide by applying a higher OLR and stoichiometric ratio (S = 0.9, Point D) to produce a hydrogen-rich biomethane blend (~65% CH_4 ~10% CO_2, ~25% H_2) which, upon CO_2 removal, could be suitable for natural gas grid injection depending on local requirements at an improved MER (0.85 L L^{-1}day^{-1}).

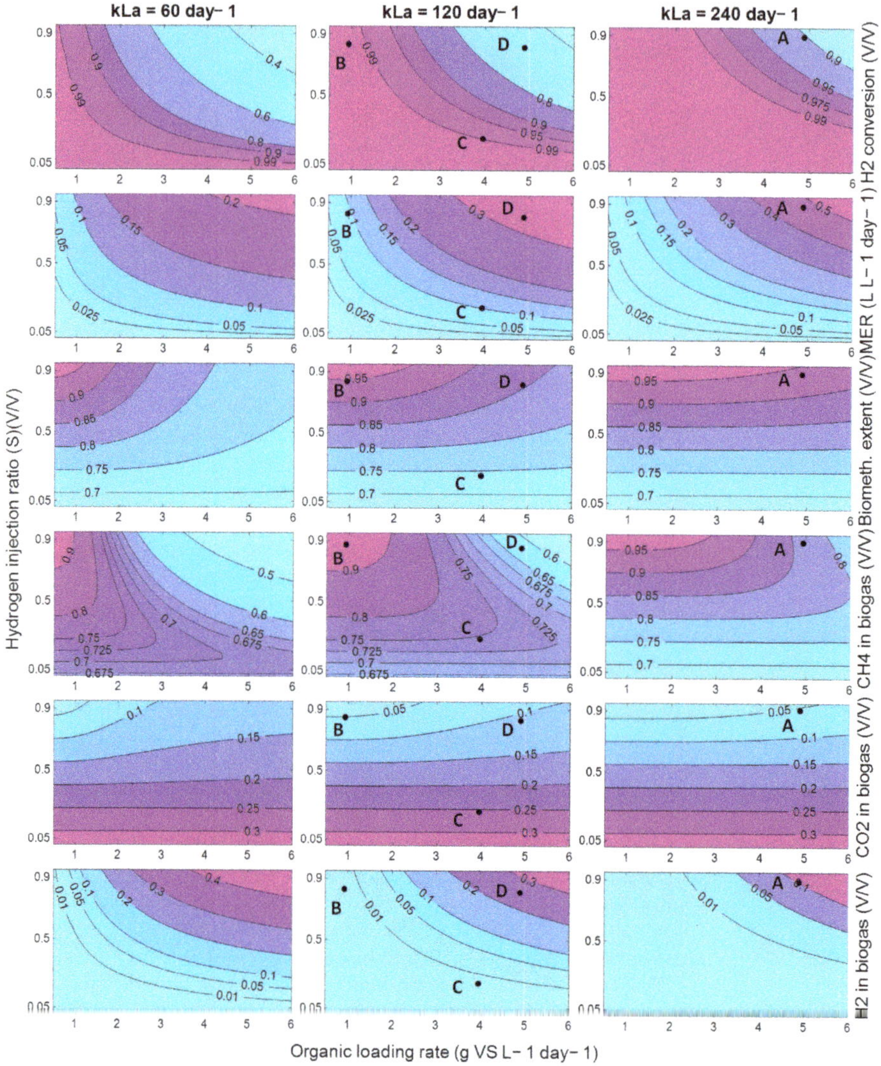

Figure 15. Contour plots of in-situ biomethanation performance using SS with a variation in the OLR and hydrogen injection ratio for k_La of 60, 120, and 240 day^{-1}. Labelled points (A, B, C, D) on the plots refer to alternative process conditions and are elaborated in the discussion.

For reference, the equivalent contours for a FW fed system are shown in Appendix A.2 Figure A1. The trends are similar, but the challenge of producing both a high quality and quantity of biomethane product gas from an in-situ system is greater for FW due to its higher biogas production and CO_2 baseline content (and therefore greater hydrogen requirement).

4. Conclusions

In this work, in-situ biomethanation, alongside the AD of sewage sludge and food waste, was successfully demonstrated at laboratory scale. Continuous feedstock and hydrogen feeding as well as the monitoring and feedback control of the hydrogen supply were implemented to emulate the envisaged full-scale implementation of this technology. The complete results dataset is made available on a data repository [39].

The performance of the lab-scale process, in terms of its capability to produce a high-concentration biomethane output gas directly from the digesters, was generally limited, mainly due to the mass-transfer of gaseous hydrogen, which was evidenced by a large residual hydrogen concentration in the produced biogas, but otherwise the biological process was stable and showed no signs of process inhibition. Increasing the biogas recirculation rate, reducing the bubble size through a sparger, and increasing the mechanical mixing improved the gas–liquid mass transfer, with the k_La estimated to be between 43–82 day^{-1}. At an OLR of 2 g VS L^{-1} day^{-1}, it was possible to achieve a MER of 0.17 and 0.23 L L^{-1}day^{-1}, a H_2 conversion of 80 and 66%, and a CH_4 content of 69 and 55% for SS and FW, respectively. The reduction in the OLR to 1 g VS L^{-1} day^{-1} allowed for an increase in biomethanation performance in terms of hydrogen conversion (94 and 87% for SS and FW, respectively) and methane content in biogas (79 and 68% for SS and FW, respectively), but at the expense of overall system productivity and the utilisation of the reactors.

To explore a broader set of operating conditions, a process model, based on a single AD reaction, the mass balance on the biomethanation reaction, and the simple treatment of gas–liquid mass transfer was developed and acceptably validated on experimental data. The exploration demonstrated the process design trade-offs that need to be made in order to have either a highly productive (high OLR and methane production) or a high-quality biomethane output gas (at low OLR) at the k_La values obtained in this experimental work. With the k_La increased up to >240 day^{-1}, a more complete in-situ biomethanation could be possible at OLR values common in large scale AD systems for SS, while for FW, even higher gas–liquid mass transfer rates would be required.

Author Contributions: Conceptualisation: A.S., D.P., S.M., and M.W; methodology; A.S., D.P., and M.W; modelling; D.P. and M.W; data-analysis: A.S., D.P., and M.W; writing—original draft preparation: A.S., D.P., and M.W; writing—review and editing: A.S., D.P., S.M., and M.W; supervision: W.N., D.P., M.P., and M.W.; project administration: W.N., D.P., M.P., and M.W; funding acquisition: W.N., M.P., and M.W. All authors have read and agreed to the published version of the manuscript.

Funding: This work was supported by the Indonesia Endowment Fund for Education (LPDP) (PhD Studentship) and the UK Engineering and Physical Sciences Research Council (EPSRC) through the IBCat H2AD project 'Biomethanisation of CO_2 in anaerobic digestion plants' (grant EP/M028208/1).

Institutional Review Board Statement: Not applicable.

Informed Consent Statement: Not applicable.

Data Availability Statement: Complete results dataset is available at the University of Sheffield data repository: https://doi.org/10.15131/shef.data.21747239.

Conflicts of Interest: The authors declare no conflict of interest.

Nomenclature

Abbreviations	
AD	Anaerobic digestion
ADM1	Anaerobic Digestion Model 1
BMP	Biochemical methane potential
CSTR	Continuous stirred-tank reactor
FW	Food waste
FW1	Duplicate 1 in food waste biomethanation experiments
FW2	Duplicate 2 in food waste biomethanation experiments
HRT	Hydraulic retention time
IA	Intermediate alkalinity
ISR	Inoculum to substrate ratio
MER	Methane evolution rate (methane from biomethanation) [L L^{-1} d^{-1}]
MFC	Mass flow controllers
O1,O2,O3	Periods in the experimental stage at different operational conditions
OLR	Organic loading rate
PA	Partial alkalinity
R1,R2,R3	Periods in the experimental stage at different recirculation rates
SS	Sewage sludge
SS1	Duplicate 1 in sewage sludge biomethanation experiments
SS2	Duplicate 2 in sewage sludge biomethanation experiments
STP	Standard temperature and pressure (273.15 K and 1 bar)
TAN	Total ammonia nitrogen
VS	Volatile solids
Symbols	
$C_{H2,g}$	Hydrogen concentration in the gas bulk phase [mol L^{-1}]
$C^{*}_{H2,l}$	Dissolved hydrogen conc. in equilibrium with gas bulk phase [mol L^{-1}]
$C_{H2,l}$	Dissolved hydrogen concentration in the liquid bulk phase [mol L^{-1}]
G_{H2_est}	Estimated stoichiometric hydrogen injection rate [mL min^{-1}]
G_{H2_CH4}	Scheduled hydrogen injection, based on CH_4 setpoint [mL min^{-1}]
G_{H2_CO2}	Scheduled hydrogen injection, based on CO_2 constraint [mL min^{-1}]
G_{H2_H2}	Scheduled hydrogen injection rate, based on H_2 constraint [mL min^{-1}]
G_{H2_pH}	Scheduled hydrogen injection rate, based on pH constraint [mL min^{-1}]
G_{H2_MFC}	Actual hydrogen injection flow rate, requested to the MFC [mL min^{-1}]
H	Henry's dimensionless constant [(mol H2 $L^{-1})_{gas}$/(mol H2 $L^{-1})_{liq}$]
$\dot{n}_{G/L}$	Molar gas–liquid transfer rate [mol d^{-1}]
p	Gas pressure [bar]
k_{CH4}	Gain parameter, of the scheduling control based on methane [-]
k_{CO2}	Gain parameter, of the scheduling control based on carbon dioxide [-]
k_{H2}	Gain parameter, of the scheduling control based on hydrogen [-]
k_{pH}	Gain parameter, of the scheduling control based on pH [-]
Q	Flow rate [L d^{-1}]
Q_{biogas}	Total biogas outflow rate [L d^{-1}]
$Q_{CH_4,AD}$	Methane flow rate from digestion of feedstock [L d^{-1}]
$Q_{CH_4,BM}$	Methane flow rate from biomethanation [L^{-1} d^{-1}]
$Q_{CO_2,AD}$	Carbon dioxide flow rate from digestion of feedstock [L d^{-1}]
$Q_{H_2,dis}$	Hydrogen dissolution rate [L^{-1} d^{-1}]
R	Ideal gas constant [L bar K^{-1} mol^{-1}]
RT_G	Gas retention time [hours]
S	Hydrogen injection ratio (injected over stoichiometric requirement) [-]
V_H	Overall gas phase volume of the system [L]
V_R	Working volume, liquid phase, of biomethanation reactor [L]
Y_{CH4}	Methane specific yield from digestion of feedstock [L g^{-1} VS]
Y_{CO2}	Carbon dioxide-specific yield from digestion of feedstock [L g^{-1} VS]
X_{H2}	Hydrogen conversion in biomethanation [-]
χ	Molar and volumetric gas fractions [-]

Appendix A.

Appendix A.1. Results of Inoculum and Biomass Characterisation and Baseline AD Testing

Table A1. Characterisation results of the inoculums used for BMP and in-situ biomethanation experiments.

Inoculum	Unit	Sewage Sludge Inoculum	Food Waste Inoculum
TS	%	3.04	3.97
VS	%	2.00	2.74
pH		7.20	7.50
Ammonia	gTAN/kg substrate	1.43	6.42
PA	$gCaCO_3/kg$	3.26	18.40
IA	$gCaCO_3/kg$	1.24	6.13
Total alkalinity	$gCaCO_3/kg$	4.51	24.52
IA/PA		0.38	0.33

Table A2. Characterisation results of the feedstocks used in the biomethanation experiments.

Feedstock	Sewage Sludge	Food Waste
TS (% wet weight)	3.96 ± 0.55	14.29 ± 0.52
VS (% wet weight)	2.79 ± 0.43	13.58 ± 0.48
Elemental Analysis		
Carbon (%TS)	39.80 ± 0.84	49.92 ± 0.31
Hydrogen (%TS)	5.98 ± 0.22	6.77 ± 0.77
Oxygen (%TS)	26.58 ± 1.51	35.27 ± 0.63
Nitrogen (%TS)	4.80 ± 0.56	3.35 ± 0.08

Table A3. Results of baseline AD experiments across all reactors.

Reactor	Methane (%)	Carbon Dioxide (%)	Specific Methane Yield (L g^{-1} VS)	Specific Carbon Dioxide yield (L g^{-1} VS)
Control SS	65.38 ± 0.01	34.02 ± 0.00	0.25 ± 0.13	0.13 ± 0.06
SS1	65.32 ± 0.00	34.26 ± 0.00	0.24 ± 0.02	0.12 ± 0.01
SS2	64.97 ± 0.01	34.22 ± 0.01	0.23 ± 0.06	0.12 ± 0.03
Control FW	59.96 ± 0.01	39.62 ± 0.01	0.44 ± 0.09	0.29 ± 0.05
FW1	59.34 ± 0.01	39.56 ± 0.01	0.42 ± 0.05	0.28 ± 0.03
FW2	59.57 ± 0.01	39.90 ± 0.01	0.41 ± 0.13	0.27 ± 0.07

Appendix A.2. In-Situ Biomethanation Results across All Experimental Periods

Table A4. Average in-situ biomethanation results for sewage sludge (SS)-fed reactors across operational periods R1–R3.

	Experimental Period	R1			R2			R3		
	Reactor	Control SS	SS1	SS2	Control SS	SS1	SS2	Control SS	SS1	SS2
Duration	days	17	17	17	12	12	12	21	21	21
OLR	gVS L_R^{-1} day^{-1}	1.91	1.88	1.92	1.93	1.89	2.03	1.87	1.99	1.97
H2 injection rate	L L_R^{-1} day^{-1}		0.81	0.80		0.85	0.85		0.85	0.87
Specific biogas production	L gVS^{-1}	0.40	0.55	0.57	0.39	0.49	0.56	0.39	0.54	0.52
Specific CH4 production	L gVS^{-1}	0.28	0.31	0.31	0.27	0.28	0.32	0.27	0.34	0.34
Specific CO2 production	L gVS^{-1}	0.12	0.09	0.09	0.12	0.08	0.10	0.08	0.05	0.04
H2 consumption rate	L L_R^{-1} day^{-1}		0.52	0.46		0.61	0.56		0.63	0.65
H2 conversion	%		64.09	58.50		71.41	66.32		73.93	75.26
Methane evolution rate (MER)	L L_R^{-1} day^{-1}		0.13	0.12		0.15	0.14		0.16	0.16

Table A4. *Cont.*

Experimental Period		R1			R2			R3		
	Reactor	Control SS	SS1	SS2	Control SS	SS1	SS2	Control SS	SS1	SS2
Gas retention time	hours		8.21	8.22		9.42	7.63		8.08	8.43
Gas–liquid transfer rate *kLa*	day^{-1}		45.8	39.8		55.8	52.5		63.0	65.2
CH4 content	vol.%	69.91	56.16	54.29	68.50	59.47	58.80	68.64	63.65	64.72
CO2 content	vol.%	29.58	16.78	16.58	29.63	15.56	16.80	28.89	15.59	14.28
H2 content	vol.%	0.07	27.91	30.24	0.11	25.28	24.63	0.24	20.23	20.38
pH		7.21	7.43	7.46	7.23	7.32	7.48	7.34	7.6	7.62
TS	wt.%	2.41	2.59	2.48	2.34	2.38	2.53	2.51	2.76	2.57
VS	wt.%	1.54	1.56	1.54	1.30	1.47	1.58	1.46	1.73	1.57
Ammonia	$gTAN\ L^{-1}$	1.61	1.55	1.49	1.46	1.36	1.56	1.46	1.56	1.52
Alkalinity ratio		0.45	0.42	0.40	0.37	0.38	0.31	0.33	0.31	0.31
Acetate	$g\ L^{-1}$	0.25	0.25	0.08	0.38	0.25	0.12	0.68	0.23	0.21
Total VFA	$g\ L^{-1}$	2.81	2.02	0.72	3.34	2.55	2.35	2.17	1.38	1.13

Table A5. Average in-situ biomethanation results for sewage sludge (SS)-fed reactors across operational periods O1–O3.

Experimental Period		O1			O2			O3		
	Reactor	Control SS	SS1	SS2	Control SS	SS1	SS2	Control SS	SS1	SS2
Duration	days	19	19	19	20	20	10	18	18	18
OLR	$gVS\ L_R^{-1}\ day^{-1}$	1.94	1.85	1.79	2.12	2.07	2.10	0.88	0.88	0.99
H2 injection rate	$L\ L_R^{-1}\ day^{-1}$		0.86	0.85		0.85	0.72		0.43	0.43
Specific biogas production	$L\ gVS^{-1}$	0.43	0.50	0.47	0.29	0.37	0.28	0.27	0.4	0.36
Specific CH4 production	$L\ gVS^{-1}$	0.30	0.32	0.33	0.21	0.24	0.19	0.20	0.31	0.29
Specific CO2 production	$L\ gVS^{-1}$	0.14	0.07	0.06	0.08	0.05	0.04	0.07	0.05	0.05
H2 consumption rate	$L\ L_R^{-1}\ day^{-1}$		0.66	0.70		0.68	0.61		0.40	0.41
H2 conversion	%		76.68	82.82		79.96	85.74		92.53	95.55
Methane evolution rate (MER)	$L\ L_R^{-1}\ day^{-1}$		0.16	0.18		0.16	0.14		0.10	0.10
Gas retention time	hours		9.42	10.91		11.89	19.75		25.06	25.11
Gas–liquid transfer rate *kLa*	day^{-1}		64.7	76.4		66.1	69.3		53.2	65.2
CH4 content	vol.%	67.90	65.36	71.49	69.92	66.22	72.05	71.1	77.62	80.90
CO2 content	vol.%	31.11	13.95	12.51	28.75	12.41	13.24	25.60	12.84	12.90
H2 content	vol.%	0.12	21.32	15.99	0.08	22.14	14.37	0.03	8.63	4.92
pH		7.74	7.78	8.01	7.43	7.59	7.24	7.57	7.78	7.62
TS	wt.%	3.02	3.19	3.37	2.59	2.37	2.72	2.48	2.24	2.31
VS	wt.%	1.63	1.75	1.87	1.24	1.31	1.45	1.29	1.25	1.24
Ammonia	$gTAN\ L^{-1}$	1.60	1.59	1.53	1.38	1.46	1.45	1.51	1.53	1.57
Alkalinity ratio		0.41	0.42	0.38	0.17	0.41	0.63	0.14	0.14	0.15
Acetate	$g\ L^{-1}$	0.94	2.11	2.94	0.95	2.00	2.25	0.31	0.23	0.40
Total VFA	$g\ L^{-1}$	2.40	3.27	4.04	2.61	3.62	3.08	1.01	0.58	0.79

Table A6. Average in-situ biomethanation results for food waste (FW)-fed reactors across operational periods R1–R3.

Experimental Period		R1			R2			R3		
	Reactor	Control FW	FW1	FW2	Control FW	FW1	FW2	Control FW	FW1	FW2
Duration	days	14	14	14	9	4	9	21	21	21
OLR	$gVS\ L_R^{-1}\ day^{-1}$	0.94	1.35	1.31	1.70	1.88	1.63	2.04	1.97	2.00
H2 injection rate	$L\ L_R^{-1}\ day^{-1}$		0.8	0.795		1.36	1.32		1.46	1.49
Specific biogas production	$L\ gVS^{-1}$	0.59	0.75	0.83	0.58	0.89	0.99	0.71	1.04	0.99

Table A6. *Cont.*

Experimental Period		R1			R2			R3		
	Reactor	Control FW	FW1	FW2	Control FW	FW1	FW2	Control FW	FW1	FW2
Specific CH4 production	L gVS^{-1}	0.38	0.40	0.42	0.36	0.42	0.48	0.45	0.50	0.53
Specific CO2 production	L gVS^{-1}	0.20	0.17	0.19	0.21	0.18	0.17	0.24	0.18	0.18
H2 consumption rate	$L\ L_R^{-1}\ day^{-1}$		0.56	0.51		0.82	0.79		0.81	0.94
H2 conversion	%		70.64	64.66		60.02	59.71		56.26	63.7
Methane evolution rate (MER)	$L\ L_R^{-1}\ day^{-1}$		0.13	0.12		0.20	0.20		0.19	0.22
Gas retention time	hours		11.12	10.15		7.83	5.29		5.43	4.57
Gas–liquid transfer rate *kLa*	day^{-1}		53.8	46.9		67.8	65.3		67.9	83.0
CH4 content	vol.%	63.83	52.75	50.86	62.45	47.96	49.67	61.82	48.45	53.18
CO2 content	vol.%	35.09	24.23	23.82	36.33	20.26	17.67	36.36	19.56	18.64
H2 content	vol.%	0.10	23.13	25.50	0.12	32.82	33.56	0.16	32.11	28.15
pH		7.6	7.66	7.63	7.71	7.86	n.a	7.77	7.94	8.16
TS	wt.%	3.71	3.78	3.63	3.24	3.24	2.97	4.08	3.50	4.34
VS	wt.%	2.73	2.80	2.73	2.38	2.29	2.24	2.82	2.46	3.14
Ammonia	gTAN L^{-1}	4.42	4.25	4.13	3.80	3.62	3.50	3.66	3.69	3.60
Alkalinity ratio		0.63	0.77	0.78	0.39	0.35	0.34	0.33	0.29	0.28
Acetate	g L^{-1}	0.64	1.31	1.53	0.47	0.53	0.51	0.37	0.34	0.48
Total VFA	g L^{-1}	1.30	2.07	2.26	0.80	0.86	0.83	0.89	0.86	0.98

Table A7. Average in-situ biomethanation results for food waste (FW)-fed reactors across operational periods O1–O3.

Experimental Period		O1			O2			O3		
	Reactor	Control FW	FW1	FW2	Control FW	FW1	FW2	Control FW	FW1	FW2
Duration	days	32	32	32	20	20	20	18	18	18
OLR	$gVS\ L_R^{-1}\ day^{-1}$	1.79	2.01	2.06	2.05	1.64	1.71	0.95	0.97	0.99
H2 injection rate	$L\ L_R^{-1}\ day^{-1}$		1.47	1.51		1.38	1.43		0.74	0.77
Specific biogas production	L gVS^{-1}	0.69	1.12	1.06	0.61	1.00	1.00	0.78	0.78	0.75
Specific CH4 production	L gVS^{-1}	0.43	0.52	0.53	0.38	0.52	0.55	0.54	0.51	0.53
Specific CO2 production	L gVS^{-1}	0.26	0.23	0.22	0.24	0.18	0.18	0.24	0.15	0.14
H2 consumption rate	$L\ L_R^{-1}\ day^{-1}$		0.73	0.84		0.90	0.96		0.64	0.69
H2 conversion	%		49.65	56.06		64.99	67.62		85.11	89.82
Methane evolution rate (MER)	$L\ L_R^{-1}\ day^{-1}$		0.18	0.21		0.22	0.24		0.16	0.17
Gas retention time	hours		3.94	3.99		6.67	5.08		12.06	12.33
Gas–liquid transfer rate *kLa*	day^{-1}		58.9	71.9		78.0	86.8		72.5	88.1
CH4 content	vol.%	61.96	46.79	50.20	61.54	54.50	56.05	67.37	66.26	70.56
CO2 content	vol.%	37.48	20.94	20.45	37.97	16.95	18.03	31.84	19.51	18.93
H2 content	vol.%	0.12	34.96	30.45	0.05	29.45	26.80	0.02	14.04	10.23
pH		7.75	8.18	8.10	7.76	8.20	8.20	7.69	8.13	8.25
TS	wt.%	3.04	2.83	2.92	2.91	2.78	2.74	2.72	2.54	2.55
VS	wt.%	2.11	1.99	2.07	2.00	1.92	1.90	1.84	1.65	1.75
Ammonia	gTAN L^{-1}	3.84	3.95	3.87	4.10	3.86	4.14	4.32	3.45	4.14
Alkalinity ratio		0.31	0.30	0.28	0.29	0.28	0.21	0.61	0.14	0.21
Acetate	g L^{-1}	0.84	0.84	0.88	2.01	1.86	0.65	4.42	0.61	0.35
Total VFA	g L^{-1}	1.19	1.15	1.17	3.11	3.02	1.12	6.88	1.20	0.74

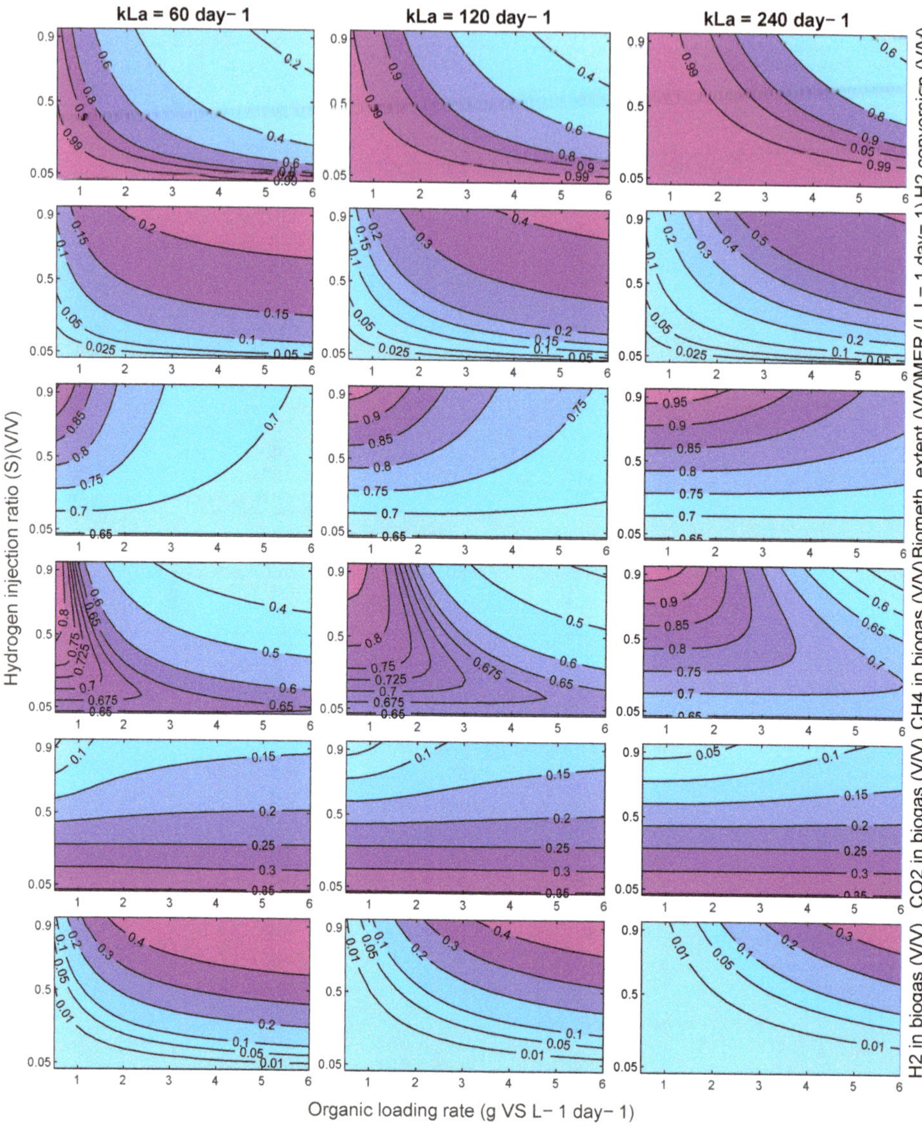

Figure A1. Contour plots of modelled in-situ biomethanation performance using FW with a variation in the OLR and hydrogen injection ratio for k_La of 60, 120, and 240 day^{-1}.

References

1. Straka, P. A Comprehensive Study of Power-to-Gas Technology: Technical Implementations Overview, Economic Assessments Methanation Plant as Auxiliary Operation of Lignite-Fired Power Station. *J. Clean. Prod.* **2021**, *311*, 127642. [CrossRef]
2. Ferry, J.G. *Methanogenesis: Ecology, Physiology, Biochemistry & Genetics*; Springer Science & Business Media: Berlin, Germany, 2012.
3. Luo, G.; Johansson, S.; Boe, K.; Xie, L.; Zhou, Q.; Angelidaki, I. Simultaneous Hydrogen Utilization and in Situ Biogas Upgrading in an Anaerobic Reactor. *Biotechnol. Bioeng.* **2012**, *109*, 1088–1094. [CrossRef] [PubMed]
4. Jee, H.S.; Yano, T.; Nishio, N.; Nagai, S. Biomethanation of H2 and CO_2 by Methanobacterium Thermoautotrophicum in Membrane and Ceramic Bioreactors. *J. Ferment. Technol.* **1987**, *65*, 413–418. [CrossRef]
5. Díaz, I.; Pérez, C.; Alfaro, N.; Fdz-Polanco, F. A Feasibility Study on the Bioconversion of CO_2 and H_2 to Biomethane by Gas Sparging through Polymeric Membranes. *Bioresour. Technol.* **2015**, *185*, 246–253. [CrossRef]

6. Luo, G.; Angelidaki, I. Hollow Fiber Membrane Based H2 Diffusion for Efficient in Situ Biogas Upgrading in an Anaerobic Reactor. *Appl. Microbiol. Biotechnol.* **2013**, *97*, 3739–3744. [CrossRef]
7. Wahid, R.; Horn, S.J. The Effect of Mixing Rate and Gas Recirculation on Biological CO_2 Methanation in Two-Stage CSTR Systems. *Biomass Bioenergy* **2021**, *144*, 105918. [CrossRef]
8. Agneessens, L.M.; Ottosen, L.D.M.; Voigt, N.V.; Nielsen, J.L.; de Jonge, N.; Fischer, C.H.; Kofoed, M.V.W. In-Situ Biogas Upgrading with Pulse H2 Additions: The Relevance of Methanogen Adaption and Inorganic Carbon Level. *Bioresour. Technol.* **2017**, *233*, 256–263. [CrossRef]
9. Tao, B.; Alessi, A.M.; Zhang, Y.; Chong, J.P.J.; Heaven, S.; Banks, C.J. Simultaneous Biomethanisation of Endogenous and Imported CO_2 in Organically Loaded Anaerobic Digesters. *Appl. Energy* **2019**, *247*, 670–681. [CrossRef]
10. Bensmann, A.; Hanke-Rauschenbach, R.; Heyer, R.; Kohrs, F.; Benndorf, D.; Reichl, U.; Sundmacher, K. Biological Methanation of Hydrogen within Biogas Plants: A Model-Based Feasibility Study. *Appl. Energy* **2014**, *134*, 413–425. [CrossRef]
11. Batstone, D.J.; Keller, J.; Angelidaki, I.; Kalyuzhnyi, S.V.; Pavlostathis, S.G.; Rozzi, A.; Sanders, W.T.; Siegrist, H.; Vavilin, V.A. The IWA Anaerobic Digestion Model No 1 (ADM1). *Water Sci. Technol.* **2002**, *45*, 65–73. [CrossRef]
12. Bassani, I.; Kougias, P.G.; Angelidaki, I. In-Situ Biogas Upgrading in Thermophilic Granular UASB Reactor: Key Factors Affecting the Hydrogen Mass Transfer Rate. *Bioresour. Technol.* **2016**, *221*, 485–491. [CrossRef] [PubMed]
13. Szuhaj, M.; Acs, N.; Tengolics, R.; Bodor, A.; Rakhely, G.; Kovacs, K.L.; Bagi, Z. Conversion of H_2 and CO_2 to CH_4 and Acetate in Fed-Batch Biogas Reactors by Mixed Biogas Community: A Novel Route for the Power-to-Gas Concept. *Biotechnol. Biofuels* **2016**, *9*, 102. [CrossRef] [PubMed]
14. Kougias, P.G.; Treu, L.; Benavente, D.P.; Boe, K.; Campanaro, S.; Angelidaki, I. Ex-Situ Biogas Upgrading and Enhancement in Different Reactor Systems. *Bioresour. Technol.* **2017**, *225*, 429–437. [CrossRef] [PubMed]
15. Jensen, M.B.; Kofoed, M.V.W.; Fischer, K.; Voigt, N.V.; Agneessens, L.M.; Batstone, D.J.; Ottosen, L.D.M. Venturi-Type Injection System as a Potential H2 Mass Transfer Technology for Full-Scale in Situ Biomethanation. *Appl. Energy* **2018**, *222*, 840–846. [CrossRef]
16. Lebranchu, A.; Blanchard, F.; Fick, M.; Pacaud, S.; Olmos, E.; Delaunay, S. Pilot-Scale Biomethanation of Cattle Manure Using Dense Membranes. *Bioresour. Technol.* **2019**, *284*, 430–436. [CrossRef] [PubMed]
17. Alfaro, N.; Fdz-Polanco, M.; Fdz-Polanco, F.; Díaz, I. Evaluation of Process Performance, Energy Consumption and Microbiota Characterization in a Ceramic Membrane Bioreactor for Ex-Situ Biomethanation of H_2 and CO_2. *Bioresour. Technol.* **2018**, *258*, 142–150. [CrossRef]
18. Kim, J.H.; Chang, W.S.; Pak, D. Factors Affecting Biological Reduction of CO_2 into CH_4 Using a Hydrogenotrophic Methanogen in a Fixed Bed Reactor. *Korean J. Chem. Eng.* **2015**, *32*, 2067–2072. [CrossRef]
19. Bassani, I.; Kougias, P.G.; Treu, L.; Angelidaki, I. Biogas Upgrading via Hydrogenotrophic Methanogenesis in Two-Stage Continuous Stirred Tank Reactors at Mesophilic and Thermophilic Conditions. *Environ. Sci. Technol.* **2015**, *49*, 12585–12593. [CrossRef]
20. Banks, C.J.; Zhang, Y.; Jiang, Y.; Heaven, S. Trace Element Requirements for Stable Food Waste Digestion at Elevated Ammonia Concentrations. *Bioresour. Technol.* **2012**, *104*, 127–135. [CrossRef]
21. Voelklein, M.A.; Rusmanis, D.; Murphy, J.D. Biological Methanation: Strategies for in-Situ and Ex-Situ Upgrading in Anaerobic Digestion. *Appl. Energy* **2019**, *235*, 1061–1071. [CrossRef]
22. Tom, Q.; Andrew, P. *Household Food Waste in the UK, 2015*; WRAP: Banbury, UK, 2017.
23. Facchin, V.; Cavinato, C.; Fatone, F.; Pavan, P.; Cecchi, F.; Bolzonella, D. Effect of Trace Element Supplementation on the Mesophilic Anaerobic Digestion of Foodwaste in Batch Trials: The Influence of Inoculum Origin. *Biochem. Eng. J.* **2013**, *70*, 71–77. [CrossRef]
24. Demirel, B.; Scherer, P. Trace Element Requirements of Agricultural Biogas Digesters during Biological Conversion of Renewable Biomass to Methane. *Biomass Bioenergy* **2011**, *35*, 992–998. [CrossRef]
25. *APHA Standard Methods for the Examination of Water and Wastewater*; American Public Health Association: Washington, DC, USA, 2005; ISBN 9780875532356.
26. Ripley, L.E.; Boyle, W.C.; Converse, J.C. Improved Alkalimetric Monitoring for Anaerobic Digestion of High-Strength Wastes. *Water Pollut. Control Fed.* **1986**, *58*, 406–411.
27. Holliger, C.; Alves, M.; Andrade, D.; Angelidaki, I.; Astals, S.; Baier, U.; Bougrier, C.; Buffière, P.; Carballa, M.; De Wilde, V.; et al. Towards a Standardization of Biomethane Potential Tests. *Water Sci. Technol.* **2016**, *74*, 2515–2522. [CrossRef]
28. Wang, B.; Nges, I.A.; Nistor, M.; Liu, J. Determination of Methane Yield of Cellulose Using Different Experimental Setups. *Water Sci. Technol.* **2014**, *70*, 599–604. [CrossRef]
29. Sastraatmaja, A. Experimental Process Development and Techno-Economic Assessment of In-Situ Biomethanation of Carbon Dioxide in Biowaste Anaerobic Digesters. Ph.D. Thesis, University of Sheffield, Sheffield, UK, 2022. Available online: https://etheses.whiterose.ac.uk/31101/ (accessed on 1 February 2023).
30. Jensen, M.B.; Ottosen, L.D.M.; Kofoed, M.V.W. H2 Gas-Liquid Mass Transfer: A Key Element in Biological Power-to-Gas Methanation. *Renew. Sustain. Energy Rev.* **2021**, *147*, 111209. [CrossRef]
31. Chow, W.L.; Chong, S.; Lim, J.W.; Chan, Y.J.; Chong, M.F. Anaerobic Co-Digestion of Wastewater Sludge: A Review of Potential Co-Substrates and Operating. *Processes* **2020**, *8*, 39. [CrossRef]

32. Grosser, A. Determination of Methane Potential of Mixtures Composed of Sewage Sludge, Organic Fraction of Municipal Waste and Grease Trap Sludge Using Biochemical Methane Potential Assays. A Comparison of BMP Tests and Semi-Continuous Trial Results. *Energy* **2018**, *143*, 488–499. [CrossRef]
33. Huiliñir, C.; Pinto-Villegas, P.; Castillo, A.; Montalvo, S.; Guerrero, L. Biochemical Methane Potential from Sewage Sludge: Effect of an Aerobic Pretreatment and Fly Ash Addition as Source of Trace Elements. *Waste Manag.* **2017**, *64*, 140–148. [CrossRef]
34. Zamanzadeh, M.; Hagen, L.H.; Svensson, K.; Linjordet, R.; Horn, S.J. Anaerobic Digestion of Food Waste-Effect of Recirculation and Temperature on Performance and Microbiology. *Water Res.* **2016**, *96*, 246–254. [CrossRef]
35. Xu, F.; Li, Y.; Ge, X.; Yang, L.; Li, Y. Anaerobic Digestion of Food Waste–Challenges and Opportunities. *Bioresour. Technol.* **2018**, *247*, 1047–1058. [CrossRef] [PubMed]
36. Voelklein, M.A.; Jacob, A.; O' Shea, R.; Murphy, J.D. Assessment of Increasing Loading Rate on Two-Stage Digestion of Food Waste. *Bioresour. Technol.* **2016**, *202*, 172–180. [CrossRef] [PubMed]
37. Buswell, A.M.; Mueller, H.F. Mechanism of Methane Fermentation. *Ind. Eng. Chem.* **1952**, *44*, 550–552. [CrossRef]
38. Guo, H.; Oosterkamp, M.J.; Tonin, F.; Hendriks, A.; Nair, R.; van Lier, J.B.; de Kreuk, M. Reconsidering Hydrolysis Kinetics for Anaerobic Digestion of Waste Activated Sludge Applying Cascade Reactors with Ultra-Short Residence Times. *Water Res.* **2021**, *202*, 117398. [CrossRef]
39. Poggio, D.; Sastraatmaja, A.; Walker, M.; Michailos, S.; Nimmo, W.; Pourkashanian, M. *Experimental Evaluation of Continuous In-Situ Biomethanation of CO_2 in Anaerobic Digesters Fed on Sewage Sludge and Food Waste, and the Influence of Hydrogen Gas-Liquid Mass Transfer*; Dataset; University of Sheffield: Sheffield, UK, 2023. [CrossRef]
40. Lecker, B.; Illi, L.; Lemmer, A.; Oechsner, H. Biological Hydrogen Methanation—A Review. *Bioresour. Technol.* **2017**, *245*, 1220–1228. [CrossRef] [PubMed]
41. Dupnock, T.L.; Deshusses, M.A. High-Performance Biogas Upgrading Using a Biotrickling Filter and Hydrogenotrophic Methanogens. *Appl. Biochem. Biotechnol.* **2017**, *183*, 488–502. [CrossRef]
42. Choi, D.; Chipman, D.C.; Bents, S.C.; Brown, R.C. A Techno-Economic Analysis of Polyhydroxyalkanoate and Hydrogen Production from Syngas Fermentation of Gasified Biomass. *Appl. Biochem. Biotechnol.* **2010**, *160*, 1032–1046. [CrossRef]
43. Yang, P.; Peng, Y.; Tan, H.; Liu, H.; Wu, D.; Wang, X.; Li, L.; Peng, X. Foaming Mechanisms and Control Strategies during the Anaerobic Digestion of Organic Waste: A Critical Review. *Sci. Total Environ.* **2021**, *779*, 146531. [CrossRef]

processes

MDPI

Article

Validation of Two Theoretically Derived Equations for Predicting pH in CO_2 Biomethanisation

Yue Zhang *, Sonia Heaven * and Charles J. Banks

Water and Environmental Engineering Group, Faculty of Engineering and Physical Sciences, University of Southampton, Southampton SO16 7QF, UK

* Correspondence: Y.Zhang@soton.ac.uk (Y.Z.); S.Heaven@soton.ac.uk (S.H.)

Abstract: CO_2 biomethanisation is a rapidly emerging technology which can contribute to reducing greenhouse gas emissions through the more sustainable use of organic feedstocks. The major technical limitation for in situ systems is that the reaction causes CO_2 depletion which drives up pH, potentially leading to instability and even digestion failure. The study aimed to test fundamentally derived predictive equations as tools to manage H_2 addition to anaerobic digesters. The methodology used data from the literature and from experimental digesters operated with excess H_2 to a point of failure and subsequent recovery. Two equations were tested: the first relating pH to CO_2 partial pressure (pCO_2), and the second extending this to include the influence of volatile fatty acids and ammonia. The first equation gave good agreement for data from studies covering a wide range of operating conditions and digester types. Where agreement was not good, this could usually be explained, and in some cases improved, using the second equation, which also showed excellent predictive performance in the experimental study. The results validated the derived equations and identified typical coefficient values for some organic feedstocks. Both equations could provide a basis for process control of CO_2 biomethanisation using routine monitoring of pH or pCO_2 with additional analysis for volatile fatty acids and total ammonia nitrogen when required.

Keywords: CO_2 biomethanisation; pH change; CO_2 partial pressure; volatile fatty acids; ammonia

Citation: Zhang, Y.; Heaven, S.; Banks, C.J. Validation of Two Theoretically Derived Equations for Predicting pH in CO_2 Biomethanisation. *Processes* **2023**, *11*, 113. https://doi.org/10.3390/pr11010113

Academic Editor: Blaž Likozar

Received: 4 December 2022
Revised: 22 December 2022
Accepted: 24 December 2022
Published: 31 December 2022

1. Introduction

The ability of certain methanogens to use hydrogen (H_2) as a reducing agent for the conversion of carbon dioxide (CO_2) into methane (CH_4) has been known to science for decades [1]. More recently, the growing realisation of the potential engineering applications has helped to move this process rapidly up through technology readiness levels from the initial laboratory research [2] to larger-scale implementation [3,4]. CO_2 biomethanisation can be carried out in various configurations. Ex situ processes are usually defined as operating on gaseous H_2 and CO_2 in dedicated reactors, with or without defined inocula, and can offer the advantages of high gas transfer rates and volumetric throughput: the main commercial systems currently in use are of this type [3–5]. In situ systems use CO_2 produced during the anaerobic digestion of organic substrates, with the addition of H_2 stimulating parts of the mixed microbial community to carry out the CO_2 biomethanisation process. Various hybrids of these two basic concepts are also possible, e.g., where additional exogenous CO_2 is provided for conversion within an in situ system [6], or where high-rate reactors are fed with organic substrates for nutrient supply and/or replenishment of the microbial population, rather than as the main source of CO_2 [7,8]. Other variants under development include those using syngas [9], zero-valent iron [10], and bioelectrochemical systems [11].

While ex situ systems are furthest along the pathway to full-scale application, all these approaches have specific advantages and features that make them suited to different applications. The urgent need for renewable hydrocarbons and for more sustainable utilisation of organic carbon has recently led to increasing interest in biomethanisation of CO_2 from waste feedstocks, as evidenced by a growing number of papers on this topic [12]. The attractions of this approach are firstly that it can enhance CH_4 yields from existing AD infrastructure, with potentially low retrofitting costs [13]. Secondly, it increases the scope for the integration of anaerobic digestion with local or grid-based renewables and other on-site technologies [14,15]. Thirdly, it can provide biogas with a high methane content, in some cases equal to that reported for single-pass ex situ reactors [12], from organic carbon inputs.

The main perceived drawback of the in situ approach is that reduction in the headspace CO_2 content induces a rise in pH [16], which if uncontrolled can lead to process instability or even digestion failure [17,18]. In full-scale digesters this is expensive and time-consuming to rectify: in the worst case, digester contents may have to be replaced with fresh inoculum, leading to extended downtime, and a digestate disposal problem [19]. As with other process innovations, widespread adoption of CO_2 biomethanisation will require companies to have confidence in their ability to control the process and to avoid any risk of instability, or to identify and deal with if for some reason it does occur. This security can be delivered by the development of effective process control tools [20,21], which ideally should be based on experimentally validated relationships that are underpinned by sound science and an in-depth understanding of the factors involved [22]. To be accepted, such relationships must also be demonstrated as robust over a wide range of conditions, including periods of dynamic change and when operations become unstable, e.g., with accumulation of volatile fatty acids (VFA) [20,22].

As a step towards this, Tao et al. [6] suggested a relationship between digester pH and partial pressure of CO_2 (pCO_2) in the headspace that would allow estimation of the minimum pCO_2 for stable operation, and thus by extension, the maximum achievable CH_4 content. The strength of this lies in the fact that it is derived from fundamental principles and requires only minimal information (digester operating temperature and a baseline value for pH and pCO_2) for calibration and use. It was developed for systems where ammonium-bicarbonate buffering is dominant, as is the case for most real-world organic feedstocks. Although, extension to phosphate, which is usually present when basal media are used, was also demonstrated [6]. However, it did not include consideration of VFA which may accumulate during CO_2 biomethanisation for various reasons, e.g., too high a H_2 partial pressure and/or too high pH without appropriate acclimatisation. The authors were also unable to explain why the maximum acceptable pH seemed to vary with different substrates, and more work is needed to clarify this.

In the current work, data gathered from an extensive review of published literature was used to investigate how well the original equation derived by Tao et al. [6] was able to predict operating pH from pCO_2, and where the fit was poor, to see whether it was possible to understand why. The original equation was modified to include consideration of the effects of total ammonia nitrogen (TAN) and VFA, and where suitable literature data were available these were used to test and compare the performance of the two equations. As the number of studies with suitable data was limited, both equations were also applied to data from a laboratory-scale CO_2 biomethanisation trial which was deliberately designed to show VFA accumulation and the onset of process instability.

The results from all parts of this work clearly demonstrated the robustness of the derived equations, and thus their potential suitability for use in process control systems based on pCO_2 and/or pH measurements, with additional monitoring of TAN and VFA when required.

2. Materials and Methods

2.1. Equations Relating pH and pCO_2

The equation derived by Tao et al. [6] relates pH to the partial pressure of CO_2 in the digester headspace

$$pH = -log_{10}\left(\frac{a \cdot p_{CO2} + \sqrt{a^2 \cdot p_{CO2}{}^2 + 4 \cdot 10^{-t} \cdot a \cdot p_{CO2}}}{2}\right) \quad (1)$$

where:

$$a = \frac{\left(10^{-pH^o}\right)^2}{10^{-t} + 10^{-pH^o}} \cdot \frac{1}{p_{CO2}{}^o} \tag{1a}$$

$$t = 0.90018 + \frac{2729.29}{T} \tag{1b}$$

and T is the digester operating temperature digester in degrees K.

The superscript ^{o} in Equation (1a) represents chosen baseline values. Symbols with this superscript are therefore constants obtained from a control digester or operating period, e.g., from conventional anaerobic digestion without CO_2 biomethanisation.

The above relationship can be taken further by considering the effect of VFA. As weak acids, these metabolic products are also capable of donating protons to NH_3 in an AD environment. A modified version of Equation A can therefore be obtained, which applies to both baseline conditions and in CO_2 biomethanisation.

At equilibrium:

$$\left[HCO_3{}^-\right] + \left[VFA^-\right] = \left[NH_4{}^+\right] \tag{2a}$$

where [ ] represents molar concentration.

Among the terms in Equation (2a)

$$\left[HCO_3{}^-\right] = \frac{K_{a1} K_H \, p_{CO2}}{10^{-pH}} \tag{2b}$$

(as in Equation (4) in Tao et al. [6]), and

$$\left[NH_4{}^+\right] = \frac{10^{-pH}}{10^{-t} + 10^{-pH}} [TAN] \tag{2c}$$

(as in Equation (7) in Tao et al. [6], but with t left as a variable rather than expressed as the constant for operation at 37 °C).

Considering that the pH of anaerobic digestion processes is commonly neutral or slightly in the alkaline range, especially under CO_2 biomethanisation:

$$\left[VFA^-\right] = [VFA] \tag{2d}$$

where [VFA] represents total VFA concentration, including both protonated and deprotonated forms.

Equation (2a) can then be converted to:

$$\frac{K_{a1} K_H \, p_{CO2}}{10^{-pH}} + [VFA] = \frac{10^{-pH}}{10^{-t} + 10^{-pH}} [TAN] \tag{2e}$$

Therefore:

$$\left(10^{-pH}\right)^2 - \frac{K_{a1} K_H p_{CO2} + [VFA] \cdot 10^{-t}}{[TAN] - [VFA]} \cdot 10^{-pH} - \frac{K_{a1} K_H p_{CO2} \cdot 10^{-t}}{[TAN] - [VFA]} = 0 \tag{2f}$$

Solving Equation (2f) gives:

$$10^{-pH} = \frac{\left(b \cdot p_{CO_2} + [VFA] \cdot 10^{-t}\right) + \sqrt{\left(b \cdot p_{CO_2} + [VFA] \cdot 10^{-t}\right)^2 - 4 \cdot ([TAN] - [VFA]) \cdot \left(-b \cdot p_{CO_2} \cdot 10^{-t}\right)}}{2 \cdot ([TAN] - [VFA])} \tag{2g}$$

where:

$$b = K_{a1} K_H = \left(\frac{10^{-pH^0}}{10^{-t} + 10^{-pH^0}} [TAN]^0 - [VFA]^0 \right) \cdot \frac{10^{-pH^0}}{p_{CO2}{}^0} \tag{2h}$$

Therefore, the relationship between pH and p_{CO2} can be expressed as in Equation B:

$$pH = -\log_{10}\frac{(b \cdot p_{CO_2} + [VFA] \cdot 10^{-t}) + \sqrt{(b \cdot p_{CO_2} + [VFA] \cdot 10^{-t})^2 - 4 \cdot ([TAN] - [VFA]) \cdot (-b \cdot p_{CO_2} \cdot 10^{-t})}}{2 \cdot ([TAN] - [VFA])} \quad (2)$$

For the remainder of this paper the expressions shown in Equations (1) and (2) are referred to as Equation A and Equation B, respectively. When the VFA concentration is zero, Equation B is identical to Equation A. It should be noted that Equation B does not apply when the molar concentration of VFA is close to or greater than that of TAN as, in this case, the ammonia produced is not sufficient to neutralise both fatty acids and carbonic acid.

2.2. Literature Data

Search terms including 'CO_2 biomethanisation' or 'CO_2 biomethanation', 'in situ biogas upgrading', and 'anaerobic digestion' in conjunction with 'H_2 addition' were used to identify papers of possible interest. This produced several thousand publications, the majority of which after a brief inspection of the title or abstract were found not to be relevant to the current study. A total of 73 papers, identified either directly in this way or from the reference lists of the selected papers, were examined in more detail. In total, 37 of these were eliminated as they did not directly address in situ conversion of CO_2 from organic feedstocks, did not include the required data on pH and pCO_2, or for other methodological reasons. Data were then extracted from the remaining 36 peer-reviewed papers and their supporting materials, with additional information obtained from the authors where possible.

The resulting data were in the form of sets of average values from different experimental periods in a study, or individual points from experimental data series. For each study or dataset considered, average values for pH, pCO_2, and where available VFA and TAN concentrations from control reactors or baseline periods were used to derive coefficients a or b for Equations A or B. Measured CO_2 concentrations were then used to predict pH for each data point, excluding the selected control or baseline period, for comparison with experimental pH values.

2.3. Digester Set-Up and Operation

Experimental work was carried out in eight 1-L digesters (designated D1-8) maintained at 37 °C in a water bath, as described in Tao et al. [23]. The digesters were mixed by 40 mm diameter 3-blade impellers driven at 200 rpm. Each digester initially received 500 mL of inoculum from a mesophilic anaerobic digester at Millbrook Wastewater Treatment Works (WWTW), Southampton, UK. The feed was co-settled primary and secondary sewage sludge collected from Budds Farm WWTW, Portsmouth, UK, in a single batch and frozen in aliquots until required, then stored under refrigeration until use. The average total solids (TS) content of the feed was 6.91% wet weight (WW) with average volatile solids (VS) of 5.69%WW. Feeding and digestate removal were carried out manually once per day to give a Hydraulic Retention Time (HRT) of 14.6 days and an organic loading rate (OLR) of 3.87 g VS L^{-1} day^{-1} unless noted. All digesters had been run under these conditions for 70 days before the start of the current trial to ensure stable and replicable operation. From day 0–54, operation of all digesters continued under the same conditions to allow the establishment of robust baseline values.

H_2 addition was achieved by dispensing the required volume into a foil-lined gas-impermeable bag using an EL-Flow Prestige mass flow controller (Bronkhorst, UK). The filled gas bag was attached to the digester immediately after the daily addition of organic feed, and the gas was bubbled up through the digestate and recirculated from the headspace at a flow rate of 8 mL min^{-1}. At the end of the daily cycle, the full gas bag was removed for measurement of gas volume and composition

In general, gas production and composition were measured daily throughout the experiment. From day 0–54 pH and VFA concentrations were measured twice a week, with

TAN, alkalinity, and solids content measured weekly. pH and VFA were measured daily between days 55–100, covering the period of H_2 addition and subsequent stabilisation, while other parameters were measured intermittently. More frequent monitoring was resumed at the end of the experimental period from day 110–125.

2.4. Analytical Methods

TS and VS were determined by Standard Method 2540 G [24]. TAN was measured using a BÜCHI K-350 Distillation Unit, with NaOH addition followed by titration of the distillate in a boric acid indicator with 0.25 N H_2SO_4. Alkalinity was measured using a SCHOTT titroline system with titration by 0.25 N H_2SO_4 to endpoints of pH 5.75 and 4.3, to allow calculation of total (TA), partial (PA), and intermediate alkalinity (IA) [25]. VFA concentrations were determined by gas chromatography (Shimadzu GC-2010) using a flame ionisation detector and a capillary column (SGE BP-21) with nitrogen as the carrier gas. Samples were acidified to 10% with formic acid and quantified against mixed standards of 50, 250, and 500 mg L^{-1} of acetic, propionic, iso-butyric, n-butyric, iso-valeric, valeric, hexanoic, and heptanoic acids. Biogas composition was determined using a MG#5 Gas Chromatograph (SRI Instruments, Torrance, CA, USA) with a thermal conductivity detector (TCD). The instrument had two linked analytical lines with CH_4 and CO_2 separated by a Porapak Q column (80/100 mesh, 6ft), and H_2 by a molecular sieve 5A column (6ft). GC calibration was conducted using standard gases supplied by BOC, UK. Gas volumes were determined in a weight-based water displacement system [26] and are reported as dry gas at a standard temperature and pressure (STP) of 0 °C and 101.325 kPa.

2.5. Performance Assessment and Statistical Analyses

Performance of Equations A and B was assessed by several measures. Heat maps were used to show the absolute value of the difference between experimental and predicted pH for single data points or for average values for an experimental period. For multiple values, in the first instance the Root Mean Square Deviation was calculated, with deviation defined here as the difference between experimental pH and predicted pH value for each point.

Where data points might be expected to show some relationship, e.g., between average values for different sets of conditions within an experiment or between individual points in a data series, regression analysis was carried out. For this purpose, the coefficient of determination (R^2) and the Root Mean Square Error (RMSE) were calculated, with error defined as the difference between the predicted pH and the line of best fit for experimental and predicted values, to take account of variation in experimental values. The slope and intercept of regression equations were also considered. T-tests were used to determine the significance of regression statistics and of differences between slopes, with values taken as significant at the 5% level.

For the purposes of assessment, the level of agreement between experimental and predicted pH was defined as good if the difference was <0.1, reasonable between 0.1–0.2, and poor if >0.2 for single data points, or for the RMSD or RMSE of multiple values. Correlation between predicted and experimental pH was defined as poor for $R^2 < 0.8$, reasonable for R^2 between 0.8–0.9, and good for $R^2 > 0.9$. These definitions were based on examination of the data and consideration of likely measurement accuracy [24,27]. The studies considered were not specifically designed for the evaluation of the relationship between pH and pCO_2 and would therefore not have taken any special steps to ensure accuracy. pH values are affected by temperature and dissolved CO_2 content, which may change after removal of the sample from the digester [19], while online in-situ measurement requires frequent calibration for accuracy [19]. As a non-combustible gas, it is also challenging to measure CO_2 accurately over the wide range of partial pressures found in CO_2 biomethanisation experiments, whether by gas chromatography (GC) or sensor [27]. The precision of the available data was also taken into account, with, e.g., pH values often only reported to 0.1 unit in AD and CO_2 biomethanisation studies.

3. Results

3.1. Application of Equation A to the Literature Data

In total, 36 studies were found with suitable data on pH and pCO_2. Most of these reported average values for monitoring parameters from periods of operation under specified conditions, often supported by graphical data. Some also provided numerical data in Supplementary Materials or on request.

The studies covered a wide range of experimental conditions and operating configurations, including different digester types (continuous stirred tank reactors (CSTR) [6,17,18,23,28–52], upflow anaerobic sludge blanket (UASB) reactors [53,54], anaerobic membrane bioreactors (AnMBR) [55,56], and anaerobic filters (AF) or trickle beds [57,58]. These were arranged in single and 2-stage systems with H_2 injection in the first [59] or second [32,36,42,43,45–47,60] stage, and with [59] or without [32,42] gas recirculation between the stages). Rates of mixing, gas addition, and recirculation were also varied both within and between studies. Some studies were carried out in pressurised systems [35,37], with gases other than H_2 (e.g., carbon monoxide [41,54] or syngas [48,49]). Different gas supply systems were used, including headspace injection [28,29,47,58], various types of bubble tube [6,23,34] or diffuser [2,32,37,42,44,46,49,50,54], venturi outlets [57], and membrane systems [30,40,48]. Trials were conducted in mesophilic [28–30,33–38,44,48,52,54–56,58] and thermophilic [17,18,31,39–47,50,53,59,60] conditions, in some cases specifically to compare these [32,49,51]. They were also performed with a variety of feedstocks including livestock manures and slurries, crop and agro-wastes, industrial and post-consumer food wastes, wastewater biosolids, and others [12].

The results of the application of Equation A to these studies are shown in Tables S1–S36 in the Supplementary Materials, in the form of a summary of key operating conditions, with heat maps showing the difference between predicted and experimental pH values, and brief comments on interesting results or trends in each study. Selected examples are discussed below.

As can be seen from the heat maps, good agreement was found under a wide range of conditions for most or all the cases reported in the majority of these studies. In the earliest work on semi-continuous thermophilic digestion of cattle slurry in CSTR [18], the predicted pH based on pCO_2 matched the experimental values exactly (see Table S1). However, it should be noted that experimental pH was only recorded to one decimal place (dp) in this trial. Table 1 shows results for a study in which a thermophilic CSTR was fed on cattle manure and whey as co-substrates, with gas supplied via a hollow fibre membrane (HFM) module and H_2 input as the main variable [40]. Values of a were derived for every case (i.e., using reported average pCO_2 and pH from each set of conditions tested as the baseline values) and used to predict the corresponding pH in all cases. The lower part of Table 1 shows the absolute value of the difference between predicted and experimental pH. As can be seen, Equation A gave a good fit both for each phase and across all experimental conditions. Values of a based on the control reactor provided the best agreement throughout. When a coefficient derived from the highest H_2 input was used, this gave a slightly higher RMSD, possibly reflecting the small accumulation of VFA seen under these conditions. A high R^2 value across a range of conditions tested, as seen here, is not an essential indicator of good performance in all cases, since some experimental designs will show differences between different phases. Yet, where it does occur, it provides additional confidence in the predictive capacity of the equation.

Table 1. Performance of Equation A for reported average values from Luo and Angelidaki [40].

		(i) No H_2	(ii) With H_2	(iii) No H_2	(iv) With H_2	(v) No H_2	(vi) With H_2	RMSD	R^2
Temp	°C	55	55	55	55	55	55	-	-
OLR	g VS L^{-1} day^{-1}	1.67	1.67	1.67	1.67	1.67	1.67	-	-
HRT	days	15	15	15	15	15	15	-	-
H_2 input	L L^{-1} day^{-1}	0	0.93	0	1.44	0	1.76	-	-
SMP_{tot}	L CH_4 g^{-1} VS	0.275	0.407	0.271	0.478	0.287	0.515	-	-
CH_4	%	54.2	78.4	53.1	90.2	55.4	96.1	-	-
TAN	M	n/r	n/r	n/r	n/r	n/r	n/r	-	-
VFA	M	0.002	0.004	0.002	0.012	0.002	0.038	-	-
pH	-	7.29	7.61	7.28	7.9	7.3	8.31	-	-
pCO_2	-	0.458	0.216	0.469	0.098	0.446	0.039	-	-
$a \times 10^8$		10.41	9.81	10.42	9.81	10.43	6.99	-	-
Difference between experimental and predicted pH based on Equation A with coefficient a(x)									
a(i)	-	-	0.02	0.00	0.02	0.00	0.12	0.06	0.997
a(ii)	-	0.02	-	0.02	0.00	0.02	0.10	0.05	0.997
a(iii)	-	0.00	0.02	-	0.02	0.00	0.12	0.06	0.997
a(iv)	-	0.02	0.00	0.02	-	0.02	0.10	0.05	1.000
a(v)	-	0.00	0.02	0.00	0.02	-	0.12	0.06	0.997
a(vi)	-	0.16	0.13	0.16	0.12	0.16	-	0.15	1.000

n/r = not reported. Lower part of Table shows heat map with white = good, grey = reasonable, black = poor agreement.

Similar levels of agreement were found in other studies using conventional single-stage CSTR, examples are given in Tables S1–S9. Equation A also performed well where stable operation was achieved in CSTR studies with the addition of carbon monoxide (Tables S10–S12) or syngas (Tables S13 and S14). One favourable factor in some of these trials may be the absence of very low pCO_2 values due to the additional CO_2 produced from CO conversion, although CO can also be inhibitory at higher concentrations [31,41].

In 2-stage CSTR systems, good agreement was found under both mesophilic and thermophilic conditions (see, e.g., Table S15). A long-term acclimatisation study in a thermophilic 2-stage CSTR initially showed good agreement between predicted and measured pH [42]. After two years of operation, agreement appeared to deteriorate: this was likely associated with a raised VFA concentration recorded at the time, linked to the high pH and low pCO_2 value (see Table S16), and thus supports the usefulness of an equation taking VFA into account. Good agreement for comparable periods of stable operation was also found for other reactor types including AnMBR (Table S17), UASB (Table S18), AF (Table S19), and trickling filter (Table S20), and in a 2-stage system consisting of a CSTR coupled to an upflow reactor (Table S21).

Equation A also performed well when applied to pressurised systems. A 35-L mesophilic CSTR fed on wastewater biosolids was operated at four different pressures [35]: after pCO_2 was adjusted for the applied pressure, the experimental and predicted pH agreed well for all but the initial lowest pressure (Table S22). No VFA accumulation was observed that might account for this, but it is possible that other factors such as acclimatisation were involved. In a trial of pulsed H_2 addition, initial pressures were not reported but experimental and predicted pH agreed well up to a H_2/CO_2 ratio of 8 [29]; deterioration above this may have been due to changes associated with higher pressure and/or to VFA accumulation (Table S23). In another pulsed H_2 trial, initial values corrected for pressure showed good agreement [28]; other results were difficult to interpret due to uncertainty over the timing of different measurements (see Table S24a,b). Another study with headspace pressurisation [33] showed good agreement within phases, although pH was only reported to 1 dp and there was relatively little variation in pCO_2 values trial (see Table S25).

Where agreement between experimental and predicted pH values was less good, the reasons for this were often clear. One common issue was VFA accumulation. In a study of mesophilic and thermophilic digestion of cheese whey, significant swings in VFA concentration were encountered, and various buffering strategies were adopted to control them [43]. Equation A gave only reasonable agreement, reflecting changing conditions between the different phases (Table S26). Thermophilic digestion of ensiled ryegrass was successfully achieved [44], but the VFA accumulation observed during periods of H_2 addition likely accounted for the poor agreement between predicted and experimental pH (see Table S27).

In some cases, there were apparent discrepancies in the data or results that could account for the poor agreement. In a study of mesophilic CSTR digestion of mixed sewage sludge [30], the pH in control and experimental digesters differed by 0.22 units during the start-up phase, before any H_2 addition had occurred, making comparisons with other phases problematic (Table S28). pH values also showed quite high standard deviations during this trial, perhaps reflecting the use of online measurement and/or variability between batches of feedstock. A study of mesophilic food waste digestion with pressurisation and H_2 addition used calculated pH values [37]; but these appeared to be incorrect, perhaps due to the use of acid dissociation and Henry's Law constants for 25 °C rather than for the operating temperature of 37 °C. However, when Equation A was applied 'in reverse' using baseline values from the pH calculations, it gave a much better agreement between predicted and measured pCO_2 (see Tables S29a,b).

In other cases, potential explanations remained speculative in the absence of supporting data. Digestion of ethanol distillery wastewater in a mesophilic AnMBR [55] showed poor agreement between predicted and experimental pH values in almost all conditions tested, although VFA accumulation was only seen at the highest H_2 loading (Table S30). However, the reactor was operated without sludge wastage, which could have caused changes in TAN or other unreported parameters during the experimental period.

Very few cases were found where there was no clear reason for poor agreement. In one study of a 2-stage mesophilic CSTR processing cattle manure and food waste, with H_2 injection in the second stage and gas recirculation between stages, pH reportedly fell on H_2 addition (Table S31). However, no factors accounting for this were identified [36].

In general, analysis of the data was able to demonstrate convincingly that Equation A performed very well whenever conditions were relatively stable and without major shifts in TAN or VFA concentrations; while also providing evidence to support the usefulness of an equation able to take changes in these parameters into account. Some general and methodological issues and the insights obtained from them are further discussed in Section 4.1 below.

3.2. Application of Equation B to the Literature Data

Equation B was designed to take account of the effect of digestate TAN and VFA concentrations. However, the majority of studies examined had no data on TAN and in many cases only limited details of VFA content, e.g., as total VFA without speciation or expressed in terms of Chemical Oxygen Demand (COD) or acetic acid (Hac) equivalent. Ten studies by other groups were found which had suitable data on both TAN and VFA, and these were used to compare the relative performance of the two equations.

Application of Equation B did not always show a clear improvement over Equation A, either because Equation A already gave good agreement and there was little change in VFA or TAN during the trials (see Tables S1, S3, S7 and S17), or because HCO_3^- and NH_4^+ remained the main buffering pair despite the presence of VFA (Table S5); or because of other issues with the data or results (Tables S14, S22 and S28). Table 2 shows results from [46] where Equation B was able to improve estimated pH by up to 0.06 units, bringing case (iv) into the good agreement range. This and other examples of improvement can be seen in Tables S3, S4, S32 and S33. In general, the results showed that Equation B performed well, and merited further testing against datasets with more striking changes in VFA and/or TAN concentrations.

Table 2. Performance of Equations A and B for reported average values from Wahid and Horn [46].

		(i)	(ii)	(iii)	(iv)	(v)	(vi)	(vii)	RMSD	R^2
Phase		1	2	3a	3b	3c	3d	3e		
H_2/CO_2		0	2	4	4	4	4	4		
Mixing	Rpm	80	80	80	120	140	170	200	-	-
Temp	°C	55	55	55	55	55	55	55	-	-
OLR	g VS L^{-1} day^{-1}	3.09	2.69	2.30	2.36	2.19	2.37	2.32	-	-
HRT	Days	20	20	20	20	20	20	20	-	-
H_2 input	L L^{-1} day^{-1}	0.0	0.8	1.6	1.7	1.7	1.7	1.7	-	-
SMP total	L CH_4 g^{-1} VS	0.217	0.246	0.305	0.331	0.411	0.321	0.315	-	-
CH_4	%	62.2	37.8	31.6	38.3	40.3	41.9	40.2	-	-
TAN	M	0.10	0.11	0.13	0.15	0.14	0.14	0.14	-	-
VFA	M	0.005	0.006	0.008	0.011	0.012	0.016	0.015	-	-
pH	-	7.86	8.07	8.25	8.37	8.41	8.43	8.41	-	-
pCO_2	-	0.378	0.207	0.137	0.124	0.119	0.116	0.120	-	-
$a \times 10^8$	-	2.85	2.82	2.43	1.81	1.63	1.56	1.62	-	-
$b \times 10^8$	-	0.255	0.277	0.278	0.233	0.193	0.162	0.171	-	-
Difference between experimental and predicted pH based on Equation A with coefficient a(x)										
a(i)	-	-	0.00	0.05	0.14	0.17	0.18	0.17	0.14	0.949
a(ii)	-	0.00	-	0.05	0.13	0.16	0.18	0.16	0.13	0.966
a(iii)	-	0.06	0.05	-	0.09	0.12	0.13	0.12	0.10	0.978
a(iv)	-	0.16	0.14	0.09	-	0.03	0.04	0.03	0.10	0.955
a(v)	-	0.19	0.17	0.12	0.03	-	0.01	0.00	0.12	0.956
a(vi)	-	0.21	0.19	0.13	0.04	0.01	-	0.01	0.13	0.958
a(vii)	-	0.19	0.18	0.12	0.03	0.00	0.01	-	0.12	0.957
Difference between experimental and predicted pH based on Equation B with coefficient b(x)										
b(i)	-	-	0.03	0.03	0.03	0.08	0.12	0.11	0.08	0.853
b(ii)	-	0.03	-	0.00	0.05	0.10	0.15	0.13	0.09	0.940
b(iii)	-	0.03	0.00	-	0.05	0.10	0.15	0.13	0.09	0.963
b(iv)	-	0.03	0.06	0.05	-	0.05	0.10	0.08	0.07	0.941
b(v)	-	0.10	0.11	0.11	0.05	-	0.05	0.03	0.08	0.929
b(vi)	-	0.16	0.17	0.16	0.10	0.05	-	0.01	0.12	0.947
b(vii)	-	0.14	0.15	0.14	0.09	0.03	0.01	-	0.11	0.942

n/r = not reported. Lower part of Table shows heat map with white = good, grey = reasonable, black = poor agreement.

3.3. Application of Equations A and B to Multi-Point Data

This part of the work was carried out to test the performance of Equations A and B with multi-point data series.

The original study that derived Equation A [6] included datasets from mesophilic CO_2 biomethanisation of food waste and from a short-term study with sewage sludge. These were re-analysed to include more datapoints and provide additional information. The results are shown in Figure 1a–d, and it can be seen that Equation A performed well in both cases. For sewage sludge the correlation coefficient R^2 was 0.991 with n = 49 (increased from 30 in [6]), $p < 0.000$, and RMSE 0.03. For food waste digestion the equivalent values were $R^2 = 0.880$, n = 299 (increased from 242), $p < 0.000$, and RMSE 0.07 indicating that most points met the good criterion: this was despite a rise in VFA concentrations when pCO_2 briefly fell below 9% [6]. The food waste digestion study included some TAN and VFA data, and Equations A and B were also applied to points where VFA measurements were available, using interpolated TAN values where necessary. However, as TAN in this experiment was around 100 times higher than VFA on a molar basis, there was little difference between the two equations (see Figure A1 in Appendix A).

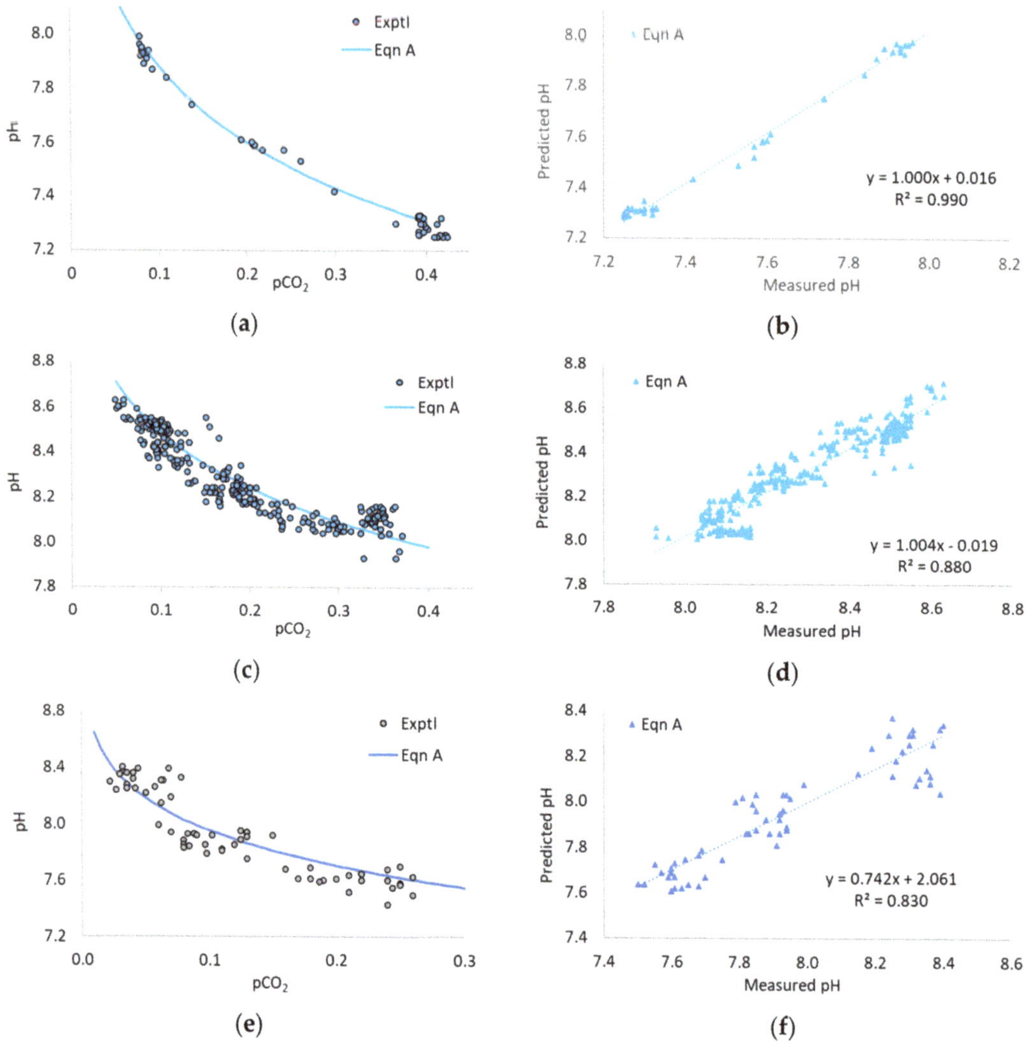

Figure 1. Experimental pCO_2 and pH values and predicted pH from Equation A for mesophilic CO_2 biomethanisation of sewage sludge (**a**,**b**) as reported in [6]; mesophilic CO_2 biomethanisation of industrial food waste (**c**,**d**) as reported in [6]; and thermophilic CO_2 biomethanisation of cattle manure (**e**,**f**) as reported in [40]. Numerical datasets were kindly provided by the authors in each case

Several researchers kindly provided additional information in the form of experimental data series from published works [17,32,39,40,50–53]. These studies were designed to test a variety of factors, so the available data were not always ideal for the current purpose: parameters such as pH and pCO_2 were not necessarily measured frequently or on the same day [32,53], or only were reported to 1 dp [40]. Data points were sometimes unevenly distributed, with a lack of results at high, intermediate, or low pCO_2 [17,32,51]. TAN and VFA concentrations were not always monitored, and when available were usually measured less frequently than pH and pCO_2, limiting the amount of data suitable for use with Equation B. In two cases, further processing was carried out to increase the number of data points available, by interpolation between measured pCO_2 values [32] and by using modelled TAN concentrations [39].

Of the eight additional datasets considered, four showed poor agreement between experimental and predicted pH values based on Equation A [17,39,50,53], in part at least for some of the reasons mentioned above. The other four showed better agreement, with R^2 values between 0.80–0.86. Figure 1e,f shows results for numerical data from the work of Luo and Angelidaki [40]: in this case Equation A gave a reasonable fit (R^2 = 0.83, n = 60, $p < 0.000$; RMSE = 0.12) for thermophilic CSTR digestion of cattle manure and whey at three H_2 loading rates across a range of pCO_2 values between 0.02–0.26. Results for data from mesophilic digestion of pig manure by Zhu et al. [51,52] were R^2 = 0.85, n = 55, $p < 0.000$, RMSE = 0.10; and from 2-stage mesophilic and thermophilic digestion of cattle manure by Bassani et al. [32] were R^2 = 0.82, n = 32, $p < 0.000$, RMSE = 0.10 and R^2 = 0.869, n = 40, $p < 0.000$, RMSE = 0.13, respectively. Graphs for these studies are given in Appendix A Figures A2–A6.

Four of the additional datasets included some data on TAN and VFA, but in each case, the application of Equation B was challenging. A further study on thermophilic co-digestion of cattle manure and whey by Luo and Angelidaki [17] had very few data points at high pCO_2 (Figure A5). TAN was measured only once in each phase, and molar VFA concentrations occasionally exceeded the average TAN value, making Equation B inapplicable for these points. In data on CO_2 biomethanisation of pig manure by Zhu et al. [50–52], Equation A had shown good results for mesophilic conditions, but TAN measurements were only available for the first part of this run (Figure A6a,b). Agreement in thermophilic conditions was poor, although there was a modest improvement in R^2 when TAN and VFA were included using Equation B (Figure A6c,d). Using VFA data and modelled TAN values from a further study on cattle manure and whey by Lovato et al. [39] also produced a slight improvement in R^2, but in some cases, VFA concentrations exceeded the estimated TAN values, and neither Equation A nor B gave good agreement between predicted and experimental pH (see Figure A4).

Two further studies with some TAN and VFA data were available from experiments using a synthetic feedstock containing phosphates [6,23]. In one of these [23], the OLR was adjusted by changing the feed concentration and thus TAN and other parameters. Equation A gave a good correlation (R^2 = 0.91) between predicted and experimental pH values, but with the predicted curve offset from the experimental data (Figure 2a). Tao et al. [6] showed that a modified approach taking account of the effect of phosphate buffering could reduce this discrepancy, but it was not possible to adapt this approach for Equation B as TAN values were already used in the phosphate adjustment. However, when Equation B was applied without any phosphate correction, it also gave a considerable improvement. Figure 2a, b shows measured and predicted pH values based on Equation A, Equation B, and the phosphate adjustment for one replicate digester at OLR 3 g COD L^{-1} day^{-1}. It can be seen that the phosphate adjustment increased the correlation coefficient and shifted the predicted pH closer to the experimental values, improving the RMSE from poor to good. Equation B (with no phosphate correction but adjusted for TAN and VFA) gave slightly a lower R^2 value than the phosphate adjustment, but a similar improvement in RMSE. Table 3 shows a heat map of the difference between predicted and experimental pH values using Equation A, Equation B, and the phosphate adjustment, for each phase of the experiment at OLR 3 g COD L^{-1} day^{-1}. Both Equation B and the phosphate adjustment were able to reduce some discrepancies although neither were able to eliminate poor results in phase 3. Table S34 gives equivalent results for all OLR in the trial with additional statistical parameters, while Figures S2–S5 show the data series graphically: it can be seen that in general both approaches performed very well, and comparison between them can help to elucidate the relative importance of phosphate buffering, and TAN and VFA concentrations in each case.

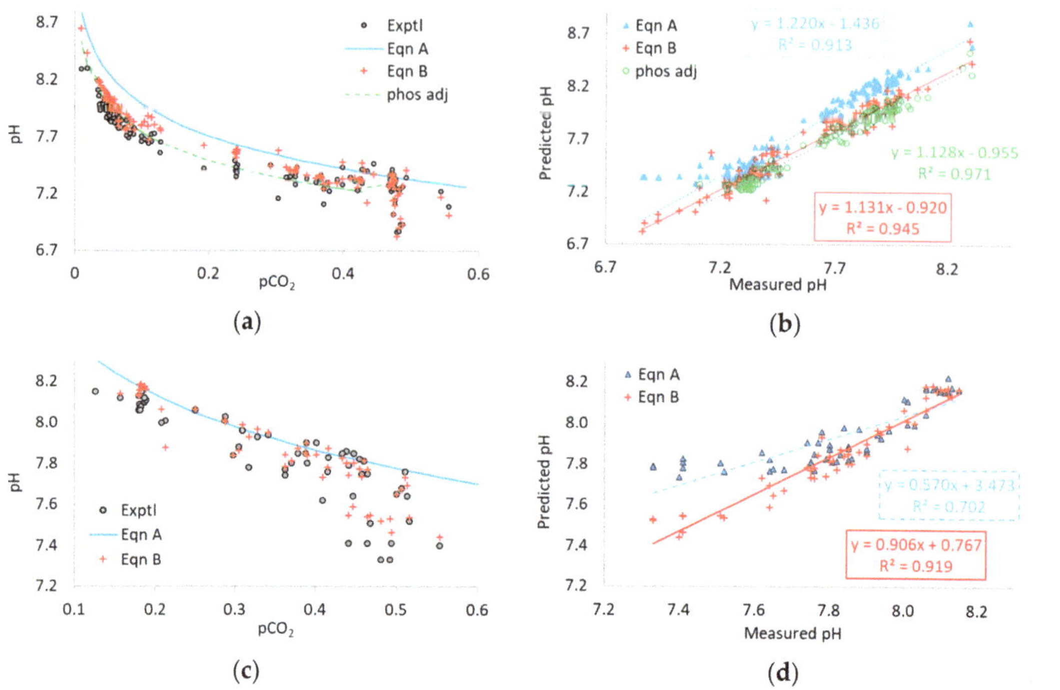

Figure 2. Experimental pCO_2 and pH values and predicted pH from Equations A and B for mesophilic CO_2 biomethanisation of phosphate-containing synthetic feed at OLR 3 g COD L^{-1} day^{-1} (**a**,**b**) as reported in [23], and for TAN concentration 3 g N L^{-1} (**c**,**d**) as reported in [6]. Numerical data was kindly provided by the authors.

Table 3. Performance of Equations A and B and phosphate adjustment for data from Tao et al. [23].

OLR 3		(i)	(ii)	(iii)	(iv)	(v)	(vi)	RMSD	R^2
Phase		1 No H_2	No H_2	2 With H_2	With H_2	3 $H_2 + CO_2$	$H_2 + CO_2$		
Temp	°C	37	37	37	37	37	37	-	-
OLR	g VS L^{-1} day^{-1}	3	3	3	3	3	3	-	-
HRT	days	15	15	15	15	15	15	-	-
H_2 input	L L^{-1} day^{-1}	0.0	0.0	2.9	2.9	2.9	7.4	-	-
SMP_{tot}	L CH_4 g^{-1} VS	0.287	0.289	0.496	0.490	0.452	0.804	-	-
CH_4	%	50.0	50.1	93.7	90.9	75.0	76.6	-	-
TAN	M	0.075	0.074	0.045	0.045	0.049	0.044	-	-
VFA	M	0.000	0.000	0.001	0.002	0.007	0.006	-	-
pH	-	7.34	7.35	8.00	7.93	7.82	7.81	-	-
pCO_2	-	0.474	0.473	0.040	0.046	0.071	0.077	-	-
$a \times 10^8$	-	9.29	9.23	22.08	23.03	19.45	18.67	-	-
$b \times 10^8$	-	0.69	0.68	0.97	1.00	0.81	0.70	-	-
Temp	°C	37	37	37	37	37	37	-	-

Table 3. *Cont.*

OLR 3		(i)	(ii)	(iii)	(iv)	(v)	(vi)	RMSD	R^2
Difference between experimental and predicted pH based on Eqn A with coefficient *a*(x)									
a(i)	-	-	0.00	0.32	0.35	0.29	0.27	0.28	0.995
a(ii)	-	0.00	-	0.33	0.35	0.29	0.28	0.28	0.996
a(iii)	-	0.37	0.37	-	0.02	0.05	0.07	0.24	1.000
a(iv)	-	0.39	0.39	0.02	-	0.07	0.09	0.25	0.998
a(iv)	-	0.31	0.32	0.05	0.07	-	0.02	0.20	0.998
a(vi)	-	0.30	0.30	0.06	0.08	0.02	-	0.19	0.998
Difference between experimental and predicted pH based on Equation B with coefficient *b*(x)									
b(i)	-	-	0.01	0.13	0.14	0.06	0.00	0.09	0.980
b(ii)	-	0.01	-	0.14	0.15	0.07	0.01	0.10	0.979
b(iii)	-	0.14	0.15	-	0.01	0.07	0.13	0.11	0.983
b(iv)	-	0.16	0.17	0.01	-	0.09	0.15	0.13	0.990
b(iv)	-	0.06	0.07	0.07	0.08	-	0.06	0.07	0.987
b(vi)	-	0.00	0.01	0.13	0.14	0.06	-	0.09	0.997
Difference between experimental and predicted pH for Equation A with *a*(x) and phosphate adjustment									
Period (i)	-	-	0.01	0.05	0.07	0.06	0.01	0.05	0.993
Period (ii)	-	0.01	-	0.06	0.08	0.07	0.01	0.06	0.993
Period (iii)	-	0.06	0.07	-	0.01	0.00	0.05	0.05	0.995
Period (iv)	-	0.09	0.10	0.03	-	0.02	0.07	0.07	0.996
Period (iv)	-	0.07	0.08	0.01	0.01	-	0.06	0.06	0.997
Period (vi)	-	0.01	0.02	0.04	0.06	0.05	-	0.04	0.999

n/r = not reported. Lower parts of Table show heat map with white = good, grey = reasonable, and black = poor agreement.

The second set of experiments using the same synthetic feed was carried out at controlled TAN concentrations [6]. At TAN 3 g N L^{-1}, the pH was mainly affected by ammonia, and the effect of the phosphate adjustment was small (see Supplementary Materials in [6]). The dataset of points in this experiment with measured VFA concentrations were re-analysed, using interpolated TAN values where necessary: results for 3 g N L^{-1} are shown in Figure 2c,d. It can be seen that Equation B gave a significant improvement in fit ($R^2 = 0.919$) compared to Equation A ($R^2 = 0.702$, with n = 54 and $p < 0.000$ in both cases). Table S35 shows heat maps for results from Equation A using average values for each phase, while Figures S6–S9 present the data series graphically: the majority of results were in good agreement, and the two trials thus provided a powerful demonstration of the equations across a range of TAN concentrations.

It is also noteworthy that the datasets in both studies using this feedstock contained some extreme/outlying points, including values from the first days of digester operation before feeding was stabilised, and well before any H_2 addition: these correspond to points with high pCO_2 and low pH in Figure 2a,c. Thus, these results are interesting as they indicate Equation B can be used for pH prediction under a wide range of conditions, not only for the purposes of controlling H_2 addition in CO_2 biomethanisation. These two studies represent a special case, as in real organic feedstocks buffering capacity is normally provided mainly by the bicarbonate-ammonium pair. The results clearly demonstrate the value of taking VFA and TAN into consideration.

Analysis of the multi-point datasets therefore gave further clear evidence of the validity of Equation A. The additional datasets obtained from other groups did not provide a completely satisfactory demonstration of Equation B, in part because the studies considered were not designed for this purpose; although the results from these and from the data of Tao et al. [6,23] strongly indicated the potential of an equation able to take VFA and TAN into account, and confirmed the need for further experimental data.

3.4. Experimental Trial of H_2 Addition to Induce Digestion Instability

3.4.1. Digestion Performance

During the period of operation under uniform conditions between days 0–54, monitoring parameters for all digesters were in good agreement, as can be seen in Figure 3.

Average values for this period are given in column (i) of Table 4. Note that specific methane production (SMP) is expressed as total CH_4 produced per unit of organic feed VS, including any CH_4 from added H_2: on days without organic feed SMP values are omitted.

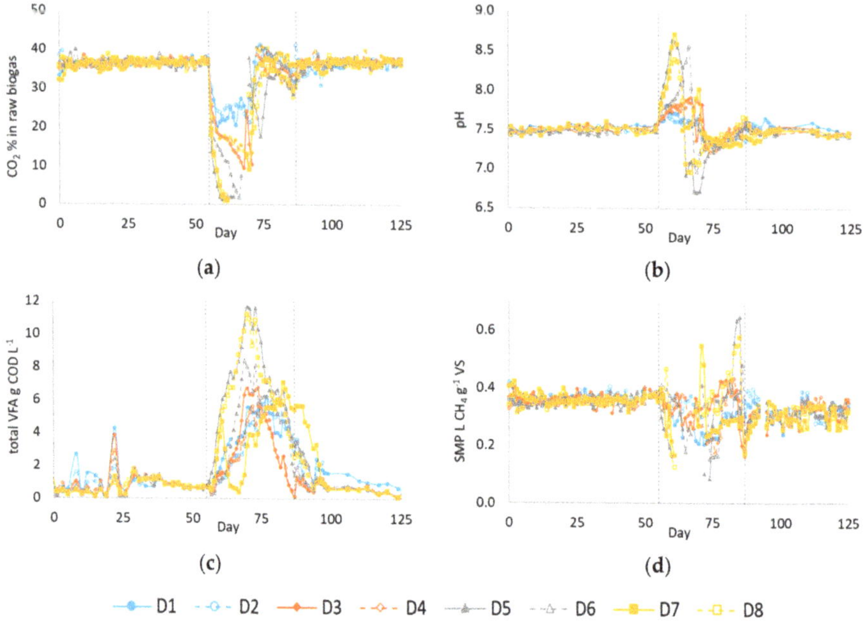

Figure 3. Values for (**a**) biogas CO_2 content, (**b**) pH, (**c**) total VFA, and (**d**) SMP in digesters D1–8 during the whole experimental period. Vertical dotted lines indicate the start of H_2 addition on day 55 and the re-establishment of full OLR in all digesters on day 87.

Table 4. Average values of monitoring parameters in D1-8 from different periods of stable operation during the experimental period.

		(i) Before H_2 Addition Day 0–54	(ii) Baseline Period 1 Day 0–24	(iii) Baseline Period 2 Day 25–54	(iv) After Recovery Day 110–125
pH	-	7.50 ± 0.04	7.48 ± 0.04	7.50 ± 0.04	7.46 ± 0.04
TAN	M	0.131 ± 0.005	0.128 ± 0.004	0.134 ± 0.003	0.133 ± 0.004
Total VFA	M	0.008 ± 0.005	0.006 ± 0.005	0.010 ± 0.004	0.006 ± 0.003
TA	g $CaCO_3$ L^{-1}	9.38 ± 0.32	9.26 ± 0.34	9.51 ± 0.22	9.37 ± 0.28
PA	g $CaCO_3$ L^{-1}	6.33 ± 0.25	6.20 ± 0.18	6.46 ± 0.22	6.40 ± 0.23
IA	g $CaCO_3$ L^{-1}	3.05 ± 0.24	3.05 ± 0.30	3.05 ± 0.17	2.97 ± 0.15
TS	%WW	3.86 ± 0.11	3.85 ± 0.11	3.88 ± 0.12	3.80 ± 0.07
VS	%WW	2.59 ± 0.09	2.59 ± 0.08	2.58 ± 0.09	2.60 ± 0.06
SMP	L CH_4 g^{-1} VS	0.358 ± 0.017	0.355 ± 0.017	0.360 ± 0.017	0.322 ± 0.027
pCH_4	-	0.616 ± 0.008	0.617 ± 0.009	0.616 ± 0.008	0.620 ± 0.009
pCO_2	-	0.365 ± 0.009	0.362 ± 0.010	0.367 ± 0.008	0.369 ± 0.008

Values after ± give standard deviation for data from all eight digesters.

From day 55 onwards, digesters were operated in pairs and received H_2 addition at the following rates: D1 and 2 0.81, D3 and 4 1.58, D5 and 6 2.34, and D7 and 8 2.82 L H_2 L^{-1} day^{-1}, corresponding to a nominal 24%, 48%, 70%, and 85% of the stoichiometric H_2 requirement for conversion of CO_2 in the produced biogas and to additional COD loadings of 0.58, 1.13, 1.67, and 2.01 g COD L^{-1} day^{-1}, respectively. These rates of addition were intentionally selected to exceed the anticipated conversion capacity of the digesters in the described operating mode, especially without acclimatisation [18].

As expected, the introduction of excess H_2 led to a rapid fall in pCO_2 (Figure 3a) accompanied by a rise in pH (Figure 3b), VFA accumulation (Figure 3c), and a fall in volumetric methane production (VMP) and in SMP (Figure 3d). In D1 and 2, and D3 and 4, H_2 addition rates were maintained until day 70 apart from a brief reduction in D1 on days 65–66. However, there were clear signs of continuous VFA accumulation, along with a decline in SMP (Figure 4c,d). H_2 addition to these digesters was stopped on day 71. In D5, D7, and D8, the initial rate of H_2 addition was maintained until day 61, by which time the pH had reached 8.61, 8.71, and 8.39 with total VFA concentrations of 6.0, 4.7, and 5.4 g COD L^{-1}, respectively. Addition rates for H_2 and organic feed were varied over the next few days (details in the experimental dataset) but any signs of recovery were temporary, and H_2 addition was stopped completely on days 74, 70, and 72, respectively. D6 initially showed a slightly lower rate of pH rise and VFA accumulation than D5, and H_2 addition continued until day 66: at this point, the pH had reached 8.6 with a total VFA concentration of 5.3 g COD L^{-1}. H_2 and organic loading rates were varied over the next few days, but without lasting recovery and H_2 addition to D6 was stopped on day 71. After H_2 addition ceased pH and VFA concentrations gradually stabilised, by day 87 the full organic loading of 3.87 g VS L^{-1} day^{-1} was resumed in all digesters. The operation continued until day 125 with less frequent monitoring until the end of the period, when parameters were again measured to provide baseline values after stabilisation. Average values for this period are shown in column (iv) of Table 4 and were close to those found at the start of the trial, indicating good stabilisation.

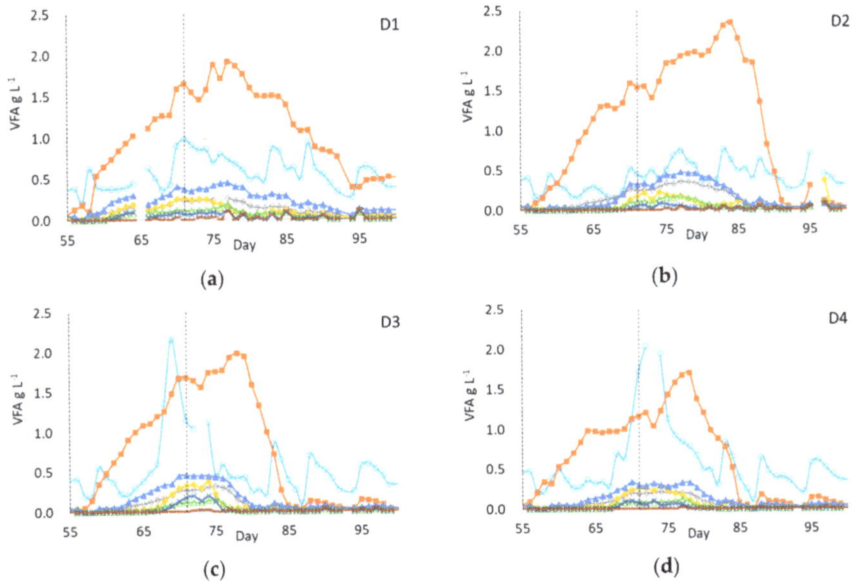

Figure 4. *Cont.*

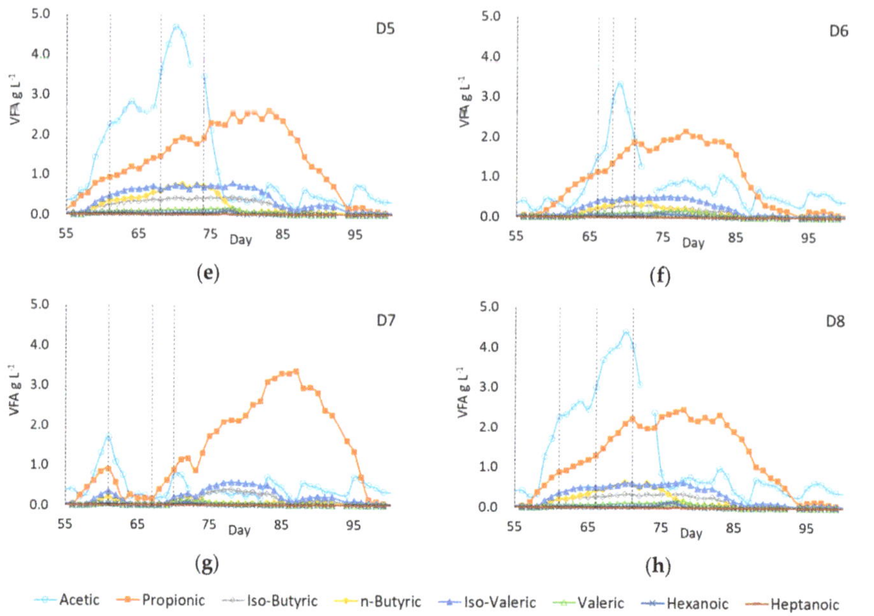

Figure 4. (**a–h**) VFA profiles between day 55−100 in digesters D1−8. All digesters initially received H_2 from day 55: vertical dotted lines indicate the start and end of subsequent periods of H_2 addition in each digester. Note the difference in *y*-axis scales for D1−4 and D5−8.

3.4.2. VFA Profiles

Figure 4 shows VFA speciation for the digesters between days 55 and 100. Following the start of H_2 injection on day 55, propionic acid accumulation occurred at similar rates in all digesters. This was expected considering the high H_2 partial pressures, which made one of the main propionate degradation pathways, to acetate, H_2, and bicarbonate/CO_2, thermodynamically unfavourable [61]. The average propionate accumulation rate was 100 mg L^{-1} day^{-1} for D1–4 over the 16 days of H_2 injection, and 140 mg L^{-1} day^{-1} for D5–8 for the first 6 days of the period. Increases in acetic acid concentrations more closely reflected the difference in H_2 injection rates, with D1 and 2 apparently showing little change in acetic acid while the average rate of acetic acid accumulation in D7 and 8 during the initial six days of H_2 injection was 310 mg L^{-1} day^{-1}. This is likely to have been due to homoacetogenesis occurring under high H_2 partial pressure.

During H_2 addition pH in D1–4 did not show such a marked rise as in D5–8, with measured values remaining below 7.9; whereas propionic acid concentrations continued to increase even after H_2 addition ceased. This indicated that the degradation of organic feed in D1–4 might also be blocked, although not as severely as in D5–8. Since sampling and pH measurement was conducted at the end of each daily feeding cycle, it is possible that pH values at some point during the cycle were higher than at the end, as reported in [6], and that this was responsible for the apparent blockage. This temporary unfavourable pH may also have caused partial inhibition of organic feed degradation during the 16 days of the H_2 injection period.

The VFA profile in D7 differed slightly from that in the other digesters, showing a fall in VFA concentrations when H_2 addition was paused on day 61, accompanied by a two-day gap in organic feeding followed by two days at reduced OLR. However, after OLR was restored and H_2 was reintroduced on day 67, propionate accumulation began and continued as in the other digesters even after H_2 addition ceased on day 70. Other VFA species also accumulated and then declined in all digesters in roughly the same order (see

Figure 5), with iso-valeric, n-butyric, and iso-butyric plateauing at around 0.5 g COD L^{-1} in most case until propionate concentrations started to fall.

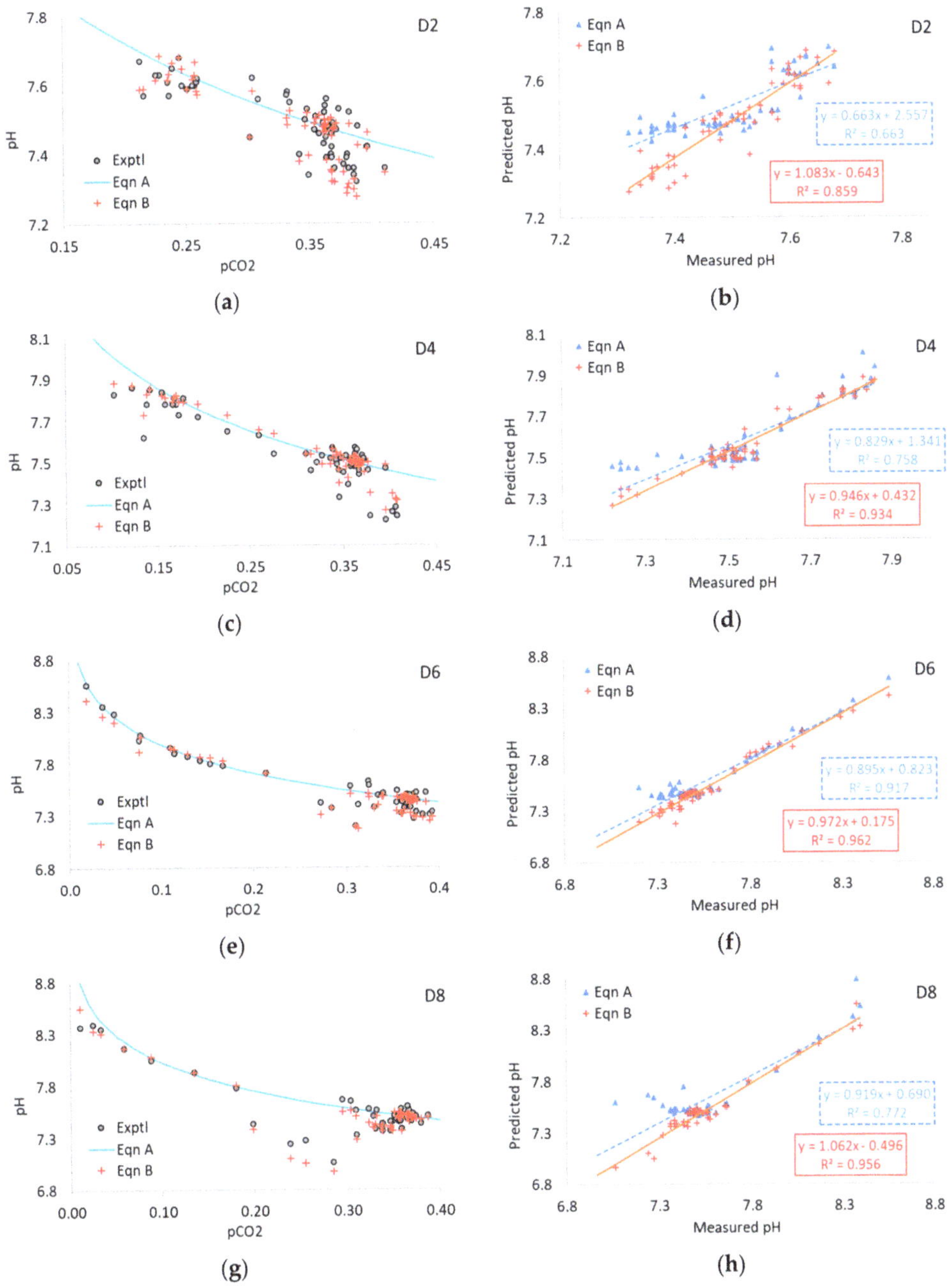

Figure 5. Experimental pCO_2 and pH values and predicted pH from Equations A and B in digesters D2 (**a**,**b**), D4 (**c**,**d**), D6 (**e**,**f**), and D8 (**g**,**h**) during mesophilic anaerobic digestion of sewage sludge at varying H_2 addition rates.

3.5. Application of Equations A and B to Experimental Data

As with the previous analysis of multi-point data series, values from the experimental work were used to evaluate the performance of Equations A and B. The dataset used consisted of points for which measured pH, pCO_2, and VFA concentrations were available, with linear interpolation of TAN values where required. Day 0–24 was initially defined as the baseline period, and average values from this (shown in column (ii) in Table 4) were used to calculate the coefficients for Equations A and B. The two equations were then applied to the data for days 25–100.

Experimental pCO_2 and pH and predicted pH values in D2, D4, D6, and D8 are shown in Figure 5, with equivalent graphs for D1, D3, D5, and D7 in Figure A7 in Appendix A. Coefficient values for Equations A and B and statistical data for all digesters are given in Table 5.

Table 5. Coefficient values and statistical parameters from the application of Equations A and B to data series from day 25–100 of experimental work.

	$a \times 10^8$	$b \times 10^8$	n	R^2 Equation A	Equation B	Slope Equation A	Equation B	Int'cept Equation A	Equation B	RMSE Equation A	Equation B
D1	8.29	0.962	50	0.513	0.877	0.492	0.856	3.86	1.09	0.06	0.04
D2	8.85	1.052	51	0.663	0.859	0.663	1.083	2.56	−0.64	0.05	0.04
D3	8.70	1.043	51	0.774	0.937	0.915	1.000	0.68	0.02	0.08	0.04
D4	8.45	1.023	50	0.758	0.934	0.829	0.946	1.34	0.43	0.08	0.04
D5	8.87	1.103	47	0.778	0.968	0.812	0.928	1.54	0.52	0.18	0.07
D6	9.18	1.118	50	0.917	0.962	0.895	0.972	0.82	0.17	0.08	0.06
D7	8.96	1.134	46	0.950	0.973	0.935	0.967	0.54	0.21	0.07	0.06
D8	8.35	1.039	45	0.772	0.956	0.919	1.062	0.69	-0.50	0.14	0.07

Note: n = number of datapoints used (i.e., where measured values were available for pCO_2, pH, and VFA with TAN interpolated where required). Results from regression analysis were significant in all cases ($p < 0.0001$).

A fairly wide range of conditions occurred during the trial, with pCO_2 values ranging from 0.41 to 0.01, pH from 6.7 to 8.7, TAN from 1.6 to 2.1 g N L^{-1}, and VFA concentrations of up to 11.2 g COD L^{-1}. Equation A provided a good fit to the experimental data during stable operation but was less able to predict pH accurately in the period of instability during and after H_2 addition when VFA concentrations were high and varying. This resulted in poor correlation coefficients ($R^2 < 0.8$) in most cases (Table 5). In contrast, the performance of Equation B was good, with R^2 values ranging from 0.859 to 0.973. In comparison with Equation A, the use of Equation B gave a marked improvement in the correlation coefficient and the RMSE for experimental versus predicted data in each case and for each replicate (see Table 4). This improvement was particularly clear in D1-5 and D8, whereas in D6 and D7 Equation A also gave relatively good results. Equation B also appeared to give a slight improvement in the slope and intercept values for graphs of predicted versus experimental pH, although the difference in slope was only significant at the 5% level in D1 and D2. In all cases the RMSE value using Equation B was <0.1, indicating that it was able to provide a reliable prediction of digester pH based on pCO_2 even in periods of rapidly changing VFA content.

As a further indication of the robustness of these equations, it can be seen from Table 4 that values for *a* and *b* obtained from the baseline period were very close for all eight reactors: substitution of the relevant coefficient from another digester made only a minimal difference to the results or to the statistical parameters shown. As an additional check, values for coefficients *a* and *b* were calculated for alternative baseline periods from day 25–54, and from day 110–125 after the final stabilisation of the digesters, based on the data in columns (iii) and (iv) of Table 3. The calculated values were closely similar to those in Table 4, and use of them again had very little impact on the performance of Equations A and B. Data points between days 110–125 were not included in the analysis shown in Table 4 and Figure 5 because of a lack of measured values for TAN in the later part of the trial. Yet, inclusion of these points using interpolated TAN values also caused only minor changes in R^2 and RMSE (details in the experimental dataset). Equation B was thus shown to be both robust and effective even in conditions of operational instability.

As noted above, in digestion of organic wastes the main buffer pair is usually HCO_3^- and NH_4^+, with VFA^- and NH_4^+ only secondary. The strategy of excess H_2 addition was very effective in pushing up VFA concentrations to create conditions of temporary instability in the digesters for testing Equation B. While the dataset would have been strengthened by more TAN measurements in the latter part of the experimental period, the trial was thus very successful in demonstrating both the impact of short-term changes in pCO_2 and VFA on digester pH; and the potential value of Equation B as a tool for process control and management in unstable operating conditions.

4. Discussion

4.1. Methodological Issues with Data in Literature Studies

The relationship between pCO_2 and pH described in Equation A was shown to work consistently and with a high predictive capability when applied under stable operating conditions. In circumstances where changes in VFA or TAN concentrations occurred, Equation B was shown to work well both for the majority of the datasets analysed, and for the dedicated experimental work in which digester stability was tested to a point of process failure. Some results were found in data from reported studies where the equations did not give reliable results; however, it is important to try and understand why these cases occurred. The first point to note is that the analysis presented used data from all relevant published studies on in-situ CO_2 biomethanisation: it did not attempt selectively to discount datasets that did not show good agreement or to dismiss reactor types or methodologies that did not conform to typical norms. It is also important to remember that the studies considered were not specifically designed for the current purpose. As a result of these factors, there were some issues with the available data for methodological and other reasons: parameters of interest were not always measured, or not measured frequently enough, or were reported in unsuitable units, or omitting relevant details of sampling or operational conditions. Examples include experiments where H_2 was added by headspace injection (see Tables S23–S25): interpretation of data from these could be challenging, as it was not always clear whether reported pH and gas composition values were measured at the same time, while headspace pressures could have varied both between treatments and during a trial.

Many studies reported average values for a given phase or set of conditions, but these could be misleading, especially when there is considerable variation during a phase. Equation A gave poor agreement when applied to average values reported in [47]. However, if the results were recalculated using the final data points, good agreement was achieved (see Table S9). Values for individual data points may thus be preferable, while the use of data series also allows the assessment of variability.

In some studies, pH was controlled by chemical addition during some or all of the experimental period [34,48,58,62], making the data difficult or unsuitable for use in this work.

In some cases, agreement between the first and subsequent phases was poor, but improved after phase 1 (see, e.g., Tables S20, S22 and S25). This may simply reflect initial acclimatisation, especially if the first phase was short; although there is also an intriguing possibility that differences could be due to changes in gas circulation and mixing. In some studies that used gas bubblers or diffusers, there was apparently no gas recirculation in the control digester or period [23,44]; whereas there is clear evidence that changes in recirculation or mixing rates can affect performance and operating parameters [17,36,46].

Despite such issues, the available data allowed a clear demonstration of performance of the equations in each case while providing some interesting insights on practical requirements for this type of research.

4.2. General Research-Related Issues

The derivation of Equation A assumes that most baseline parameters remain applicable throughout the operating period, and changes in pH are driven primarily by pCO_2. As both the literature data and the experimental results showed, in practice this is not always so. Changes may occur over short periods, as in the current study where excess H_2 induced rapid VFA accumulation, or, as a result of longer-term processes such as washout and acclimatisation. CO_2 biomethanisation studies are often not run under the same conditions for three HRT for each phase or during start-up (see e.g., [31,49,51,60]). This is a reasonable approach where the main aim is to assess gas transfer and utilisation or changes in the microbial community, which happen on relatively rapid timescales. Yet, the stabilisation of digestate parameters can take longer and may appear as a drift in values between phases. For example, when tested on data from mesophilic cattle slurry digestion in a 140-L helically-mixed CSTR [38], Equation A gave good results (Table S8). However, there was a slight downward trend in agreement through the trial, which might have been associated with incremental shifts in TAN or VFA during acclimatisation, since the total duration for all phases was around 2.5 HRT. Unfortunately, these parameters were not reported.

Experimental design is an important factor here. Changes in baseline values during a trial due, e.g., due to acclimatisation or feedstock variability may be easy to distinguish where there is a control reactor. If TAN and/or VFA changes in the same way in both control and experimental reactors then agreement within each phase may be good, even if that between phases is not (e.g., Table S25). However, the identification of such changes is more difficult when different conditions are run in series without controls.

There may also be specific reasons for differences in TAN between control and experimental digesters within a phase, such as changes in microbial biomass associated with growth in the hydrogenotrophic methanogenic population during in situ CO_2 biomethanisation [23], or toxicity effects from syngas or other inhibitors. These effects may help to explain patterns where agreement across several phases is good for control reactors but poor for experimental ones (e.g., Table S36). In such cases, results from control digesters do not resolve the problem, although it may still be possible to observe trends in the results.

The original studies by Luo et al. [17,18,40,41] were run with control reactors, at relatively short HRTs of 15 days, in conditions that allowed steady stable operation, and therefore produced good agreement in all cases: see Tables S1, S2, S3 and S10. Studies of this type can give high R^2 values for prediction across multiple conditions but, as noted earlier, this is not a requirement for demonstrating the good performance of the equations tested if there are other reasons for variability across phases.

Several of the above examples add further support to the practical value of an equation that can deal with changes over both long and short timescales. The present work demonstrated that Equation B was able to deal with acute short-term changes, such as the rapid VFA accumulation in the current trial. It could also accommodate longer-term changes and shifts in TAN: the evidence for this is slightly more tentative due to the relative scarcity of datasets with TAN measurements but can be seen, e.g., in Tables S7, S22, S34 and S35. Equation B also performed well when there were changes in both TAN and VFA (e.g., Tables S4 and S32). As noted above, Equation B is thus applicable in cases where pH is affected by other factors in addition to changes in pCO_2 induced by CO_2 biomethanisation.

4.3. Considerations for Large-Scale Operation

In this work pCO_2 values from literature and experimental data were used to predict pH, and the results were then compared with reported pH values. pH was selected as the output because this is widely used as an indicator of stability in day-to-day operation. However, the same relationships can of course be used to predict pCO_2 from pH, and monitoring of either parameter in conjunction with the use of these equations could thus provide a useful basis for process control in CO_2 biomethanisation.

Equation A allows calculation of pH values from two parameters (digester temperature and pCO_2), with a further two (TAN and VFA) required for Equation B. Calibration can be achieved by measurement of these parameters under one set of conditions, which does not necessarily have to be baseline operation (e.g., without H_2 addition in the case of CO_2 biomethanisation), as long as these are representative of a stable operating mode. In contrast, alternative models to predict pH such as ADM1 [63] generally require a large number of input values for calibration. These more complex models also predict many other performance and stability parameters, but their use in real-time control is still uncommon, although interest in this application is expanding [20,22].

Variations in the feedstock can cause changes in total Kjeldahl nitrogen and TAN, and while these can be controlled in laboratory experiments they also affect commercial AD plants [20]. Even with relatively consistent feedstocks such as sewage sludges, biosolids concentration may vary over time as a result, e.g., of changes in dewatering practices. AD operators at WWTW can accommodate this by changing the applied OLR or HRT if required, but TAN concentrations will still vary and may affect the ammonium-bicarbonate equilibrium, pCO_2 and pH and thus the coefficients for Equation A and B. At commercial sites with a single digester, operation without a 'control' reactor represents the normal situation: these sites may also have less staff time available for digester management, so simple robust process control systems for CO_2 biomethanisation are likely to be particularly valuable in this case [21]. At multi-digester sites it may be possible to process all of the biogas generated on the site in one digester retrofitted for CO_2 biomethanisation [6]; in this case the performance of other digesters could provide control data indicating any changes due, e.g., to feedstock variation.

Use of a relationship which takes account of the effect of changes in TAN and VFA concentrations may appear to require increased monitoring compared to current practice at many AD plants [27]. However, as noted earlier, the objective in full-scale operation is normally to remain within an envelope of safe operating conditions. As Equations A and B are equivalent when no VFA is present, Equation A can be used when conditions are stable. Analysis for TAN and VFA is relatively straightforward, and in normal circumstances TAN values do not change very rapidly. Monitoring of additional parameters can therefore be performed infrequently or not at all when an operation is stable, and introduced or stepped up only if some disturbance makes this necessary, meaning the requirement for more parameters need not be burdensome [21].

No detailed analysis was made to compare the values of coefficients *a* or *b* derived for different sets of feedstocks or operating conditions. However, it is interesting to note that four studies on thermophilic (55 °C) CSTR digestion of cattle manure obtained from three different sources gave values for *a* of around 2–3 $\times$ 10^{-8} when operating stably (Tables S1, S15, S16 and S32); values for mesophilic (37 and 40 °C) operation were slightly higher at around 5–6 $\times$ 10^{-8} (Tables S8 and S15). Three independent studies of sewage sludge digestion in mesophilic (37 °C) CSTR gave *a* values of 11–15 $\times$ 10^{-8} (Tables S6, S10 and S38). Crop and agro-wastes and food waste feedstocks, as defined in [12], were more diverse, forestalling any direct comparison. These different values of *a* reflect both the digestate TAN concentration, and the effect of other physico-chemical characteristics of the digestate on the Henry's and acid disassociation constants; when the VFA concentration is zero, *b* is simply equal to *a* divided by the molar TAN concentration. While it is easy to obtain values for *a* or *b* from any suitable set of baseline parameters for a given digester, the idea that there may be typical values of these coefficients which could be applied generally and/or that can be linked to other digestate properties is of interest and worth consideration in further work

5. Conclusions

In the current work, the performance of the two equations was well demonstrated across a wide range of operating conditions and data sources.

Equation A provides a relationship between two simple parameters, often routinely measured, which showed good agreement with experimental data when applied under stable operating conditions. This approach therefore appears suitable as a basis for control of H_2 addition in CO_2 biomethanisation of organic substrates.

Equation A performed less well for digesters undergoing dynamic change or experiencing instability, as indicated, e.g., by VFA accumulation. In these situations, the use of Equation B incorporating VFA and TAN concentrations was generally able to improve prediction, giving RMSD values of <0.1 pH units: this was clearly demonstrated by experimental work in which unstable conditions were deliberately induced to test the equations' performance. Both TAN and VFA are relatively straightforward to monitor and when they are included along with pCO_2 as in the derived Equation B, the range of application can be extended from stable operation to cover both dynamic change, and crisis management and recovery if digester operation becomes unstable.

As both Equation A and B are derived from fundamental principles, the discrepancies that are seen in some cases between predicted and experimental data most likely reflect limitations in typical measurement accuracy and/or experimental design. This and the fact that the equations performed well when applied to data from a wide range of plant configurations and substrates gives strong support to the view that they are suitable for application on real-world operational sites.

Establishing the required coefficient values for each equation is a simple process requiring only basic data from a period of stable operation. Once these are determined for a particular waste/digester combination, the use of this approach offers a reliable means of optimising the biomethanisation potential of a plant without the risk of exceeding critical values for pH or pCO_2. This should provide the industry with the confidence to adopt this emerging biotechnology, which could significantly increase the efficiency of utilisation of carbon from organic substrates. The research thus further strengthens the case for promoting in situ CO_2 biomethanisation as a means of maximising the value of existing infrastructure and resources in the waste and agri-food sectors.

Supplementary Materials: The following supporting information can be downloaded at: https://www.mdpi.com/article/10.3390/pr11010113/s1. [64,65].

Author Contributions: Development of equations, Y.Z.; Conceptualisation of study, C.J.B., Y.Z. and S.H.; literature review and data analysis, S.H. and Y.Z.; supervision of experimental work, C.J.B., S.H. and Y.Z.; analysis of experimental data: Y.Z. and S.H.; writing—original draft preparation, S.H.; writing—review and editing, Y.Z. and C.J.B.; funding acquisition, C.J.B., S.H. and Y.Z. All authors have read and agreed to the published version of the manuscript.

Funding: This research was supported by the IBCat H2AD project (EP/M028208/1) funded by the Engineering and Physical Sciences Research Council (EPSRC). The APC were funded by MDPI Topic Anaerobic Digestion Processes.

Data Availability Statement: The experimental data in this study are openly available in https://doi.org/10.5258/SOTON/D2488. For other data requests please contact the corresponding authors.

Acknowledgments: The authors would like to thank the following for the provision of additional numerical datasets: Ilaria Bassani, Istituto Italiano di Tecnologia, Genova, Italy; Giovanna Lovato, Instituto Mauá di Tecnologia, São Caetano do Sul, Brazil; Gang Luo, Department of Environmental Science and Engineering, Fudan University, Shanghai, China; Bing Tao, formerly of the University of Southampton, Southampton, UK; Xianpu Zhu, Shanghai Jiao Tong University, Shanghai, China. The authors also gratefully acknowledge the help and support of the Specialist Technical Staff of the Environmental Laboratories at the University of Southampton, UK.

Conflicts of Interest: The authors declare no conflict of interest. The funders had no role in the design of the study, namely in the collection, analyses, or interpretation of data; in the writing of the manuscript; or in the decision to publish the results.

Appendix A

Figures A1–A7 show results for the application of Equation A and where possible Equation B to datasets associated with [6,17,32,39,40,50–53], and with the experimental work in the current trial. Numerical datasets associated with the published literaturewere kindly provided by the authors in each case. Note that graphs do not include baseline values used for deriving coefficients *a* and *b*.

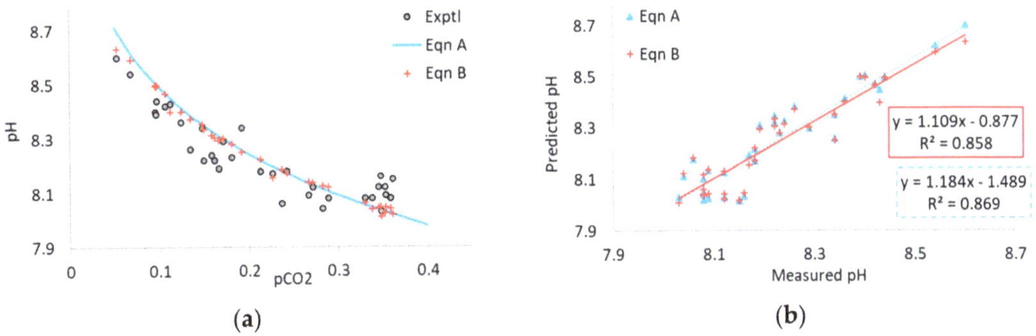

Figure A1. (**a**) Measured pCO_2 and measured and predicted pH values and (**b**) predicted pH values from Equations A and B for mesophilic CO_2 biomethanisation of industrial food waste for points with measured VFA and measured or interpolated TAN associated with Tao et al. [6]. See Figure 1c,d in the main text for results from Equation A for the full dataset, and Table S5 for operating conditions and average results for each phase.

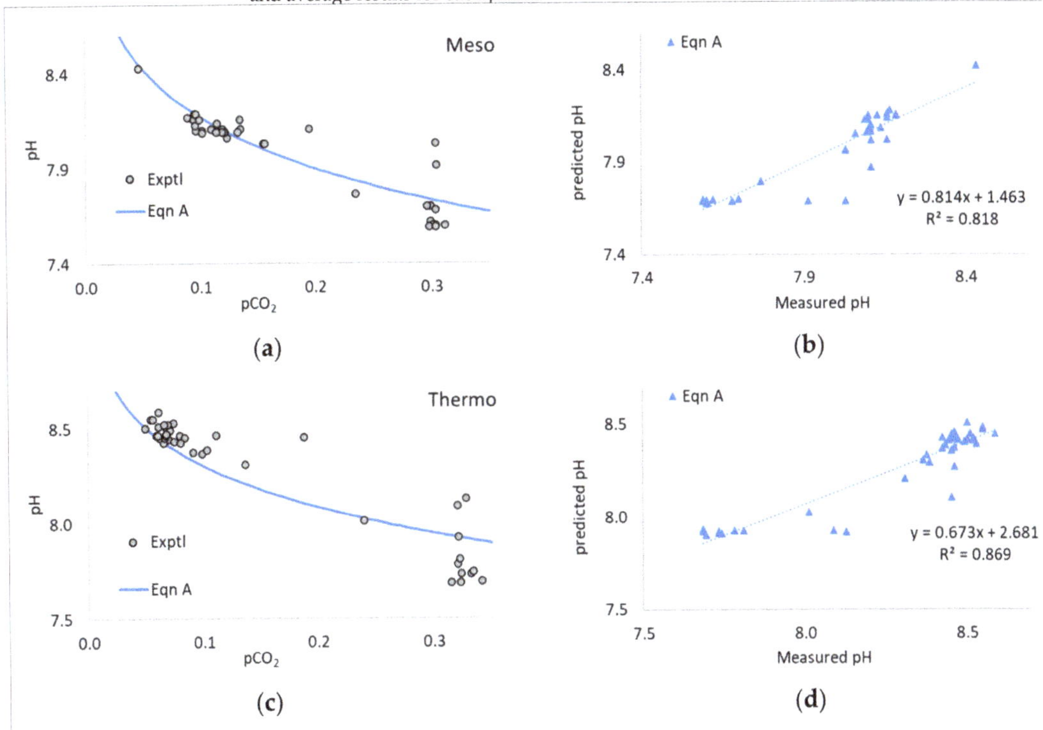

Figure A2. Experimental pCO_2 and pH values and predicted pH from Equation A for 2-stage mesophilic (**a**,**b**) and thermophilic (**c**,**d**) CO_2 biomethanisation of cattle manure associated with Bassani et al. [32]. See Table S15 for operating conditions and average results for each phase.

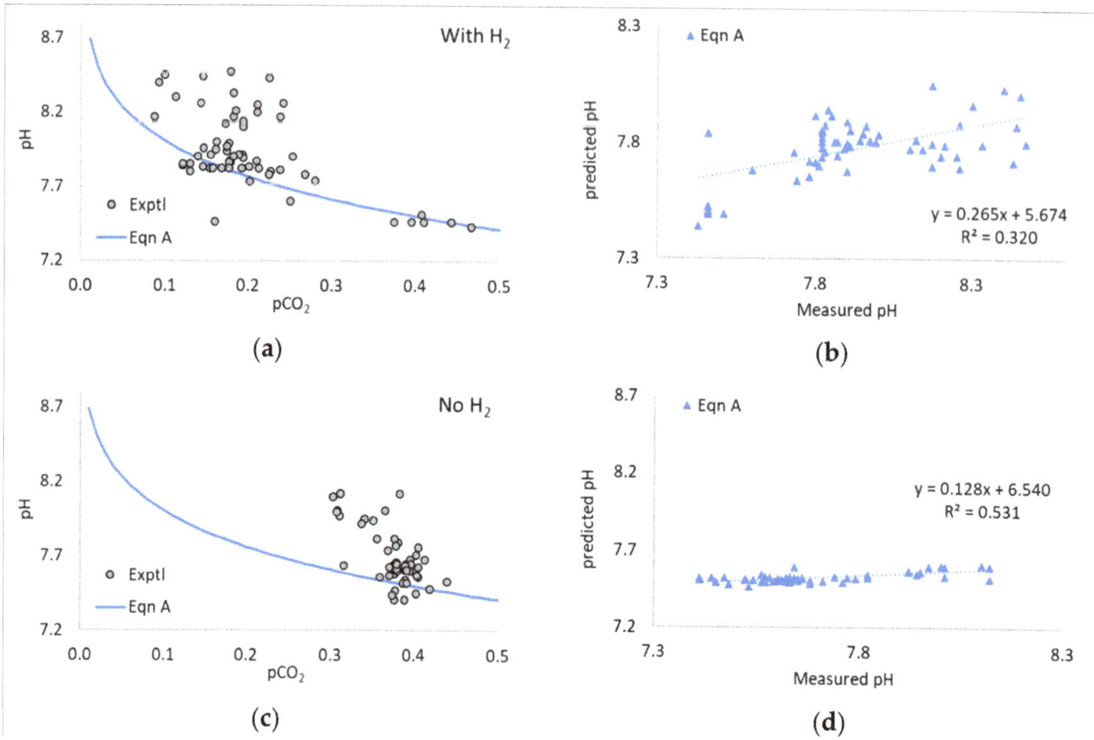

Figure A3. Experimental pCO_2 and pH values and predicted pH from Equation A for thermophilic UASB fed on potato starch wastewater with (**a**,**b**) and without (**c**,**d**) H_2 addition associated with Bassani et al. [53]. See Table S18 for operating conditions and average results for each phase.

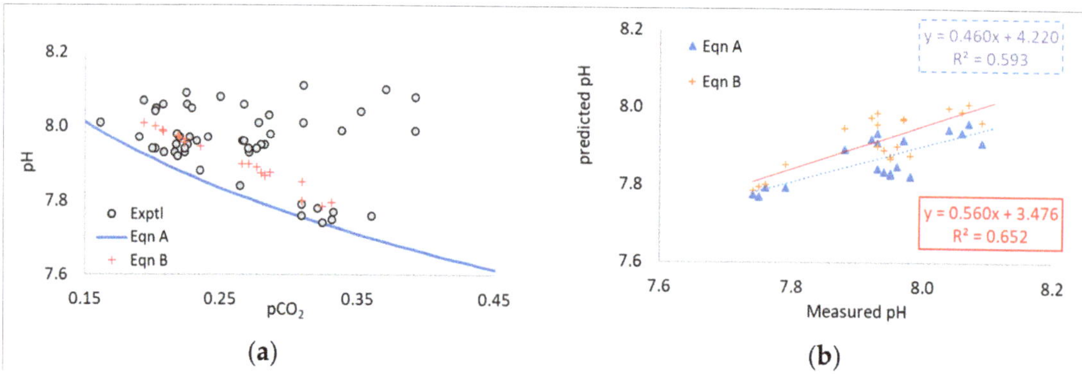

Figure A4. (**a**) Measured pCO_2 and measured and predicted pH values and (**b**) predicted pH values from Equations A and B for thermophilic CO_2 biomethanisation of cattle manure and whey using modelled TAN values associated with Lovato et al. [39]. See Table S7 for operating conditions and average results for each phase.

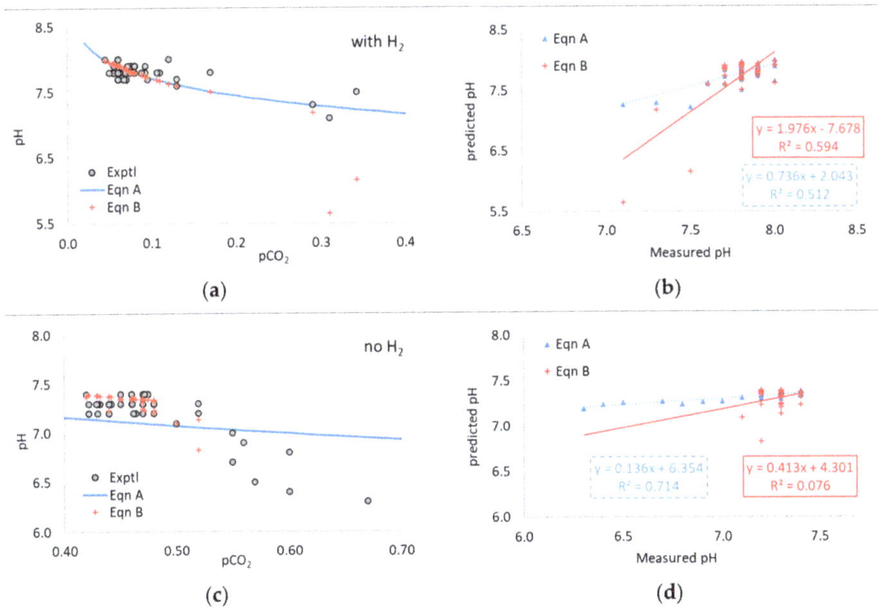

Figure A5. Experimental pCO_2 and pH values and predicted pH from Equations A and B for thermophilic CO_2 biomethanisation of cattle manure and whey with (**a**,**b**) and without (**c**,**d**) H_2 addition associated with Luo and Angelidaki [17]. See Table S3 for operating conditions and average results for each phase.

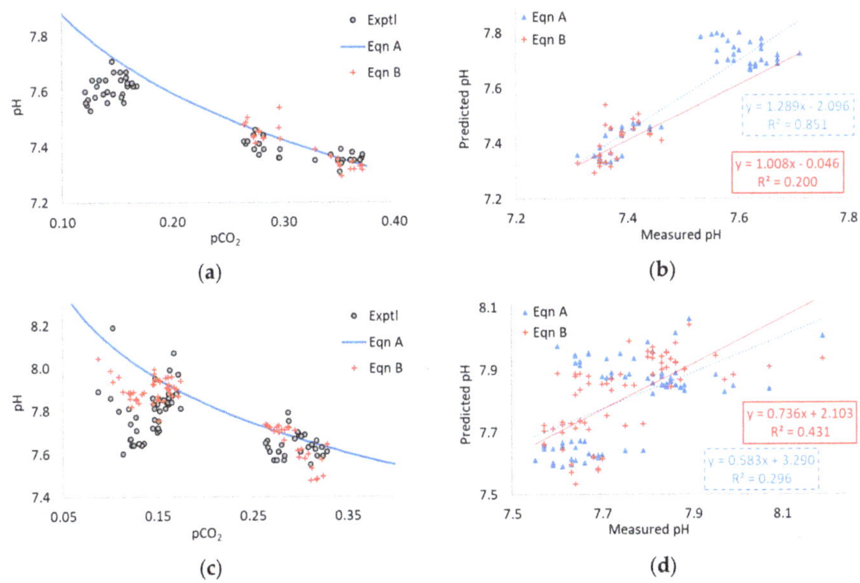

Figure A6. Experimental pCO_2 and pH values and predicted pH from Equation A for mesophilic (**a**,**b**) and thermophilic (**c**,**d**) CO_2 biomethanisation of pig manure associated with Zhu et al. [50–52]. See Table S4 for operating conditions and average results for each phase.

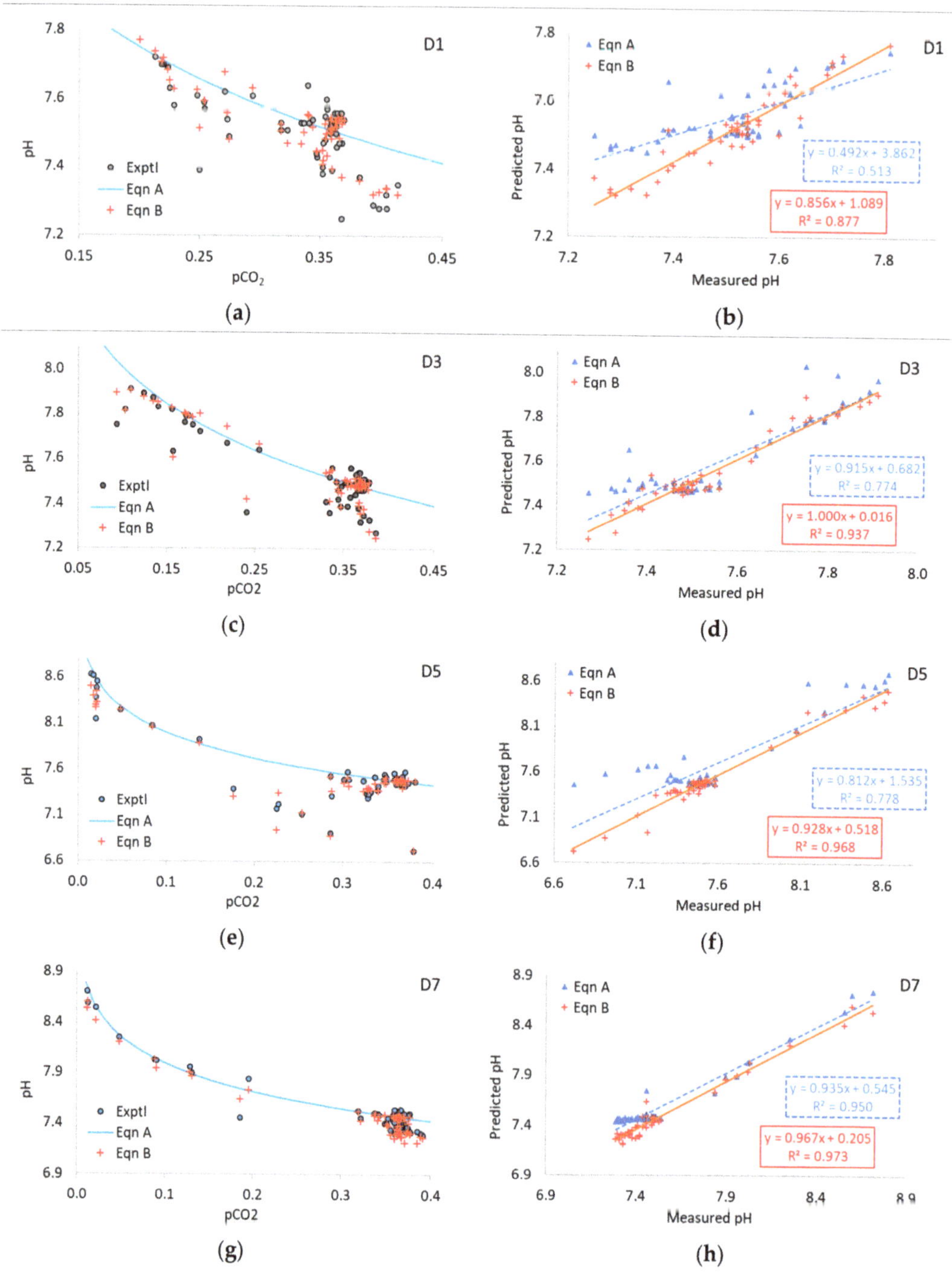

Figure A7. Experimental pCO_2 and pH values and predicted pH from Equations A and B in digesters D1 (**a**,**b**), D3 (**c**,**d**), D5 (**e**,**f**), and D7 (**g**,**h**) during mesophilic anaerobic digestion of sewage sludge at varying H_2 addition rates. For even-number digesters, see Figure 5 in the main text.

References

1. Zeikus, J. The biology of methanogenic bacteria. *Bacteriol. Rev.* **1977**, *41*, 514–541. [CrossRef] [PubMed]
2. Luo, G.; Angelidaki, I. Integrated biogas upgrading and hydrogen utilization in an anaerobic reactor containing enriched hydrogenotrophic methanogenic culture. *Biotechnol. Bioeng.* **2012**, *109*, 2729–2736. [CrossRef] [PubMed]
3. Rafrafi, Y.; Laguillaumie, L.; Dumas, C. Biological Methanation of H_2 and CO_2 with Mixed Cultures: Current Advances, Hurdles and Challenges. *Waste Biomass Valorization* **2020**, *12*, 5259–5282. [CrossRef]
4. Zavarkó, M.; Imre, A.R.; Pörzse, G.; Csedő, Z. Past, Present and Near Future: An Overview of Closed, Running and Planned Biomethanation Facilities in Europe. *Energies* **2021**, *14*, 5591. [CrossRef]
5. Aryal, N.; Kvist, T.; Ammam, F.; Pant, D.; Ottosen, L.D. An overview of microbial biogas enrichment. *Bioresour. Technol.* **2018**, *264*, 359–369. [CrossRef]
6. Tao, B.; Zhang, Y.; Heaven, S.; Banks, C.J. Predicting pH rise as a control measure for integration of CO_2 biomethanisation with anaerobic digestion. *Appl. Energy* **2020**, *277*, 115535. [CrossRef]
7. Andreides, D.; Stransky, D.; Bartackova, J.; Pokorna, D.; Zabranska, J. Syngas biomethanation in countercurrent flow trickle-bed reactor operated under different temperature conditions. *Renew. Energy* **2022**, *199*, 1329–1335. [CrossRef]
8. Ghofrani-Isfahani, P.; Tsapekos, P.; Peprah, M.; Kougias, P.; Zervas, A.; Zhu, X.; Yang, Z.; Jacobsen, C.S.; Angelidaki, I. Ex-Situ biogas upgrading in thermophilic trickle bed reactors packed with micro-porous packing materials. *Chemosphere* **2022**, *296*, 133987. [CrossRef]
9. Paniagua, S.; Lebrero, R.; Muñoz, R. Syngas biomethanation: Current state and future perspectives. *Bioresour. Technol.* **2022**, *358*, 127436. [CrossRef]
10. Dong, D.; Choi, O.K.; Lee, J.W. Influence of the continuous addition of zero valent iron (ZVI) and nano-scaled zero valent iron (nZVI) on the anaerobic biomethanation of carbon dioxide. *Chem. Eng. J.* **2022**, *430*, 132233. [CrossRef]
11. Geppert, F.; Liu, D.; van Eerten-Jansen, M.; Weidner, E.; Buisman, C.; Ter Heijne, A. Bioelectrochemical power-to-gas: State of the art and future perspectives. *Trends Biotechnol.* **2016**, *34*, 879–894. [CrossRef]
12. Bywater, A.; Heaven, S.; Zhang, Y.; Banks, C.J. Potential for Biomethanisation of CO_2 from Anaerobic Digestion of Organic Wastes in the United Kingdom. *Processes* **2022**, *10*, 1202. [CrossRef]
13. D'Silva, T.C.; Isha, A.; Chandra, R.; Vijay, V.K.; Subbarao, P.M.V.; Kumar, R.; Chaudhary, V.P.; Singh, H.; Khan, A.A.; Tyagi, V.K. Enhancing methane productionh in anaerobic digestion through hydrogen assisted pathways—A state-of-the-art review. *Renew. Sustain. Energy Rev.* **2021**, *151*, 111536. [CrossRef]
14. Lai, C.S.; McCulloch, M.D. Sizing of stand-alone solar PV and storage system with anaerobic digestion biogas power plants. *IEEE Trans. Ind. Electron.* **2016**, *64*, 2112–2121. [CrossRef]
15. Wang, P.; Wang, J.; Jin, R.; Li, G.; Zhou, M.; Xia, Q. Integrating biogas in regional energy systems to achieve near-zero carbon emissions. *Appl. Energy* **2022**, *322*, 119515. [CrossRef]
16. Angelidaki, I.; Treu, L.; Tsapekos, P.; Luo, G.; Campanaro, S.; Wenzel, H.; Kougias, P.G. Biogas upgrading and utilization: Current status and perspectives. *Biotechnol. Adv.* **2018**, *36*, 452–466. [CrossRef]
17. Luo, G.; Angelidaki, I. Co-digestion of manure and whey for in situ biogas upgrading by the addition of H_2: Process performance and microbial insights. *Appl. Microbiol. Biotechnol.* **2013**, *97*, 1373–1381. [CrossRef]
18. Luo, G.; Johansson, S.; Boe, K.; Xie, L.; Zhou, Q.; Angelidaki, I. Simultaneous hydrogen utilization and in situ biogas upgrading in an anaerobic reactor. *Biotechnol. Bioeng.* **2012**, *109*, 1088–1094. [CrossRef]
19. Drosg, B. *Process Monitoring in Biogas Plants*; IEA Bioenergy: Paris, France, 2013.
20. Gaida, D.; Wolf, C.; Bongards, M. Feed control of anaerobic digestion processes for renewable energy production: A review. *Renew. Sustain. Energy Rev.* **2017**, *68*, 869–875. [CrossRef]
21. Nguyen, D.; Gadhamshetty, V.; Nitayavardhana, S.; Khanal, S.K. Automatic process control in anaerobic digestion technology: A critical review. *Bioresour. Technol.* **2015**, *193*, 513–522. [CrossRef]
22. Wade, M.J. Not just numbers: Mathematical modelling and its contribution to anaerobic digestion processes. *Processes* **2020**, *8*, 888. [CrossRef]
23. Tao, B.; Alessi, A.M.; Zhang, Y.; Chong, J.P.; Heaven, S.; Banks, C.J. Simultaneous biomethanisation of endogenous and imported CO_2 in organically loaded anaerobic digesters. *Appl. Energy* **2019**, *247*, 670–681. [CrossRef]
24. APHA. *Standard Methods for the Examination of Water and Wastewater*; American Public Health Association: Washington, DC, USA, 2005.
25. Ripley, L.E.; Boyle, W.C.; Converse, J.C. Improved alkalimetric monitoring for anaerobic digestion of high-strength wastes. *J. (Water Pollut. Control Fed.)* **1986**, *58*, 406–411.
26. Walker, M.; Zhang, Y.; Heaven, S.; Banks, C. Potential errors in the quantitative evaluation of biogas production in anaerobic digestion processes. *Bioresour. Technol.* **2009**, *100*, 6339–6346. [CrossRef] [PubMed]
27. Wu, D.; Li, L.; Zhao, X.; Peng, Y.; Yang, P.; Peng, X. Anaerobic digestion: A review on process monitoring. *Renew. Sustain. Energy Rev.* **2019**, *103*, 1–12. [CrossRef]
28. Agneessens, L.M.; Ottosen, L.D.M.; Andersen, M.; Olesen, C.B.; Feilberg, A.; Kofoed, M.V.W. Parameters affecting acetate concentrations during in-situ biological hydrogen methanation. *Bioresour. Technol.* **2018**, *258*, 33–40. [CrossRef] [PubMed]
29. Agneessens, L.M.; Ottosen, L.D.M.; Voigt, N.V.; Nielsen, J.L.; de Jonge, N.; Fischer, C.H.; Kofoed, M.V.W. In-situ biogas upgrading with pulse H_2 additions: The relevance of methanogen adaption and inorganic carbon level. *Bioresour. Technol.* **2017**, *233*, 256–263. [CrossRef] [PubMed]
30. Alfaro, N.; Fdz-Polanco, M.; Fdz-Polanco, F.; Díaz, I. H_2 addition through a submerged membrane for in-situ biogas upgrading in the anaerobic digestion of sewage sludge. *Bioresour. Technol.* **2019**, *280*, 1–8. [CrossRef]

31. Andreides, D.; Pokorna, D.; Zabranska, J. Assessing the syngas biomethanation in anaerobic sludge digestion under different syngas loading rates and homogenisation. *Fuel* **2022**, *320*, 123929. [CrossRef]
32. Bassani, I.; Kougias, P.G.; Treu, L.; Angelidaki, I. Biogas upgrading via hydrogenotrophic methanogenesis in two-stage continuous stirred tank reactors at mesophilic and thermophilic conditions. *Environ. Sci. Technol.* **2015**, *49*, 12585–12593. [CrossRef]
33. Corbellini, V.; Catenacci, A.; Malpei, F. Hydrogenotrophic biogas upgrading integrated into WWTPs: Enrichment strategy. *Water Sci. Technol.* **2019**, *79*, 759–770. [CrossRef] [PubMed]
34. Corbellini, V.; Feng, C.; Bellucci, M.; Catenacci, A.; Stella, T.; Espinoza-Tofalos, A.; Malpei, F. Performance Analysis and Microbial Community Evolution of In Situ Biological Biogas Upgrading with Increasing H_2/CO_2 Ratio. *Archaea* **2021**, *2021*, 8894455. [CrossRef] [PubMed]
35. Díaz, I.; Fdz-Polanco, F.; Mutsvene, B.; Fdz-Polanco, M. Effect of operating pressure on direct biomethane production from carbon dioxide and exogenous hydrogen in the anaerobic digestion of sewage sludge. *Appl. Energy* **2020**, *280*, 115915. [CrossRef]
36. Khan, A.; Akbar, S.; Okonkwo, V.; Smith, C.; Khan, S.; Shah, A.A.; Adnan, F.; Ijaz, U.Z.; Ahmed, S.; Badshah, M. Enrichment of the hydrogenotrophic methanogens for, in-situ biogas up-gradation by recirculation of gases and supply of hydrogen in methanogenic reactor. *Bioresour. Technol.* **2022**, *345*, 126219. [CrossRef]
37. Kim, S.; Mostafa, A.; Im, S.; Lee, M.-K.; Kang, S.; Na, J.-G.; Kim, D.-H. Production of high-calorific biogas from food waste by integrating two approaches: Autogenerative high-pressure and hydrogen injection. *Water Res.* **2021**, *194*, 116920. [CrossRef]
38. Lebranchu, A.; Blanchard, F.; Fick, M.; Pacaud, S.; Olmos, E.; Delaunay, S. Pilot-scale biomethanation of cattle manure using dense membranes. *Bioresour. Technol.* **2019**, *284*, 430–436. [CrossRef]
39. Lovato, G.; Alvarado-Morales, M.; Kovalovszki, A.; Peprah, M.; Kougias, P.G.; Rodrigues, J.A.D.; Angelidaki, I. In-situ biogas upgrading process: Modeling and simulations aspects. *Bioresour. Technol.* **2017**, *245*, 332–341. [CrossRef]
40. Luo, G.; Angelidaki, I. Hollow fiber membrane based H_2 diffusion for efficient in situ biogas upgrading in an anaerobic reactor. *Appl. Microbiol. Biotechnol.* **2013**, *97*, 3739–3744. [CrossRef]
41. Luo, G.; Wang, W.; Angelidaki, I. Anaerobic digestion for simultaneous sewage sludge treatment and CO biomethanation: Process performance and microbial ecology. *Environ. Sci. Technol.* **2013**, *47*, 10685–10693. [CrossRef]
42. Treu, L.; Kougias, P.; de Diego-Díaz, B.; Campanaro, S.; Bassani, I.; Fernández-Rodríguez, J.; Angelidaki, I. Two-year microbial adaptation during hydrogen-mediated biogas upgrading process in a serial reactor configuration. *Bioresour. Technol.* **2018**, *264*, 140–147. [CrossRef]
43. Treu, L.; Tsapekos, P.; Peprah, M.; Campanaro, S.; Giacomini, A.; Corich, V.; Kougias, P.G.; Angelidaki, I. Microbial profiling during anaerobic digestion of cheese whey in reactors operated at different conditions. *Bioresour. Technol.* **2019**, *275*, 375–385. [CrossRef] [PubMed]
44. Voelklein, M.; Rusmanis, D.; Murphy, J. Biological methanation: Strategies for in-situ and ex-situ upgrading in anaerobic digestion. *Appl. Energy* **2019**, *235*, 1061–1071. [CrossRef]
45. Wahid, R.; Horn, S.J. Impact of operational conditions on methane yield and microbial community composition during biological methanation in in situ and hybrid reactor systems. *Biotechnol. Biofuels* **2021**, *14*, 1–15. [CrossRef] [PubMed]
46. Wahid, R.; Horn, S.J. The effect of mixing rate and gas recirculation on biological CO_2 methanation in two-stage CSTR systems. *Biomass Bioenergy* **2021**, *144*, 105918. [CrossRef]
47. Wahid, R.; Mulat, D.G.; Gaby, J.C.; Horn, S.J. Effects of H_2:CO_2 ratio and H_2 supply fluctuation on methane content and microbial community composition during in-situ biological biogas upgrading. *Biotechnol. Biofuels* **2019**, *12*, 1–15. [CrossRef]
48. Wang, W.; Xie, L.; Luo, G.; Zhou, Q.; Angelidaki, I. Performance and microbial community analysis of the anaerobic reactor with coke oven gas biomethanation and in situ biogas upgrading. *Bioresour. Technol.* **2013**, *146*, 234–239. [CrossRef]
49. Yang, Z.; Liu, Y.; Zhang, J.; Mao, K.; Kurbonova, M.; Liu, G.; Zhang, R.; Wang, W. Improvement of biofuel recovery from food waste by integration of anaerobic digestion, digestate pyrolysis and syngas biomethanation under mesophilic and thermophilic conditions. *J. Clean. Prod.* **2020**, *256*, 120594. [CrossRef]
50. Zhu, X.; Cao, Q.; Chen, Y.; Sun, X.; Liu, X.; Li, D. Effects of mixing and sodium formate on thermophilic in-situ biogas upgrading by H_2 addition. *J. Clean. Prod.* **2019**, *216*, 373–381. [CrossRef]
51. Zhu, X.; Chen, L.; Chen, Y.; Cao, Q.; Liu, X.; Li, D. Differences of methanogenesis between mesophilic and thermophilic in situ biogas-upgrading systems by hydrogen addition. *J. Ind. Microbiol. Biotechnol.* **2019**, *46*, 1569–1581. [CrossRef]
52. Zhu, X.; Chen, L.; Chen, Y.; Cao, Q.; Liu, X.; Li, D. Effect of H_2 addition on the microbial community structure of a mesophilic anaerobic digestion system. *Energy* **2020**, *198*, 117368. [CrossRef]
53. Bassani, I.; Kougias, P.G.; Angelidaki, I. In-situ biogas upgrading in thermophilic granular UASB reactor: Key factors affecting the hydrogen mass transfer rate. *Bioresour. Technol.* **2016**, *221*, 485–491. [CrossRef] [PubMed]
54. Jing, Y.; Campanaro, S.; Kougias, P.; Treu, L.; Angelidaki, I.; Zhang, S.; Luo, G. Anaerobic granular sludge for simultaneous biomethanation of synthetic wastewater and CO with focus on the identification of CO-converting microorganisms. *Water Res.* **2017**, *126*, 19–28. [CrossRef] [PubMed]
55. Deschamps, L.; Imatoukene, N.; Lemaire, J.; Mounkaila, M.; Filali, R.; Lopez, M.; Theoleyre, M.-A. In-situ biogas upgrading by bio-methanation with an innovative membrane bioreactor combining sludge filtration and H_2 injection. *Bioresour. Technol.* **2021**, *337*, 125444. [CrossRef]
56. Hafuka, A.; Fujino, S.; Kimura, K.; Oshita, K.; Konakahara, N.; Takahashi, S. In-situ biogas upgrading with H_2 addition in an anaerobic membrane bioreactor (AnMBR) digesting waste activated sludge. *Sci. Total Environ.* **2022**, *828*, 154573. [CrossRef] [PubMed]

57. Illi, L.; Lecker, B.; Lemmer, A.; Müller, J.; Oechsner, H. Biological methanation of injected hydrogen in a two-stage anaerobic digestion process. *Bioresour. Technol.* **2021**, *333*, 125126. [CrossRef]
58. Thapa, A.; Park, J.-G.; Yang, H.-M.; Jun, H.-B. In-situ biogas upgrading in an anaerobic trickling filter bed reactor treating a thermal post-treated digestate. *J. Environ. Chem. Eng.* **2021**, *9*, 106780. [CrossRef]
59. Corbellini, V.; Kougias, P.G.; Treu, L.; Bassani, I.; Malpei, F.; Angelidaki, I. Hybrid biogas upgrading in a two-stage thermophilic reactor. *Energy Convers. Manag.* **2018**, *168*, 1–10. [CrossRef]
60. Andreides, D.; Quispe, J.I.B.; Bartackova, J.; Pokorna, D.; Zabranska, J. A novel two-stage process for biological conversion of syngas to biomethane. *Bioresour. Technol.* **2021**, *327*, 124811. [CrossRef]
61. Scholten, J.C.; Conrad, R. Energetics of syntrophic propionate oxidation in defined batch and chemostat cocultures. *Appl. Environ. Microbiol.* **2000**, *66*, 2934–2942. [CrossRef]
62. Fontana, A.; Kougias, P.G.; Treu, L.; Kovalovszki, A.; Valle, G.; Cappa, F.; Morelli, L.; Angelidaki, I.; Campanaro, S. Microbial activity response to hydrogen injection in thermophilic anaerobic digesters revealed by genome-centric metatranscriptomics. *Microbiome* **2018**, *6*, 194. [CrossRef]
63. Batstone, D.J.; Keller, J.; Angelidaki, I.; Kalyuzhnyi, S.; Pavlostathis, S.; Rozzi, A.; Sanders, W.; Siegrist, H.; Vavilin, V. The IWA anaerobic digestion model no 1 (ADM1). *Water Sci. Technol.* **2002**, *45*, 65–73. [CrossRef] [PubMed]
64. Heaven, S.; Zhang, Y.; Bywater, A.; Banks, C. Dataset for 'Potential for Biomethanisation of CO2 from Anaerobic Digestion of Organic Wastes in the UK'. 2022. Available online: https://eprints.soton.ac.uk/467644/1/processes_10_01202_1_.pdf (accessed on 6 June 2022).
65. Palù, M.; Peprah, M.; Tsapekos, P.; Kougias, P.; Campanaro, S.; Angelidaki, I.; Treu, L. In-situ biogas upgrading assisted by bioaugmentation with hydrogenotrophic methanogens during mesophilic and thermophilic co-digestion. *Bioresour. Technol.* **2022**, *348*, 126754. [CrossRef] [PubMed]

processes

Communication

Toward the Transition of Agricultural Anaerobic Digesters into Multiproduct Biorefineries

David Bolzonella *, **Davide Bertasini, Riccardo Lo Coco, Miriam Menini, Fabio Rizzioli, Anna Zuliani, Federico Battista, Nicola Frison, Aleksandra Jelic and Giovanna Pesante**

Department of Biotechnology, University of Verona, Strada Le Grazie 15, 37134 Verona, Italy
* Correspondence: david.bolzonella@univr.it; Tel.:+ 39-045-8027965

Abstract: Anaerobic digestion allows for the proper management of agro-waste, including manure. Currently, more than 18,000 anaerobic digestion plants are under operation in EU, 80% of which are employed in the rural context. Tariff schemes for power generation from biogas produced during anaerobic digestion of agricultural feedstocks in Germany, Italy and Austria are coming to an end and new approaches are needed to exploit the existing infrastructures. Digesters in the rural context can be implemented and modified to be transformed into sustainable multi-feedstock and multi-purpose biorefineries for the production of energy, nutrients, proteins, bio-chemicals such as carboxylic acids, polyesters and proteins. This paper describes how the transition of agricultural anaerobic digesters into multi-products biorefineries can be achieved and what are the potential benefits originating from the application of a pilot scale platform able to treat cow manure and other crop residues while producing volatile fatty acids, polyhydroxyalkanoates, microbial protein material, hydrogen, methane and a concentrated liquid stream rich in nitrogen, potassium and phosphorus.

Keywords: agricultural waste; anaerobic digestion; biobased products; biogas; nutrients; polyhydroxyalkanoates; proteins; volatile fatty acids

Citation: Bolzonella, D.; Bertasini, D.; Lo Coco, R.; Menini, M.; Rizzioli, F.; Zuliani, A.; Battista, F.; Frison, N.; Jelic, A.; Pesante, G. Toward the Transition of Agricultural Anaerobic Digesters into Multiproduct Biorefineries. *Processes* **2023**, *11*, 415. https://doi.org/10.3390/pr11020415

Academic Editors: Sonia Heaven, Sigrid Kusch-Brandt and Charles Banks

Received: 21 December 2022
Revised: 24 January 2023
Accepted: 25 January 2023
Published: 30 January 2023

1. Introduction

The European Union is an international leader in the bioeconomy sector: Germany, France and Italy are the greatest global producers of crops, meat, fish and processed food, which determine the production of large amounts of residual organic streams, mainly livestock manure. The total European production of livestock effluents accounts for 1.4 billion tonnes per year, with France, Germany and Italy producing 276, 197 and 103 million tonnes per year, respectively [1]. The production of agricultural residues derived from crop cultivation accounted for 21 million tonnes in 2020 in the European Union [2]. These streams, collectively named agro-waste, are a rich untapped source of carbon and nutrients and should not be disposed of massively but rather properly valorized within the circular economy concept with the potential to support human well-being and saving resources [3].

Anaerobic Digestion (AD) is a robust technology for the management and valorization of agro-waste in the rural context: in fact, AD enables the bioconversion of the organic matter present in manure, residual crops, and other residual organic streams of the food processing chain while recovering biogas for power or biomethane production [4,5] and a digestate containing considerable quantities of valuable nutrients [6].

Because of its benefits and the incentives schemes applied in different Countries, AD is largely diffused in Europe [3]: more than 18,000 plants with an installed capacity of 8000 MWel are globally under operation in Europe according to the European Biogas Association [7]. In 2021, the EU produced 3 billion m^3 of biomethane [8]. However, the EU production should increase to 35 billion m^3 by 2030, to be able to comply to the new EU communication COM/2022/108, implementing the "RE. PowerEU" plan.

As reported above, besides biogas, AD plants generate also digestate, a sort of renewable fertilizer [3,6]. The use of digestate provides macro (nitrogen, phosphorus, potassium, etc.) and micro (cobalt, selenium, etc.) nutrients to the soil. Moreover, it determines the supplementation of stable carbon to the fields, thus increasing the carbon sink capability of soils [9].

In the last 20 years, the number of AD plants has increased considerably in Western and Central Europe because of favorable incentive schemes dedicated to the supporting of power and biomethane generation [10]. However, the incentive schemes for power generation in Germany, Italy and Austria are rapidly coming to an end, therefore the use and exploitation of rural anaerobic digesters should be reconsidered.

In perspective, an obvious upgrade of this technology is the transformation of anaerobic digesters into a multi-feedstock, multipurpose and multiproduct biorefinery able to valorize agro-waste into high added value biobased products such as nutrients and chemicals, thus creating social and economic benefits at local level [11–13].

In recent years, several studies showed the possibility to generate biofuels such as hydrogen and methane, biobased chemicals such as polyhydroxyalkanoates, produce microalgae, bacteria or yeast, convert CO_2 and recover nutrients from organic wastes.

It is therefore possible to imagine anaerobic digestion at the center of a future biorefinery approach where agro-wastes and food processing waste are converted into high added value biobased products other than biofuels. This new bioeconomy approach is crucial for the rural renaissance of Europe.

In this framework, the Horizon Europe project AgriLoop will study the upcycling of agro- and food-processing waste and byproducts to high value biobased products. A pilot scale platform biorefinery will be operated where cow manure, crops silage and other agro-waste will be transformed into biofuels, carboxylic acids, and polyesters, while nutrients will be recovered in concentrated forms.

In this paper, we will provide evidence of this global approach, here implemented in a pilot scale biorefinery able to convert agro-waste into valuable products working in a real environment. For the purpose of this study, agricultural wastes, by-products and co-products are defined as plant or animal residues that are not (or not further processed into) food or feed. This large amount of undervalued biomass is further called agro-waste as it is intended to be fully integrated as resource in optimised cascading biorefinery and therefore revoked as waste, to become multiple levers of the transition away from a fossil-based carbon-intensive economy.

2. Materials and Methods

The general approach proposed in this study envisages that livestock effluents, crop silage and other agro-waste are transformed into biofuels and biobased products such as volatile fatty acids (VFAs) and polyhydroxyalkanoates (PHAs), while nutrients and microbial proteins are recovered. All these activities will be carried out treating real substrates and in a real environment: the pilot scale platform is placed in "La Torre" farm, Isola della Scala, Verona, in the northeast of Italy.

This biorefinery platform will allow for the treatment of different types of feedstocks and their bioconversion into biobased products.

2.1. Pilot Plant Description

The biorefinery plant is an upgrade and modification of the one presented by Righetti and colleagues [13], first developed in the framework of the project NoAW, No Agro-Waste-innovative approaches to turn agricultural waste into ecological and economic assets (http://noaw2020.eu/), accessed on 20 December 2022, and it can be divided into the following main areas:

- agro-waste pre-treatment and preparation;
- acidogenic fermentation where volatile fatty acid production and hydrogen are produced;

- mixed microbial biomass selection and PHA storing;
- PHA extraction and purification;
- anaerobic digestion of the solid part of the fermentation effluent;
- nutrient recovery;
- microbial protein production.

The overall approach of the biorefinery and the obtainable products are schematically shown in Figure 1.

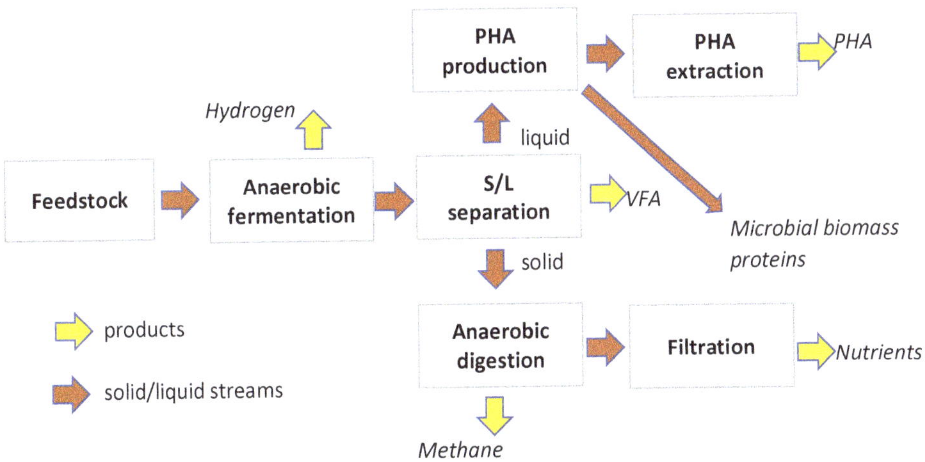

Figure 1. Scheme of the unit operations involved in the studied pilot scale process.

The 2 m^3 storage tank for feedstock mixing and preparation is equipped with a weighing system for the control of the amount of material fed into the system. The tank has a mixer to homogenize the substrates. Mixed material is then ground by means of a special centrifuge pump model TM451 (Vidotto Dissipatori, Eraclea, Italy, Italy) and pumped to the anaerobic acidogenic reactor. The fermentation unit is a 5 m^3 unit with an adjustable volume. The fermenter is temperature controlled with warm water and can operate both in mesophilic and thermophilic conditions.

The operational conditions applied to the reactor are shown in Table 1.

Table 1. Operational conditions of the anaerobic acidogenic fermenter and anaerobic digester.

Parameter	Anaerobic Fermenter	Anaerobic Digester
Hydraulic retention time (HRT)	3–5 days	25–35 days
Organic loading rate (OLR)	15–25 kgCOD/m^3 day	2.5–4.5 kgCOD/m^3 day

The fermentation process operates in mesophilic conditions with an average organic loading rate in the range of 15–25 kgCOD/m^3 d, a desired OLR of approximately 18 kgCOD/m^3 d and a hydraulic retention time of 3 to 5 days to improve the conversion of particulate COD into short chain VFAs to be fed to the PHA bioreactor. The operational pH of the anaerobic fermenter is determined by the characteristics of fed substrates and it is typically around 5.5. Operational conditions such as OLR can be set in order to tune the abundance of different short chain VFAs in the liquid phase.

The fermenter effluent is treated in a horizontal screw press unit (Sepcom, WAM Group, Modena, Italy) for the solid/liquid separation. The screw press treatment capacity is 5 m^3/h.

The liquid fraction obtained through solid/liquid separation is then pre-filtered at 500 micron and then micro-filtered in a rotating ceramic unit. The permeate rich in short chain VFAs is then sent to the mixed microbial biomass selection unit.

The first bioreactor is a 2 m^3 volume system for the selection of PHA-accumulating microorganism, while the second reactor, 1 m^3 in volume, is dedicated to the accumulation of PHAs.

Both bioreactors can operate in batch or continuous mode.

The sludge rich in PHAs produced in the second bioreactor, after settling and concentration, is quenched with sulfuric acid and sent to a chemical extraction unit.

The separated solid fraction obtained from the screw press is sent to a mesophilic anaerobic digester designed to operate at an optimal organic loading rate of 4 $kgVS/m^3$ d and with a hydraulic retention time of approximately 30 days. The anaerobic digester has a working volume of 1 m^3. The operational pH of the anaerobic digester is typically self-buffered around 8. Table 1 reports the expected ranges of operation for both OLR and HRT.

The anaerobic digestion effluent undergoes to solid/liquid separation and the liquid stream is sent to a screening and filtration unit with mesh of 250 μm and then to a double stage filtration unit where ultra-filtration is followed by reverse osmosis for nutrient recovery and concentration. The permeate (clean water) is used to dilute the treated feedstock in the initial storage tank. The ultra-filtration unit has a ceramic disc filter with a porosity of 0.22 μm, with a maximum flow of 10 L/h. The reverse osmosis unit has a membrane porosity of less than 1 nm, with a maximum flow of 200 L/h.

2.2. Analysis

In order to carry out the mass balances for the system and for each operation unit, the feedstocks and the effluents of the reactors were monitored once or more times per week through the analysis of total and volatile solids (TS and TVS), chemical oxygen demand (COD), Total Kjeldahl Nitrogen (TKN) and total phosphorus (TP) concentrations. The process stability parameters for the anaerobic fermenter and digester, namely pH, VFAs content and speciation, total and partial alkalinity and ammonia (NH_3-N) were regularly checked daily. All the analyses, except for the VFAs, were carried out in accordance with the standard methods for water and wastewater analysis [14].

The biogas flowrate was measured by a volumetric counter (Ritter Gmbh, Schwabmünchen, Germany), while biogas composition in terms of CH_4 and CO_2 concentrations was measured using a portable gas monitor (GAS Counter BIOGAS 5000), equipped with a hydraulic guard to remove H_2S. Hydrogen content was determined by a gas chromatograph (GC Agilent Technology 6890 N) equipped with the column HP-PLOT MOLESIEVE, 30 × 0.53 mm ID × 25 μm film, using a thermal conductivity detector and an argon as gas carrier.

2.3. Mass Balance Equation and Yields of the Pilot Plant

The mass balance of the pilot scale plant is based on the chemical characteristics of the influent and effluent streams and the relative flowrates. The general equation for mass balances for a given chemical compound considers influent and effluent mass, generated and consumed mass and a term for accumulation according to the general formula:

$$M\ influent - M\ effluent + M\ generated - M\ consumed = M\ accumulated\ in\ the\ system. \quad (1)$$

The general equation can be then applied considering flowrate and concentration for any chemical compound according to the following equation:

$$V \cdot \frac{dC}{dt} = Q_{in} \cdot C_{in} - Q_{out} \cdot C_{out} - V \cdot R_c + V \cdot R_p, \quad (2)$$

where

Q_{in} is influent flowrate, m^3 per day;

C_{in} is influent concentration of a given compound, kg per m^3;
C_{out} is effluent concentration of a given compound, kg per m^3;
Q_{out} is effluent flowrate, m^3 per day;
C is compound concentration in the bioreactor, kg per m^3;
V is volume of the bioreactor, m^3;
Rp is specific production rate in the system, kg per m^3 per day;
Rc is specific consumption rate in the system, kg per m^3 per day.

In steady state conditions, the concentration of a given component, C, is constant and Equation (2) equals zero.

Yields for a given product are determined considering the quantity of product per time unit, typically per day, over the quantity of substrate entering the system at the same time.

The mass balance and yields of the system discussed below are determined according to the description reported here.

3. Expected Results and Discussion

As described above, existing AD plants can be converted into multipurpose biorefineries able to produce a portfolio of different biobased products.

In this study, we will consider the conversion of a mix of cow manure, straw and residual crops into different bioproducts:

- hydrogen and methane,
- short chain VFAs,
- PHAs,
- nutrients,
- microbial proteins.

The scheme of the general approach is shown in Figure 1.

Specific details for the different biobased products are illustrated and discussed below.

3.1. *Biofuels: Hydrogen and Methane Production*

Hydrogen and methane have gained particular interest in the scientific community in recent years due to the need of decarbonisation to fight climate change and to the increasing price of natural gas. This has led to the further development and implementation of the two-stage anaerobic digestion technology, a biological process which allows for the simultaneous production of hydrogen and methane from organic residues. The two anaerobic reactors treating agro-waste are dedicated to dark fermentation and methanogenesis, respectively. The first step consists in hydrogen and carbon dioxide production by the acidogenic microorganisms, through the hydrolysis of complex organic compounds into soluble macro-molecules and then into short chain VFAs. During the methanogenic step, in the second reactor, VFAs are degraded to carbon dioxide and methane, while part of the carbon dioxide is bio-converted to methane through its reaction with hydrogen [15]. The resulting gaseous blend can be upgraded to eliminate undesired compounds such as hydrogen sulphide, ammonia and carbon dioxide [16] in order to obtain the hydrogen–methane blend bio-hythane.

In the AgriLoop project, we will operate the two reactors as described in Table 1 and we expect yields of approximately 20 L H_2 and 180 L CH_4 per kgVS treated. Besides the gaseous effluents, the process produces digestate, a mix of solid and liquid parts.

3.2. *Volatile Fatty Acids Production*

VFAs are short chain carboxylic acids from acetate (C1) to caproate (C6), usually derived from fossil sources. The biological production of VFAs is receiving great attention to favour the transition from a linear to a circular economy model. Through anaerobic fermentation, agro-waste can be converted into valuable VFAs with high economic value, ranging from 800 to 2500 €/tonne [17]. In our project, agro-waste will be converted into VFAs for the synthesis of PHAs and single cell proteins (SCP, see Section 3.5).

The VFAs yields depend on the chemical and physical nature of the organic substrates: in this case, the mixed material will be composed of both hard-to-degrade organic material (straw) and easily biodegradable cellulose and hemi-cellulose, therefore expected yields are in the range of 0.2–0.3 gVFA/gVS. These values are in line with literature data: agro-food wastes rich in simplest compounds, for example, winery waste such as lees, having high content of glucose, fructose and ethanol, have typical VFAs yields of 0.5–0.7 gVFA/gCOD; olive pomace, rich in lipids and recalcitrant lignocellulosic compounds, have lower yields [18]. In order to increase the process yield, lignocellulosic materials represented by rice, corn, wheat straws and by-products from energy crops are usually pre-treated under steam explosion to favour the degradation of the long and complex carbon chains into smaller organic fractions, which can be easily converted by the mixed microbial culture [19], increasing the VFAs yield to 0.45–0.55 gVFA/gCOD [20].

In this specific study, chemical pre-treatments to increase bioconversion of recalcitrant organic matter will be applied; in particular, the use of soda will be tested.

In particular, crop residues and manure with straw will be added with soda at 36% to match up a final concentration in the pre-treatment vessel of 1% and digested overnight at environment temperature.

In terms of concentration, VFAs in the fermentation broth are usually in the range of 15–40 g/L [21] depending on both bioconversion capability and dilution. To be exploitable for industrial purpose, VFAs need to be separated from the other organic compounds and concentrated in order to have a minimal concentration level of 100 g/L [22].

Several techniques such as adsorption on solid matrix [23], distillation and evaporation [22], extraction by solvent [24], electrodialysis [25] and pressure-driven membrane processes [26] have been implemented for VFAs recovery. Traditional recovery techniques by solvents and distillation are known to be high in cost and energy, but are still largely applied in order to separate VFA from fermentation medium. It was demonstrated that the maximum VFA recovery by this technique falls in the range of 61–98% [27]. Filtration techniques, such as nanofiltration and reverse osmosis, are becoming attractive alternatives to traditional recovery techniques, being capable of recovering over 90% of the VFAs content from the fermentation medium [28].

VFA recovery has other advantages if applied directly on the liquid phase of the fermenter: it can enhance anaerobic fermentation because (i) it shifts the balance toward VFAs biosynthesis, (ii) it alleviates product-induced inhibition, and (iii) it maintains stable alkalinity levels, allowing further waste degradation [29].

In order to obtain a functional cascade of integrated membrane separation processes that allow the recovery, purification and optimal concentration of VFAs from waste-derived AD effluents, it is essential to have pre-treatment processes able to deeply remove suspended solids and colloids. To ensure the long-term application of a membrane filtration system, it is necessary to take into account the factors that influence filtration operations, such as pH and solids content (TS and TVS) [28]. The solids content is the limiting factor for the application of specific membranes, since it can cause damage to the membrane itself. Consequently, pre-treatments such as solid/liquid separations with a centrifuge or vibrating screen must be applied [17].

In the AgriLoop project, because of the bio-recalcitrant nature of treated agro-waste, mainly due to straw, expected yields are low, in the range of 0.1–0.2 kgVFA per kg of VS fed to the anaerobic fermenter, while the fermentation broth will show VFA concentrations in the range of 10–15 g/L. Several recovery techniques will be tested at lab scale with a preference for membrane contactors.

3.3. *Polyhydroxyalkanoates Production*

Polyhydroxyalkanoates are a family of thermoplastic polyesters of hydroxyacid (HA) monomers connected by an ester bond [18]. They can be produced from different renewable sources by bacterial fermentation, as a form of intracellular carbon and energy storage. The process of PHA production by mixed microbial culture applied in this biorefinery platform is conducted in three independent stages [30]. In the first step, the complex organic substrate is fermented to obtain a stream rich in VFAs; the second step is based on the natural principles of selection and competition of PHA-accumulating microorganisms against microorganisms that are unable to accumulate PHAs, by applying transient conditions in sequential batch reactors (SBR) [30]. It has been shown that conditions of external substrate excess (feast) and limitation (famine) select microbial populations with an increased capacity to store PHAs [31]. Finally, in the third phase, when the cells reach maximum PHAs content, they are harvested and sent to downstream extraction processes [18]. The downstream processes for extracting the polymer from the cells are among the most important factors affecting the overall cost of PHAs production, and they represent the bottleneck of the whole process. So far, the most studied methods for the recovery of PHAs can be grouped into two categories: solvent extraction and digestion of the non-polymeric material. Solvent extraction is the predominant method for PHAs extraction, used when high polymer purity is desired [32]. It typically involves soaking the PHA-containing biomass in an appropriate solvent or solvent mixture to dissolve the granules, followed by the addition of a precipitating agent to recover the polymers in crystalline form [33]. In the dissolution step, chlorinated solvents are typically used, e.g., chloroform and dichloromethane. For the precipitation step, methanol and ethanol are used [33]. Solvent extraction in large-scale applications is generally not an environmentally friendly method due to the harmful characteristics of the most used solvents, especially chlorinated solvents such as chloroform, and the massive amounts of antisolvent used (approximately 10 volumes per volume of PHAs solution) in order to precipitate the polymer. However, PHA-producing microorganisms can accumulate these polymers in quantities of up to 70% g PHA/gVSS [18], therefore it is possible to remove the small fraction of non-PHA cellular mass (NPCM) surrounding the PHA rather than extracting the PHA by solubilizing it in a suitable solvent. For this reason, several methods have been developed to release the PHAs granules by solubilizing the surrounding NPCM. Various chemical compounds can be used: sodium hypochlorite, surfactants, acid and alkaline compounds. All these chemicals can be used in the aqueous phase, thus avoiding the energy consumption required to dry the biomass when solvent extraction is applied [33].

Although the mass balance indicates higher productions, the expected final PHA productions for this pilot scale biorefinery process can reach a maximum of 2 kg PHA per day with a yield estimated in 0.01 kgPHA per kgVS in the feedstock because of the complexity of the biomass treated in this specific context, mainly due to straw, and because of the set-up of the extraction and purification phase.

3.4. Nutrient Recovery

The main by-product of the AD process is represented by a solid-liquid mixture, the digestate, which contains the stabilized organic matter not converted into biogas and considerable amounts of macro- and micro-nutrients. Digestate is now receiving great attention from the agricultural community because it is rich in nitrogen and phosphorous, and because of the high costs of fossil fertilizers. For example, when livestock effluents are treated, TKN concentration can reach values in the range of 2.5–9 g per kg of fresh matter with an ammonium content exceeding 50–60% *w*/*w* [17]. Agricultural digestate have also a good content of phosphorous, typically in the range of 0.5–1.5 g per kg fresh matter, a small part of which is in the soluble form.

There are currently several technological options for digestate treatment and nutrients recovery available on the market. All the options require a common step of solid/liquid separation of the agricultural digestate. These two phases, solid and liquid, can be then further processed to obtain concentrated nutrient streams minimizing the transport costs [16,34].

The most developed technologies at industrial scales for nitrogen and phosphorus recovery are ammonia stripping and struvite precipitation, respectively, which allow both an average removal and recovery yields of 90–95%. However, they have very high environmental and energetical costs [17]. For this reason, new technologies are emerging to reduce the use of chemicals, such as pressure-driven membranes, ultrafiltration and reverse osmosis (RO).

Membrane processes are physical-based technologies, where the agricultural digestate is treated under sequential solid/liquid separations. The solid fraction from each stage is called concentrate or retentate, while the liquid phase is named permeate. Usually, the first separation stage consists in a preliminary removal of the coarse solids by means of a centrifuge or a vibrating screen. Then, the liquid is treated in several steps: (i) microfiltration (MF), having membrane with pore size >0.1 μm under a trans-membrane pressure of 0.1–3 bar, (ii) ultrafiltration (UF, pore size >0.001 μm, pressure 2–10 bar) and a (iii) RO (pore size <1 nm, pressure 10–100 bar), to obtain a nutrient-rich retentate and water as permeate. In this way, it is possible to obtain a fertilizer rich in N and P (8.2–12.0 kg/tonne TN; 5.6–10.4 kg/tonne P_2O_5) [19,20].

In this study, we will treat the liquid fraction of digestate obtained after solid/liquid separation via a screw press in a train of operation consisting of filtering/sieving, microfiltration and reverse osmosis to obtain a retentate characterized by relatively high concentrations of nitrogen, potassium and phosphorus, as well as other important minerals such as iron, magnesium and sulfur, with the characteristics of distillation vinasse which can be used as soil amendment. Expected typical concentrations for nitrogen are in the range of 7–10 gN/L [35] while phosphorus is mainly recovered in the form of fine microfiltered sludge before the RO unit.

3.5. Proteins Production

Due to the growing human population, expected to reach 10 billion by 2050 [36], proteins are a sought-after resource not only as a source of food or animal feed, but also as the starting material for other high-value products such as protein hydrolysates, biostimulants for agriculture, wood adhesives, flocculants, surfactants and protein-based plastics [37]. Animal- and vegetal-based proteins are at present mainly produced through low-efficiency processes with a high environmental impact in terms of greenhouse gas emissions, land and water use, and soil and air pollution [38,39]. However, AD plants are an underexploited source of proteins, which could be used in place of traditional protein sources, with a lower ecological footprint. AD plants, which at present are either run at a cost or in need of subsidies, could be turned into profitable or at least neutral-cost establishments, once equipped for the upcycling of proteins.

Proteins can be recovered from AD plants through different routes as described in Figure 2:

1. The process of acidogenesis generates volatile fatty acids (VFA), which can easily be utilised as a source of carbon for the growth of different types of microorganisms. These are unicellular bacteria, fungi, yeasts and algae, which are rich in proteins and appear in the market under the name of microbial proteins (MP) or single cell proteins (SCP) once they have been dried and processed [40]. At present, established technologies are already producing microbial proteins at competitive conversion rates, which are marketed with the required certifications and patents [41].
2. CH_4, CO_2 and H_2 produced by methanogenesis are generally turned into thermal or electrical energy. However, these gases can be upcycled into microbial protein production by methanotrophic and hydrogenotrophic microbial strains, while the digestate can also be utilised as a source of nitrogen in the form of ammonia [42,43]. In the AD context, the CO_2 used as feedstock for bacterial strains can be generated from the upgrading of biogas into biomethane, or from the CO_2 emitted as a bioproduct of the combustion of biogas.
3. A more straightforward route for the recycling of proteins in AD plants is the utilisation of the microbial biomass produced in the digesters directly as MP source. This biomass containing mixed microbial cultures can be richer in protein than pure cultures, and has many applications, mainly in the animal feed sector [44].

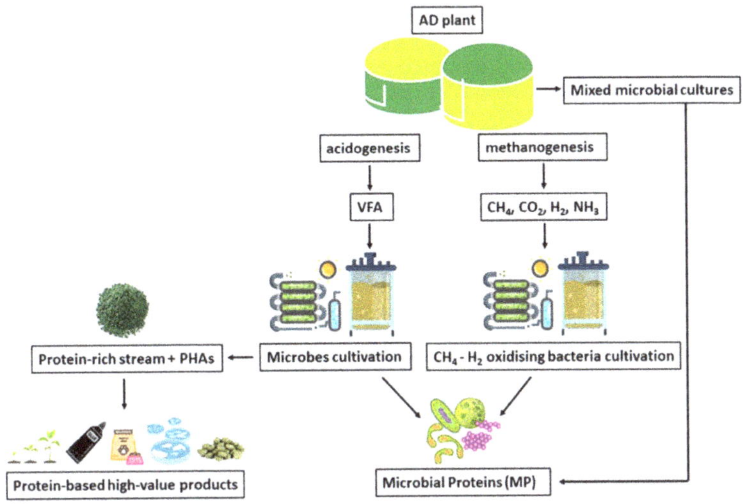

Figure 2. Schematic representation of the routes through which proteins can be recovered from AD plants.

In this study, route 1 is followed: specifically, VFA generated in the acidogenic anaerobic fermenter is used to produce PHA-accumulating microbial biomass which can be employed for feeding purposes, especially in the aquaculture sector. Expected yields are high and interesting with microbial biomass production in the range of 0.3–0.5 kg VSS per kg COD consumed.

3.6. Expected Mass Balance for the Overall Biorefinery Process

The overall biorefinery process allows for the recovery of biofuels (hydrogen and methane, VFA, PHA and nutrients). PHA-rich microbial mass can be used as protein microbial material instead of as a source of polyesters.

Considering the feeding of 1 m^3 per day at 20% dry matter, the expected productions of hydrogen, methane, VFAs, PHAs, nitrogen and phosphorus are those reported in Table 2.

Table 2. Expected quantitative products obtained from the overall AgriLoop process.

Product	Expected Production per Day
Hydrogen	2 m^3
Methane	18 m^3
Volatile fatty acids (short chain)	40 kg
Polyhydroxyalkanoates	Up to 8 kg
Nitrogen	0.6 kg as N
Phosphorus	0.05 kg as P

The overall feedstock is anaerobically fermented and can produce some 40 kg of short chain fatty acids, where relative concentrations of acetic, propionic, butyric and valeric acids can be tuned using different operational conditions in terms of HRT and OLR. During the anaerobic process, up to 2 m^3 of hydrogen is produced.

Produced VFAs can be considered a product itself, or, after solid/liquid separation and further purification, can be fed to the PHA-accumulating bioreactors. If transferred to this section, VFAs can produce up to 20 kg of microbial biomass containing up to 8 kg of PHAs. Because of the recovery procedure, the PHAs effectively recovered will be clearly lower than the calculated mass.

The solid fraction recovered after the solid/liquid separation step, equivalent to 60% of the total solids in the feedstock, is sent to the anaerobic digestion unit where up to 18 m^3 methane can be produced. Anaerobic digestate will be then treated to separate solids and recover nutrients from the liquid phase via screening, microfiltration and ultrafiltration: 0.6 kg and 0.05 kg of nitrogen and phosphorus, respectively, can be recovered.

Table 2 summarizes the expected quantities of products.

4. Conclusions

Currently, more than 18,000 anaerobic digestion units are under operation in EU, 80% of which are employed in the agricultural sector. Because of the need for the sustainable production of biobased products from agro-waste and since the tariff schemes are rapidly approaching their end in several countries, there is the urgent need for a reconfiguration of these plants. In particular, these can be transformed into biorefineries able to transform agro-waste into valuable bio-based products.

In this study, a complete biorefinery platform for the treatment of agro-waste (especially cow manure) was designed and implemented at pilot scale in a real farm environment: biofuels (methane and hydrogen), VFAs, PHAs or microbial mass proteins were the main products. The high flexibility of the process allows for the different production of biofuels or biochemicals depending on the operator choices.

Considering the treatment of 1 m^3 of slurry at 20% dry matter every day, for biofuels, up to 2 and 18 m^3 of hydrogen and methane, respectively, can be produced. Then, 40 kg of VFA can be produced and up to 8 kg PHA after extraction and purification. Moreover, 0.6 kg and 0.05 kg of N and P, respectively, can be recovered in a concentrated form, easy to transported.

The practical set up of this platform will be further studied and its applicability will be analyzed especially through the environmental, social and economic benefits and sustainability performance that will be quantitatively assessed during the AgriLoop project implementation.

Author Contributions: Conceptualization, D.B. (David Bolzonella); data curation; writing—original draft preparation, writing—review and editing, D.B. (David Bolzonella), D.B. (Davide Bertasini), R.L.C., M.M., F.R., A.Z., F.B., N.F., A.J. and G.P.; funding acquisition, D.B. (David Bolzonella). All authors have read and agreed to the published version of the manuscript.

Funding: The authors acknowledge financial support from the Project AgriLoop. This project has received funding from the European Union's Horizon Europe research and innovation programme under grant agreement No. 101081776. Views and opinions expressed are however those of the authors only and do not necessarily reflect those of the European Union. Neither the European Union nor the granting authority can be held responsible for them.

Data Availability Statement: Not applicable.

Conflicts of Interest: The authors declare no conflict of interest. The funders had no role in the design of the study; in the collection, analyses, or interpretation of data; in the writing of the manuscript; or in the decision to publish the results.

References

1. Köninger, J.; Lugato, E.; Panagos, P.; Kochupillai, M.; Orgiazzi, A.; Briones, M.J. Manure management and soil biodiversity: Towards more sustainable food systems in the EU. *Agric. Syst.* **2021**, *194*, 103251. [CrossRef]
2. Eurostat. Crop Production in EU Standard Humidity. 2020. Available online: https://ec.europa.eu/eurostat/databrowser/explore/all/all_themes (accessed on 14 December 2022).
3. Gontard, N.; Sonesson, U.; Birkved, M.; Majone, M.; Bolzonella, D.; Celli, A.; Angellier-Coussy, H.; Jang, G.-W.; Verniquet, A.; Broeze, J. A research challenge vision regarding management of agricultural waste in a circular bio-based economy. *Crit. Rev. Environ. Sci. Technol.* **2018**, *48*, 614–654. [CrossRef]
4. Pöschl, M.; Ward, S.; Owende, P. Evaluation of energy efficiency of various biogas production and utilization pathways. *Appl. Energy* **2010**, *87*, 3305–3321. [CrossRef]
5. Wall, D.M.; O'Kiely, P.; Murphy, J.D. The potential for biomethane from grass and slurry to satisfy renewable energy targets. *Bioresour. Technol.* **2013**, *149*, 425–431. [CrossRef] [PubMed]

6. Arthurson, V. Closing the global energy and nutrient cycles through application of biogas residue to agricultural land-potential benefits and drawbacks. *Energies* **2009**, *2*, 226–242. [CrossRef]
7. EBA. *EBA Statistical Report 2021*; EBA: Brussells, Belgium, 2021.
8. Alberici, S.; Grimme, W.; Toop, G. *Biomethane Production Potential in the EU. Feasibility of REPowerEU 2030 targets, Production Potentials in the Member States and Outlook to 2050*; Guidehouse: Utrecht, The Netherlands, 2022; p. 35.
9. Minasny, B.; Malone, B.P.; McBratney, A.B.; Angers, D.A.; Arrouays, D.; Chambers, A.; Chaplot, V.; Chen, Z.-S.; Cheng, K.; Das, B.S. Soil carbon 4 per mille. *Geoderma* **2017**, *292*, 59–86. Available online: https://www6.inrae.fr/4p1000science (accessed on 15 November 2022). [CrossRef]
10. del Pablo-Romero, M.; Sánchez-Braza, A.; Salvador-Ponce, J.; Sánchez-Labrador, N. An overview of feed-in tariffs, premiums and tenders to promote electricity from biogas in the EU-28. *Renew. Sustain. Energy Rev.* **2017**, *73*, 1366–1379. [CrossRef]
11. Battista, F.; Frison, N.; Bolzonella, D. Energy and nutrients' recovery in anaerobic digestion of agricultural biomass: An Italian perspective for future applications. *Energies* **2019**, *12*, 3287. [CrossRef]
12. Surendra, K.; Sawatdeenarunat, C.; Shrestha, S.; Sung, S.; Khanal, S.K. Anaerobic digestion-based biorefinery for bioenergy and biobased products. *Ind. Biotechnol.* **2015**, *11*, 103–112. [CrossRef]
13. Righetti, E.; Nortilli, S.; Fatone, F.; Frison, N.; Bolzonella, D. A multiproduct biorefinery approach for the production of hydrogen, methane and volatile fatty acids from agricultural waste. *Waste Biomass Valorization* **2020**, *11*, 5239–5246. [CrossRef]
14. APHA. *Standard Methods for the Examinations of Water and Wastewater*; American Public Health Association: Washington, DC, USA, 1998.
15. Bolzonella, D.; Battista, F.; Cavinato, C.; Gottardo, M.; Micolucci, F.; Lyberatos, G.; Pavan, P. Recent developments in biohythane production from household food wastes: A review. *Bioresour. Technol.* **2018**, *257*, 311–319. [CrossRef] [PubMed]
16. IEA. *Outlook for Biogas and Biomethane: Prospects for Organic Growth*; IEA: Paris, France, 2020.
17. Rizzioli, F.; Bertasini, D.; Bolzonella, D.; Frison, N.; Battista, F. A critical review on the techno-economic feasibility of nutrients recovery from anaerobic digestate in the agricultural sector. *Sep. Purif. Technol.* **2022**, 122690. [CrossRef]
18. Gottardo, M.; Bolzonella, D.; Tuci, G.A.; Valentino, F.; Majone, M.; Pavan, P.; Battista, F. Producing volatile fatty acids and polyhydroxyalkanoates from foods by-products and waste: A review. *Bioresour. Technol.* **2022**, *306*, 127716. [CrossRef]
19. Battista, F.; Bolzonella, D. Some critical aspects of the enzymatic hydrolysis at high dry-matter content: A review. *Biofuel Bioprod. Biorefin.* **2018**, *12*, 711–723. [CrossRef]
20. Alvaro, A.G.; Palomar, C.R.; Redondo, D.H.; Munoz, R.; de Godos Crespo, I. Potential Transformation of Cereal By-Products into Volatile Fatty Acids through Anaerobic Digestion. *SSRN Electron. J.* 2022. Available online: https://ssrn.com/abstract=4127756 (accessed on 15 November 2022).
21. Strazzera, G.; Battista, F.; Garcia, N.H.; Frison, N.; Bolzonella, D. Volatile fatty acids production from food wastes for biorefinery platforms: A review. *J. Environ. Manag.* **2018**, *226*, 278–288. [CrossRef]
22. Rizzioli, F.; Battista, F.; Bolzonella, D.; Frison, N. Volatile fatty acid recovery from anaerobic fermentate: Focusing on adsorption and desorption performances. *Ind. Eng. Chem. Res.* **2021**, *60*, 13701–13709. [CrossRef]
23. Reyhanitash, E.; Kersten, S.R.; Schuur, B. Recovery of volatile fatty acids from fermented wastewater by adsorption. *ACS Sustain. Chem. Eng.* **2017**, *5*, 9176–9184. [CrossRef]
24. Katikaneni, S.P.; Cheryan, M. Purification of fermentation-derived acetic acid by liquid-liquid extraction and esterification. *Ind. Eng. Chem. Res.* **2002**, *41*, 2745–2752. [CrossRef]
25. Strathmann, H. Electrodialysis, a mature technology with a multitude of new applications. *Desalination* **2010**, *264*, 268–288. [CrossRef]
26. Zacharof, M.-P.; Mandale, S.J.; Williams, P.M.; Lovitt, R.W. Nanofiltration of treated digested agricultural wastewater for recovery of carboxylic acids. *J. Clean. Prod.* **2016**, *112*, 4749–4761. [CrossRef]
27. Atasoy, M.; Owusu-Agyeman, I.; Plaza, E.; Cetecioglu, Z. Bio-based volatile fatty acid production and recovery from waste streams: Current status and future challenges. *Bioresour. Technol.* **2018**, *268*, 773–786. [CrossRef] [PubMed]
28. Pervez, M.N.; Mahboubi, A.; Uwineza, C.; Zarra, T.; Belgiorno, V.; Naddeo, V.; Taherzadeh, M.J. Factors influencing pressure-driven membrane-assisted volatile fatty acids recovery and purification-A review. *Sci. Total Environ.* **2022**, *817*, 152993. [CrossRef] [PubMed]
29. Ramos-Suarez, M.; Zhang, Y.; Outram, V. Current perspectives on acidogenic fermentation to produce volatile fatty acids from waste. *Rev. Environ. Sci. Biotechnol.* **2021**, *20*, 439–478. [CrossRef]
30. Kourmentza, C.; Plácido, J.; Venetsaneas, N.; Burniol-Figols, A.; Varrone, C.; Gavala, H.N.; Reis, M.A. Recent advances and challenges towards sustainable polyhydroxyalkanoate (PHA) production. *Bioengineering* **2017**, *4*, 55. [CrossRef] [PubMed]
31. Albuquerque, M.; Martino, V.; Pollet, E.; Avérous, L.; Reis, M. Mixed culture polyhydroxyalkanoate (PHA) production from volatile fatty acid (VFA)-rich streams: Effect of substrate composition and feeding regime on PHA productivity, composition and properties. *J. Biotechnol.* **2011**, *151*, 66–76. [CrossRef] [PubMed]
32. Raza, Z.A.; Abid, S.; Banat, I.M. Polyhydroxyalkanoates: Characteristics, production, recent developments and applications. *Int. Biodeterior. Biodegrad.* **2018**, *126*, 45–56. [CrossRef]
33. Pagliano, G.; Galletti, P.; Samorì, C.; Zaghini, A.; Torri, C. Recovery of polyhydroxyalkanoates from single and mixed microbial cultures: A review. *Front. Bioeng. Biotechnol.* **2021**, *9*, 624021. [CrossRef]

34. Fuchs, W.; Drosg, B. Assessment of the state of the art of technologies for the processing of digestate residue from anaerobic digesters. *Water Sci. Technol.* **2013**, *67*, 1984–1993. [CrossRef]
35. Van Puffelen, J.L.; Brienza, C.; Regelink, I.C.; Sigurnjak, I.; Adani, F.; Meers, E.; Schoumans, O.F. Performance of a full-scale processing cascade that separates agricultural digestate and its nutrients for agronomic reuse. *Sep. Purif. Technol.* **2022**, *297*, 121501. [CrossRef]
36. UN. *World Population Prospects 2022: Summary of Results*; United Nations—Department of Economic and Social Affairs, Population Division: New York, NY, USA, 2022.
37. Yadav, B.; Chavan, S.; Atmakuri, A.; Tyagi, R.D.; Drogui, P. A review on recovery of proteins from industrial wastewaters with special emphasis on PHA production process: Sustainable circular bioeconomy process development. *Bioresour. Technol.* **2020**, *317*, 124006. [CrossRef]
38. Pesante, G.; Zuliani, A.; Cannone, E.; Greco, F.; Tesoriero, C.; Vettori, A.; Frison, N. Biological conversion of agricultural residues into microbial proteins for aquaculture using PHA-producing mixed microbial cultures. *J. Clean. Prod.* **2022**, *378*, 134554. [CrossRef]
39. Pikaar, I.; Matassa, S.; Bodirsky, B.L.; Weindl, I.; Humpenöder, F.; Rabaey, K.; Boon, N.; Bruschi, M.; Yuan, Z.; van Zanten, H. Decoupling livestock from land use through industrial feed production pathways. *Environ. Sci. Technol.* **2018**, *52*, 7351–7359. [CrossRef]
40. Ravindra, A.P. Value-added food Single cell protein. A review. *Biotechnol. Adv.* **2000**, *18*, 459–479. [CrossRef]
41. Ritala, A.; Häkkinen, S.T.; Toivari, M.; Wiebe, M.G. Single cell protein—State-of-the-art, industrial landscape and patents 2001–2016. *Front. Microbiol.* **2017**, *8*, 2009. [CrossRef] [PubMed]
42. Verbeeck, K.; De Vrieze, J.; Pikaar, I.; Verstraete, W.; Rabaey, K. Assessing the potential for up-cycling recovered resources from anaerobic digestion through microbial protein production. *Microb. Biotechnol.* **2021**, *14*, 897–910. [CrossRef]
43. Matassa, S.; Boon, N.; Verstraete, W. Resource recovery from used water: The manufacturing abilities of hydrogen-oxidizing bacteria. *Water Res.* **2015**, *68*, 467–478. [CrossRef] [PubMed]
44. Vethathirri, R.S.; Santillan, E.; Wuertz, S. Microbial community-based protein production from wastewater for animal feed applications. *Bioresour. Technol.* **2021**, *341*, 125723. [CrossRef]

MDPI
St. Alban-Anlage 66
4052 Basel
Switzerland
Tel. +41 61 683 77 34
Fax +41 61 302 89 18
www.mdpi.com

Processes Editorial Office
E-mail: processes@mdpi.com
www.mdpi.com/journal/processes

www.ingramcontent.com/pod-product-compliance
Lightning Source LLC
Chambersburg PA
CBHW061612220926
51850CB00005B/485